T0310177

LQ Dynamic Optimization and Differential Games

LQ Dynamic Optimization and Differential Games

Jacob Engwerda
Tilburg University, The Netherlands

JOHN WILEY & SONS, LTD

Copyright © 2005   John Wiley & Sons Ltd, The Atrium, Southern Gate, Chichester,
West Sussex PO19 8SQ, England

Telephone   (+44) 1243 779777

Email (for orders and customer service enquiries): cs-books@wiley.co.uk
Visit our Home Page on www.wiley.com

All Rights Reserved. No part of this publication may be reproduced, stored in a retrieval system or transmitted in
any form or by any means, electronic, mechanical, photocopying, recording, scanning or otherwise, except
under the terms of the Copyright, Designs and Patents Act 1988 or under the terms of a licence issued
by the Copyright Licensing Agency Ltd, 90 Tottenham Court Road, London W1T 4LP, UK, without the
permission in writing of the Publisher. Requests to the Publisher should be addressed to the Permissions
Department, John Wiley & Sons Ltd, The Atrium, Southern Gate, Chichester, West Sussex PO19 8SQ,
England, or emailed to permreq@wiley.co.uk, or faxed to (+44) 1243 770620.

Designations used by companies to distinguish their products are often claimed as trademarks. All brand names
and product names used in this book are trade names, service marks, trademarks or registered trademarks of their
respective owners. The Publisher is not associated with any product or vendor mentioned in this book.

This publication is designed to provide accurate and authoritative information in regard to the
subject matter covered. It is sold on the understanding that the Publisher is not engaged in rendering professional
services. If professional advice or other expert assistance is required, the services
of a competent professional should be sought.

Other Wiley Editorial Offices

John Wiley & Sons Inc., 111 River Street, Hoboken, NJ 07030, USA

Jossey-Bass, 989 Market Street, San Francisco, CA 94103-1741, USA

Wiley–VCH Verlag GmbH, Boschstr. 12, D-69469 Weinheim, Germany

John Wiley & Sons Australia Ltd, 33 Park Road, Milton, Queensland 4064, Australia

John Wiley & Sons (Asia) Pte Ltd, 2 Clementi Loop #02-01, Jin Xing Distripark, Singapore 129809

John Wiley & Sons Canada Ltd, 22 Worcester Road, Etobicoke, Ontario, Canada M9W 1L1

Wiley also publishes its books in a variety of electronic formats. Some content that appears in print may not be
available in electronic books.

British Library Cataloguing in Publication Data

A catalogue record for this book is available from the British Library

ISBN-13 978-0-470-01524-7 (HB)
ISBN-10 0-470-01524-1 (HB)

Typeset in 10/12pt Times by Thomson Press (India) Limited, New Delhi
Printed and bound in Great Britain by Antony Rowe Ltd, Chippenham, Wiltshire
This book is printed on acid-free paper responsibly manufactured from sustainable forestry
in which at least two trees are planted for each one used for paper production.

Contents

Preface

Interaction between processes at various levels takes place everywhere around us. Either in industry, economics, ecology or on a social level, in many places processes influence each other. Particularly in those cases where people can affect the outcome of a process, the question arises how do people come to take a particular action.

To obtain a clearer view of this question within mathematics the paradigms of optimal control theory and game theory evolved, and dynamic game theory resulted as a merging of both these topics. This theory brings together the issues of optimizing behavior, the presence of multiple agents, enduring consequences of decisions and robustness with respect to variability in the environment. Within this field, linear quadratic differential games developed and these play an important role for three reasons: first, many applications of differential game theory fall into this category, secondly, there are many analytical results available and, thirdly, efficient numerical solution techniques can be used to solve these games.

Going through the literature, one can find many instances of results dealing with linear quadratic differential games. Unfortunately, there is no textbook focusing on this subject and giving a rigorous self-contained treatment of this theory. This textbook intends on the one hand to fill this gap and on the other hand to show the relevance of this theory by illustrating the theoretical concepts and results by means of simple economics examples.

Given this background, the organization of the book was chosen as follows. The last four chapters of the book deal with differential games. Since the theoretical development of these chapters sets out some knowledge on the one-player case and dynamic optimization techniques, these chapters are preceded by a rigorous treatment of this one-player case, the so-called regular linear quadratic control problem, and a chapter on dynamic optimization theory. To tackle these issues, however, one needs to be familiar with some preliminary work on linear algebra and dynamical systems. For that reason those subjects are dealt with first. The first chapter gives some historical developments and an outline of the book.

Having worked for more than two years on this book I ask myself whether it was worth all the trouble. Was it worth spending so much spare time and research time to produce something from which all one can hope is that it will be digestible and does not contain too many flaws and, on the other hand, does not contain a number of closely related interesting issues leaving the reader with information without the theory to back it up? The main motivation for writing this book was in the hope that it somehow might contribute a small amount to people's understanding and that some people will appreciate this book and use it for this purpose...

Last, but not least, I would like to thank the people who have contributed, most of them unknowingly, to this book. First of all my wife, **Carine**, and children, **Elsemiek**, **Ton** and **Heiko**, for offering me the chance to use my spare time for doing this. Next, I am indebted to Hans Schumacher for many discussions over the past decade concerning various issues on dynamic games, and to both him and Malo Hautus for the use of their excellent course notes on dynamic optimization. Furthermore Arie Weeren, Rudy Douven and Bram van den Broek contributed indirectly to the sections on differential games through their Ph.D. theses. Finally, Hendri Adriaens is acknowledged for his technical support on LaTEX and Tomasz Michalak for his proof-reading of an early version of this book.

Answers to the exercises included in this book can be found at http://www.wiley. com/go/engwerda

Notation and symbols

Matrix (Operations)

x^T	transpose of the vector x
$\| x \|_2$	length $x^T x$ of vector x
0	zero; zero vector; zero matrix
I	identity matrix
A^{-1}	inverse of matrix A
A^T	transpose of matrix A
$\text{adj}(A)$	adjoint of matrix A
$\det(A)$	determinant of matrix A
$\text{rank}(A)$	rank of matrix A
$\text{trace}(A)$	trace of matrix A
$\sigma(A)$	spectrum of matrix A
$\sigma(A|_S)$	spectrum of matrix A restricted to the invariant subspace S
$A_i(b)$	matrix A with ith column replaced by b
A_{ji}	matrix A with row j and column i deleted
$\text{Ker } A$	null space of A
$N(A)$	null space of A
$\text{Im } A$	column space of A
$R(A)$	column space of A
$\text{diag}\{A,B\}$	diagonal matrix with diagonal entries A and B
$\dim(S)$	dimension of subspace S
E_λ	eigenspace corresponding with eigenvalue λ
E_λ^g	generalized eigenspace corresponding with eigenvalue λ
E^c	center subspace
E^s	stable subspace
E^u	unstable subspace
$\text{vec}(A)$	vector of stacked columns of matrix A
$A \oplus B$	Kronecker product of matrices A and B
$S \oplus V$	direct sum subspaces S and V
$S^\perp$	orthogonal complement of subspace S
$\mathbb{C}_0^-$	set of complex numbers with non-positive real part
$\mathbb{C}_0^+$	set of complex numbers with non-negative real part
$\mathbb{C}^n$	set of complex vectors with n entries
$\mathbb{C}^{n\times m}$	set of $n \times m$ matrices with complex entries

C^1	set of continuous differentiable functions
CD	set of continuous functions satisfying some differentiability properties (definition 3.4)
Γ_i	set of control functions of player i
Γ^{aff}	set of affine functions of the state variable
$\mathcal{F}$	set of all stabilizing feedback matrices
L_2	set of all measurable Lebesgue square integrable functions on $[0, \infty)$
$L_{2,loc}$	set of all measurable functions that are square integrable over all finite intervals $[0, T]$
$\mathcal{M}^{inv}$	set of M-invariant subspaces
$\mathbb{N}$	set of natural numbers
$\mathcal{P}^{pos}$	set of n-dimensional M-invariant graph subspaces
$\mathbb{Q}$	set of rational numbers
$\mathbb{R}$	set of real numbers
$\mathbb{R}^n$	set of vectors with n real entries
$\mathbb{R}^{n \times m}$	set of $n \times m$ matrices with real entries
Σ_d^N	set of bargaining problems (see Chapter 6)
$\mathcal{U}$	set of control functions for which the differential equation has a solution in the extended sense
$\mathcal{U}_{re}$	$\{u \in \mathcal{U} \mid J(u)$ exists as a finite number and $\lim_{t \to \infty} x(t)$ exists $\}$
$\mathcal{U}_s$	set of locally square integrable control functions yielding a stable closed-loop system

Miscellaneous

ARE	algebraic Riccati equation
δ_{ij}	Kronecker δ, $\delta_{ij} = 1$ if $i = j$, $\delta_{ij} = 0$ otherwise
$\delta f(x_0, h)$	Fréchet/Gateaux differential of f at x_0 with increment h
∂f	Fréchet derivative of f
$\partial_i f$	ith Fréchet partial derivative of f
$E(S, d)$	egalitarian bargaining solution
inf	infimum
$I(S, d)$	ideal point
$K(S, d)$	Kalai–Smorodinsky bargaining solution
lim	limit
$N(S, d)$	Nash bargaining solution
$0(h)$	higher-order terms in h
$\Pi_{i=1}^n p_i$	product of the n p_i variables
RDE	Riccati differential equation
sup	supremum
$\|z\|$	modulus, $\sqrt{x^2 + y^2}$, of complex number $z = x + iy$
Re(z)	real part, x, of complex number $z = x + iy$
$x \succ y$	vector inequality $x_i > y_i$, $i = 1, \ldots, N$

1

Introduction

1.1 Historical perspective

Dynamic game theory brings together four features that are key to many situations in economics, ecology and elsewhere: optimizing behavior, the presence of multiple agents/players, enduring consequences of decisions and robustness with respect to variability in the environment.

To deal with problems which have these four features the dynamic game theory methodology splits the modeling of the problem into three parts. One part is the modeling of the environment in which the agents act. To obtain a mathematical model of the agents' environment a set of differential or difference equations is usually specified. These equations are assumed to capture the main dynamical features of the environment. A characteristic property of this specification is that these dynamic equations mostly contain a set of so-called 'input' functions. These input functions model the effect of the actions taken by the agents on the environment during the course of the game. In particular, by viewing 'nature' as a separate player in the game who can choose an input function that works against the other player(s), one can model worst-case scenarios and, consequently, analyze the robustness of the 'undisturbed' game solution.

A second part is the modeling of the agents' objectives. Usually the agents' objectives are formalized as cost/utility functions which have to be minimized. Since this minimization has to be performed subject to the specified dynamic model of the environment, techniques developed in optimal control theory play an important role in solving dynamic games. In fact, from a historical perspective, the theory of dynamic games arose from the merging of static game theory and optimal control theory. However, this merging cannot be done without further reflection. This is exactly what the third modeling part is about. To understand this point it is good to summarize the rudiments of static games.

Most research in the field of static game theory has been – and is being – concentrated on the normal form of a game. In this form all possible sequences of decisions of each player are set out against each other. So, for example, for a two-player game this results in a matrix structure. Characteristic for such a game is that it takes place in one moment of time: all players make their choice once and simultaneously and, dependent on the choices made, each player receives his payoff. In such a formulation important issues like the order of play in the decision process, information available to the players at the time of their decisions, and the evolution of the game are suppressed, and this is the reason this branch of game theory is usually classified as 'static'. In case the agents act in a dynamic environment these issues are, however, crucial and need to be properly specified before one can infer what the outcome of the game will be. This specification is the third modeling part that characterizes the dynamic game theory methodology.

In this book we study a special class of dynamic games. We study games where the environment can be modeled by a set of linear differential equations and the objectives can be modeled as functions containing just affine quadratic terms. Concerning the information structure of the game we will basically describe two cases: the 'open-loop' and the 'linear feedback' case. A proper introduction of these notions is postponed until the relevant chapters later on.

The popularity of these so-called linear quadratic differential games is caused on the one hand by practical considerations. To some extent these kinds of differential games are analytically and numerically solvable. If one leaves this track, one easily gets involved in the problem of solving sets of nonlinear partial differential equations, and not many of these equations can be solved analytically. Even worse, when the number of state variables is more than two in these equations, a numerical solution is in general hard to obtain. On the other hand this linear quadratic problem setting naturally appears if the agents' objective is to minimize the effect of a small perturbation of their nonlinear optimally controlled environment. By solving a linear quadratic control problem, and using the optimal actions implied by this problem, players can avoid most of the additional cost incurred by this perturbation (section 5.1).

So, linear quadratic differential games are a subclass of dynamic games. As already indicated above, optimal control techniques play an important role in solving dynamic games and, in fact, optimal control theory is one of the roots of dynamic game theory. For these reasons the first part of this book (Chapters 2–5) gives, broadly speaking, an introduction to the basics of the theory of dynamic optimization. To appreciate this theory we next provide a short historical overview of its development.

To outline the field, optimal control theory is defined as the subject of obtaining optimal (i.e. minimizing or maximizing) solutions and developing numerical algorithms for one-person single-objective dynamic decision problems.

Probably the first recorded feedback control application is the water clock invented by the Greek Ktesibios around 300 BC in Alexandria, Egypt. This was definitely a successful design as similar clocks were still used around 1260 AD in Baghdad. The theory on optimal control has its roots in the calculus of variations. The Greek Pappus of Alexandria[1] already posed 300 AD the isoperimetric problem, i.e. to find a closed plane curve of a given length which encloses the largest area, and concluded that this was a circle. Remarkably, the most essential contribution towards its rigorous proof was only given in 1841 by Steiner[2]. Some noteworthy landmarks in between are the derivation by

---

[1]Pappus, ±290–±350, born in Alexandria (Egypt), was the last of the great Greek geometers. He is sometimes called the founding father of projective geometry.

[2]Steiner, 1796–1863, was a Swiss mathematician who first went to school at the age of 18. He made very significant contributions to projective geometry.

Fermat[3] around 1655 of the sine law of refraction (as proposed by Snell[4]) using the principle that light always follows the shortest possible path; the publication of Newton's Principia[5] in 1687 in which he analyses the motion of bodies in resisting and non-resisting media under the action of centripetal forces; and the formulation and solution in 1696 of the 'Brachistochrone' problem by Johann Bernoulli[6], i.e. to find the curve along which a particle uses the minimal time to slide between two points A and B, if the particle is influenced by gravitational forces only – this curve turns out to be a cycloid. The name calculus of variation was introduced by Euler[7] in 1766 and both he (in 1744) and Lagrange[8] (1759, 1762 and 1766) contributed substantially to the development of this theory. Both their names are attached to the famous Euler–Lagrange differential equation that an optimal function and its derivative have to satisfy in order to solve a dynamic optimization problem. This theory was further developed in the nineteenth century by Hamilton[9] in his papers on general methods in dynamics (1834,1835), and in the lectures by Weierstrass[10] at the University of Berlin during 1860–1890 and Jacobi[11] who carried out important research in partial differential equations and used this theory to analyze equations describing the motion of dynamical systems. Probably the first mathematical model to describe plant behavior for control purposes is due to Maxwell[12], who in 1868 used differential equations to explain instability problems encountered with James Watt's flyball governor (later on used to regulate the speed of steam-engine vehicles). 'Fore-runners' of modern optimal control theory, associated with the maximum principle, are Valentine (1937), McShane[13] (1939), Ambartsumian (1943) and Hesteness (1949) (the expanded version of this work was later published in 1966). Particularly in the 1930s and

---

[3]Fermat, 1601–1665, was a French lawyer who is famous for his mathematical contributions to the algebraic approach to geometry and number theory.

[4]Snell, 1580–1626, was a Dutch mathematician/lawyer who made significant contributions to geodesy and geometric optics.

[5]Newton, 1643–1727, born in England, was famous for the contributions he made in the first half of his career to mathematics, optics, physics and astronomy. The second half of his life he spent as a government official in London.

[6]Johann Bernoulli, 1667–1748, was a Swiss mathematician who made significant contributions to analysis.

[7]Euler, 1707–1783, a Swiss mathematician/physician was one of the most prolific writers on mathematics of all time. He made decisive and formalistic contributions to geometry, calculus, number theory and analytical mechanics.

[8]Lagrange, 1736–1813, was a Sardinian mathematician who made substantial contributions in various areas of physics, the foundation of calculus, dynamics, probability and number theory.

[9]Hamilton, 1805–1865, an Irish mathematician/physician, introduced the characteristic function and used this to study dynamics. Moreover, he introduced and studied the algebra of quaternions that play an important role in mathematical physics.

[10]Weierstrass, 1815–1897, was a German mathematician. The standards of rigour Weierstrass set in his courses strongly affected the mathematical world and for that reason he is sometimes called the father of modern analysis.

[11]Jacobi, 1804–1851, was a German mathematician, who was prepared (but not allowed) to enter university when he was 12. He was a very prolific writer in many fields of mathematics. His contributions to the theory of partial differential equations and determinants are well-known.

[12]Maxwell, 1831–1879, a Scottish mathematician/physicist, published his first paper when he was only 14. His most well-known contributions are about electricity and magnetism and his kinetic theory of gases.

[13]McShane, 1904–1989, was an American mathematician, well-known for his work in the calculus of variations, ballistics, integration theory and stochastic differential equations.

1940s frequency domain methods and Laplace transformation techniques were used to study control problems. At the beginning of the second half of the twentieth century further theoretical impetus was provided on the one hand by the Russian Pontrjagin[14], around 1956, with the development of the so-called maximum principle. This culminated in the publication of his book (1961). On the other hand, around 1950, the Americans Bellman and Isaacs started the development of the dynamic programming principle[15] which led to the publication of the books by Bellman (1956) and Bellman and Dreyfus (1962).

Furthermore, it was recognized in the late 1950s that a state space approach could be a powerful tool for the solution of, in particular, linear feedback control problems. The main characteristics of this approach are the modeling of systems using a state space description, optimization in terms of quadratic performance criteria, and incorporation of Kalman–Bucy optimal state reconstruction theory. The significant advantage of this approach is its applicability to control problems involving multi-input multi-output systems and time-varying situations (for example Kwakernaak and Sivan, 1972). A historical overview on early control theory can be found in Neustadt (1976).

Progress in stochastic, robust and adaptive control methods from the 1960s onwards, together with the development of computer technology, have made it possible to control much more accurately dynamical systems which are significantly more complex. In Bushnell (1996), one can find a number of papers, including many references, concerning the more recent history of control theory.

For more or less the same reasons we covered the historical background of optimal control theory we next trace back some highlights of game theory, ending the overview more or less at the time the theory of dynamic games emerged. This avoids the challenge of providing an overview of its most important recent theoretical developments.

The first static cooperative game problem reported seems to date back to 0–500 AD. In the Babylonian Talmud, which serves as the basis of Jewish religious, criminal and civil law, the so-called marriage contract problem is discussed. In this problem it is specified that when a man who has three wives dies, wives receive 100, 200 and 300, respectively. However, it also states that if the estate is only worth 100 all three receive the same amount, if it is worth 200 they receive 50, 75 and 75, respectively, and if it is worth 300 they receive a proportional amount, i.e. 50, 100 and 150, respectively. This problem puzzled Talmudic scholars for two millennia. It was only recognized in 1985 that the solution presented by the Talmud can be interpreted using the theory of co-operative games. Some landmarks in this theory are a book on probability theory written by

---

[14]Pontrjagin, 1908–1988, a Russian mathematician who due to an accident was left blind at 14. His mother devoted herself to help him succeed to become a mathematician. She worked for years in fact as his private secretary, reading scientific works aloud to him, writing in the formulae in his manuscripts, correcting his work, though she had no mathematical training and had to learn to read foreign languages. He made important contributions to topology, algebra and, later on, control theory.

[15]Both scientists worked in the late 1940s and early 1950s at the Research and New Development (RAND) Corporation in Santa Monica, California, USA. During presentations and discussions at various seminars held at RAND at that time the dynamic programming principle probably arose as a principle to solve dynamic optimization problems. From the discussions later on (Breitner, 2002) it never became clear whether just one or both scientists should be considered as the founding father(s) of this principle.

Montmort[16] in 1708 in which he deals in a systematic way with games of chance. In 1713 Waldegrave, inspired by a card game, provided the first known minimax mixed strategy solution to a two-person game. Cournot[17] (1838) discusses for the first time the question of what equilibrium price might result in the case of two producers who sell an identical product (duopoly). He utilizes a solution concept that is a restricted version of the non-cooperative Nash equilibrium. The theory of cooperation between economic agents takes its origin from economic analysis. Probably Edgeworth[18] and Pareto[19] provided the first definitions of a cooperative outcome. Edgeworth (1881) proposed the contract curve as a solution to the problem of determining the outcome of trading between individuals, whereas Pareto (1896) introduced the notion of efficient allocation. Both used the formalism of ordinal utility theory. Zermelo[20] (1913) presented the first theorem of game theory which asserts that chess is a strictly determined game, i.e. assuming that only an a priori fixed number of moves is allowed, either (i) white has a strategy which always wins; or (ii) white has a strategy which always at least draws, but no strategy as in (i); or (iii) black has a strategy which always wins. Fortunately, no one knows which of the above is actually the case. There are games where the assumption that each player may choose from only a finite number of actions seems to be inappropriate. Consider, for example, the 'princess and the monster' game (see Foreman (1977) and Başar and Olsder (1999)). In this game, which is played in a completely dark room, there is a monster who wants to catch a princess. The moment both bump into each other the game terminates. The monster likes to catch the princess as soon as possible, whereas the princess likes to avoid the monster as long as possible. In this game the optimal strategies cannot be deterministic. For, if the monster were to have a deterministic optimal strategy, then the princess would be able to calculate this strategy. This would enable her to determine the monster's path and thereby to choose for herself a strategy such that she avoids the monster forever. Therefore, an optimal strategy for the monster (if it exists) should have random actions, so that his strategy cannot be predicted by the princess. Such a strategy is called mixed. Borel[21] published from 1921–1927 five notes in which he gave the first modern formulation of a mixed strategy along with finding the minimax[22] solution for

---

[16]Montmort, 1678–1719, was a French mathematician.

[17]Cournot, 1801–1877, a French mathematician was the pioneer of mathematical economics.

[18]Edgeworth, 1845-1926, an Irish economist/mathematician was well-known for his work on utility theory and statistics.

[19]Pareto, 1848–1923, was an Italian economist/sociologist who studied classics and engineering in Turin. After his studies in 1870 he worked for a couple of years in industry. From 1889 onwards he began writing numerous polemical articles against the Italian government and started giving public lectures. In 1893 he succeeded Walras at the University of Lausanne (Switzerland). Famous are his Manual of Political Economy (1906), introducing modern microeconomics, and his Trattato di Sociologia Generale (1916), explaining how human action can neatly be reduced to residue (non-logical sentiments) and derivation (afterwards justifications).

[20]Zermelo, 1871–1953, was a German mathematician/physician famous for his work on axiomatic set theory.

[21]Borel, 1871–1956, was a French mathematician/politician famous for his work on measure theory. In the second half of his life he embarked on a political career and became Minister of the Navy.

[22]The idea of a minimax solution is that a player wants to choose that action which minimizes the maximum risk that can occur due to the actions of his opponent(s).

two-person games with three or five possible strategies. Von Neumann[23] (1928) considered two-person zero-sum games. These are games where the revenues of one player are costs for the other player. He proved that every two-person zero-sum game with finitely many pure strategies for each player is determined and introduces the extensive form of a game. The extensive form of a game basically involves a tree structure with several nodes and branches, providing an explicit description of the order of play and the information available to each player at the time of his decision. The game evolves from the top of the tree to the tip of one of its branches. The case that players decide to cooperate in order to maximize their profits was considered by Zeuthen[24] (1930). If players agree to cooperate in order to maximize their profits the question arises as to what extent efforts will be used to maximize the individual profit of each single player. This is called the bargaining problem. Zeuthen (1930) proposes a solution to the bargaining problem which Harsanyi[25] (1956) showed is equivalent to Nash's bargaining solution. A distinction between the order of play was introduced by von Stackelberg[26] (1934) within the context of economic competition. He distinguished between the first mover, called the leader, and the second mover, called the follower. The idea is that the follower can observe the move of the leader and subsequently act. The seminal work of Von Neumann and Morgenstern[27] (1944) presents the two-person zero-sum theory; the notion of a cooperative game with transferable utility, its coalitional form and stable sets; and axiomatic utility theory. In their book they argue that economics problems may be analyzed as games. Once all irrelevant details are stripped away from an economics problem, one is left with an abstract decision problem – a game. The book led to an era of intensive game theory research. The next cornerstone in static game theory was set by Nash from 1950–1953 in four papers. He proved the existence of a strategic equilibrium for non-cooperative games – the Nash equilibrium – and proposed the 'Nash program', in which he suggested approaching the study of cooperative games via their reduction to non-cooperative form (1950a,1951). A Nash equilibrium was defined as a strategy combination – consisting of one stategy for each player – with the property that no player can gain (in terms of utility) by unilaterally deviating from it. Hence this equilibrium solution is self-enforcing. That is, it is an optimal solution for each player as long as his opponent players stick to their recommendations. Unfortunately, it turns out that a game may have more than one such Nash equilibrium. Since not all equilibria are in general equally attractive, refinement criteria for selecting among multiple equilibria were proposed later on (for example van Damme, 1991). Another, still relevant, issue associated with this non-uniqueness is how one can determine numerically all refined Nash equilibria (for example Peeters, 2002). In his two papers on bargaining theory (Nash, 1950b,1953), Nash founded axiomatic bargaining theory, proved the existence of the Nash bargaining solution and provided the first execution of the Nash program. In

---

[23]Von Neumann, 1903–1957, was a Hungarian mathematician who was a pioneer in various fields including quantum mechanics, algebra, applied mathematics and computer science.

[24]Zeuthen, 1888–1959, was a Danish economist known for his work on general equilibrium theory, bargaining and monopolistic competition.

[25]Harsanyi, 1920–2000, was a Hungarian pharmacist/philosopher/economist well-known for his contributions to game theory.

[26]Von Stackelberg, 1905–1946, was a German economist.

[27]Morgenstern, 1902–1976, was an Austrian economist.

axiomatic bargaining theory one tries to identify rules that yield a fair sharing of the benefits of cooperation (see also Chapter 6). From this time onwards the theoretical developments in game theory rapidly increased. By consulting the latest textbooks in this field (for example Tijs (1981) or Fudenberg and Tirole (1991)) or the website of Walker, (2001) one can get an impression of these developments.

Probably as a spin-off of all these new ideas in control and game theory in the late 1940s and early 1950s the first dynamic game models came to light. The initial drive towards the development of this theory was provided by Isaacs during his stay at the RAND Corporation from 1948–1955. He wrote the first paper on games of pursuit (Isaacs, 1951). In this paper the main ideas for the solution of two-player, zero-sum dynamic games of the pursuit–evasion type are already present. He furthered these ideas (1954–1955) and laid a basis for the theory of dynamic games within the framework of two-person zero-sum games. This theory was first discussed by Berkovitz (1961). The first contribution on nonzero-sum games seems to date back to Case (1967). A nice historic overview on the early years of differential games and Isaacs' contributions to this field can be found in Breitner (2002).

The first official papers on dynamic games were published in a special issue of the Annals of Mathematics Studies edited by Dresher, Tucker and Wolfe (1957). In a special section of this volume entitled 'Games with a continuum of moves' Berkovitz, Fleming and Scarf published their papers (Berkovitz and Fleming, 1957; Fleming, 1957; Scarf, 1957). On the Russian side early official contributions on dynamic games have been published by Kelendzeridze (1961), Petrosjan (1965), Pontrjagin (1961) and Zelikin and Tynyanskij (1965). The books written by Isaacs (1965) and Blaquiere, Gerard and Leitmann (1969) document the theoretical developments of dynamic game theory during its first two decades. The historical development of the theory since the late 1960s is documented by the works of, for example, Friedman (1971), Leitmann (1974), Krasovskii and Subbotin (1988), Mehlmann (1988), Başar and Olsder (1999) and Haurie (2001). Particularly in Başar and Olsder (1999) one can find at the end of each chapter a section where relevant historical remarks are included concerning the subjects discussed in that chapter.

Current applications of differential games range from economics, financial engineering, ecology and marketing to the military. The work of Dockner *et al.* (2000) provides an excellent comprehensive, self-contained survey of the theory and applications of differential games in economics and management science. The proceedings and the associated Annals of the International Symposia on Dynamic Games and Applications held every other year (for example Petrosjan and Zenkevich, 2002) document the development of both theory and applications over the last 20 years.

We conclude this section by presenting a historical outline of the development of the theory on non-cooperative linear quadratic differential games. As already indicated the linear quadratic differential games constitute a subclass of differential games. Starr and Ho might be called the founding fathers of this theory. Ho, Bryson and Baron (1965) analyzed the particular class of pursuit–evasion games, and the results were later put into a rigorous framework by Schmitendorf (1970). With their paper, Starr and Ho (1969) generalized the zero-sum theory developed by Isaacs. Using the Hamilton–Jacobi theory they provided a sufficient condition for existence of a linear feedback Nash equilibrium for a finite-planning horizon.

Lukes (1971) showed that, if the planning horizon in the game is chosen to be sufficiently small, the game always has – for every initial state of the system a unique linear feedback Nash equilibrium. Moreover, this equilibrium can be computed by solving a set of so-called feedback Nash Riccati differential equations. Papavassilopoulos and Cruz (1979) show that if the set of strategy spaces is restricted to analytic functions of the current state and time, then the Nash equilibrium is unique, if it exists. Bernhard (1979) considers the zero-sum game with the additional restriction that the final state should lie in some prespecified linear subspace. Mageirou (1976) considered the infinite-horizon zero-sum game. An important point demonstrated by this paper and made more explicit by Jacobson (1977) is that the strategy spaces should be clearly defined before one can derive equilibria. Papavassilopoulos, Medanić and Cruz (1979) discussed parametric conditions under which the coupled set of algebraic feedback Nash Riccati equations has a solution. If these conditions are met the infinite-horizon game has at least one feedback Nash equilibrium. Papavassilopoulos and Olsder (1984) demonstrate that an infinite-horizon game may have either none, a unique or multiple feedback Nash equilibria even though every finite-horizon version of it has a unique feedback Nash equilibrium. In particular they present a sufficient condition under which the set of feedback Nash Riccati differential equations has a solution. This last result was general-ized by Freiling, Jank and Abou-Kandil (1996). Weeren, Schumacher and Engwerda (1999) give an asymptotic analysis of the regular finite-planning two-player scalar game. They show that this game always has a unique equilibrium but that the convergence of the equilibrium actions depends on the scrap value. Three different convergence schemes may occur and the equilibrium actions always converge in this regular case to a solution of the infinite-horizon game. The number of equilibria for the scalar N-player infinite-horizon game was studied in Engwerda (2000b). For the two-player case parameter conditions under which this game has a unique equilibrium were derived in Lockwood (1996) and Engwerda (2000a).

The problem of calculating the solutions of the feedback Riccati differential equations is adressed in Cruz and Chen (1971) and Ozgüner and Perkins (1977). Iterative algorithms to calculate a stabilizing solution of the algebraic feedback Nash equations were deve-loped by Krikelis and Rekasius (1971), Tabak (1975), Mageirou (1977) and Li and Gajic (1994). A disadvantage of these algorithms is that they depend on finding good initial conditions and provide just one solution (if they converge) of the equations. In Engwerda (2003) an algorithm based on determining the eigenstructure of a certain matrix was presented to calculate the whole set of stabilizing solutions in case the system is scalar.

Under the assumption that the planning horizon is not too long, Friedman (1971) showed that the game will have a unique open-loop Nash equilibrium (see also Starr (1969)). For an arbitrary finite-planning horizon length Lukes and Russel (1971) presented a sufficient condition for the existence and uniqueness of open-loop equilibria in terms of the invertibility of a Hilbert space operator. Eisele (1982) used this latter approach to show that if this operator is not invertible the game either has no open-loop equilibrium solution or an infinite number of solutions, this depending on the initial state of the system. Feucht (1994) reconsidered the open-loop problem for a general indefinite cost function and studied in particular its relationship with the associated set of open-loop Riccati differential equations.

Analytic solutions of these Riccati differential equations have been studied in Simaan and Cruz (1973), Abou-Kandil and Bertrand (1986), Jódar and Abou-Kandil (1988,1989), Jódar (1990), Jódar and Navarro (1991a,b), Jódar, Navarro and Abou-Kandil (1991), and Abou-Kandil, Freiling and Jank (1993). These results were generalized by Feucht (1994).

The basic observation made in these references is that this set of differential equations constitute an ordinary (non-symmetric, high-order) Riccati differential equation and its solution (Reid, 1972) can thus be analyzed as the solution of a set of linear differential equations (see also Abou-Kandil *et al.*, 2003). If some additional parametric assumptions are made the set of coupled equations reduces to one single (non-symmetric, low-order) Riccati differential equation. An approximate solution is derived in Simaan and Cruz (1973) and Jódar and Abou-Kandil (1988). Scalzo (1974) showed that if the controls are constrained to take values in compact convex subsets the finite-horizon game always has an open-loop equilibrium. This is irrespective of the duration of the game. Engwerda and Weeren (1994) and Engwerda (1998a) studied both the limiting behavior of the finite-planning horizon solutions as the final time approaches infinity and the infinite-horizon case using a variational approach. An algorithm to calculate all equilibria for the infinite-horizon game was provided in Engwerda (1998b). Kremer (2002) used the Hilbert space approach to analyze the infinite-horizon game and showed in particular that similar conclusions hold in this case as those obtained by Eisele.

To model uncertainty basically two approaches have been taken in literature (but see also Bernhard and Bellec (1973) and Broek, Engwerda and Schumacher (2003) for a third approach). Usually either a stochastic or a worst-case approach is taken. A stochastic approach, for example, is taken in Kumar and Schuppen (1980), Başar (1981), Bagchi and Olsder (1981) and Başar and Li (1989). A worst-case approach (see, for example, the seminal work by Başar and Bernhard (1995)) is taken, in a cooperative setting, by Schmitendorf (1988). The non-cooperative open-loop setting is dealt with by Kun (2001) and the linear feedback setting by Broek, Engwerda and Schumacher (2003).

Applications in economics are reported in various fields. In, for example, in industrial organization by Fershtman and Kamien (1987), Reynolds (1987), Tsutsui and Mino (1990), Chintagunta (1993), Jørgensen and Zaccour (1999, 2003); in exhaustible and renewable resources by Hansen, Epple and Roberts (1985), Mäler and de Zeeuw (1998) and Zeeuw and van der Ploeg (1991); in interaction between monetary and fiscal authorities by Pindyck (1976), Kydland (1976), Hughes-Hallett (1984), Neese and Pindyck (1984), Tabellini (1986), Petit (1989), Hughes-Hallett and Petit (1990), van Aarle, Bovenberg and Raith (1995), Engwerda, van Aarle and Plasmans (1999, 2002) van Aarle *et al.* (2001) and van Aarle, Engwerda and Plasmans (2002); in international policy coordination by Miller and Salmon (1985a,b), Cohen and Michel (1988), Curie, Holtham and Hughes-Hallett (1989), Miller and Salmon (1990) and Neck and Dockner (1995); and in monetary policy games by Obstfeld (1991) and Lockwood and Philippopoulos (1994). In this context it should be mentioned that various discrete-time macroeconomic game models have been estimated and analyzed in literature (e.g. the SLIM model developed by Douven and Plasmans (1996) and the Optgame model developed by Neck *et al.* (2001).

In literature information structures different from the ones that are considered in this book have also been investigated. For instance Foley and Schmitendorf (1971) and Schmitendorf (1970) considered the case that one player has open-loop information and the other player uses a linear-feedback strategy. Furthermore Başar (1975,1977 or, for its discrete-time counterpart, 1974) showed that finite-planning horizon linear quadratic differential games also permit multiple nonlinear Nash equilibria if at least one of the players has access to the current and initial state of the system (the dynamic information case). Finally, we should point out the relationship between the optimal control of stochastic linear systems with an exponential performance criterion and zero-sum differential games, which enables a stochastic interpretation of worst-case design of linear systems (Jacobson (1973) and Broek, Engwerda and Schumacher (2003a) or Klompstra (2000) in a discrete-time framework).

1.2 How to use this book

This book is a self-contained introduction to linear quadratic differential games. The book is introductory, but not elementary. It requires some basic knowledge of mathematical analysis, linear algebra and ordinary differential equations. The last chapter also assumes some elementary knowledge of probability theory. The topics covered in the various chapters can be followed up to a large extent in related literature. In particular the sections entitled 'Notes and references' which end each chapter can be regarded as pointers to the sources consulted and related items that are not mentioned in this book.

This book is intended to be used either as a textbook by students in the final year of their studies or as a reference work. The material is written in such a way that most of the material can be read without consulting the mathematical details. Lengthy proofs are provided, in most cases, in the appendix to each chapter. Broadly speaking, the book consists of two parts: the first part (Chapters 2–5) is about dynamic optimization with, as a special case, the linear quadratic control problem. The second part (Chapters 6–9) is about linear quadratic differential games. So, this book could be used to teach a first semester introductory course on dynamic optimization and a second semester course on linear quadratic differential games. Throughout this book the theory is illustrated by examples which are often taken from the field of economics. These examples should help the reader to understand the presented theory.

1.3 Outline of this book

A summary of each chapter is given below. Note that some of the statements in this section are not precise. They hold under certain assumptions which are not explicitly stated. Readers should consult the corresponding chapters for the exact results and conditions.

Chapter 2 reviews some basic linear algebra which in some instances goes beyond the introductory level. To fully understand the different dynamics of systems that can occur over time, it is convenient to cover the arithmetic of complex numbers. For that reason we introduce this arithmetic in a seperate section. This analysis is used to introduce complex eigenvalues of a real square $n \times n$ matrix A. We show that each eigenvector has a generalized eigenspace. By choosing a basis for each of these generalized eigenspaces in an appropriate way we then obtain a basis for $\mathbb{R}^n$. With respect to this basis matrix A has the Jordan canonical structure. Since this Jordan canonical form has a diagonal structure it is a convenient way of analyzing the dynamics of a linear system (Chapter 3). Algebraic Riccati equations play a crucial role in this book. Therefore, we introduce and discuss a number of their elementary properties in the second part of Chapter 2. In this chapter we focus on the Riccati equation that is associated with the one-player linear quadratic control problem. We show that the solutions of this Riccati equation can be obtained by determining the eigenstructure of this equation with the associated so-called Hamiltonian matrix. The so-called stabilizing solutions of Riccati equations play a crucial role later on. We show that the Riccati equation considered in Chapter 2 always has at most one stabilizing solution. Furthermore, we show that 'under some conditions' the Riccati equation has a stabilizing solution if and only if the associated Hamiltonian matrix has no eigenvalues on the imaginary axis.

Chapter 3 reviews some elementary theory on dynamical systems. The dynamics of systems over time are described in this book by sets of differential equations. Therefore the first question which should be answered is whether such equations always have a unique solution. In its full generality the answer to this question is negative. Therefore we review in Chapter 3 some elementary theory that provides us with sufficient conditions to conclude that a set of differential equations has a unique solution. To that end we first consider systems generated by a set of linear differential equations. Using the Jordan canonical form we show that such systems always have a unique solution.

For systems described by a set of nonlinear differential equations we recall some fundamental existence results from literature. A disadvantage of these theorems is that they are often useless if one is considering the existence of solutions for a dynamical system that is subject to control. This is because in most applications one would like to allow the control function to be a discontinuous function of time – a case which does not fit into the previous framework. For that reason the notion of a solution to a set of differential equations is extended. It turns out that this new definition of a solution is sufficient to study optimal control problems with discontinuous control functions.

Stability of dynamical systems plays an important role in convergence analyses of equilibrium strategies. For that reason we give in sections 3.3 and 3.4 an outline of how the behavior of, in particular planar, dynamical systems can be analyzed. Section 3.5 reviews some system theoretical concepts: controllability, stabilizability, observability and detectability. In section 3.6 we specify the standard linear quadratic framework that is used throughout this book. We show how a number of problems can be reformulated into this framework. Finally, we present in the last section of this chapter a number of examples of linear quadratic differential games which should help to motivate the student to study this book.

Chapter 4 deals with the subject of how to solve optimal control problems. The first section deals with the optimization of functions. The rest of the sections deal with dynamic optimization problems. As an introduction we derive the Euler–Lagrange conditions. Then, we prove Pontrjagin's maximum principle. Since the maximum principle only provides a set of necessary conditions which must be satisfied by the optimal solution, we also present some sufficient conditions under which one can conclude that a solution that satisfies the maximum principle conditions is indeed optimal.

Next, we prove the basic theorem of dynamic programming which gives us the optimal control of the problem, provided some conditions are met. It is shown how the maximum principle and dynamic programming are related.

Chapter 5 studies the regular linear quadratic control problem. The problem is called regular because we assume that every control effort is disliked by the control designer. We consider the indefinite problem setting, i.e. in our problem setting we do not make assumptions about preferences of the control designer with respect to the sign of deviations from the state variable from zero. The problem formulation allows for both a control designer who likes some state variables becoming as large as possible and for a control designer who is keen on keeping them as small as possible. For both a finite- and infinite-planning horizon we derive necessary and sufficient conditions for the existence of a solution for this control problem. For the infinite-planning horizon setting this is done with the additional assumption that the closed-loop system must be stabilized by the chosen control. These existence conditions are phrased in terms of solvability conditions on Riccati equations. Moreover, conditions are provided under which the finite-planning horizon solution converges. We show that, generically, this solution will converge to the solution of the infinite-planning horizon problem.

Chapter 6 is the first chapter on differential games. It considers the case that players cooperate to achieve their goals in which case, in general, a curve of solutions results. Each of the solutions on this curve (the Pareto frontier) has the property that it cannot be improved by all the players simultaneously. We show for our linear quadratic setting how solutions of this Pareto frontier can be determined. Moreover, we show how the whole Pareto frontier can be calculated if all the individual players want to avoid the state variables deviating from zero. This can be done by solving a parameterized linear quadratic control problem.

Given this cooperative mode of play from the players the question arises as to how they will coordinate their actions or, to put it another way, which solution on the Pareto frontier will result. In section 6.2 we present a number of outcomes that may result. Different outcomes are obtained as a consequence of the fact that the sought solution satisfies different desired properties. For some of these outcomes we indicate how they can be calculated numerically.

Chapter 7 considers the case that the players do not cooperate to realize their goals. Furthermore, the basic assumption in this section is that the players have to formulate their actions as soon as the system starts to evolve and these actions cannot be changed once the system is running. Under these assumptions we look for control actions (Nash equilibrium actions) that are such that no player can improve his position by a unilateral deviation from such a set of actions. Given this problem setting we derive in section 7.2, for a finite-planning horizon, a both necessary and sufficient condition under which, for every initial state, there exists a Nash equilibrium. It turns out that if an equilibrium exists, it is unique. Moreover we show that during some time interval a Nash equilibrium exists if and only if some Riccati differential equation has a solution. A numerical algorithm is provided to calculate the unique Nash equilibrium actions.

For the infinite-planning horizon case things are more involved. In this case, if an equilibrium exists at all, it will in general not be unique. That is, in most cases there will exist an infinite number of Nash equilibrium actions. We show that if the equilibrium actions should permit a feedback synthesis, then the game has an equilibrium if and only if a set of coupled algebraic Riccati equations has a stabilizing solution; but, also in this case, there may exist an infinite number of equilibrium actions. A numerical algorithm is provided to calculate these equilibrium actions. A necessary and sufficient condition is given under which the game has a unique equilibrium. This equilibrium always permits a feedback synthesis.

Finally, we show that generically the finite-planning horizon equilibrium actions converge. If convergence takes place they usually converge to the actions implied by the infinite-planning horizon solution which stabilizes the system most. The chapter concludes with elaborating the scalar case and providing some examples from economics.

Chapter 8 also considers the non-cooperative mode of play. However, the basic assumption in this section is that the players know the exact state of the system at every point in time and, furthermore, they use linear functions of this state as a means of control to realize their goals. For a finite-planning horizon we show that this game has for every initial state a linear feedback Nash equilibrium if and only if a set of coupled Riccati differential equations has a symmetric set of solutions. Moreover, this equilibrium is unique and the equilibrium actions are a linear function of these Riccati solutions.

For the infinite-planning horizon case things are even more involved. In this case the game has for every initial state a Nash equilibrium if and only if a set of coupled algebraic Riccati equations has a set of symmetric solutions which, if they are simultaneously used to control the system, stabilize it. We elaborate the scalar case and show, in particular, that the number of equilibria may range from zero to $2^N - 1$. A computational algorithm is provided to calculate all Nash equilibrium actions. For the non-scalar case it is shown that there are games which have an infinite number of equilibrium actions.

Finally we show that in the two-player scalar game, if deviations of the state variable from zero are penalized, the solution of the finite-planning horizon game converges if the planning horizon converges. This solution always converges to a solution of the infinite-planning horizon game. However, different from the open-loop case, the converged solution now depends crucially on the scrap values used by both players.

Chapter 9 is the last chapter on non-cooperative games. It considers what effect uncertainty has on the equilibrium actions assuming that players are aware of the fact that they have to control a system characterized by dynamic quasi-equilibrium. That is, up to now we assumed that optimization takes place with no regard to possible deviations. It can safely be assumed, however, that agents follow a different strategy in reality. If an accurate model can be formed for all of the system, it will in general be complicated and difficult to handle. Moreover, it may be unwise to optimize on the basis of a model which is too detailed, in view of possible changes in dynamics that may take place in the course of time and that may be hard to predict. It makes more sense for agents to work on the basis of a relatively simple model and to look for strategies that are robust with respect to deviations between the model and reality.

We consider two approaches to model such situations. One is based on a stochastic approach. The other is based on the introduction of a malevolent deterministic disturbance input and the specification of how each player will cope with his aversion against this input.

We show that the equilibrium actions from Chapter 8 are also equilibrium actions for the stochastic counterparts of the games we study in this chapter. For the deterministic approach we see a more diverse pattern of consequences. Equilibrium actions may cease to exist for the adapted game, whereas opposite results are possible too. That is, a game which at first did not have an equilibrium may now have one or more equilibria. Sufficient existence conditions for such, so-called soft-constrained, Nash equilibria are provided. These conditions are formulated in terms of whether certain Riccati (in)equalities have an appropriate solution. For the scalar case, again, an algorithm is provided to calculate all soft-constrained Nash equilibria.

Finally, we show that the deterministic approach also facilitates the so-called linear exponential gaussian stochastic interpretation. That is, by considering a stochastic framework with gaussian white noise and players considering some exponential cost

function, the same equilibrium actions result. This result facilitates a stochastic interpretation of worst-case design and vice versa.

1.4 Notes and references

For the historical survey in this chapter we extensively used the MacTutor History of
Mathematics Archive from the University of St. Andrews in Scotland (2003) and the
outline of the history of game theory by Walker (2001). Furthermore, the paper by
Breitner (2002) was used for a reconstruction of the early days of dynamic game theory.

2

Linear algebra

This chapter reviews some basic linear algebra including material which in some instances goes beyond the introductory level. A more detailed treatment of the basics of linear algebra can for example, be found in Lay (2003), whereas Lancaster and Tismenetsky (1985) provide excellent work for those who are interested in more details at an advanced level. We will outline the most important basic concepts in linear algebra together with some theorems we need later on for the development of the theory. A detailed treatment of these subjects and the proofs of most of these theorems is omitted, since they can be found in almost any textbook that provides an introduction to linear algebra. The second part of this chapter deals with some subjects which are, usually, not dealt with in an introduction to linear algebra. This part provides more proofs because either they cannot easily be found in the standard linear algebra textbooks or they give an insight into the understanding of problems which will be encountered later on in this book.

2.1 Basic concepts in linear algebra

Let $\mathbb{R}$ denote the set of real numbers and $\mathbb{C}$ the set of complex numbers. For those who are not familiar with the set of complex numbers, a short introduction to this set is given in section 2.3. Furthermore, let $\mathbb{R}^n$ be the set of vectors with n entries, where each entry is an element of $\mathbb{R}$. Now let $x_1, \ldots, x_k \in \mathbb{R}^n$. Then an element of the form $\alpha_1 x_1 + \cdots + \alpha_k x_k$ with $\alpha_i \in \mathbb{R}$ is a **linear combination** of $x_1, \ldots, x_k$. The set of all linear combinations of $x_1, x_2, \ldots, x_k \in \mathbb{R}^n$, called the **Span** of $x_1, x_2, \ldots, x_k$, constitutes a **linear subspace** of $\mathbb{R}^n$. That is, with any two elements in this set the sum and any scalar multiple of an element also belong to this set. We denote this set by Span $\{x_1, x_2, \ldots, x_k\}$.

A set of vectors $x_1, x_2, \ldots, x_k \in \mathbb{R}^n$ are called **linearly dependent** if there exists $\alpha_1, \ldots, \alpha_k \in \mathbb{R}$, not all zero, such that $\alpha_1 x_1 + \cdots + \alpha_k x_k = 0$; otherwise they are said to be **linearly independent**.

LQ Dynamic Optimization and Differential Games   J. Engwerda
© 2005 John Wiley & Sons, Ltd

Let S be a subspace of $\mathbb{R}^n$, then a set of vectors $\{b_1, b_2 \ldots, b_k\}$ is called a **basis** for S if this set of vectors are linearly independent and $S = \text{Span}\{b_1, b_2, \ldots, b_k\}$.

Example 2.1

Consider the vectors $e_1 := \begin{bmatrix} 1 \\ 0 \end{bmatrix}$ and $e_2 := \begin{bmatrix} 0 \\ 1 \end{bmatrix}$ in $\mathbb{R}^2$. Then, both $\{e_1, e_2\}$ and $\left\{ \begin{bmatrix} 1 \\ 2 \end{bmatrix}, \begin{bmatrix} -1 \\ 1 \end{bmatrix} \right\}$ are a basis for $\mathbb{R}^2$. The set $\{e_1, e_2\}$ is called the **standard basis** for $\mathbb{R}^2$.

□

So, a basis for a subspace S is not unique. However, all bases for S have the same number of elements. This number is called the **dimension** of S and is denoted by $\dim(S)$. In the above example the dimension of $\mathbb{R}^2$ is 2.

Next we consider the problem under which conditions are two vectors perpendicular. First the **(Euclidean) length** of a vector x is introduced which will be denoted by $\| x \|_2$. If $x = \begin{bmatrix} \alpha_1 \\ \alpha_2 \end{bmatrix}$ then, using the theorem of Pythagoras, the length of x is $\| x \|_2 = \sqrt{\alpha_1^2 + \alpha_2^2}$.

Using induction it is easy to verify that the length of a vector $x = \begin{bmatrix} \alpha_1 \\ \vdots \\ \alpha_n \end{bmatrix} \in \mathbb{R}^n$ is $\| x \|_2 = \sqrt{\alpha_1^2 + \cdots + \alpha_n^2}$. Introducing the superscript T for **transposition** of a vector, i.e. $x^T = [\alpha_1 \cdots \alpha_n]$, we can rewrite this result in shorthand as $\| x \|_2 = \sqrt{x^T x}$. Now, two vectors x and y are perpendicular if they enclose an angle of $90°$. Using Pythagoras theorem again we conclude that two vectors x and y are perpendicular if and only if the length of the hypotenuse $\| x - y \|^2 = \| x \|^2 + \| y \|^2$. Or, rephrased in our previous terminology: $(x - y)^T (x - y) = x^T x + y^T y$. Using the elementary vector calculation rules and the fact that the transpose of a scalar is the same scalar (i.e. $x^T y = y^T x$) straightforward calculation shows that the following theorem holds.

Theorem 2.1

Two vectors $x, y \in \mathbb{R}^n$ are perpendicular if and only if $x^T y = 0$.   □

Based on this result we next introduce the concept of orthogonality. A set of vectors $\{x_1, \ldots, x_n\}$ are mutually **orthogonal** if $x_i^T x_j = 0$ for all $i \neq j$ and **orthonormal** if $x_i^T x_j = \delta_{ij}$. Here δ_{ij} is the **Kronecker delta function** with $\delta_{ij} = 1$ for $i = j$ and $\delta_{ij} = 0$ for $i \neq j$. More generally, a collection of subspaces $S_1, \ldots, S_k$ are mutually orthogonal if $x^T y = 0$ whenever $x \in S_i$ and $y \in S_j$, for $i \neq j$.

The **orthogonal complement** of a subspace S is defined by

$$S^\perp := \{y \in \mathbb{R}^n | y^T x = 0 \text{ for all } x \in S\}.$$

A set of vectors $\{u_1, u_2, \ldots, u_k\}$ is called an **orthonormal basis** for a subspace $S \subset \mathbb{R}^n$ if they form a basis of S and are orthonormal. Using the orthogonalization procedure of Gram–Schmidt it is always possible to extend such a basis to a full orthonormal basis $\{u_1, u_2, \ldots, u_n\}$ for $\mathbb{R}^n$. This procedure is given in the following theorem.

Theorem 2.2 (Gram–Schmidt)[1]

Given a basis $\{b_1, b_2, \ldots, b_m\}$ for a subspace S of $\mathbb{R}^n$, define

$$v_1 := b_1$$

$$v_2 := b_2 - \frac{b_2^T v_1}{v_1^T v_1} v_1$$

$$v_3 := b_3 - \frac{b_3^T v_1}{v_1^T v_1} v_1 - \frac{b_3^T v_2}{v_2^T v_2} v_2$$

$$\vdots$$

$$v_m := b_m - \frac{b_m^T v_1}{v_1^T v_1} v_1 - \frac{b_m^T v_2}{v_2^T v_2} v_2 - \cdots - \frac{b_m^T v_{m-1}}{v_{m-1}^T v_{m-1}} v_{m-1}.$$

Then $\left\{ \frac{v_1}{\|v_1\|}, \frac{v_2}{\|v_2\|}, \ldots, \frac{v_m}{\|v_m\|} \right\}$ is an orthonormal basis for S. Furthermore,

$$\text{Span}\{v_1, v_2, \ldots, v_k\} = \text{Span}\{b_1, b_2, \ldots, b_k\}. \qquad \square$$

In the above sketched case, with $u_i = \frac{v_i}{\|v_i\|}$,

$$S^\perp = \text{Span}\{u_{k+1}, \ldots, u_n\}.$$

Since $\{u_1, \ldots, u_n\}$ form a basis for $\mathbb{R}^n$, $\{u_{k+1}, \ldots, u_n\}$ is called an **orthonormal completion** of $\{u_1, u_2, \ldots, u_k\}$.

An ordered array of mn elements $a_{ij} \in \mathbb{R}$, $i = 1, \ldots, n;\ j = 1, \ldots, m$, written in the form

$$A = \begin{bmatrix} a_{11} & a_{12} & \cdots & a_{1m} \\ a_{21} & a_{22} & \cdots & a_{2m} \\ \vdots & \vdots & \vdots & \vdots \\ a_{n1} & a_{n2} & \cdots & a_{nm} \end{bmatrix}$$

is said to be an $n \times m$-**matrix** with entries in $\mathbb{R}$. The set of all $n \times m$-matrices will be denoted by $\mathbb{R}^{n \times m}$. If $m = n$, A is called a **square matrix**. With matrix $A \in \mathbb{R}^{n \times m}$ one can associate the linear map $x \to Ax$ from $\mathbb{R}^m \to \mathbb{R}^n$. The **kernel** or **null space** of A is defined by

$$\ker A = N(A) := \{x \in \mathbb{R}^m | Ax = 0\},$$

and the **image** or **range** of A is

$$\text{Im } A = R(A) := \{y \in \mathbb{R}^n | y = Ax, x \in \mathbb{R}^m\}.$$

---

[1]Gram was a Danish mathematician who lived from 1850–1916 and worked in the insurance business. Schmidt was a German mathematician who lived from 1876–1959. Schmidt reproved the orthogonalization procedure from Gram in a more general context in 1906. However, Gram was not the first to use this procedure. The procedure seems to be a result of Laplace and it was essentially used by Cauchy in 1836.

One can easily verify that $\ker A$ is a subspace of $\mathbb{R}^m$ and $\operatorname{Im} A$ is a subspace of $\mathbb{R}^n$. Furthermore, we recall the following fundamental result.

Theorem 2.3

Let $A \in \mathbb{R}^{n \times m}$. Then,

1. $\dim(\ker A) + \dim(\operatorname{Im} A) = m$.

2. $\dim(\operatorname{Im} A) = \dim((\ker A)^{\perp})$. $\square$

Let a_i, $i = 1, \ldots, m$, denote the columns of matrix $A \in \mathbb{R}^{n \times m}$, then

$$\operatorname{Im} A = \operatorname{Span}\{a_1, \ldots, a_m\}.$$

The **rank** of a matrix A is defined by $\operatorname{rank}(A) = \dim(\operatorname{Im} A)$, and thus the rank of a matrix is just the number of independent columns in A. One can show that $\operatorname{rank}(A) = \operatorname{rank}(A^T)$. Consequently, the rank of a matrix also coincides with the number of independent rows in A. A matrix $A \in \mathbb{R}^{n \times m}$ is said to have **full row rank** if $n \leq m$ and $\operatorname{rank}(A) = n$. Equally, it is said to have **full column rank** if $m \leq n$ and $\operatorname{rank}(A) = m$. A full rank square matrix is called a **nonsingular** or **invertible** matrix, otherwise it is called singular. The following result is well-known.

Theorem 2.4

Let $A \in \mathbb{R}^{n \times m}$ and $b \in \mathbb{R}^n$. Then, $Ax = b$ has

1. at most one solution x if A has full column rank;

2. at least one solution x if A has full row rank;

3. a solution if and only if $\operatorname{rank}([A|b]) = \operatorname{rank}(A)$;

4. a unique solution if A is invertible. $\square$

If a matrix A is invertible one can show that the matrix equation $AX = I$ has a unique solution $X \in \mathbb{R}^{n \times n}$. Here I is the $n \times n$ **identity matrix** with entries $e_{ij} := \delta_{ij}$, $i, j = 1, \ldots, n$, and δ_{ij} is the Kronecker delta. Moreover, this matrix X also satisfies the matrix equation $XA = I$. Matrix X is called the **inverse** of matrix A and the notation A^{-1} is used to denote this inverse.

A notion that is useful to see whether or not a square $n \times n$ matrix A is singular is the **determinant** of A, denoted by $\det(A)$. The next theorem lists some properties of determinants.

Theorem 2.5

Let $A, B \in \mathbb{R}^{n \times n}$; $C \in \mathbb{R}^{n \times m}$; $D \in \mathbb{R}^{m \times m}$; and $0 \in \mathbb{R}^{m \times n}$ be the matrix with all entries zero. Then,

1. $\det(AB) = \det(A)\det(B)$;

2. A is invertible if and only if $\det(A) \neq 0$;

3. If A is invertible, $\det(A^{-1}) = \frac{1}{\det(A)}$;

4. $\det(A) = \det(A^T)$;

5. $\det\left(\begin{bmatrix} A & C \\ 0 & D \end{bmatrix}\right) = \det(A)\det(D)$.   □

Next we present Cramer's rule to calculate the inverse of a matrix. This way of calculating the inverse is sometimes helpful in theoretical calculations, as we will see later on.

For any $n \times n$ matrix A and any $b \in \mathbb{R}^n$, let $A_i(b)$ be the matrix obtained from A by replacing column i by the vector b

$$A_i(b) := [a_1 \cdots a_{i-1}\ b\ a_{i+1} \cdots a_n].$$

Theorem 2.6 (Cramer)[2]

Let A be an invertible $n \times n$ matrix. For any $b \in \mathbb{R}^n$, the unique solution x of $Ax = b$ has entries given by

$$x_i = \frac{\det A_i(b)}{\det A}, \quad i = 1, 2, \ldots, n.$$ □

Cramer's rule leads easily to a general formula for the inverse of an $n \times n$ matrix A. To see this, notice that the *jth* column of A^{-1} is a vector x that satisfies

$$Ax = e_j$$

where e_j is *jth* column of the identity matrix. By Cramer's rule,

$$x_{ij} := [(i,j) - \text{entry of } A^{-1}] = \frac{\det A_i(e_j)}{\det A}. \tag{2.1.1}$$

Let A_{ji} denote the submatrix of A formed by deleting row j and column i from matrix A. An expansion of the determinant down column i of $A_i(e_j)$ shows that

$$\det A_i(e_j) = (-1)^{i+j}\det A_{ji} =: C_{ji}.$$

---

[2]Cramer was a well-known Swiss mathematician who lived from 1704–1752. He showed this rule in an appendix of his book (Cramer, 1750). However, he was not the first one to give this rule. The Japanese mathematician Takakazu and the German mathematician Leibniz had considered this idea already in 1683 and 1693, respectively, long before a separate theory of matrices was developed.

By (2.1.1), the (i,j)–entry of A^{-1} is C_{ji} divided by $\det A$. Thus

$$A^{-1} = \frac{1}{\det} \begin{bmatrix} C_{11} & C_{21} & \cdots & C_{n1} \\ C_{12} & C_{22} & \cdots & C_{n2} \\ \vdots & \vdots & \vdots & \vdots \\ C_{1n} & C_{2n} & \cdots & C_{nn} \end{bmatrix} \qquad (2.1.2)$$

The entries C_{ij} are the so-called **cofactors** of matrix A. Notice that the subscripts on C_{ji} are the reverse of the entry number (i,j) in the matrix. The matrix of cofactors on the right-hand side of (2.1.2) is called the **adjoint** of A, denoted by $\text{adj}\,A$. The next theorem simply restates (2.1.2).

Theorem 2.7

Let A be an invertible $n \times n$ matrix. Then

$$A^{-1} = \frac{1}{\det(A)}\,\text{adj}\,(A). \qquad (2.1.3)$$

$\square$

Example 2.2

If

$$A = \begin{bmatrix} 2 - \lambda & 1 & -1 \\ 0 & 2 - \lambda & 0 \\ 1 & -1 & 3 - \lambda \end{bmatrix},$$

then

$$\text{adj}(A) = \begin{bmatrix} \det\begin{bmatrix} 2 - \lambda & 0 \\ -1 & 3 - \lambda \end{bmatrix} & -\det\begin{bmatrix} 1 & -1 \\ -1 & 3 - \lambda \end{bmatrix} & \det\begin{bmatrix} 1 & -1 \\ 2 - \lambda & 0 \end{bmatrix} \\ -\det\begin{bmatrix} 0 & 0 \\ 1 & 3 - \lambda \end{bmatrix} & \det\begin{bmatrix} 2 - \lambda & -1 \\ 1 & 3 - \lambda \end{bmatrix} & -\det\begin{bmatrix} 2 - \lambda & -1 \\ 0 & 0 \end{bmatrix} \\ \det\begin{bmatrix} 0 & 2 - \lambda \\ 1 & -1 \end{bmatrix} & -\det\begin{bmatrix} 2 - \lambda & 1 \\ 1 & -1 \end{bmatrix} & \det\begin{bmatrix} 2 - \lambda & 1 \\ 0 & 2 - \lambda \end{bmatrix} \end{bmatrix}$$

$$= \begin{bmatrix} \lambda^2 - 5\lambda + 6 & \lambda - 2 & -\lambda + 2 \\ 0 & \lambda^2 - 5\lambda + 7 & 0 \\ \lambda - 2 & -\lambda + 3 & \lambda^2 - 4\lambda + 4 \end{bmatrix}.$$

Notice that all entries of the adjoint matrix are polynomials with a degree that does not exceed 2. Furthermore, $\det(A) = -\lambda^3 + 7\lambda^2 - 17\lambda + 14$, which is a polynomial of degree 3. $\square$

If the entries of A are a_{ij}, $i, j = 1, \ldots, n$, the **trace** of matrix A is defined as $\text{trace}(A)$ $:= \sum_{i=1}^{n} a_{ii}$. The next properties are well-known.

Theorem 2.8

1. $\text{trace}(\alpha A) = \alpha \, \text{trace} \, (A)$, $\forall A \in \mathbb{R}^{n \times n}$ and $\alpha \in \mathbb{R}$;

2. $\text{trace}(A + B) = \text{trace} \, (A) + \text{trace} \, (B)$, $\forall A, B \in \mathbb{R}^{n \times n}$;

3. $\text{trace}(AB) = \text{trace} \, (BA)$, $\forall A \in \mathbb{R}^{n \times m}$, $B \in \mathbb{R}^{m \times n}$. $\qquad\qquad\square$

2.2 Eigenvalues and eigenvectors

Let $A \in \mathbb{R}^{n \times n}$, then $\lambda \in \mathbb{R}$ is called an **eigenvalue** of A if there exists a vector $x \in \mathbb{R}^n$, different from zero, such that $Ax = \lambda x$. If such a scalar λ and corresponding vector x exist, the vector x is called an **eigenvector**. If A has an eigenvalue λ it follows that there exists a nonzero vector x such that $(A - \lambda I)x = 0$. Stated differently, matrix $A - \lambda I$ is singular. So, according to Theorem 2.5, λ is an eigenvalue of matrix A if and only if $\det(A - \lambda I) = 0$. All vectors in the null space of $A - \lambda I$ are then the eigenvectors corresponding to λ. As a consequence we have that the set of eigenvectors corresponding with an eigenvalue λ forming a subspace. This subspace is called the **eigenspace** of λ and we denote this subset by E_λ. So, to find the eigenvalues of a matrix A we have to find those values λ for which $\det(A - \lambda I) = 0$. Since $p(\lambda) := \det(A - \lambda I)$ is a polynomial of degree n, $p(\lambda)$ is called the **characteristic polynomial** of A. The set of roots of this polynomial is called the **spectrum** of A and is denoted by $\sigma(A)$.

An important property of eigenvectors corresponding to different eigenvalues is that they are always independent.

Theorem 2.9

Let $A \in \mathbb{R}^{n \times n}$ and λ_1, λ_2 be two different eigenvalues of A with corresponding eigenvectors x_1 and x_2, respectively. Then $\{x_1, x_2\}$ are linearly independent.

Proof

Assume $x_2 = \mu x_1$, for some nonzero scalar $\mu \in \mathbb{R}$. Then $0 = A(x_2 - \mu x_1) = Ax_2 - \mu A x_1 = \lambda_2 x_2 - \mu \lambda_1 x_1 = \lambda_2 \mu x_1 - \mu \lambda_1 x_1 = \mu(\lambda_2 - \lambda_1)x_1 \neq 0$, due to the stated assumptions. So this yields a contradiction and therefore our assumption that x_2 is a multiple of x_1 must be incorrect. $\qquad\qquad\square$

Example 2.3

1. Consider matrix $A_1 = \begin{bmatrix} -1 & -3 \\ 2 & 4 \end{bmatrix}$. The characteristic polynomial of A_1 is

$$p(\lambda) = \det(A_1 - \lambda I) = (\lambda - 1)(\lambda - 2).$$

So, $\sigma(A_1) = \{1, 2\}$. Furthermore, $E_1 = N(A_1 - I) = \{\alpha \begin{bmatrix} 3 \\ -2 \end{bmatrix}, \alpha \in \mathbb{R}\}$ and $E_2 = N(A_1 - 2I) = \{\alpha \begin{bmatrix} -1 \\ 1 \end{bmatrix}, \alpha \in \mathbb{R}\}$. .

2. Consider matrix $A_2 = \begin{bmatrix} -6 & 4 \\ -4 & 2 \end{bmatrix}$. The characteristic polynomial of A_2 is $(\lambda + 2)^2$. So, $\sigma(A_2) = \{-2\}$. Furthermore, $E_{-2} = N(A_2 + 2I) = \{\alpha \begin{bmatrix} 1 \\ 1 \end{bmatrix}, \alpha \in \mathbb{R}\}$.

3. Consider matrix $A_3 = \begin{bmatrix} 3 & 0 \\ 0 & 3 \end{bmatrix}$. The characteristic polynomial of A_3 is $(\lambda - 3)^2$. So, $\sigma(A_3) = \{3\}$. Furthermore, $E_3 = N(A_3 - 3I) = \mathbb{R}^2$.

4. Consider matrix $A_4 = \begin{bmatrix} 3 & 1 \\ -2 & 1 \end{bmatrix}$. The characteristic polynomial of A_4 is $(\lambda^2 - 4\lambda + 5)$. This polynomial has no real roots. So, matrix A_4 has no real eigenvalues. □

The above example illustrates a number of properties that hold in the general setting too (Lancaster and Tismenetsky, 1985).

Theorem 2.10

Any polynomial $p(\lambda)$ can be factorized as the product of different linear and quadratic terms, i.e.

$$p(\lambda) = c(\lambda - \lambda_1)^{n_1}(\lambda - \lambda_2)^{n_2} \ldots (\lambda - \lambda_k)^{n_k}(\lambda^2 + b_{k+1}\lambda + c_{k+1})^{n_{k+1}} \ldots (\lambda^2 + b_r\lambda + c_r)^{n_r},$$

for some scalars c, λ_i, b_i and c_i. Here, for $i \neq j$, $\lambda_i \neq \lambda_j$ and $\begin{bmatrix} b_i \\ c_i \end{bmatrix} \neq \begin{bmatrix} b_j \\ c_j \end{bmatrix}$ and the quadratic terms do not have real roots. Furthermore, $\sum_{i=1}^{k} n_i + 2\sum_{i=k+1}^{r} n_i = n$. □

The power index n_i appearing in the factorization with the factor $\lambda - \lambda_i$ is called the **algebraic multiplicity** of the eigenvalue λ_i. Closely related to this number is the so-called **geometric multiplicity** of the eigenvalue λ_i, which is the dimension of the corresponding eigenspace E_{λ_i}. In Example 2.3 we see that for every eigenvalue the geometric multiplicity is smaller than its algebraic multiplicity. For instance, for A_1 both multiplicities are 1 for both eigenvalues, whereas for A_2 the geometric multiplicity of the eigenvalue -2 is 1 and its algebraic multiplicity is 2. This property holds in general.

Theorem 2.11

Let λ_i be an eigenvalue of A. Then its geometric multiplicity is always smaller than (or equal to) its algebraic multiplicity.

Proof

Assume λ_1 is an eigenvalue of A and $\{b_1, \ldots, b_k\}$ is a set of independent eigenvectors that span the corresponding eigenspace. Extend this basis of the eigenspace to a basis

$\{b_1, \ldots, b_k, b_{k+1}, \ldots, b_n\}$ for $\mathbb{R}^n$ (using the Gram–Schmidt procedure, for example). Let S be the $n \times n$ matrix with columns $b_1, \ldots, b_n$. Then

$$
\begin{aligned}
AS &= (Ab_1 \ldots Ab_k \; Ab_{k+1} \ldots Ab_n) \\
&= (\lambda_1 b_1 \ldots \lambda_1 b_k \; Ab_{k+1} \ldots Ab_n) \\
&= S \begin{bmatrix} \lambda_1 I_k & D \\ 0 & E_{n-k} \end{bmatrix},
\end{aligned}
$$

where I_k denotes the $k \times k$ identity matrix and E_{n-k} a square $(n-k) \times (n-k)$ matrix. Then right-multiplying this relation by S^{-1}, we have $A = S \begin{bmatrix} \lambda_1 I_k & D \\ 0 & E_{n-k} \end{bmatrix} S^{-1}$. Therefore

$$
\begin{aligned}
\det(A - \lambda I) &= \det\left(S \begin{bmatrix} \lambda_1 I_k & D \\ 0 & E_{n-k} \end{bmatrix} S^{-1} - \lambda I \right) \\
&= \det(S) \det\left(\begin{bmatrix} \lambda_1 I_k & D \\ 0 & E_{n-k} \end{bmatrix} - \lambda I \right) \det(S^{-1}) \\
&= \det\left(\begin{bmatrix} (\lambda_1 - \lambda) I_k & D \\ 0 & E_{n-k} - \lambda I_{n-k} \end{bmatrix} \right) \\
&= (\lambda_1 - \lambda)^k \det(E_{n-k} - \lambda I_{n-k}). \qquad \square
\end{aligned}
$$

It turns out that if, for an eigenvalue λ_1, there holds a strict inequality between its geometric and algebraic multiplicity, there is a natural way to extend the eigenspace towards a larger subspace whose dimension has the corresponding algebraic multiplicity. This larger subspace is called the **generalized eigenspace** of the eigenvalue λ_1 and is given by $E^g_{\lambda_1} := N(A - \lambda_1 I)^n$. In fact, there exists a minimal index $p \le n$ for which $E^g_{\lambda_1} = N(A - \lambda_1 I)^p$. This, follows from the property that $N(A - \lambda_1 I)^p \subset N(A - \lambda_1 I)^{p+1}$, for all $p = 1, 2, \ldots$, and the fact that whenever $N(A - \lambda_1 I)^p = N(A - \lambda_1 I)^{p+1}$ also $N(A - \lambda_1 I)^{p+1} = N(A - \lambda_1 I)^{p+2}$ (see Exercises).

2.3 Complex eigenvalues

In the previous subsection we saw that the characteristic polynomial of an $n \times n$ matrix involves a polynomial of degree n that can be factorized as the product of different linear and quadratic terms (see Theorem 2.10). Furthermore, it is not possible to factorize any of these quadratic terms as the product of two linear terms. Without loss of generality such a quadratic term can be written as

$$
p(\lambda) = \lambda^2 - 2a\lambda + b^2 + a^2. \tag{2.3.1}
$$

Next introduce the symbol i to denote the square root of -1. So, by definition

$$
i := \sqrt{-1}.
$$

Using this notation the equation $p(\lambda) = 0$ has two solutions, i.e.

$$\lambda_j = \frac{2a \pm \sqrt{4a^2 - 4(a^2 + b^2)}}{2} = a \pm \sqrt{i^2 b^2} = a \pm bi, \ j = 1, 2.$$

Or, stated differently, $p(\lambda)$ in equation (2.3.1) has, with this notation, the two square roots

$$\lambda_1 = a + bi \text{ and } \bar{\lambda}_1 = a - bi. \tag{2.3.2}$$

An expression z of the form

$$z = x + yi$$

where x and y are real numbers and i is the formal symbol satisfying the relation $i^2 = -1$ is called a **complex number**. x is called the **real part** of z and y the **imaginary part** of z. Since any complex number $x + iy$ is uniquely determined by the numbers x and y one can visualize the set of all complex numbers as the set of all points (x, y) in the plane $\mathbb{R}^2$, as in Figure 2.1. The horizontal axis is called the **real axis** because the points $(x, 0)$ on it

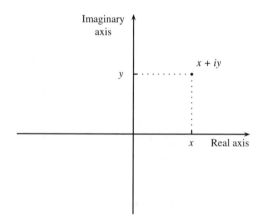

Figure 2.1 The complex plane $\mathbb{C}$

correspond to the real numbers. The vertical axis is the **imaginary axis** because the points $(0, y)$ on it correspond to the **pure imaginary numbers** of the form $0 + yi$, or simply yi. Given this representation of the set of complex numbers it seems reasonable to introduce the **addition** of two complex numbers just like the addition of two vectors in $\mathbb{R}^2$, i.e.

$$(x_1 + y_1 i) + (x_2 + y_2 i) := (x_1 + x_2) + (y_1 + y_2)i.$$

Note that this rule reduces to ordinary addition of real numbers when y_1 and y_2 are zero. Furthermore, by our definition of $i^2 = -1$, we have implicitly also introduced the operation of **multiplication** of two complex numbers. This operation is defined by

$$(x_1 + y_1 i)(x_2 + y_2 i) := x_1 x_2 + x_1 y_2 i + y_1 x_2 i + y_1 y_2 i^2 = (x_1 x_2 - y_1 y_2) + (x_1 y_2 + x_2 y_1)i.$$

Note that this operation also induces a multiplication rule of two vectors in $\mathbb{R}^2$ as

$$\begin{bmatrix} x_1 \\ y_1 \end{bmatrix} * \begin{bmatrix} x_2 \\ y_2 \end{bmatrix} := \begin{bmatrix} x_1 x_2 - y_1 y_2 \\ x_1 y_2 + x_2 y_1 \end{bmatrix}.$$

We will not, however, elaborate this point.

From equation (2.3.2) it is clear that closely related to the complex number $z = x + yi$ is the complex number $x - yi$. The complex number $x - yi$ is called the **conjugate** of z and denote it by $\bar{z}$ (read as 'z bar'). So, the conjugate of a complex number z is obtained by reversing the sign of the imaginary part.

Example 2.4

The conjugate of $z = -2 - 4i$ is $\bar{z} = -2 + 4i$, that is $\overline{-2 - 4i} = -2 + 4i$. □
Geometrically, $\bar{z}$ is the mirror image of z in the real axis (see Figure 2.2).

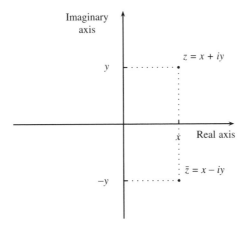

Figure 2.2 The conjugate of a complex number z

The **absolute value** or **modulus** of a complex number $z = x + yi$ is the length of the associated vector $\begin{bmatrix} x \\ y \end{bmatrix}$ in $\mathbb{R}^2$. That is, the absolute value of z is the real number $|z|$ defined by

$$|z| = \sqrt{x^2 + y^2}.$$

This number $|z|$ coincides with the square root of the product of z with its conjugate $\bar{z}$, i.e. $|z| = \sqrt{z\bar{z}}$.

We now turn to the **division** of complex numbers. The objective is to define devision as the inverse of multiplication. Thus, if $z \neq 0$, then the definition of $\frac{1}{z}$ is the complex number w that satisfies

$$wz = 1. \tag{2.3.3}$$

Obviously, it is not a priori clear that for every complex number $z \neq 0$ there always exists a unique number w satisfying this relationship. The next theorem states that this complex number w always exists and gives an explicit representation of this number.

Theorem 2.12

If $z \neq 0$, then equation (2.3.3) has a unique solution, which is

$$w = \frac{1}{|z|^2}\bar{z}.$$

Proof

Let $z = a + bi$ and $w = x + yi$. Then equation (2.3.3) can be written as

$$(x + yi)(a + bi) = 1,$$

or

$$ax - by - 1 + (bx + ay)i = 0.$$

Therefore, the equation (2.3.3) has a unique solution if and only if the next set of two equations has a unique solution x, y

$$ax - by - 1 = 0$$
$$bx + ay = 0.$$

Or, stated differently, the equation

$$\begin{bmatrix} a & -b \\ b & a \end{bmatrix}\begin{bmatrix} x \\ y \end{bmatrix} = \begin{bmatrix} 1 \\ 0 \end{bmatrix}$$

has a unique solution. Note that $\det\left(\begin{bmatrix} a & -b \\ b & a \end{bmatrix}\right) = a^2 + b^2 \neq 0$. So, the solution $\begin{bmatrix} x \\ y \end{bmatrix}$ is unique. It is easily verified that $x = \frac{a}{a^2+b^2}$ and $y = \frac{-b}{a^2+b^2}$ satisfy the equation, from which the stated result follows straightforwardly. $\qquad\square$

Example 2.5

If $z = 3 + 4i$, then $\frac{1}{z} = \frac{1}{25}(3 - 4i)$. The complex number $z = \frac{2-i}{3+4i}$ can be written in the standard form $z = \frac{1}{25}(2 - i)(3 - 4i) = \frac{1}{25}(2 - 11i) = \frac{2}{25} - \frac{11}{25}i$. $\qquad\square$

Theorem 2.13 lists some useful properties of the complex conjugate. The proofs are elementary and left as an exercise to the reader.

Theorem 2.13

For any complex numbers z_1 and z_2

1. $\overline{z_1 + z_2} = \bar{z}_1 + \bar{z}_2$
2. $\overline{z_1 z_2} = \bar{z}_1 \bar{z}_2$ (and consequently $\overline{z_1/z_2} = \bar{z}_1/\bar{z}_2$).
3. $\bar{\bar{z}}_1 = z_1$. $\qquad\square$

Just as vectors in $\mathbb{R}^n$ and matrices in $\mathbb{R}^{n\times m}$ are defined, one can define vectors in $\mathbb{C}^n$ and matrices in $\mathbb{C}^{n\times m}$ as vectors and matrices whose entries are now complex numbers. The operations of addition and (scalar) multiplication are defined in the same way. Furthermore for a matrix Z with elements z_{ij} from $\mathbb{C}$ the **complex conjugate** $\bar{Z}$ of Z is defined as the matrix obtained from Z by changing all its entries to their complex conjugates. In other words, the entries of $\bar{Z}$ are $\bar{z}_{ij}$.

Example 2.6

Let z be any complex number, $z_1 = \begin{bmatrix} 1+i \\ 2-3i \end{bmatrix}$, $z_2 = \begin{bmatrix} 2-4i \\ -3+2i \end{bmatrix}$, $A_1 = \begin{bmatrix} 1+i & i \\ 2-3i & 2 \end{bmatrix}$ and $A_2 = \begin{bmatrix} 4+2i & 1-i \\ -2-i & 1 \end{bmatrix}$. Then,

$$z_1 + z_2 = \begin{bmatrix} 3-3i \\ -1-i \end{bmatrix}, A_1 + A_2 = \begin{bmatrix} 5+3i & 1 \\ -4i & 3 \end{bmatrix}, z\begin{bmatrix} 2-4i \\ -3+2i \end{bmatrix} = \begin{bmatrix} z(2-4i) \\ z(-3+2i) \end{bmatrix},$$

$$zA_1 = \begin{bmatrix} z(1+i) & zi \\ z(2-3i) & 2z \end{bmatrix}, A_1z_1 = \begin{bmatrix} (1+i)^2 + 3 + 2i \\ (2-3i)(1+i) + 4 - 6i \end{bmatrix} = \begin{bmatrix} 3+4i \\ 9-7i \end{bmatrix} \text{ and,}$$

$$A_1A_2 = \begin{bmatrix} (1+i)(4+2i) + i(-2-i) & (1+i)(1-i) + i \\ (2-3i)(4+2i) + 2(-2-i) & (2-3i)(1-i) + 2 \end{bmatrix} = \begin{bmatrix} 3+4i & 2+i \\ 10-10i & 1-5i \end{bmatrix}.$$

□

Any vector $z \in \mathbb{C}^n$ can be written as $z = x + yi$, where $x, y \in \mathbb{R}^n$. Similarly any matrix $Z \in \mathbb{C}^{n\times m}$ can be written as $Z = A + Bi$, where $A, B \in \mathbb{R}^{n\times m}$.

The eigenvalue–eigenvector theory already developed for matrices in $\mathbb{R}^{n\times n}$ applies equally well to matrices with complex entries. That is, a complex scalar λ is called a **complex eigenvalue** of a complex matrix $Z \in \mathbb{C}^{n\times n}$ if there is a nonzero complex vector z such that

$$Zz = \lambda z.$$

Before we elaborate this equation, we first generalize the notion of determinant to complex matrices. The definition coincides with the definition of the determinant of a real matrix. That is, the **determinant of a complex matrix** $Z = [z_{ij}]$ is

$$\det Z = z_{11}\det Z_{11} - z_{12}\det Z_{12} + \cdots + (-1)^{n+1}z_{1n}\det Z_{1n},$$

where Z_{ij} denotes the submatrix of Z formed by deleting row i and column j and the determinant of a complex number z is z. One can now copy the theory used in the real case, to derive the results of Theorem 2.5, and show that these properties also apply in the complex case. In particular, one can also show that in the complex case a square matrix Z is nonsingular if and only if its determinant differs from zero. This result will be used to analyze the eigenvalue problem in more detail. To be self-contained, this result is now shown. Its proof requires the basic facts that, like in the real case, adding a multiple of one row to another row of matrix Z, or adding a multiple of one column to another column of Z does not change the determinant of matrix Z. Taking these and the fact that

$\det\left(\begin{bmatrix} A & C \\ 0 & D \end{bmatrix}\right) = \det(A)\det(D)$ for granted (see Exercises) the next fundamental property on the existence of solutions for the set of linear equations $Zz = 0$ is proved.

Theorem 2.14

Let $Z \in \mathbb{C}^{n \times n}$. Then the set of equations

$$Zz = 0$$

has a complex solution $z \neq 0$ if and only if $\det Z = 0$.

Proof

Let $Z = A + iB$ and $z = x + iy$, with $A, B \in \mathbb{R}^{n \times n}$ and $x, y \in \mathbb{R}^n$. Then,

$$Zz = Ax - By + i(Bx + Ay).$$

Therefore $Zz = 0$ has a unique solution if and only if the next set of equations are uniquely solvable for some vectors $x, y \in \mathbb{R}^n$:

$$Ax - By = 0 \text{ and } Bx + Ay = 0.$$

Since this is a set of equations with only real entries, Theorem 2.4 and 2.5 can be used to conclude that this set of equations has a solution different from zero if and only if

$$\det\left(\begin{bmatrix} A & -B \\ B & A \end{bmatrix}\right) = 0.$$

Since adding multiples of one row to another row and adding multiples of one column to another column does not change the determinant of a matrix

$$\det\left(\begin{bmatrix} A & -B \\ B & A \end{bmatrix}\right) = \det\left(\begin{bmatrix} I & iI \\ 0 & I \end{bmatrix}\right)\det\left(\begin{bmatrix} A & -B \\ B & A \end{bmatrix}\right)\det\left(\begin{bmatrix} I & -iI \\ 0 & I \end{bmatrix}\right).$$

(Note that the above mentioned addition operations can indeed be represented in this way.) Spelling out the right-hand side of this equation yields

$$\det\left(\begin{bmatrix} A & -B \\ B & A \end{bmatrix}\right) = \det\left(\begin{bmatrix} A+iB & 0 \\ B & A-iB \end{bmatrix}\right).$$

So, $\det\left(\begin{bmatrix} A & -B \\ B & A \end{bmatrix}\right) = \det(A + iB)\det(A - iB)$. Therefore, $Zz = 0$ has a solution different from zero if and only if $\det(A + iB)\det(A - iB) = 0$. Since $w := \det(A + iB)$ is a complex number and $\det(A - iB) = \bar{w}$ (see Exercises) it follows that $\det(A + iB)\det(A - iB) = 0$ if and only if $w\bar{w} = |w|^2 = 0$, i.e. $w = 0$, which proves the claim. $\qquad\square$

Corollary 2.15

$\lambda \in \mathbb{C}$ is an eigenvalue of A if and only if $\det(A - \lambda I) = 0$. Moreover, all $z \in \mathbb{C}^n (\neq 0)$ satisfying $(A - \lambda I)z = 0$ are eigenvectors corresponding to λ. $\qquad \square$

Example 2.7

(see also Example 2.3 part **4**) Let $A_4 = \begin{bmatrix} 3 & 1 \\ -2 & 1 \end{bmatrix}$. Its characteristic polynomial is

$\det(A_4 - \lambda I) = \lambda^2 - 4\lambda + 5$. The complex roots of this equation are $\lambda_1 = 2 + i$ and

$\lambda_2 = 2 - i$. The eigenvectors corresponding to λ_1 are $N(A_4 - (2 + i)I) = \{\alpha \begin{bmatrix} 1 \\ -1 + i \end{bmatrix},$

$\alpha \in \mathbb{C}\}$. The eigenvectors corresponding to λ_2 are $N(A_4 - (2 - i)I) = \{\alpha \begin{bmatrix} 1 \\ -1 - i \end{bmatrix},$

$\alpha \in \mathbb{C}\}$. $\qquad \square$

From Example 2.7 we see that with $\lambda_1 = 2 + i$ being an eigenvalue of A_4, its conjugate $2 - i$ is also an eigenvalue of A_4. This property is, of course, something one would expect given the facts that the characteristic polynomial of a matrix with real entries can be factorized as a product of linear and quadratic terms, and equation (2.3.2).

Theorem 2.16

Let $A \in \mathbb{R}^{n \times n}$. If $\lambda \in \mathbb{C}$ is an eigenvalue of A and z a corresponding eigenvector, then $\bar{\lambda}$ is also an eigenvalue of A and $\bar{z}$ a corresponding eigenvector.

Proof

By definition x and λ satisfy $Az = \lambda z$. Taking the conjugate on both sides of this equation gives $\overline{Az} = \overline{\lambda z}$. Using Theorem 2.13 and the fact that A is a real matrix, so that its conjugate is matrix A again, yields $A\bar{z} = \bar{\lambda}\bar{z}$. Therefore, by definition, $\bar{z}$ is an eigenvector corresponding with the eigenvalue $\bar{\lambda}$. $\qquad \square$

Theorem 2.17 below, shows that whenever $A \in \mathbb{R}^{n \times n}$ has a complex eigenvalue, then A has a so-called two-dimensional invariant subspace (see Section 2.5 for a formal introduction to this notion) a property that will be used in the next section.

Theorem 2.17

Let $A \in \mathbb{R}^{n \times n}$. If $\lambda = a + bi$ $(a, b \in \mathbb{R}, b \neq 0)$ is a complex eigenvalue of A and $z = x + iy$, with $x, y \in \mathbb{R}^n$, a corresponding eigenvector, then A has a two-dimensional invariant subspace $S = \text{Im}[x \, y]$. In particular:

$$AS = S \begin{bmatrix} a & b \\ -b & a \end{bmatrix}.$$

Proof

Theorem 2.16 shows that both

$$Az = \lambda z \text{ and } A\bar{z} = \bar{\lambda}\bar{z}.$$

Writing out both equations yields

$$Ax + iAy = ax - by + i(bx + ay)$$
$$Ax - iAy = ax - by - i(bx + ay).$$

Adding and subtracting both equations, respectively, gives the next two equations

$$Ax = ax - by \tag{2.3.4}$$
$$Ay = bx + ay. \tag{2.3.5}$$

or, stated differently,

$$A[x\ y] = [Ax\ Ay] = [ax - by\ bx + ay] = [x\ y]\begin{bmatrix} a & b \\ -b & a \end{bmatrix}.$$

So what is left to be shown is that $\{x, y\}$ are linearly independent. To show this, assume that $y = \mu x$ for some real $\mu \neq 0$. Then from equations (2.3.4) and (2.3.5) we get

$$Ax = (a - b\mu)x \tag{2.3.6}$$
$$Ax = \frac{1}{\mu}(b + a\mu)x. \tag{2.3.7}$$

According to equation (2.3.6), x is a real eigenvector corresponding to the real eigenvalue $a - b\mu$; whereas according to equation (2.3.7) x is a real eigenvector corresponding to the real eigenvalue $\frac{1}{\mu}(b + a\mu)$. However, according to Theorem 2.9 eigenvectors corresponding to different eigenvalues are always linearly independent. So, the eigenvalues $a - b\mu$ and $\frac{1}{\mu}(b + a\mu)$ must coincide, but this implies $\mu^2 = -1$, which is not possible.   □

Example 2.8

1. (see also Example 2.7) Let $A = \begin{bmatrix} 3 & 1 & 0 \\ -2 & 1 & 0 \\ 0 & 0 & 1 \end{bmatrix}$. The characteristic polynomial of A is $(\lambda^2 - 4\lambda + 5)(\lambda - 1)$. So, A has one real root $\lambda_1 = 1$ and two complex roots. The complex roots of this equation are $\lambda_2 = 2 + i$ and $\lambda_3 = 2 - i$. The eigenvectors corresponding to λ_2 are $N(A - (2 + i)I) = \{\alpha \begin{bmatrix} 1 \\ 1 - i \\ 0 \end{bmatrix}, \alpha \in \mathbb{C}\}$. The real part of this eigenvector is $x := \begin{bmatrix} 1 \\ -1 \\ 0 \end{bmatrix}$ and the imaginary part of this eigenvector is $y := \begin{bmatrix} 0 \\ 1 \\ 0 \end{bmatrix}$. Therefore, A has a two-dimensional invariant subspace consisting of $S = \text{Im} \begin{bmatrix} 1 & 0 \\ -1 & 1 \\ 0 & 0 \end{bmatrix}$. Indeed, with $2 + i =: a + bi$ we have (see Theorem 2.17)

$$AS = \begin{bmatrix} 2 & 1 \\ -3 & 1 \\ 0 & 0 \end{bmatrix} = \begin{bmatrix} 1 & 0 \\ -1 & 1 \\ 0 & 0 \end{bmatrix} \begin{bmatrix} 2 & 1 \\ -1 & 2 \end{bmatrix} = S \begin{bmatrix} a & b \\ -b & a \end{bmatrix}.$$

2. Let $A = \begin{bmatrix} 2 & 2 & -3 \\ 0 & -1 & 0 \\ 3 & 1 & 2 \end{bmatrix}$. The characteristic polynomial of A is $(\lambda^2 - 4\lambda + 13)$

$(\lambda + 1)$. So, A has one real root $\lambda_1 = -1$ and two complex roots. The complex roots of this equation are $\lambda_2 = 2 + 3i$ and $\lambda_3 = 2 - 3i$. The eigenvectors correspond-

ing to λ_2 are $N(A - (2 + 3i)I) = \left\{ \alpha \begin{bmatrix} 1 \\ 0 \\ -i \end{bmatrix}, \ \alpha \in \mathbb{C} \right\}$. Therefore, A has a two-dimen-

sional invariant subspace consisting of $S = \text{Im} \begin{bmatrix} 1 & 0 \\ 0 & 0 \\ 0 & -1 \end{bmatrix}$. Verification shows that

indeed with $a = 2$ and $b = 3$, $AS = S \begin{bmatrix} a & b \\ -b & a \end{bmatrix}$.   □

2.4 Cayley–Hamilton theorem

A theorem that is often used in linear algebra is the Cayley–Hamilton theorem. In the next section this theorem will be used to derive the Jordan canonical form.

To introduce and prove this Cayley–Hamilton theorem let $p(\lambda)$ be the characteristic polynomial of A, that is

$$p(\lambda) = \det(A - \lambda I) = \lambda^n + a_1 \lambda^{n-1} + \cdots + a_n.$$

According to equation (2.1.3), $\det(B)I = B\text{adj}(B) = \text{adj}(B)B$. So, with $B := A - \lambda I$, we obtain the identity

$$p(\lambda)I = (A - \lambda I)\text{adj}(A - \lambda I) \tag{2.4.1}$$
$$= \text{adj}(A - \lambda I)(A - \lambda I). \tag{2.4.2}$$

By a straightforward spelling out of the right-hand side of the next equation, one can verify that $p(\lambda)I$ can be rewritten as

$$p(\lambda)I = -(A - \lambda I)\left[I\lambda^{n-1} + (A + a_1 I)\lambda^{n-2} + \cdots + (A^{n-1} + a_1 A^{n-2} + \cdots + a_{n-1}I) \right] +$$
$$A^n + a_1 A^{n-1} + \cdots + a_{n-1}A + a_n I. \tag{2.4.3}$$

Now, if $p(\lambda) = \lambda^n + a_1 \lambda^{n-1} + \cdots + a_n$, let $p(A)$ be defined as the matrix formed by replacing each power of λ in $p(\lambda)$ by the corresponding power of A (with $A^0 = I$). That is,

$$p(A) = A^n + a_1 A^{n-1} + \cdots + a_n I. \tag{2.4.4}$$

Comparing the right-hand sides of equations (2.4.1) and (2.4.3) it follows that the next equality holds

$$(A - \lambda I)\text{adj}(A - \lambda I) = -(A - \lambda I)Q(\lambda) + p(A), \tag{2.4.5}$$

where we used the shorthand notation

$$Q(\lambda) := I\lambda^{n-1} + (A + a_1 I)\lambda^{n-2} + \cdots + (A^{n-1} + a_1 A^{n-2} + \cdots + a_{n-1}I).$$

Next, rewrite equation (2.4.5) as

$$(A - \lambda I)\{\mathrm{adj}(A - \lambda I) + Q(\lambda)\} = p(A). \qquad (2.4.6)$$

Using equation (2.4.2), equation (2.4.6) can be rewritten as

$$\mathrm{adj}(A - \lambda I) + Q(\lambda) = \frac{\mathrm{adj}(A - \lambda I)}{p(\lambda)} p(A). \qquad (2.4.7)$$

Using the definition of the adjoint matrix, it is easily verified that every entry of the adjoint of $A - \lambda I$ is a polynomial with a degree that does not exceed $n - 1$ (see also Example 2.2). Therefore, the left-hand side of equation (2.4.7) is a polynomial matrix function. Furthermore, since the degree of the characteristic polynomial $p(\lambda)$ of A is n,

$$\frac{\mathrm{adj}(A - \lambda I)}{p(\lambda)} = P_0 \frac{1}{p(\lambda)} + P_1 \frac{\lambda}{p(\lambda)} + \cdots + P_{n-1} \frac{\lambda^{n-1}}{p(\lambda)}$$

is a strict rational function of λ. Therefore, the left-hand side and right-hand side of equation (2.4.7) coincide only if both sides are zero. That is,

$$\frac{\mathrm{adj}(A - \lambda I)}{p(\lambda)} p(A) = 0. \qquad (2.4.8)$$

In particular this equality holds for $\lambda = \bar{\lambda}$, where $\bar{\lambda}$ is an arbitrary number that is not an eigenvalue of matrix A. But then, $\mathrm{adj}(A - \bar{\lambda} I)$ is invertible. So, from equation (2.4.8) it follows that $p(A) = 0$. This proves the Cayley–Hamilton theorem.

Theorem 2.18 (Cayley–Hamilton)[3]

Let $A \in \mathbb{R}^{n \times n}$ and let $p(\lambda)$ be the characteristic polynomial of A. Then

$$p(A) = 0.$$

□

Example 2.9

Reconsider the matrices A_i, $i = 1, \ldots, 4$ from Example 2.3. Then straightforward calculations show that $(A_1 - I)(A_1 - 2I) = 0$; $(A_2 + 2I)^2 = 0$; $(A_3 - 3I)^2 = 0$ and $A_4^2 - 4A_4 + 5I = 0$, respectively.

□

---

[3]Cayley was an English lawyer/mathematician who lived from 1821–1895. His lecturer was the famous Irish mathematician Hamilton who lived from 1805–1865 (see also Chapter 1).

Next, assume that the characteristic polynomial $p(\lambda)$ of A is factorized as in Theorem 2.10. That is

$$p(\lambda) = p_1(\lambda)p_2(\lambda)\cdots p_k(\lambda)p_{k+1}(\lambda)\cdots p_r(\lambda), \qquad (2.4.9)$$

with $p_i(\lambda) = (\lambda - \lambda_i)^{n_i}, i = 1, \ldots, k$, and $p_i(\lambda) = (\lambda^2 + b_i\lambda + c_i)^{n_i}, i = k+1, \ldots, r$. The next lemma shows that the null spaces of $p_i(A)$ do not have any points in common (except for the zero vector). Its proof is provided in the Appendix to this chapter (section 2.10).

Lemma 2.19

Let $p_i(A)$ be as described above. Then

$$\ker p_i(A) \cap \ker p_j(A) = \{0\}, \text{ if } i \neq j. \qquad \square$$

The next lemma is rather elementary but, nevertheless, gives a useful result.

Lemma 2.20

Let $A, B \in \mathbb{R}^{n \times n}$. If $AB = 0$ then

$$\dim(\ker A) + \dim(\ker B) \geq n.$$

Proof

Note that $\operatorname{Im} B \subset \ker A$. Consequently,

$$\dim(\ker A) \geq \dim(\operatorname{Im} B).$$

Therefore, using Theorem 2.3,

$$\dim(\ker B) + \dim(\ker A) \geq \dim(\ker B) + \dim(\operatorname{Im} B) = n. \qquad \square$$

We are now able to prove the next theorem which is essential to construct the Jordan Canonical form of a square matrix A in the next section.

Theorem 2.21

Let the characteristic polynomial of matrix A be as described in equation (2.4.9). Assume that the set $\{b_{j1}, \ldots, b_{jm_j}\}$ forms a basis for $\ker p_j(A)$, with $p_j(A)$ as before, $j = 1 \ldots, r$. Then,

1. the set of vectors $\{b_{11}, \ldots, b_{1m_1}, \ldots, b_{r1}, \ldots, b_{rm_r}\}$ forms a basis for $\mathbb{R}^n$;

2. $m_i = n_i, i = 1, \ldots, k$, and $m_i = 2n_i, i = k+1, \ldots, r$. That is, the algebraic multiplicities of the real eigenvalues coincide with the dimension of the corresponding

generalized eigenspaces; and the dimension of the generalized eigenspaces of the complex eigenvalues are twice their algebraic multiplicities.

Proof

1. First construct for the nullspace of each $p_j(A)$ a basis $\{b_{j1}, \ldots, b_{jm_j}\}$. From Lemma 2.19 it is clear that all the vectors in the collection $\{b_{11}, \ldots, b_{1m_1}, \ldots, b_{r1}, \ldots, b_{rm_r}\}$ are linearly independent. Since all vectors are an element of $\mathbb{R}^n$ this implies that the number of vectors that belong to this set is smaller than or equal to the dimension of $\mathbb{R}^n$, i.e. n.

   On the other hand it follows straightforwardly by induction from Lemma 2.20 that, since according Cayley–Hamilton's theorem $p_1(A)p_2(A) \cdots p_r(A) = 0$, the sum of the dimensions of the nullspaces of $p_i(A)$ should be at least n. Therefore, combining both results, we conclude that this sum should be exactly n.

2. Since the dimension of the nullspace of $p_1(A)$ is m_1, the dimension of the nullspace of $A - \lambda_1 I$ is at most m_1. So A has at most m_1 independent eigenvectors corresponding with the eigenvalue λ_1. A reasoning similar to Theorem 2.11 then shows that the characteristic polynomial $p(\lambda)$ of A can be factorized as $(\lambda - \lambda_1)^{m_1} h(\lambda)$, where $h(\lambda)$ is a polynomial of degree $n - m_1$. Since by assumption $p(\lambda) = (\lambda - \lambda_1)^{n_1} h_2(\lambda)$, where the polynomial $h_2(\lambda)$ does not contain the factor $\lambda - \lambda_1$, it follows that $m_i \le n_i$, $i = 1, \ldots, k$. In a similar way one can show that $m_i \le 2n_i$, $i = k+1, \ldots, r$. Since according to Theorem 2.10

$$\sum_{i=1}^{k} n_i + 2 \sum_{i=k+1}^{r} n_i = n$$

it is clear that this equality can only hold if $m_i = n_i$, $i = 1, \ldots, k$, and $m_i = 2n_i$, $i = k+1, \ldots, r$. $\qquad\square$

2.5 Invariant subspaces and jordan canonical form

Let λ be an eigenvalue of $A \in \mathbb{R}^{n \times n}$, and x be a corresponding eigenvector. Then $Ax = \lambda x$ and $A(\alpha x) = \lambda(\alpha x)$ for any $\alpha \in \mathbb{R}$. Clearly, the eigenvector x defines a one-dimensional subspace that is invariant with respect to pre-multiplication by A since $A^k x = \lambda^k x, \forall k$. In general, a subspace $S \subset \mathbb{R}^n$ is called A-**invariant** if $Ax \in S$ for every $x \in S$. In other words, S is A-invariant means that the image of S under A is contained in S, i.e. Im $AS \subset S$. Examples of A-invariant subspaces are the trivial subspace $\{0\}$, $\mathbb{R}^n$, $\ker A$ and Im A.

A-invariant subspaces play an important role in calculating solutions of the so-called algebraic Riccati equations. These solutions constitute the basis for determining various equilibria as we will see later on. A-invariant subspaces are intimately related to the generalized eigenspaces of matrix A. A complete picture of all A-invariant subspaces is in fact provided by considering the so-called Jordan canonical form of matrix A. It is a well-known (but nontrivial) result in linear algebra that any square matrix $A \in \mathbb{R}^{n \times n}$ admits a **Jordan canonical** representation.

To grasp the idea of the Jordan form, first consider the case that A has n different eigenvalues λ_i, $i = 1, \ldots, n$. Let x_i, $i = 1, \ldots, n$, be the corresponding eigenvectors. From Theorem 2.9 it then follows straightforwardly that $\{x_1, x_2, \ldots, x_n\}$ are linearly independent and in fact constitute a basis for $\mathbb{R}^n$. Now let matrix $S := [x_1 \ x_2 \ \ldots \ x_n]$. Since this matrix is full rank its inverse exists. Then $AS = [Ax_1 \ Ax_2 \ \ldots \ Ax_n] =$

$$[\lambda_1 x_1 \ \lambda_2 x_2 \ \cdots \ \lambda_n x_n] = SJ_1, \text{ where } J_1 = \begin{bmatrix} \lambda_1 & & & \\ & \lambda_2 & & \\ & & \ddots & \\ & & & \lambda_n \end{bmatrix}. \text{ So if matrix } A \text{ has } n$$

different eigenvalues it can be factorized as $A = SJ_1 S^{-1}$. Notice that this same procedure can be used to factorize matrix A as long as matrix A has n independent eigenvectors. The fact that the eigenvalues all differed is not crucial for this construction.

Next consider the case that A has an eigenvector x_1 corresponding to the eigenvalue λ_1, and that the generalized eigenvectors x_2 and x_3 are obtained through the following equations:

$$(A - \lambda_1 I)x_1 = 0$$
$$(A - \lambda_1 I)x_2 = x_1$$
$$(A - \lambda_1 I)x_3 = x_2.$$

From the second equation we see that $Ax_2 = x_1 + \lambda_1 x_2$ and from the third equation that $Ax_3 = x_2 + \lambda_1 x_3$. With $S := [x_1 \ x_2 \ x_3]$ we therefore have $AS = [Ax_1 \ Ax_2 \ Ax_3] =$

$$[\lambda_1 x_1 \ x_1 + \lambda_1 x_2 \ x_2 + \lambda_1 x_3] = SJ_2 \text{ with } J_2 = \begin{bmatrix} \lambda_1 & 1 & 0 \\ 0 & \lambda_1 & 1 \\ 0 & 0 & \lambda_1 \end{bmatrix}. \text{ In particular, if } A \in \mathbb{R}^{3 \times 3}$$

and $\{x_1, x_2, x_3\}$ are linearly independent, we conclude that S is invertible and therefore A can be factorized as $A = SJ_2 S^{-1}$.

For the general case, assume that matrix A has an eigenvector x_1 corresponding with the eigenvalue λ_1 and that the generalized eigenvectors $x_2, \ldots, x_k$ are obtained through the following equations:

$$(A - \lambda_1 I)x_1 = 0$$
$$(A - \lambda_1 I)x_2 = x_1$$
$$\vdots$$
$$(A - \lambda_1 I)x_k = x_{k-1}.$$

Now, let S be an A-invariant subspace that contains x_2. Then x_1 should also be in S. This, since from the second equality above $Ax_2 = \lambda_1 x_2 + x_1$, and, as both Ax_2 and $\lambda_1 x_2$ belong to S, x_1 has to belong to this subspace too. Similarly it follows that if x_3 belongs to some A-invariant subspace S, x_2 and, consequently, x_1 have also to belong to this subspace S. So, in general, we observe that if some x_i belongs to an A-invariant subspace S all its 'predecessors' have to be in it too.

Furthermore note that $(A - \lambda_1 I)^k x_k = 0$. That is,

$$x_k \in \ker(A - \lambda_1 I)^k. \tag{2.5.1}$$

An A-invariant subspace S is called a **stable invariant subspace** if all the (possibly complex) eigenvalues of A constrained to S have negative real parts.

Lemma 2.22

Consider a set of vectors $\{x_1, \ldots, x_p\}$ which are recursively obtained as follows:

$$(A - \lambda_1 I)x_1 = 0; \quad (A - \lambda_1 I)x_{i+1} = x_i, \ i = 1, \ldots, p - 1. \tag{2.5.2}$$

Then, with $S := [x_1 \cdots x_p]$,

$$AS = S\Lambda, \tag{2.5.3}$$

where $\Lambda \in \mathbb{R}^{p \times p}$ is the matrix $\begin{bmatrix} \lambda_1 & 1 & & & \\ & \lambda_1 & \ddots & & \\ & & \ddots & \ddots & \\ & & & \lambda_1 & 1 \\ & & & & \lambda_1 \end{bmatrix}$. Moreover, S has full column rank.

Proof

From the reasoning above, the first part of the lemma is obvious. All that is left to be shown is that the set of vectors $\{x_1, \ldots, x_p\}$ are linearly independent. To prove this we consider the case $p = 3$. The general case can be proved similarly. So, assume that

$$\mu_1 x_1 + \mu_2 x_2 + \mu_3 x_3 = 0. \tag{2.5.4}$$

Then also

$$\begin{aligned} 0 &= (A - \lambda_1 I)(\mu_1 x_1 + \mu_2 x_2 + \mu_3 x_3) \\ &= \mu_2 x_1 + \mu_3 x_2, \end{aligned} \tag{2.5.5}$$

and, consequently,

$$\begin{aligned} 0 &= (A - \lambda_1 I)(\mu_2 x_1 + \mu_3 x_2) \\ &= \mu_3 x_1. \end{aligned} \tag{2.5.6}$$

Since $x_i \neq 0$ it is clear from equations (2.5.4) and (2.5.5) that necessarily $\mu_i = 0$, $i = 1, 2, 3$. That is $\{x_1, x_2, x_3\}$ are linearly independent. $\square$

Since in $\mathbb{R}^n$ one can find at most n linearly independent vectors, it follows from Lemma 2.22 that the set of vectors defined in equation (2.5.2) has at most n elements. A sequence of vectors $\{x_1, \ldots, x_p\}$ that are recursively defined by equation (2.5.2) and for which the equation $(A - \lambda_1 I)x_{p+1} = x_p$ has no solution, is called a **Jordan chain** generated by x_1. So a Jordan chain, denoted by $J(x_1)$, is the maximal sequence of vectors generated by equation (2.5.2) starting with the eigenvector x_1. The following property holds.

Proposition 2.23

Consider the generalized eigenspace E_1^g corresponding to the eigenvalue λ_1. Let $\{b_1, \ldots, b_k\}$ constitute a basis for the corresponding eigenspace $\ker(A - \lambda_1 I)$ and $J(b_i)$ the Jordan chain generated by the basis vector b_i. Then, $\{J(b_1), \ldots, J(b_k)\}$ constitutes a basis for E_1^g. Moreover, with $S := [J(b_1), \ldots, J(b_k)]$, equation (2.5.3) holds.

Proof

From Lemma 2.22 it is obvious that for each b_i, $J(b_i)$ are a set of linearly independent vectors. To show that, for example, the set of vectors $\{J(b_1), J(b_2)\}$ are linearly independent too, assume that some vector y_k in the Jordan chain $J(b_2)$ is in the span of $J(b_1)$. Denoting $J(b_1) = \{b_1, x_2, \ldots, x_r\}$ and $J(b_2) = \{b_2, y_2, \ldots, y_s\}$, then there exist μ_i, not all zero, such that

$$y_k = \mu_1 b_1 + \mu_2 x_2 + \cdots + \mu_r x_r.$$

So,

$$b_2 = (A - \lambda_1 I)^{k-1} y_k$$
$$= \mu_1 (A - \lambda_1)^{k-1} b_1 + \cdots + \mu_r (A - \lambda_1)^{k-1} x_r.$$

Using the definition of the Jordan chain we conclude that if $k > r$ all vectors on the right-hand side of this equation become zero. So $b_2 = 0$, which clearly contradicts our assumption that b_2 is a basis vector. On the other hand, if $k \leq r$, the equation reduces to

$$b_2 = \mu_k b_1 + \mu_{k+1} x_2 + \cdots + \mu_r x_{r-k+1}. \tag{2.5.7}$$

From this we see, by premultiplying both sides of this equation on the left by $(A - \lambda_1 I)$ again, that

$$0 = (A - \lambda_1 I)b_2$$
$$= \mu_{k+1}(A - \lambda_1 I)x_2 + \cdots + \mu_r (A - \lambda_1 I)x_{r-k+1}$$
$$= \mu_{k+1} b_1 + \cdots + \mu_r x_{r-k}.$$

However, according Lemma 2.22 the vectors in the Jordan chain $J(b_1)$ are linearly independent, so $\mu_{k+1} = \cdots = \mu_r = 0$; but this implies according to equation (2.5.7) that $b_2 = \mu_k b_1$, which violates our assumption that $\{b_1, b_2\}$ are linearly independent. This shows that in general $\{J(b_1), \ldots, J(b_k)\}$ are a set of linearly independent vectors.

Furthermore we have from equation (2.5.1) and its preceding discussion that all vectors in $J(b_i)$ belong to E_1^g. So, what is left to be shown is that the number of these independent vectors equals the dimension of E_1^g. To that purpose we show that every vector in E_1^g belongs to a suitably chosen Jordan chain for some vector $x_1 \in \ker(A - \lambda_1 I)$. Let $x \in E_1^g$. Then $(A - \lambda_1)^{n_1} x = 0$ (where n_1 is the algebraic multiplicity of λ_1) or, equivalently,

$$(A - \lambda_1 I)(A - \lambda_1 I)^{n_1-1} x = 0.$$

Let $x_1 := (A - \lambda_1 I)^{n_1-1} x$. Then,

$$(A - \lambda_1 I) x_1 = 0, \text{ where } (A - \lambda_1 I)^{n_1-1} x = x_1.$$

Next, introduce $x_2 := (A - \lambda_1 I)^{n_1-2} x$. Then, $(A - \lambda_1 I) x_2 = x_1$. Repeating this process we see that $\{x_1, \ldots, x_{n_1-1}, x\}$ satisfy

$$(A - \lambda_1 I) x_1 = 0, \ (A - \lambda_1 I) x_i = x_{i-1}, \ i = 1, \ldots, n_1.$$

So, x belongs to a Jordan chain for some $x_1 \in \ker(A - \lambda_1 I)$. From this it is easily verified that $x \in \mathrm{Span}\{J(b_1), \ldots, J(b_k)\}$. Which completes the proof. □

Example 2.10

Suppose a matrix A has the following form

$$A = [x_1 \ x_2 \ x_3 \ x_4]^{-1} \begin{bmatrix} \lambda_1 & 1 & & \\ & \lambda_1 & & \\ & & \lambda_2 & \\ & & & \lambda_3 \end{bmatrix} [x_1 \ x_2 \ x_3 \ x_4].$$

Then it is easy to verify that

$$\begin{aligned} S_1 &= \mathrm{Span}\{x_1\}; \ S_{12} = \mathrm{Span}\{x_1, x_2\}; \ S_{123} = \mathrm{Span}\{x_1, x_2, x_3\}; \\ S_3 &= \mathrm{Span}\{x_3\}; \ S_{13} = \mathrm{Span}\{x_1, x_3\}; \ S_{124} = \mathrm{Span}\{x_1, x_2, x_4\}; \\ S_4 &= \mathrm{Span}\{x_4\}; \ S_{14} = \mathrm{Span}\{x_1, x_4\}; \ \text{and } S_{34} = \mathrm{Span}\{x_3, x_4\} \end{aligned}$$

are all A-invariant subspaces. □

Example 2.11

Assume that matrix A has a (possibly complex) eigenvalue λ_1 and that its geometric multiplicity is 2. Then A has infinitely many invariant subspaces.

For, in the case of $\lambda_1 \in \mathbb{R}$, let $\{x_1, x_2\}$ be a basis for the nullspace of $A - \lambda_1 I$. Then any linear combination of these basis vectors is an A-invariant subspace, because $A(\alpha_1 x_1 + \alpha_2 x_2) = \alpha_1 A x_1 + \alpha_2 A x_2 = \alpha_1 \lambda_1 x_1 + \alpha_2 \lambda_2 x_2 = \lambda_1(\alpha_1 x_1 + \alpha_2 x_2)$. Similarly, if $\lambda = a + bi$ (with $b \neq 0$) is a complex number, with geometric multiplicity 2, there exist four linearly independent vectors x_i, y_i, $i = 1, 2$ such that

$$A[x_i \ y_i] = [x_i \ y_i] \begin{bmatrix} a & b \\ -b & a \end{bmatrix}.$$

Simple calculation shows then that for arbitrary α_i the subspace $\mathrm{Im}[\alpha_1 x_1 + \alpha_2 x_2, \alpha_1 y_1 + \alpha_2 x_2]$ is A-invariant too. $\qquad\square$

Below we will use the notation $\mathbf{diag}\{A, B\}$ for square, but not necessarily the same sized matrices A and B, to denote the partitioned matrix $\begin{bmatrix} A & 0 \\ 0 & B \end{bmatrix}$.

Theorem 2.24 (Jordan canonical form)[4]

For any square matrix $A \in \mathbb{R}^{n \times n}$ there exists a nonsingular matrix T such that

$$A = TJT^{-1}$$

where $J = \mathrm{diag}\{J_1, J_2, \ldots, J_r\}$ and J_i either has one of the following real ($\mathbb{R}$), real extended ($\mathbb{R}E$), complex ($\mathbb{C}$) or complex extended ($\mathbb{C}E$) forms

$$i) \ J_{\mathbb{R}} = \begin{bmatrix} \lambda_i & & & & \\ & \lambda_i & & & \\ & & \ddots & & \\ & & & \lambda_i & \\ & & & & \lambda_i \end{bmatrix}; \quad ii) \ J_{\mathbb{R}E} = \begin{bmatrix} \lambda_i & 1 & & & \\ & \lambda_i & \ddots & & \\ & & \ddots & \ddots & \\ & & & \lambda_i & 1 \\ & & & & \lambda_i \end{bmatrix}$$

$$iii) \ J_{\mathbb{C}} = \begin{bmatrix} a_i & b_i & & & \\ -b_i & a_i & & & \\ & & \ddots & & \\ & & & a_i & b_i \\ & & & -b_i & a_i \end{bmatrix}; \quad iv) \ J_{\mathbb{C}E} = \begin{bmatrix} C_i & I & & \\ & C_i & \ddots & \\ & & \ddots & \ddots \\ & & & C_i & I \\ & & & & C_i \end{bmatrix}$$

Here $\{\lambda_i, \ i = 1, \ldots, r\}$ are the distinct real eigenvalues of A, $C_i = \begin{bmatrix} a_i & b_i \\ -b_i & a_i \end{bmatrix}$, $i = k+1, \ldots, r$, and I is the 2×2 identity matrix.

Proof

Let E_1^g be the generalized eigenspace corresponding to the eigenvalue λ_1, which has an algebraic multiplicity m_1. Then, by definition

$$E_1^g = \ker(A - \lambda_1 I)^{m_1} = \ker(A - \lambda_1 I)^{m_1 + i}, \quad i \geq 0.$$

So,

$$(A - \lambda_1 I)^{m_1}(A - \lambda_1 I)E_1^g = 0.$$

---

[4]Jordan was a famous French engineer/mathematician who lived from 1838–1922.

From which we conclude that

$$(A - \lambda_1 I)E_1^g \subset E_1^g.$$

Consequently, introducing the matrix S_1 such that $E_1^g = \operatorname{Im} S_1$, $(A - \lambda_1 I)S_1 = S_1 V_1$ for some matrix V_1. Or, stated differently,

$$AS_1 = S_1(\lambda_1 I + V_1).$$

So, assuming the characteristic polynomial of A is factorized as in Theorem 2.10, we conclude that there exist matrices S_i and V_i, $i = 1, \ldots, r$, such that

$$A[S_1 \cdots S_r] = [S_1 \cdots S_r]\operatorname{diag}\{\lambda_1 I + V_1, \ldots, \lambda_r I + V_r\}.$$

This proves the first part of the theorem. What is left to be shown is that for each generalized eigenspace we can find a basis such that with respect to this basis

$$AS_i = S_i J_i,$$

where J_i has either one of the four representations $J_{\mathbb{R}}$, $J_{\mathbb{RE}}$, $J_{\mathbb{C}}$ or $J_{\mathbb{CE}}$. However, this was basically shown in Proposition 2.23. In the case where the eigenvalue λ_1 is real and the geometric multiplicity of λ_1 coincides with its algebraic multiplicity, $J_i = J_{\mathbb{R}}$. In the case where its geometric multiplicity is one and its algebraic multiplicity is larger than one, $J_i = J_{\mathbb{RE}}$. If the geometric multiplicity of λ_1 is larger than one but differs from its algebraic multiplicity, a mixture of both forms $J_{\mathbb{R}}$ and $J_{\mathbb{RE}}$ occurs. The exact form of this mixture depends on the length of the Jordan chains of the basis vectors chosen in $\ker(A - \lambda_1 I)$. The corresponding results $J_i = J_{\mathbb{C}}$ and $J_i = J_{\mathbb{CE}}$, in the case where $\lambda_1 = a_i + b_i i$, $b_i \neq 0$, is a complex root, follow similarly using the result of Theorem 2.17.   □

In the above theorem the numbers a_i and b_i in the boxes $C_i = \begin{bmatrix} a_i & b_i \\ -b_i & a_i \end{bmatrix}$ come from the complex roots $a_i + b_i i$, $i = k+1, \ldots, r$, of matrix A. Note that the characteristic polynomial of C_i is $\lambda^2 + 2a_i\lambda + a_i^2 + b_i^2$.

An immediate consequence of this theorem is the following corollary.

Corollary 2.25

Let $A \in \mathbb{R}^{n \times n}$. Then,

1. All A-invariant subspaces can be constructed from the Jordan canonical form.

2. Matrix A has a finite number of invariant subspaces if and only if all geometric multiplicities of the (possibly complex) eigenvalues are one.

Proof

1. Let $\operatorname{Im} S$ be an arbitrarily chosen k-dimensional A-invariant subspace. Then there exists a matrix Λ such that

$$\operatorname{Im} AS = \operatorname{Im} S\Lambda.$$

Let $\Lambda = TJT^{-1}$, where J is the Jordan canonical form corresponding to Λ. Then $\operatorname{Im} AST = \operatorname{Im} STJ$. However, since T is invertible, $\operatorname{Im} ST = \operatorname{Im} S$. So, $\operatorname{Im} AS = \operatorname{Im} SJ$. This implies that the columns of matrix S are either eigenvectors or generalized eigenvectors of A.

2. From part 1 it follows that all A-invariant subspaces can be determined from the Jordan canonical form of A. In Example 2.11 we already showed that if there is an eigenvalue which has a geometric multiplicity larger than one, there exist infinitely many invariant subspaces. So what remains to be shown is that if all real eigenvalues have exactly one corresponding eigenvector and with each pair of conjugate eigenvalues there corresponds exactly one two-dimensional invariant subspace, then there will only exist a finite number of A-invariant subspaces.

However, under these assumptions it is obvious that the invariant subspaces corresponding with the eigenvalues are uniquely determined. So, there are only a finite number of such invariant subspaces. This implies that there are also only a finite number of combinations possible for these subspaces, all yielding additional A-invariant subspaces. Therefore, all together there will be only a finite number of invariant subspaces. □

From the above corollary it is clear that with each A-invariant subspace V one can associate a part of the spectrum of matrix A. We will denote this part of the spectrum by $\sigma(A|_V)$.

Now, generically, a polynomial of degree n will have n distinct (possibly complex) roots. Therefore if one considers an arbitrary matrix $A \in \mathbb{R}^{n \times n}$ its Jordan form is most of the times a combination of the first, $J_\mathbb{R}$, and third, $J_\mathbb{C}$, Jordan form.

Example 2.12 (generic Jordan canonical form)

If $A \in \mathbb{R}^{n \times n}$ has n distinct (possibly complex) eigenvalues then its Jordan form is

$$
J = \begin{bmatrix}
\lambda_1 & & & & & & & & & & \\
& \lambda_2 & & & & & & & & & \\
& & \ddots & & & & & & & & \\
& & & \lambda_{k-1} & & & & & & & \\
& & & & \lambda_k & & & & & & \\
& & & & & a_{k+1} & b_{k+1} & & & & \\
& & & & & -b_{k+1} & a_{k+1} & & & & \\
& & & & & & & \ddots & & & \\
& & & & & & & & a_r & b_r & \\
& & & & & & & & -b_r & a_r &
\end{bmatrix},
$$

where all numbers appearing in this matrix differ. If A has only real roots the numbers a_i, b_i disappear and $k = n$. □

Example 2.13

Consider matrix $A = \begin{bmatrix} 3 & 1 & 0 \\ -2 & 1 & 0 \\ 4 & 5 & -1 \end{bmatrix}$. The characteristic polynomial of A is

$(\lambda + 1)(\lambda^2 - 4\lambda + 5)$. Therefore it has one real eigenvalue, -1, and two complex eigenvalues, $2 + i$ and $2 - i$. Consequently the Jordan canonical form of A is

$J = \begin{bmatrix} -1 & 0 & 0 \\ 0 & 2 & 1 \\ 0 & -1 & 2 \end{bmatrix}$. $\qquad\qquad\qquad\qquad\qquad\qquad\qquad\qquad$ □

2.6 Semi-definite matrices

We start this section with the formal introduction of the transposition operation. The **transpose** of a matrix $A \in \mathbb{R}^{n \times m}$, denoted by A^T, is obtained by interchanging the rows and columns of matrix A. In more detail, if

$$A = \begin{bmatrix} a_{11} & a_{12} & \cdots & a_{1n} \\ a_{21} & a_{22} & \cdots & a_{2n} \\ \vdots & \vdots & \vdots & \vdots \\ a_{m1} & a_{m2} & \cdots & a_{mn} \end{bmatrix} \text{ then } A^T = \begin{bmatrix} a_{11} & a_{21} & \cdots & a_{m1} \\ a_{12} & a_{22} & \cdots & a_{m2} \\ \vdots & \vdots & \vdots & \vdots \\ a_{1n} & a_{2n} & \cdots & a_{mn} \end{bmatrix}.$$

If a matrix equals its transpose, i.e. $A = A^T$, the matrix is called **symmetric**. One important property of symmetric matrices is that a symmetric matrix has no complex eigenvalues. Furthermore, eigenvectors corresponding to different eigenvalues are perpendicular.

Theorem 2.26

Let $A \in \mathbb{R}^{n \times n}$ be symmetric. Then,

1. if λ is an eigenvalue of A, then λ is a real number;

2. if λ_i, $i = 1, 2$, are two different eigenvalues of A and x_i, $i = 1, 2$, corresponding eigenvectors, then $x_1^T x_2 = 0$.

Proof

1. Let x be an eigenvector corresponding to λ. According to Theorem 2.16 $\bar{\lambda}$ is then an eigenvalue of A too and $\bar{x}$ is a corresponding eigenvector. Therefore $x^T A \bar{x} = x^T \bar{\lambda} \bar{x} = \bar{\lambda} x^T \bar{x}$. On the other hand, since $x^T A \bar{x}$ is a scalar and A is symmetric, $x^T A \bar{x} = (x^T A \bar{x})^T = \bar{x}^T A x = \lambda \bar{x}^T x$. Since $\bar{x}^T x = x^T \bar{x} \in \mathbb{R}$, different from zero, we conclude that $\bar{\lambda} = \lambda$. So $\lambda \in \mathbb{R}$.

2. Consider $p := x_1^T A x_2$. Then, on the one hand, $p = \lambda_2 x_1^T x_2$. On the other hand, $p = p^T = x_2^T A x_1 = \lambda_1 x_2^T x_1 = \lambda_1 x_1^T x_2$. So comparing both results we conclude that $\lambda_2 x_1^T x_2 = \lambda_1 x_1^T x_2$. Since $\lambda_1 \neq \lambda_2$ it follows that $x_1^T x_2 = 0$. $\qquad$ □

Next note that $\ker(A) = \ker(A^2)$ if A is symmetric since, assuming that $x \in \ker(A^2)$ but $x \notin \ker(A)$, then, since $Ax \neq 0$, also $x^T A^T A x \neq 0$. However, from the symmetry of A it follows that $x^T A^T A x = x^T A^2 x = 0$, which contradicts our previous result. Therefore, the generalized eigenspaces of a symmetric matrix coincide with its eigenspaces. Now, for every eigenspace E_i of A one can construct an orthonormal basis using the Gram–Schmidt orthogonalization procedure. Since, according to Theorem 2.26, eigenvectors corresponding to different eigenvalues are always perpendicular, we obtain in this way an orthonormal basis of eigenvectors $\{u_1, \ldots, u_n\}$ for $\mathbb{R}^n$. Then with $U := [u_1, u_2, \cdots u_n]$, $AU = UJ$, where matrix J is a diagonal matrix whose diagonal entries are the eigenvalues of matrix A. It is easily verified that $U^T U = I$. So, by definition, the inverse of U is U^T. A matrix U which has the property that $U^T U = I$ is called an **orthonormal** matrix. This is because all the columns of this matrix U are orthogonal and have length (norm) one. So summarizing we have the following theorem.

Theorem 2.27

If A is symmetric, then there exists an orthonormal matrix U and a diagonal matrix $J = \text{diag}(\lambda_i)$ such that

$$A = UJU^T.$$

The ith-column of U is an eigenvector corresponding to the eigenvalue λ_i. $\qquad\square$

A square symmetric matrix A is said to be **positive definite (semi-definite)**, denoted by $A > 0$ $(A \geq 0)$, if $x^T A x > 0$ (≥ 0) for all $x \neq 0$. A is called **negative definite (semi-definite)**, denoted by $A < 0$ $(A \leq 0)$, if $x^T A x < 0$ (≤ 0) for all $x \neq 0$.

Theorem 2.28

Let A be a symmetric matrix. Then, $A > 0$ (≥ 0) if and only if all eigenvalues λ_i of A satisfy $\lambda_i > 0$ (≥ 0).

Proof

Consider the decomposition $A = UJU^T$ from Theorem 2.27. Choose $x = u_i$, where u_i is the ith-column of U. Then $x^T A x = \lambda_i$ and the result is obvious.
The converse statement is left as an exercise for the reader. $\qquad\square$

2.7 Algebraic Riccati equations

In this subsection we study the so-called **algebraic Riccati[5] equation** or ARE for short. AREs have an impressive range of applications, such as linear quadratic optimal control,

---

[5]Count Jacopa Francesco Riccati (1676–1754) studied the differential equation $\dot{x}(t) + t^{-n} x^2(t) - n t^{m+n-1} = 0$, where m and n are constants (Riccati, 1724). Since then, these kind of equations have been extensively studied in literature. See Bittanti (1991) and Bittanti, Laub and Willems (1991) for an historic overview of the main issues evolving around the Riccati equation.

stability theory, stochastic filtering and stochastic control, synthesis of linear passive networks and robust optimal control. In this book we will see that they also play a central role in the determination of equilibria in the theory of linear quadratic differential games.

Let A, Q and R be real $n \times n$ matrices with Q and R symmetric. Then an algebraic Riccati equation in the $n \times n$ matrix X is the following quadratic matrix equation:

$$A^T X + XA + XRX + Q = 0. \tag{2.7.1}$$

The above equation can be rewritten as

$$[I \ X] \begin{bmatrix} Q & A^T \\ A & R \end{bmatrix} \begin{bmatrix} I \\ X \end{bmatrix} = 0.$$

From this we infer that the image of matrix $[I \ X]$ is orthogonal to the image of $\begin{bmatrix} Q & A^T \\ A & R \end{bmatrix} \begin{bmatrix} I \\ X \end{bmatrix}$. Or, stated differently, the image of $\begin{bmatrix} Q & A^T \\ A & R \end{bmatrix} \begin{bmatrix} I \\ X \end{bmatrix}$ belongs to the orthogonal complement of the image of matrix $[I \ X]$. It is easily verified that the orthogonal complement of the image of matrix $[I \ X]$ is given by the image of $\begin{bmatrix} -X \\ I \end{bmatrix}$.

Therefore, ARE has a solution if and only if there exists a matrix $\Lambda \in \mathbb{R}^{n \times n}$ such that

$$\begin{bmatrix} Q & A^T \\ A & R \end{bmatrix} \begin{bmatrix} I \\ X \end{bmatrix} = \begin{bmatrix} -X \\ I \end{bmatrix} \Lambda.$$

Premultiplication of both sides from the above equality with the matrix $\begin{bmatrix} 0 & I \\ -I & 0 \end{bmatrix}$ yields then

$$\begin{bmatrix} A & R \\ -Q & -A^T \end{bmatrix} \begin{bmatrix} I \\ X \end{bmatrix} = \begin{bmatrix} I \\ X \end{bmatrix} \Lambda.$$

Or, stated differently, the symmetric solutions X of ARE can be obtained by considering the invariant subspaces of matrix

$$H := \begin{bmatrix} A & R \\ -Q & -A^T \end{bmatrix}. \tag{2.7.2}$$

Theorem 2.29 gives a precise formulation of this observation.

Theorem 2.29

Let $V \subset \mathbb{R}^{2n}$ be an n-dimensional invariant subspace of H, and let $X_1, X_2 \in \mathbb{R}^{n \times n}$ be two real matrices such that

$$V = \mathrm{Im} \begin{bmatrix} X_1 \\ X_2 \end{bmatrix}.$$

If X_1 is invertible, then $X := X_2 X_1^{-1}$ is a solution to the Riccati equation (2.7.1) and $\sigma(A + RX) = \sigma(H|_V)$. Furthermore, the solution X is independent of the specific choice of the basis of V.

Proof

Since V is an H-invariant subspace, there is a matrix $\Lambda \in \mathbb{R}^{n \times n}$ such that

$$\begin{bmatrix} A & R \\ -Q & -A^T \end{bmatrix} \begin{bmatrix} X_1 \\ X_2 \end{bmatrix} = \begin{bmatrix} X_1 \\ X_2 \end{bmatrix} \Lambda.$$

Post-multiplying the above equation by X_1^{-1} we get

$$\begin{bmatrix} A & R \\ -Q & -A^T \end{bmatrix} \begin{bmatrix} I \\ X \end{bmatrix} = \begin{bmatrix} I \\ X \end{bmatrix} X_1 \Lambda X_1^{-1}. \tag{2.7.3}$$

Now pre-multiply equation (2.7.3) by $[-X \; I]$ to get

$$[-X \; I] \begin{bmatrix} A & R \\ -Q & -A^T \end{bmatrix} \begin{bmatrix} I \\ X \end{bmatrix} = [-X \; I] \begin{bmatrix} I \\ X \end{bmatrix} X_1 \Lambda X_1^{-1}.$$

Rewriting both sides of this equality

$$-XA - A^T X - XRX - Q = 0,$$

which shows that X is indeed a solution of equation (2.7.1). Some rewriting of equation (2.7.3) also gives

$$A + RX = X_1 \Lambda X_1^{-1};$$

therefore, $\sigma(A + RX) = \sigma(\Lambda)$. However, by definition, Λ is a matrix representation of the map $H|_V$, so $\sigma(A + RX) = \sigma(H|_V)$. Next notice that any other basis spanning V can be represented as

$$\begin{bmatrix} X_1 \\ X_2 \end{bmatrix} P = \begin{bmatrix} X_1 P \\ X_2 P \end{bmatrix}$$

for some nonsingular P. The final conclusion follows then from the fact that $(X_2 P)(X_1 P)^{-1} = X_2 X_1^{-1}$. □

The converse of Theorem 2.29 also holds.

Theorem 2.30

If $X \in \mathbb{R}^{n \times n}$ is a solution to the Riccati equation (2.7.1), then there exist matrices $X_1, X_2 \in \mathbb{R}^{n \times n}$, with X_1 invertible, such that $X = X_2 X_1^{-1}$ and the columns of $\begin{bmatrix} X_1 \\ X_2 \end{bmatrix}$ form a basis of an n-dimensional invariant subspace of H.

Proof

Define $\Lambda := A + RX$. Multiplying this by X and using equation (2.7.1) gives

$$X\Lambda = XA + XRX = -Q - A^T X.$$

Write these two relations as

$$\begin{bmatrix} A & R \\ -Q & -A^T \end{bmatrix} \begin{bmatrix} I \\ X \end{bmatrix} = \begin{bmatrix} I \\ X \end{bmatrix} \Lambda.$$

Hence, the columns of $\begin{bmatrix} I \\ X \end{bmatrix}$ span an n-dimensional invariant subspace of H, and defining $X_1 := I$, and $X_2 := X$ completes the proof. $\qquad\square$

Note

Matrix H is called a **Hamiltonian matrix**. It has a number of nice properties. One of them is that whenever $\lambda \in \sigma(H)$, then also $-\lambda \in \sigma(H)$. That is, the spectrum of a Hamiltonian matrix is symmetric with respect to the imaginary axis. This fact is easily established by noting that with $J := \begin{bmatrix} 0 & -I \\ I & 0 \end{bmatrix}$, $H = -JH^T J^{-1}$. So, $p(\lambda) = \det(H - \lambda I) = \det(-JH^T J^{-1} - \lambda I) = \det(-JH^T J^{-1} - \lambda JJ^{-1}) = \det(-H^T - \lambda I) = \det(H^T + \lambda I) = \det(H + \lambda I) = p(-\lambda)$. $\qquad\square$

From Theorems 2.29 and 2.30 it will be clear why the Jordan canonical form is so important in this context. As we saw in the previous section, the Jordan canonical form of a matrix H can be used to construct all invariant subspaces of H. So, all solutions of equation (2.7.1) can be obtained by considering all n-dimensional invariant subspaces $V = \text{Im} \begin{bmatrix} X_1 \\ X_2 \end{bmatrix}$ of equation (2.7.2), with $X_i \in \mathbb{R}^{n \times n}$, that have the additional property that X_1 is invertible. A subspace V that satisfies this property is called a **graph subspace** (since it can be 'visualized' as the graph of the map: $x \rightarrow X_2 X_1^{-1} x$).

Example 2.14

(see also Example 2.3) Let

$$A = \begin{bmatrix} -1 & -3 \\ 2 & 4 \end{bmatrix}, \quad R = \begin{bmatrix} 2 & 0 \\ 0 & 1 \end{bmatrix}, \quad \text{and } Q = \begin{bmatrix} 0 & 0 \\ 0 & 0 \end{bmatrix}.$$

Then

$$H = \begin{bmatrix} -1 & -3 & 2 & 0 \\ 2 & 4 & 0 & 1 \\ 0 & 0 & 1 & -2 \\ 0 & 0 & 3 & -4 \end{bmatrix}.$$

The eigenvalues of H are $1, 2, -1, -2$, and the corresponding eigenvectors are

$$v_1 = \begin{bmatrix} 3 \\ -2 \\ 0 \\ 0 \end{bmatrix}, v_2 = \begin{bmatrix} -1 \\ 1 \\ 0 \\ 0 \end{bmatrix}, v_{-1} = \begin{bmatrix} -13 \\ 4 \\ 6 \\ 6 \end{bmatrix} \text{ and } v_{-2} = \begin{bmatrix} -33 \\ 5 \\ 24 \\ 36 \end{bmatrix}.$$

All solutions of the Riccati equation are obtained by combinations of these vectors as follows.

1. Consider $\text{Span}\{v_1, v_2\} =: \begin{bmatrix} X_1 \\ X_2 \end{bmatrix}$. Then $X_1 = \begin{bmatrix} 3 & -1 \\ -2 & 1 \end{bmatrix}$, and $X_2 = \begin{bmatrix} 0 & 0 \\ 0 & 0 \end{bmatrix}$, which yields

$$X := X_2 X_1^{-1} = \begin{bmatrix} 0 & 0 \\ 0 & 0 \end{bmatrix} \text{ and } \sigma(A + RX) = \{1, 2\}.$$

2. Consider $\text{Span}\{v_1, v_{-1}\} =: \begin{bmatrix} X_1 \\ X_2 \end{bmatrix}$. Then $X_1 = \begin{bmatrix} 3 & -13 \\ -2 & 4 \end{bmatrix}$, and $X_2 = \begin{bmatrix} 0 & 6 \\ 0 & 6 \end{bmatrix}$, which yields

$$X := X_2 X_1^{-1} = \frac{-1}{7} \begin{bmatrix} 6 & 9 \\ 6 & 9 \end{bmatrix} \text{ and } \sigma(A + RX) = \{1, -1\}.$$

3. Let $\text{Span}\{v_1, v_{-2}\} =: \begin{bmatrix} X_1 \\ X_2 \end{bmatrix}$. Then $X_1 = \begin{bmatrix} 3 & -33 \\ -2 & 5 \end{bmatrix}$, and $X_2 = \begin{bmatrix} 0 & 24 \\ 0 & 36 \end{bmatrix}$, which yields

$$X := X_2 X_1^{-1} = \frac{-1}{51} \begin{bmatrix} 48 & 72 \\ 72 & 108 \end{bmatrix} \text{ and } \sigma(A + RX) = \{1, -2\}.$$

4. Let $\text{Span}\{v_2, v_{-1}\} =: \begin{bmatrix} X_1 \\ X_2 \end{bmatrix}$. Then $X_1 = \begin{bmatrix} -1 & -13 \\ 1 & 4 \end{bmatrix}$, and $X_2 = \begin{bmatrix} 0 & 6 \\ 0 & 6 \end{bmatrix}$, which yields

$$X := X_2 X_1^{-1} = \frac{-2}{3} \begin{bmatrix} 1 & 1 \\ 1 & 1 \end{bmatrix} \text{ and } \sigma(A + RX) = \{2, -1\}.$$

5. Let $\text{Span}\{v_2, v_{-2}\} =: \begin{bmatrix} X_1 \\ X_2 \end{bmatrix}$. Then $X_1 = \begin{bmatrix} -1 & -33 \\ 1 & 5 \end{bmatrix}$, and $X_2 = \begin{bmatrix} 0 & 24 \\ 0 & 36 \end{bmatrix}$, which yields

$$X := X_2 X_1^{-1} = \frac{-1}{7} \begin{bmatrix} 6 & 6 \\ 9 & 9 \end{bmatrix} \text{ and } \sigma(A + RX) = \{2, -2\}.$$

6. Let $\text{Span}\{v_{-1}, v_{-2}\} =: \begin{bmatrix} X_1 \\ X_2 \end{bmatrix}$. Then $X_1 = \begin{bmatrix} -13 & -33 \\ 4 & 5 \end{bmatrix}$, and $X_2 = \begin{bmatrix} 6 & 24 \\ 6 & 36 \end{bmatrix}$, which yields

$$X := X_2 X_1^{-1} = \frac{1}{67} \begin{bmatrix} 66 & -114 \\ -114 & -270 \end{bmatrix} \text{ and } \sigma(A + RX) = \{-1, -2\}. \qquad \square$$

A natural question that arises is whether one can make any statements on the number of solutions of the algebraic Riccati equation (2.7.1). As we already noted, there is a one-to-one relationship between the number of solutions and the number of graph subspaces of matrix H. So, this number can be estimated by the number of invariant subspaces of matrix H. From the Jordan canonical form, Theorem 2.24 and more in particular Corollary 2.25, we see that if all eigenvalues of matrix H have a geometric multiplicity of one then H has only a finite number of invariant subspaces. So, in those cases the algebraic Riccati equation (2.7.1) will have either no, or at the most a finite number, of solutions. That there indeed exist cases where the equation has an infinite number of solutions is illustrated in the next example.

Example 2.15

Let

$$A = R = \begin{bmatrix} 1 & 0 \\ 0 & 1 \end{bmatrix} \text{ and } Q = \begin{bmatrix} 0 & 0 \\ 0 & 0 \end{bmatrix}.$$

Then

$$H = \begin{bmatrix} 1 & 0 & 1 & 0 \\ 0 & 1 & 0 & 1 \\ 0 & 0 & -1 & 0 \\ 0 & 0 & 0 & -1 \end{bmatrix}.$$

The eigenvalues of H are $\{1, 1, -1, -1\}$ and the corresponding eigenvectors are

$$v_{11} = \begin{bmatrix} 1 \\ 0 \\ 0 \\ 0 \end{bmatrix}, v_{12} = \begin{bmatrix} 0 \\ 1 \\ 0 \\ 0 \end{bmatrix}, v_{-11} = \begin{bmatrix} -1 \\ 0 \\ 2 \\ 0 \end{bmatrix} \text{ and } v_{-12} = \begin{bmatrix} 0 \\ -1 \\ 0 \\ 2 \end{bmatrix}.$$

The set of all one-dimensional H-invariant subspaces, from which all other H-invariant subspaces can be determined, is given by

$$b_1 = \begin{bmatrix} 1 \\ 0 \\ 0 \\ 0 \end{bmatrix}, b_2 = \begin{bmatrix} 0 \\ 1 \\ 0 \\ 0 \end{bmatrix}, b_3 = \begin{bmatrix} x \\ y \\ 0 \\ 0 \end{bmatrix}, b_4 = \begin{bmatrix} -1 \\ 0 \\ 2 \\ 0 \end{bmatrix}, b_5 = \begin{bmatrix} 0 \\ -1 \\ 0 \\ 2 \end{bmatrix} \text{ and } b_6 = \begin{bmatrix} -p \\ -q \\ 2p \\ 2q \end{bmatrix}.$$

All solutions of the Riccati equation are obtained by combinations of these vectors. This yields the next solutions X of equation (2.7.1)

1. $\begin{bmatrix} 0 & 0 \\ 0 & 0 \end{bmatrix}$ yielding $\sigma(A + RX) = \{1, 1\}$.

2. $\begin{bmatrix} -2 & \alpha \\ 0 & 0 \end{bmatrix}$; $\begin{bmatrix} -2 & 0 \\ \alpha & 0 \end{bmatrix}$; $\begin{bmatrix} 0 & \alpha \\ 0 & -2 \end{bmatrix}$; $\begin{bmatrix} 0 & 0 \\ \alpha & -2 \end{bmatrix}$; and $\frac{2}{\alpha - \beta} \begin{bmatrix} -\alpha & \frac{\alpha\beta}{\gamma} \\ -\gamma & \beta \end{bmatrix}$ yielding $\sigma(A + RX) = \{-1, 1\}$.

3. $\begin{bmatrix} -2 & 0 \\ 0 & -2 \end{bmatrix}$ yielding $\sigma(A + RX) = \{-1, -1\}$. $\qquad\qquad$ □

Up to this point, we have said nothing about the structure of the solutions given by Theorems 2.29 and 2.30. In Chapters 5 and 6 we will see that only solutions that are symmetric and for which $\sigma(A + RX) \subset \mathbb{C}^-$ (the so-called **stabilizing solutions**) will interest us. From the above example we see that there is only one stabilizing solution and that this solution is symmetric. This is not a coincidence as the next theorem shows. In fact the property that there will be at most one stabilizing solution is already indicated by our Note following Theorem 2.30. For, if matrix H has n different eigenvalues in $\mathbb{C}^-$, then it also has n different eigenvalues in $\mathbb{C}^+$. So, there can exist at most one appropriate invariant subspace of H in that case. To prove this observation in a more general context, we use another well-known lemma.

Lemma 2.31 (Sylvester)[6]

Consider the Sylvester equation

$$AX + XB = C \tag{2.7.4}$$

where $A \in \mathbb{R}^{n \times n}$, $B \in \mathbb{R}^{m \times m}$ and $C \in \mathbb{R}^{n \times m}$ are given matrices. Let $\{\lambda_i, \ i = 1, \ldots, n\}$ be the eigenvalues (possibly complex) of A and $\{\mu_j, \ j = 1, \ldots, m\}$ the eigenvalues (possibly complex) of B. There exists a unique solution $X \in \mathbb{R}^{n \times m}$ if and only if $\lambda_i(A) + \mu_j(B) \neq 0$, $\forall i = 1, \ldots, n$ and $j = 1, \ldots, m$.

Proof

First note that equation (2.7.4) is a linear matrix equation. Therefore, by rewriting it as a set of linear equations one can use the theory of linear equations to obtain the conclusion. For readers familiar with the Kronecker product and its corresponding notation we will provide a complete proof. Readers not familiar with this material are referred to the literature (see, for example, Zhou, Doyle and Glover (1996)). Using the Kronecker product, equation (2.7.4) can be rewritten as

$$(B^T \oplus A)\text{vec}(X) = \text{vec}(C).$$

This is a linear equation and therefore it has a unique solution if and only if matrix $B^T \oplus A$ is nonsingular; or put anotherway, matrix $B^T \oplus A$ has no zero eigenvalues. Since the eigenvalues of $B^T \oplus A$ are $\lambda_i(A) + \mu_j(B^T) = \lambda_i(A) + \mu_j(B)$, the conclusion follows. $\qquad$ □

An immediate consequence of this lemma is the following corollary.

---

[6]English mathematician/actuary/lawyer/poet who lived from 1814–1897. He did important work on matrix theory and was a friend of Cayley.

Corollary 2.32 (Lyapunov)[7]

Consider the so-called Lyapunov equation

$$AX + XA^T = C \tag{2.7.5}$$

where $A, C \in \mathbb{R}^{n \times n}$ are given matrices. Let $\{\lambda_i, \ i = 1, \ldots, n\}$ be the eigenvalues (possibly complex) of A. Then equation (2.7.5) has a unique solution $X \in \mathbb{R}^{n \times n}$ if and only if $\lambda_i(A) + \lambda_j(A) \neq 0$, $\forall i, j = 1, \ldots, n$. $\qquad \square$

Example 2.16

Consider the matrix equation

$$AX + XA = C,$$

where $\sigma(A) \subset \mathbb{C}^-$. This matrix equation has a unique solution for every choice of matrix C. This is because the sum of any two eigenvalues of matrix A will always be in $\mathbb{C}^-$ and thus differ from zero. $\qquad \square$

Example 2.17

Consider the matrix equation

$$A_1 X + XA_2 = C,$$

with A_i as in Example 2.3, and C an arbitrarily chosen 2×2 matrix, that is: $A_1 = \begin{bmatrix} -1 & -3 \\ 2 & 4 \end{bmatrix}$, and $A_2 = \begin{bmatrix} -6 & 4 \\ -4 & 2 \end{bmatrix}$. According to Example 2.3 the eigenvalues of A_1 are $\{1, 2\}$ and A_2 has only one eigenvalue $\{-2\}$. Therefore, according to Lemma 2.31, there exist matrices C for which the above equation has no solution and also matrices C for which the equation has an infinite number of solutions. There exists no matrix C for which the equation has exactly one solution. $\qquad \square$

The next theorem states the important result that if the algebraic Riccati equation (2.7.1) has a stabilizing solution then it is unique and, moreover, symmetric.

Theorem 2.33

The algebraic Riccati equation (2.7.1) has at most one stabilizing solution. This solution is symmetric.

---

[7]Russian mathematician who lived from 1857–1918. Schoolfriend of Markov and student of Chebyshev. He did important work on differential equations, potential theory, stability of systems and probability theory.

Proof

First, we prove the uniqueness property. To that end assume that X_1, X_2 are two symmetric solutions of equation (2.7.1). Then

$$A^T X_i + X_i A + X_i R X_i + Q = 0, \ i = 1, 2.$$

After subtracting the equations, we find

$$A^T (X_1 - X_2) + (X_1 - X_2)A + X_1 R X_1 - X_2 R X_2 = 0.$$

Using the symmetry of matrix X_1 this equation can be rewritten as

$$(A + R X_1)^T (X_1 - X_2) + (X_1 - X_2)(A + R X_2)^T = 0. \tag{2.7.6}$$

The above equation is a Sylvester equation. Since by assumption both $A + R X_1$ and $A + R X_2$ have all their eigenvalues in $\mathbb{C}^-$, $A + R X_1$ and $-(A + R X_2)$ have no eigenvalues in common. Thus, it follows from Lemma 2.31 that the above equation (2.7.6) has a unique solution $X_1 - X_2$. Obviously, $X_1 - X_2 = 0$ satisfies the equation. So $X_1 = X_2$, which proves the uniqueness result.

Next, we show that if equation (2.7.1) has a stabilizing solution X, then this solution will be symmetric. To that end we first note that, with the definition of J and H as before, the matrix JH defined by

$$JH := \begin{bmatrix} 0 & -I \\ I & 0 \end{bmatrix} H = \begin{bmatrix} Q & A^T \\ A & R \end{bmatrix}$$

is a symmetric matrix.

Now, let X solve the Riccati equation. Then, according Theorem 2.30, $X = X_2 X_1^{-1}$, where

$$H \begin{bmatrix} X_1 \\ X_2 \end{bmatrix} = \begin{bmatrix} X_1 \\ X_2 \end{bmatrix} \Lambda, \ \text{with } \sigma(\Lambda) \subset \mathbb{C}^- \text{ and } X_1 \text{ invertible.}$$

This implies that

$$[X_1^T \ X_2^T] JH \begin{bmatrix} X_1 \\ X_2 \end{bmatrix} = [X_1^T \ X_2^T] J \begin{bmatrix} X_1 \\ X_2 \end{bmatrix} \Lambda.$$

Since the left-hand side of this equation is symmetric, we conclude that the right-hand side of this equation has to satisfy

$$(X_2^T X_1 - X_1^T X_2)\Lambda = \Lambda^T (X_1^T X_2 - X_2^T X_1).$$

Or, stated differently,

$$(X_2^T X_1 - X_1^T X_2)\Lambda + \Lambda^T (X_2^T X_1 - X_1^T X_2) = 0.$$

This is a Lyapunov equation. Since, by assumption, the eigenvalues of $\Lambda \subset \mathbb{C}^-$ it is obvious from Corollary 2.32 that this equation has a unique solution $X_2^T X_1 - X_1^T X_2$. Obviously, 0 satisfies the equation. Hence, $X_1^T X_2 = X_2^T X_1$. But, then

$$
\begin{aligned}
X &= X_2 X_1^{-1} \\
&= X_1^{-T} X_1^T X_2 X_1^{-1} \\
&= X_1^{-T} X_2^T X_1 X_1^{-1} \\
&= X_1^{-T} X_2^T \\
&= X^T.
\end{aligned}
$$

$\square$

Apart from the fact that the stabilizing solution, X_s, of the algebraic Riccati equation (if it exists) is characterized by its uniqueness, there is another characteristic property. It is the **maximal solution** of equation (2.7.1). That is, every other solution X of equation (2.7.1) satisfies $X \leq X_s$. Notice that maximal (and minimal) solutions are unique if they exist.

To prove this property we first show a lemma whose proof uses the concept of the exponential of a matrix. The reader not familiar with this, can think of e^{At} like the scalar exponential function e^{at}. A formal treatment of this notion is given in section 3.1.

Lemma 2.34

If $Q \leq 0$ and A is stable, the Lyapunov equation

$$
AX + XA^T = Q \tag{2.7.7}
$$

has a unique semi-positive definite solution X.

Proof

Since A is stable we immediately infer from Corollary 2.32 that equation (2.7.7) has a unique solution X. To show that $X \geq 0$, consider $V(t) := \frac{d}{dt}(e^{A^T t} X e^{At})$. Using the product rule of differentiation and the fact that $\frac{de^{At}}{dt} = Ae^{At} = e^{At}A$ (see section 3.1), we have

$$
V(t) = e^{A^T t}(A^T X + XA)e^{At} = e^{A^T t} Q e^{At}. \tag{2.7.8}
$$

Since A is stable e^{At} converges to zero if t becomes arbitrarily large (see section 3.1 again). Consequently, since $e^{A.0} = I$, and the operation of integration and differentiation 'cancel out' we obtain on the one hand that

$$
\int_0^\infty V(t)dt = 0 - X, \tag{2.7.9}
$$

and on the other hand, using equation (2.7.8), that

$$
\int_0^\infty V(t)dt = \int_0^\infty e^{A^T t} Q e^{At} dt \leq 0. \tag{2.7.10}
$$

Combining equations (2.7.9) and (2.7.10) yields then that $X \geq 0$.

$\square$

Proposition 2.35

If the algebraic Riccati equation (2.7.1) has a stabilizing solution X_s and X is another solution of equation (2.7.1), then $X \leq X_s$.

Proof

Since both X_s and X satisfy equation (2.7.1)

$$(A - SX_s)^T (X_s - X) + (X_s - X)(A - SX_s) =$$
$$-A^T X - XA + (A^T X_s + X_s A - X_s SX_s) - X_s SX_s + X_s SX + XSX_s =$$
$$-XSX_s - X_s SX_s + X_s SX + XSX_s =$$
$$-(X - X_s)S(X - X_s) \leq 0.$$

Since, by assumption, $A - SX_s$ is stable Lemma 2.34 gives that $X_s - X \geq 0$. □

This maximality property has also led to iterative procedures to compute the stabilizing solution (see, for example, Zhou, Doyle and Glover (1996).

Next we provide a necessary condition on the spectrum of H from which one can conclude that the algebraic Riccati equation has a stabilizing solution. Lancaster and Rodman (1995) have shown that this condition together with a condition on the associated so-called matrix sign function are both necessary and sufficient to conclude that equation (2.7.1) has a real, symmetric, stabilizing solution.

Theorem 2.36

The algebraic Riccati equation (2.7.1) has a stabilizing solution only if H has no eigenvalues on the imaginary axis.

Proof

Let X be the stabilizing solution of equation (2.7.1). Simple manipulations, using equation (2.7.1), yield

$$\begin{bmatrix} I & 0 \\ -X & I \end{bmatrix} H \begin{bmatrix} I & 0 \\ X & I \end{bmatrix} = \begin{bmatrix} A + RX & R \\ 0 & -(A + RX)^T \end{bmatrix}.$$

As $\begin{bmatrix} I & 0 \\ -X & I \end{bmatrix}^{-1} = \begin{bmatrix} I & 0 \\ X & I \end{bmatrix}$, we conclude from the above identity that

$$\sigma(H) = \sigma(A + RX) \cup \sigma(-(A + RX)).$$

Since X is a stabilizing solution, $\sigma(A + RX)$ is contained in $\mathbb{C}^-$ and $\sigma(-(A + RX))$ in $\mathbb{C}^+$. Hence, H does not have eigenvalues on the imaginary axis. □

The next example illustrates that the above mentioned condition on the spectrum of matrix H in general is not enough to conclude that the algebraic Riccati equation will have a solution.

Example 2.18

Let $A = \begin{bmatrix} -1 & 0 \\ 0 & 1 \end{bmatrix}$, $R = \begin{bmatrix} 1 & 0 \\ 0 & 0 \end{bmatrix}$ and $Q = \begin{bmatrix} 0 & 0 \\ 0 & 0 \end{bmatrix}$. Then it is readily verified that the eigenvalues of H are $\{-1, 1\}$. A basis for the corresponding eigenspaces are

$$b_1 = \begin{bmatrix} 1 \\ 0 \\ 0 \\ 0 \end{bmatrix}, b_2 = \begin{bmatrix} 0 \\ 0 \\ 0 \\ 1 \end{bmatrix}; \text{ and } b_3 = \begin{bmatrix} 0 \\ 1 \\ 0 \\ 0 \end{bmatrix}, b_4 = \begin{bmatrix} 1 \\ 0 \\ 2 \\ 0 \end{bmatrix}, \text{ respectively.}$$

Consequently, H has no graph subspace associated with the eigenvalues $\{-1, -1\}$. $\square$

We conclude this section by providing a sufficient condition under which the algebraic Riccati equation (2.7.1) has a stabilizing solution. The next theorem shows that the converse of Theorem 2.36 also holds, under some additional assumptions on matrix R. One of these assumptions is that the matrix pair (A, R) should be stabilizable. A more detailed treatment of this notion is given in section 3.5. For the moment it is enough to bear in mind that the pair (A, R) is called stabilizable if it is possible to steer any initial state x_0 of the system

$$\dot{x}(t) = Ax(t) + Ru(t), \ x(0) = x_0,$$

towards zero using an appropriate control function $u(.)$. The proof of this theorem can be found in the Appendix to this chapter.

Theorem 2.37

Assume that (A, R) is stabilizable and R is positive semi-definite. Then the algebraic Riccati equation (2.7.1) has a stabilizing solution if and only if H has no eigenvalues on the imaginary axis. $\square$

2.8   Notes and references

Good references for a book with more details on linear algebra (and in particular section 2.5) is Lancaster and Tismenetsky (1985) and Horn and Johnson (1985).

For section 2.7 the book by Zhou, Doyle and Glover (1996) in particular chapters 2 and 13, has been consulted. For a general treatment (that is without the positive definiteness assumption) of solutions of algebraic Riccati equations an appropriate reference is the book by Lancaster and Rodman (1995). Matrix H in equation (2.7.2) has a special structure which is known in literature as a Hamiltonian structure. Due to this structure, in particular the eigenvalues and eigenvectors of this matrix have some nice properties. Details on this can be found, for example, in chapter 7 of Lancaster and Rodman (1995). A geometric classification of all solutions can also be found in Kučera (1991). The eigenvector solution method for finding the solutions of the algebraic Riccati equation was popularized in the optimal control literature by MacFarlane (1963) and Potter (1966). Methods and references on how to evercome numerical difficulties with this approach can be found, for example, in Laub (1991). Another numerical approach to calculate the stabilizing solution of the algebraic Riccati equation is based on the Newton – Raphson method. This gives rise to an iterative procedure to calculate this solution and has first been formalized by Kleinman (1968).

2.9 Exercises

1. Consider

$$v_1 = \begin{bmatrix} 1 \\ 2 \\ 3 \end{bmatrix}; \ v_2 = \begin{bmatrix} 3 \\ 2 \\ 1 \end{bmatrix}; \ v_3 = \begin{bmatrix} 1 \\ 1 \\ 1 \end{bmatrix} \text{ and } v_4 = \begin{bmatrix} -1 \\ 0 \\ 1 \end{bmatrix}.$$

(a) Show that $\{v_1, v_2\}$ are linearly independent.

(b) Show that $\{v_1, v_2, v_3\}$ are linearly dependent.

(c) Does a set of vectors v_i exist such that they constitute a basis for $\mathbb{R}^3$?

(d) Determine Span$\{v_2, v_3, v_4\}$. What is the dimension of the subspace spanned by these vectors?

(e) Determine the length of vector v_3.

(f) Determine all vectors in $\mathbb{R}^3$ that are perpendicular to v_4.

2. Consider

$$S := \text{Span} \left\{ \begin{bmatrix} 1 \\ 0 \\ 1 \\ 0 \end{bmatrix}, \begin{bmatrix} 0 \\ 0 \\ 1 \\ 1 \end{bmatrix}, \begin{bmatrix} 1 \\ 1 \\ 1 \\ 0 \end{bmatrix} \right\}.$$

(a) Use the orthogonalization procedure of Gram–Schmidt to find an orthonormal basis for S.

(b) Find a basis for $S^{\perp}$.

3. Consider

$$A = \begin{bmatrix} 1 & 0 & 1 \\ 1 & 1 & 2 \\ 1 & -1 & 0 \\ 0 & 1 & 1 \end{bmatrix}.$$

(a) Determine Ker A and Im A.

(b) Determine rank(A), dim(Ker A) and dim$((\text{Ker } A)^{\perp})$.

(c) Determine all $b \in \mathbb{R}^4$ for which $Ax = b$ has a solution x.

(d) If the equation $Ax = b$ in item (c) has a solution x, what can you say about the number of solutions?

4. Let $A \in \mathbb{R}^{n \times m}$ and S be a linear subspace. Show that the following sets are linear subspaces too.

(a) $S^{\perp}$;

(b) Ker A;

(c) Im A.

5. Consider the system
$$\dot{x}(t) = Ax(t), \ x(0) = x_0$$
$$y(t) = Cx(t),$$

where $x(.) \in \mathbb{R}^n$ and $y(.) \in \mathbb{R}^m$. Denote by $y(t, x_0)$ the value of y at time t induced by the initial state x_0. Consider for a fixed t_1 the set
$$V(t_1) := \{x_0 | y(t, x_0) = 0, \ t \in [0, t_1]\}.$$

Show that $V(t_1)$ is a linear subspace (see also Theorem 3.2).

6. Use Cramer's rule to determine the inverse of matrix
$$A = \begin{bmatrix} 1 & 0 & 1 \\ 0 & 1 & 0 \\ 1 & -1 & 0 \end{bmatrix}.$$

7. Consider
$$v_1^T = [1, \ 2, \ 3, \ 0, \ -1, \ 4] \text{ and } v_2^T = [1, \ -1, \ 1, \ 3, \ 4, \ 1].$$

Let $A := v_1 v_2^T$.
(a) Determine trace$(13 * A)$.
(b) Determine trace$(2 * A + 3 * I)$.
(c) Determine trace$(A * A^T)$.

8. Determine for each of the following matrices
$$A = \begin{bmatrix} 1 & 1 \\ 0 & 2 \end{bmatrix}; \ B = \begin{bmatrix} 2 & 0 \\ 1 & 2 \end{bmatrix}; \ C = \begin{bmatrix} 1 & 1 \\ 1 & 1 \end{bmatrix}; \ D = \begin{bmatrix} 1 & 0 & 0 \\ 0 & 1 & 0 \\ 1 & 1 & 2 \end{bmatrix},$$

(a) the characteristic polynomial,
(b) the eigenvalues,
(c) the eigenspaces,
(d) for every eigenvalue both its geometric and algebraic multiplicity,
(e) for every eigenvalue its generalized eigenspace.

9. Assume that $A = SBS^{-1}$. Show that matrix A and B have the same eigenvalues.

10. Let $A \in \text{TR}^{u \times u}$

(a) Show that for any $p \geq 1$, $N(A^p) \subset N(A^{p+1})$.
(b) Show that whenever $N(A^k) = N(A^{k+1})$ for some $k \geq 1$ also $N(A^p) = N(A^{p+1})$, for every $p \geq k$.
(c) Show that $N(A^p) = N(A^n)$, for every $p \geq n$.

11. Determine for the following complex numbers
$$z_1 = 1 + 2i; \ z_2 = 2 - i; \text{ and } z_3 = -1 - 2i,$$

$\bar{z}_i$, $\| z_i \|$, and $\frac{1}{z_i}$, respectively. Plot all these numbers on one graph in the complex plane.

12. Show that for any complex numbers z_1 and z_2

(a) $\overline{z_1 + z_2} = \bar{z}_1 + \bar{z}_2$,

(b) $\overline{z_1 z_2} = \bar{z}_1 \bar{z}_2$ and $\overline{z_1/z_2} = \bar{z}_1/\bar{z}_2$,

(c) $\bar{\bar{z}}_1 = z_1$.

13. Show that for any complex vector $z \in \mathbb{C}^n |z| = \sqrt{\bar{z}^T z}$.

14. (a) Show that for any complex matrix $Z = \begin{bmatrix} z_1 \\ \vdots \\ z_n \end{bmatrix}$, where $z_i = [z_{i1}, \ldots, z_{in}]$, $i = 1, \ldots, n+1$,

$$\det \begin{bmatrix} z_1 + \lambda z_{n+1} \\ z_2 \\ \vdots \\ z_n \end{bmatrix} = \det Z + \lambda \det \begin{bmatrix} z_{n+1} \\ z_2 \\ \vdots \\ z_n \end{bmatrix}.$$

(b) Consider in (a) a matrix Z for which $z_1 = z_2$. Show that $\det Z = 0$.

(c) Show, using (a) and (b), that if matrix Z_1 is obtained from matrix Z by adding some multiple of the second row of matrix Z to its first row, $\det Z_1 = \det Z$.

(d) Show, using (a), that if the first row of matrix Z is a zero row, $\det Z = 0$.

15. Let $w := \det(A + iB)$, where $A, B \in \mathbb{R}^{n \times n}$. Show that $\det(A - iB) = \bar{w}$. (Hint: consider $\overline{A + iB}$; the result follows then directly by a simple induction proof using the definition of a determinant and Theorem 2.13.)

16. Starting from the assumption that the results shown in Exercise 14 also hold when we consider an arbitrary row of matrix Z, show that

$$\det\left(\begin{bmatrix} A & C \\ 0 & D \end{bmatrix} \right) = \det(A)\det(D).$$

17. Determine for each of the following matrices

$$A = \begin{bmatrix} 0 & 1 \\ -1 & 0 \end{bmatrix}; \ B = \begin{bmatrix} 1 & 1 \\ -1 & 1 \end{bmatrix}; \ C = \begin{bmatrix} 1 & -1 & 0 \\ 1 & 1 & 0 \\ 1 & 1 & -1 \end{bmatrix}; \text{ and } D = \begin{bmatrix} 1 & -1 & 1 & 0 \\ 1 & 3 & 0 & 1 \\ 0 & 0 & 1 & -1 \\ 0 & 0 & 2 & 3 \end{bmatrix},$$

(a) the characteristic polynomial,

(b) the (complex) eigenvalues,

(c) the (complex) eigenspaces,

(d) for every eigenvalue both its geometric and algebraic multiplicity,

(e) for every eigenvalue its generalized eigenspace.

18. Show that $\ker A$ and $\operatorname{Im} A$ are A-invariant subspaces.

19. Consider

$$A = \begin{bmatrix} 0 & -2 & 0 \\ 1 & 2 & 0 \\ 1 & 2 & 1 \end{bmatrix}.$$

(a) Show that the characteristic polynomial of A is $p(\lambda) = -\lambda^3 + 3\lambda^2 - 4\lambda + 2$.
(b) Show that $-A^3 + 3A^2 - 4A + 2I = 0$.
(c) Show that $p(\lambda) = (1 - \lambda)(\lambda^2 - 2\lambda + 2)$.
(d) Determine $N_1 := \mathrm{Ker}(I - A)$ and $N_2 := \mathrm{Ker}(A^2 - 2A + 2I)$.
(e) Show that $N_1 \cap N_2 = \{0\}$.
(f) Show that any set of basis vectors for N_1 and N_2, respectively, form together a basis for $\mathbb{R}^3$.

20. Consider the matrices

$$A = \begin{bmatrix} 2 & 1 & -1 \\ 0 & 2 & 0 \\ 0 & 0 & 1 \end{bmatrix} \text{ and } B = \begin{bmatrix} 3 & 0 & 0 \\ 1 & 3 & 0 \\ 0 & 0 & 3 \end{bmatrix}.$$

(a) Determine the (generalized) eigenvectors of matrix A and B.
(b) Determine the Jordan canonical form of matrix A and B.

21. Determine the Jordan canonical form of the following matrices

$$A = \begin{bmatrix} 0 & 2 \\ -1 & 3 \end{bmatrix}; \ B = \begin{bmatrix} 0 & 1 \\ -1 & -2 \end{bmatrix}; \ C = \begin{bmatrix} 3 & 1 \\ -2 & 1 \end{bmatrix}.$$

22. Determine the Jordan canonical form of the matrices

$$A = \begin{bmatrix} 0 & 0 & -1 \\ 0 & 2 & 0 \\ -1 & 0 & 0 \end{bmatrix}; \ B = \begin{bmatrix} -1 & 0 & 0 \\ -1 & 1 & 1 \\ 0 & 0 & -1 \end{bmatrix}; \ C = \begin{bmatrix} 2 & -1 & 0 \\ 0 & 0 & -1 \\ -1 & 1 & 1 \end{bmatrix};$$

$$D = \begin{bmatrix} 2 & -2 & 2 \\ 0 & 2 & 0 \\ -1 & 2 & 0 \end{bmatrix}.$$

23. Determine for each matrix in Exercise 22 all its invariant subspaces.

24. Factorize the matrices below as $U^T D U$, where U is an orthonormal matrix and D a diagonal matrix.

$$A = \begin{bmatrix} 3 & 4 \\ 4 & -3 \end{bmatrix}; B = \begin{bmatrix} 0 & 0 & -1 \\ 0 & 2 & 0 \\ -1 & 0 & 0 \end{bmatrix} \ C = \begin{bmatrix} 5 & 2 & 1 \\ 2 & 2 & -2 \\ 1 & -2 & 5 \end{bmatrix}.$$

25. Let A be a symmetric matrix. Show that $A > 0$ (≥ 0) if all eigenvalues λ_i of A satisfy $\lambda_i > 0$ (≥ 0).

26. Verify which matrices in Exercise 24 are positive definite and positive semi-definite, respectively.

27. Determine all the solutions of the following algebraic Riccati equations. Moreover, determine for every solution $\sigma(A + RX)$.

(a) $\begin{bmatrix} 1 & 1 \\ 0 & 1 \end{bmatrix}^T X + X \begin{bmatrix} 1 & 1 \\ 0 & 1 \end{bmatrix} + X \begin{bmatrix} 1 & 0 \\ 0 & 0 \end{bmatrix} X + \begin{bmatrix} 0 & 0 \\ 0 & 1 \end{bmatrix} = 0.$

(b) $\begin{bmatrix} 0 & 2 \\ -1 & 3 \end{bmatrix}^T X + X \begin{bmatrix} 0 & 2 \\ -1 & 3 \end{bmatrix} + X \begin{bmatrix} 1 & 0 \\ 0 & 1 \end{bmatrix} X = 0.$

28. Verify whether the following matrix equations have a unique solution.

(a) $\begin{bmatrix} 1 & 1 \\ 0 & -3 \end{bmatrix} X + X \begin{bmatrix} 1 & 2 \\ 0 & 2 \end{bmatrix} = \begin{bmatrix} 1 & 2 \\ 2 & 3 \end{bmatrix}.$

(b) $\begin{bmatrix} 1 & 2 \\ 0 & 2 \end{bmatrix} X + X \begin{bmatrix} -1 & 0 \\ 2 & 3 \end{bmatrix} = \begin{bmatrix} 1 & 2 \\ 2 & 3 \end{bmatrix}.$

(c) $\begin{bmatrix} 1 & 1 \\ -1 & 1 \end{bmatrix} X + X \begin{bmatrix} -4 & 5 \\ -2 & 2 \end{bmatrix} = \begin{bmatrix} 1 & 2 \\ 2 & 3 \end{bmatrix}.$

29. Determine all solutions of the matrix equation

$$\begin{bmatrix} 1 & 0 \\ 0 & 2 \end{bmatrix} X + X \begin{bmatrix} -1 & 0 \\ 0 & 3 \end{bmatrix} = C,$$

where $C = \begin{bmatrix} 1 & 2 \\ 2 & 3 \end{bmatrix}$ and $C = \begin{bmatrix} 0 & 4 \\ 3 & 10 \end{bmatrix}$, respectively. Can you find a matrix C for which the above equation has a unique solution?

2.10 Appendix

Proof of Lemma 2.19

We distinguish three cases: (1) $i, j \in \{1, \ldots, k\}$; (2) $i \in \{1, \ldots, k\}$ and $j \in \{k+1, \ldots, r\}$; and (3) $i, j \in \{k+1, \ldots, r\}$. The proofs for all three cases are more or less the same (this is not surprising, because if we could have allowed for more advanced complex arithmetic all the cases could have been dealt with as one case).

(1) Assume without loss of generality $y \in \ker p_1(A)$. Then there exists an index $1 \leq m \leq n_1$ such that $y \in \ker(A - \lambda_1 I)^m$ but $y \notin \ker(A - \lambda_1 I)^{m-1}$ where, by definition, $(A - \lambda_1 I)^0 := I$ (see Exercises). Then for an arbitrary index $i \geq 1$

$$(A - \lambda_1 I)^{m-1}(A - \lambda_2 I)^i y = (A - \lambda_1 I)^{m-1}(A - \lambda_1 I + (\lambda_1 - \lambda_2)I)^i y$$

$$= (A - \lambda_1 I)^{m-1} \left\{ \sum_{j=0}^{i} \binom{i}{j} (\lambda_1 - \lambda_2)^{i-j} (A - \lambda_1 I)^j \right\} y$$

$$= (\lambda_1 - \lambda_2)^i (A - \lambda_1 I)^{m-1} y$$

$$\neq 0.$$

Consequently $(A - \lambda_2 I)^i y$ must differ from zero. So, $y \notin \ker(A - \lambda_2 I)^i$, and thus in particular $y \notin \ker p_2(A)$.

(2) Using the same notation as in **(1)** now

$$(A - \lambda_1 I)^{m-1}(A^2 + b_{k+1}A + c_{k+1}I)^i y = (A - \lambda_1 I)^{m-1}((A - \lambda_1 I)^2 + (2\lambda_1 + b_{k+1})(A - \lambda_1 I) +$$
$$(\lambda_1^2 + b_{k+1}\lambda_1 + c_{k+1})I)^i y$$
$$= ((\lambda_1^2 + b_{k+1}\lambda_1 + c_{k+1})I)^i (A - \lambda_1 I)^{m-1} y$$
$$\neq 0,$$

from which in a similar way the conclusion results.

(3) Again assume that m is such that $y \in \ker(A^2 + b_{k+1}A + c_{k+1}I)^m$ and $y \notin \ker(A^2 + b_{k+1}A + c_{k+1}I)^{m-1}$. Moreover, assume $y \in \ker(A^2 + b_{k+2}A + c_{k+2}I)^i$. Then

$$0 = (A^2 + b_{k+1}A + c_{k+1}I)^{m-1}(A^2 + b_{k+2}A + c_{k+2}I)^i y$$
$$= (A^2 + b_{k+1}A + c_{k+1}I)^{m-1}(A^2 + b_{k+1}A + c_{k+1}I + (b_{k+2} - b_{k+1})A + (c_{k+2} - c_{k+1})I)^i y$$
$$= (A^2 + b_{k+1}A + c_{k+1}I)^{m-1}((b_{k+2} - b_{k+1})A + (c_{k+2} - c_{k+1})I)^i y$$
$$= ((b_{k+2} - b_{k+1})A + (c_{k+2} - c_{k+1})I)^i (A^2 + b_{k+1}A + c_{k+1}I)^{m-1} y.$$

So, $v := (A^2 + b_{k+1}A + c_{k+1}I)^{m-1} y \in \ker((b_{k+2} - b_{k+1})A + (c_{k+2} - c_{k+1})I)^i$.

Next assume that $b_{k+2} = b_{k+1}$ and thus $c_{k+2} \neq c_{k+1}$ (otherwise $p_{k+1}(\lambda) = p_{k+2}(\lambda)$ holds). Then $\ker((b_{k+2} - b_{k+1})A + (c_{k+2} - c_{k+1})I)^i = 0$. So in that case $v = 0$ and $y \in \ker(A^2 + b_{k+1}A + c_{k+1}I)^{m-1}$ which contradicts our assumption on y. So $b_{k+2} \neq b_{k+1}$, and we conclude that $v \in \ker(A + \frac{c_{k+2} - c_{k+1}}{b_{k+2} - b_{k+1}}I)$ **(i)**.

On the other hand we have

$$0 = (A^2 + b_{k+1}A + c_{k+1}I)^{m-1}(A^2 + b_{k+2}A + c_{k+2}I)^i y$$
$$= (A^2 + b_{k+2}A + c_{k+2}I)^i (A^2 + b_{k+1}A + c_{k+1}I)^{m-1} y.$$

Therefore, also $v \in \ker(A^2 + b_{k+2}A + c_{k+2}I)^i$ **(ii)**.

However, according to **(2)** (i) and (ii) cannot both occur simultaneously, from which the assertion now readily follows. ∎

Proof of Theorem 2.37

From Theorem 2.36 we conclude the necessity of the condition. So what is left to be shown is that if H has no eigenvalues on the imaginary axis, equation (2.7.1) has a stabilizing solution.

From the Note following Theorem 2.30 it is clear that, since H has no eigenvalues on the imaginary axis, H has an n-dimensional stable invariant subspace. That is, there exists a matrix Λ, with $\sigma(\Lambda) \subset \mathbb{C}^-$, and a full column rank matrix

$$V = \text{Im}\begin{bmatrix} X_1 \\ X_2 \end{bmatrix},$$

such that

$$\begin{bmatrix} A & R \\ -Q & -A^T \end{bmatrix} \begin{bmatrix} X_1 \\ X_2 \end{bmatrix} = \begin{bmatrix} X_1 \\ X_2 \end{bmatrix} \Lambda. \tag{2.10.1}$$

As was shown in the proof of Theorem 2.33, we have under these conditions that

$$X_2^T X_1 = X_1^T X_2. \tag{2.10.2}$$

Then, according to Theorem 2.29, $X := X_2 X_1^{-1}$ is the unique stabilizing solution of equation (2.7.1) provided X_1 is invertible. So, what is left to be shown is that from our assumptions on R it follows that X_1 is invertible. To that end we first show that if matrix W is a full column rank matrix, such that $\mathrm{Im}\, W := \ker X_1$, then there exists a square matrix P such that

$$\Lambda W = WP. \tag{2.10.3}$$

For that purpose consider the first n equations of equation (2.10.1). That is,

$$AX_1 + RX_2 = X_1 \Lambda. \tag{2.10.4}$$

Pre- and post-multiplying this equation (2.10.4) by $W^T X_2^T$ and W, respectively, yields

$$W^T X_2^T A X_1 W + W^T X_2^T R X_2 W = W^T X_2^T X_1 \Lambda W.$$

From $X_1 W = 0$ and equation (2.10.2) it follows then that $W^T X_2^T R X_2 W = 0$. Since R is semi-positive definite we conclude that

$$R X_2 W = 0. \tag{2.10.5}$$

Using this it follows by post-multiplying equation (2.10.4) with W that $X_1 \Lambda W = 0$. Or, stated differently, $\mathrm{Im}\Lambda W \subset \ker X_1$. Obviously, this can be rephrased as in equation (2.10.3).

Now, let λ be an eigenvalue of P and y a corresponding eigenvector. Then, from equation (2.10.3),

$$\Lambda Wy = \lambda Wy. \tag{2.10.6}$$

That is, λ is an eigenvalue of Λ and Wy a corresponding eigenvector. Since Λ is a stable matrix, we conclude that $\lambda \in \mathbb{C}^-$. Next consider the second n-equations of equation (2.10.1)

$$-QX_1 - A^T X_2 = X_2 \Lambda.$$

Post-multiplying this equation by Wy gives

$$-QX_1 Wy - A^T X_2 Wy = X_2 \Lambda Wy.$$

Using equation (2.10.6) and the fact that $X_1 W = 0$, we infer from this equation that

$$(A^T + \lambda I)X_2 Wy = 0. \tag{2.10.7}$$

Since $\begin{bmatrix} X_1 \\ X_2 \end{bmatrix}$ is a full column rank matrix, and $\begin{bmatrix} X_1 \\ X_2 \end{bmatrix} Wy = \begin{bmatrix} 0 \\ X_2 Wy \end{bmatrix}$, it follows that $X_2 Wy \neq 0$ if $Wy \neq 0$. So, from equations (2.10.5) and (2.10.7) we conclude that if $\ker X_1$ is nonempty, then there exists a vector $X_2 Wy \neq 0$ and a $\lambda \in \mathbb{C}^-$ such that

$$(A^T + \lambda I)X_2 Wy = 0 \text{ and } RX_2 Wy = 0.$$

However, this implies that the equation

$$\begin{bmatrix} A^T - \mu I \\ R \end{bmatrix} X_2 Wy = 0$$

holds for some $\mu \in \mathbb{C}^+$ and $X_2 Wy \neq 0$. That is, the matrix

$$[A - \mu I \ R]$$

does not have a full row rank for some $\mu \in \mathbb{C}^+$. But this implies, according to Theorem 3.20, that the matrix pair (A, R) is not stabilizable, which contradicts our assumption. So, our assumption that $\ker X_1 \neq 0$ must be wrong. $\qquad \square$

3

Dynamical systems

A linear quadratic differential game studies situations involving two or more decision makers (individuals, organizations or governments). Decision makers are called the players in the game. These players often have partly conflicting interests and make individual or collective decisions. In a linear differential game the basic assumption is that all players can influence a number of variables which are crucial in realizing their goals and that these variables change over time due to external (natural) forces. These variables are called the state variables of the system. It is assumed that the movement over time of these state variables can be described by a set of linear differential equations in which the direct impact of the players' actions is in an additive linear way. Consequently, the extent to which the players succeed in realizing their goals depends on the actions of the other players. Obviously, if one player has information on the action that another player will take, he can incorporate this information into the decision making about his own action. Therefore, information plays a crucial role in the design of optimal actions for the players. So, summarizing, to analyze linear quadratic differential games one first has to introduce systems of differential equations, performance criteria and information sets.

This provides the framework for this chapter. First, to become more familiar with dynamical systems, linear systems are introduced. As already mentioned above, basically these systems are assumed to describe the motion of the game over time. Some results on the existence of trajectories satisfying the associated set of differential equations are outlined.

The optimal behavior of the involved players will be formalized mainly in terms of the optimization of a quadratic performance criterion. The associated optimization problems give rise to the analysis of sets of nonlinear differential equations. It is a well-known result that not every set of differential equations has a unique solution. Therefore, this issue is dealt with in a separate section. In particular the consequences of this observation for our situation, where players are free to manipulate the state dynamics, are discussed.

The analysis of sets of nonlinear differential equations is a delicate matter. Usually one starts this analysis by a study of the local behavior of trajectories near so-called equilibrium points of this set. This is the subject covered in section 3.3. An algorithm on how to proceed in case the set consists of just two differential equations is provided in section 3.4. To understand the potential evolution of linear systems a geometric point of view is very helpful. For that reason in section 3.5 the notions of controllability, observability, stabilizability and detectability are defined and various algebraic and geometric characterizations of these notions are summarized.

At the end of this chapter the standard framework of the control problem is discussed and this will be analysed throughout the book. In particular it is shown how various problem settings can be reformulated into this framework and some illustrative examples are presented. As already indicated above, the information players have on the game crucially affects the outcome of the game. Therefore, this chapter concludes with a discussion about the presumed information the players have on the game.

3.1 Description of linear dynamical systems

Let a finite dimensional **linear time invariant dynamical system** be described by the following set of constant coefficient differential equations:

$$\dot{x}(t) = Ax(t) + B_1 u_1(t) + B_2 u_2(t) + \cdots + B_N u_N(t), \ x(t_0) = x_0. \tag{3.1.1}$$

Here $x(t) \in \mathbb{R}^n$ is called the system **state**, x_0 the **initial condition** of the system, $u_i(t) \in \mathbb{R}^{m_i}$ contains the m_i control input variables that are chosen by player i, $i = 1, 2, \cdots, N$, and N indicates the number of players. The A and B_i are appropriately dimensioned real constant matrices. We assume that the state of the system coincides with the output of the system. That is, we assume that all players observe the complete state of the system. In case each player controls a single variable ($m_i = 1$, $i = 1, \cdots, N$) and the state is scalar too the system is called a single-input single-output system, or a scalar system.

Stating an equation in the form (3.1.1) raises the question whether solutions exist to such a differential equation. In fact, this question is far from trivial and in general negative. Fortunately, however, for a large class of input functions $u_i(.)$ a unique solution to equation (3.1.1) does exist. This is the case if all input functions $u_i(.)$ are, for example, continuous. The rest of this section will be used to justify (or at least to argue the plausibility of) this claim. The result presented here has a global character. That is, we will show that on the whole real line the equation (3.1.1) has a unique solution if, for example, the $u_i(.)$ functions are continuous. In the next section we will consider nonlinear systems of differential equations. Nonlinear differential equations will appear later on, for example in finding Nash equilibria when the planning horizon is finite. We shall state general conditions under which a set of nonlinear differential equations has a unique solution. However, these are local conditions. That is, from these conditions one can only conclude that there will exist a unique solution in some open interval around t_0. So, under those conditions all one knows is that a solution will exist for the differential equations for some limited time period.

Before considering the differential equation (3.1.1) we first treat a special case, that is

$$\dot{x}(t) = Ax(t), \ x(t_0) = x_0. \tag{3.1.2}$$

Definition 3.1

Let $A \in \mathbb{R}^{n \times n}$ and $t \in \mathbb{R}$ be fixed. Then the **exponential** e^{At} is defined as

$$e^{At} := \sum_{k=0}^{\infty} \frac{A^k t^k}{k!},$$

where $k! = k(k-1) \cdots 1$, with $0! = 1$. □

Without giving a formal proof notice that for a fixed t_1, the above series is absolutely and uniformly convergent for all $t \in [t_0, t_1]$ (due to the term $k!$ in the denominator, see Perko (2001)). So, e^{At} is well-defined. Next, it will be shown that the initial value problem (3.1.2) has the unique solution for all $t \in \mathbb{R}$

$$x(t) = e^{At} x_0. \tag{3.1.3}$$

Notice the similarity in the form of the above solution (3.1.3) and the solution $x(t) = e^{at} x_0$ of the elementary first-order differential equation $\dot{x} = ax$ with initial condition $x(t_0) = x_0$. Skipping a formal proof again (see again Perko (2001) for details) it follows that

$$\frac{d}{dt} e^{At} = \frac{d}{dt} \left(I + \frac{At}{1!} + \frac{A^2 t^2}{2!} + \frac{A^3 t^3}{3!} + \cdots \right)$$

$$= 0 + A + \frac{2A^2 t}{2!} + \frac{3A^3 t^2}{3!} + \cdots$$

$$= A \left(\sum_{k=0}^{\infty} \frac{A^k t^k}{k!} \right)$$

$$= A e^{At}.$$

Lemma 3.1

Let $A \in \mathbb{R}^{n \times n}$. Then

$$\frac{d}{dt} e^{At} = A e^{At}. \qquad \square$$

This result leads then to the next fundamental uniqueness and existence result for the differential equation (3.1.2).

Theorem 3.2

Let $A \in \mathbb{R}^{n \times n}$. Then for a given $x_0 \in \mathbb{R}^n$, the initial value problem

$$\dot{x} = Ax, \quad x(t_0) = x_0$$

has a unique solution given by $x(t) = e^{At} x_0$.

Proof

Using Lemma 3.1 it follows by straightforward substitution that $x(t) = e^{At}x_0$ satisfies the initial value problem. To see that it is the unique solution, assume that there is another solution $y(t)$ that satisfies the initial value problem. Next, consider

$$z(t) := e^{-At}y(t).$$

Then, using the product rule of differentiation and Lemma 3.1,

$$\frac{d}{dt}z(t) = \frac{de^{-At}}{dt}y(t) + e^{-At}\frac{dy(t)}{dt}$$
$$= -Ae^{-At}y(t) + e^{-At}(Ay(t))$$
$$= 0.$$

From the definition of e^{-At} it follows straightforwardly that $e^{-At}A = Ae^{-At}$ (that is the matrices e^{-At} and A commute). So, $\frac{d}{dt}z(t) = 0$ and, consequently, $z(t)$ must be constant. By taking $t = 0$ in the definition of $z(t)$ we see that $z(t) = y(0)$. It is easily verified (from Definition 3.1 again) that whenever two matrices A and B commute, then $e^{A+B} = e^A e^B$. In particular it follows from this that the inverse of matrix e^A is e^{-A}. Therefore $y(t) = e^{At}z(t) = e^{At}y(0)$; but since $y(0) = x_0$, we conclude that $y(t)$ and $x(t)$ coincide.   □

Example 3.1

1. Let $A_1 = \begin{bmatrix} \lambda & 0 \\ 0 & \mu \end{bmatrix}$. Recalling the power series expansion of $e^x = 1 + x + \frac{x^2}{2!} + \frac{x^3}{3!} + \cdots$, it follows straightforwardly from the definition of e^A that $e^{A_1t} = \begin{bmatrix} e^{\lambda t} & 0 \\ 0 & e^{\mu t} \end{bmatrix}$. So, the unique solution of equation (3.1.2) is $x(t) = \begin{bmatrix} e^{\lambda t} & 0 \\ 0 & e^{\mu t} \end{bmatrix} x_0$.

2. Let $A_2 = \begin{bmatrix} \lambda & 1 & 0 \\ 0 & \lambda & 1 \\ 0 & 0 & \lambda \end{bmatrix}$. Again straightforward calculation shows that $e^{A_2t} = e^{\lambda t}\begin{bmatrix} 1 & t & \frac{t^2}{2!} \\ 0 & 1 & t \\ 0 & 0 & 1 \end{bmatrix}$. From this one immediately obtains the solution of equation (3.1.2).

3. Let $A_3 = \begin{bmatrix} a & b \\ -b & a \end{bmatrix}$. Then, $e^{A_3t} = e^{at}\begin{bmatrix} \cos(bt) & \sin(bt) \\ -\sin(bt) & \cos(bt) \end{bmatrix}$. This can be directly verified by comparing the entries obtained in the calculation of e^{A_3t} with the power series expansion of $\sin(x) = \frac{x}{1!} - \frac{x^3}{3!} + \frac{x^5}{5!} - \frac{x^7}{7!} + \cdots$ and of $\cos(x) = 1 - \frac{x^2}{2!} + \frac{x^4}{4!} - \frac{x^6}{6!} + \frac{x^8}{8!} - \cdots$. Again, then the solution of equation (3.1.2) is obvious.   □

In fact the cases considered in the above example are representative of the solution of equation (3.1.2) in general. To show this we will use the Jordan canonical form of matrix A. As a preliminary, we have the following lemma.

Lemma 3.3

Let $A = S \operatorname{diag}\{J_i\}S^{-1}$. Then

$$e^{At} = S \operatorname{diag}\{e^{J_i t}\}S^{-1}.$$

Proof

Consider the Definition 3.1 of e^{At}. From this

$$e^{At} = \sum_{k=0}^{\infty} \frac{A^k t^k}{k!} = \sum_{k=0}^{\infty} \frac{(S \operatorname{diag}\{J_i\}S^{-1})^k t^k}{k!} = \sum_{k=0}^{\infty} \frac{S(\operatorname{diag}\{J_i\})^k t^k S^{-1}}{k!}$$

$$= S\left[\sum_{k=0}^{\infty} \frac{(\operatorname{diag}\{J_i\})^k t^k}{k!}\right] S^{-1} = S \operatorname{diag}\{e^{J_i t}\}S^{-1}. \qquad \square$$

Theorem 3.4

Let $A = S \operatorname{diag}\{J_1, J_2, \cdots, J_k\}S^{-1}$, where J_i has either one of the Jordan canonical forms $J_{\mathbb{R}}$, $J_{\mathbb{R}E}$, $J_{\mathbb{C}}$ or $J_{\mathbb{C}E}$. Then,

$$e^{At} = S \operatorname{diag}\{e^{J_1 t}, e^{J_2 t}, \cdots, e^{J_k t}\}S^{-1},$$

where $e^{J_i t}$ has one of the following forms:

$$(i)\ e^{J_{\mathbb{R}} t} = e^{\lambda_i t} \begin{bmatrix} 1 & & & & \\ & 1 & & & \\ & & \ddots & & \\ & & & 1 & \\ & & & & 1 \end{bmatrix} ; \quad (ii)\ e^{J_{\mathbb{R}E} t} = e^{\lambda_i t} \begin{bmatrix} 1 & t & \frac{t^2}{2!} & & \frac{t^{n_i-1}}{(n_i-1)!} \\ & 1 & \ddots & & \frac{t^{n_i-2}}{(n_i-2)!} \\ & & \ddots & \ddots & \\ & & & 1 & t \\ & & & & 1 \end{bmatrix} ;$$

$$(iii)\ e^{J_{\mathbb{C}} t} = e^{at} \begin{bmatrix} R_i & & & & \\ & R_i & & & \\ & & \ddots & & \\ & & & R_i & \\ & & & & R_i \end{bmatrix} ; \quad (iv)\ e^{J_{\mathbb{C}E} t} = e^{at} \begin{bmatrix} R & Rt & R\frac{t^2}{2!} & & R\frac{t^{n_i-1}}{(n_i-1)!} \\ & R & \ddots & & R\frac{t^{n_i-2}}{(n_i-2)!} \\ & & \ddots & \ddots & \\ & & & R & Rt \\ & & & & R \end{bmatrix} .$$

Here the matrices $J_{\mathbb{R}}$, $J_{\mathbb{R}E}$, $J_{\mathbb{C}}$ and $J_{\mathbb{C}E}$ are as denoted in Theorem 2.24, on Jordan's canonical form, and matrix R is the rotation matrix $\begin{bmatrix} \cos(bt) & \sin(bt) \\ -\sin(bt) & \cos(bt) \end{bmatrix}$.

Proof

Straightforward calculations show that the exponential of the Jordan canonical matrix forms J_R, J_{RE}, J_C and J_{CE} is as stated above, respectively. The final result follows then directly from Lemma 3.3.

Example 3.2

Let $A_1 = \begin{bmatrix} 8 & 7 & -13 \\ 1 & -1 & 2 \\ 7 & 4 & -9 \end{bmatrix}$. The characteristic polynomial of A is $-\lambda^3 - 2\lambda^2 - 5\lambda + 26$,

which can be factorized as $-(\lambda - 2)(\lambda^2 + 4\lambda + 13)$. So, A has a real eigenvalue 2 and the

complex eigenvalues $-2 \pm 3i$. An eigenvector corresponding to 2 is $\begin{bmatrix} 1 \\ 1 \\ 1 \end{bmatrix}$ and an

eigenvector corresponding to $-2 + 3i$ is $\begin{bmatrix} 1+i \\ -i \\ 1 \end{bmatrix}$. Therefore, with $S = \begin{bmatrix} 1 & 1 & 1 \\ 1 & 0 & -1 \\ 1 & 1 & 0 \end{bmatrix}$,

the solution $x(t)$ of the differential equation $\dot{x} = Ax$; $x(0) = x_0$, is

$$x(t) = S \begin{bmatrix} e^{2t} & 0 & 0 \\ 0 & e^{-2t}\cos(3t) & e^{-2t}\sin(3t) \\ 0 & -e^{-2t}\sin(3t) & e^{-2t}\cos(3t) \end{bmatrix} S^{-1} x_0. \qquad \square$$

Next, consider the nonhomogeneous linear differential equation

$$\dot{x}(t) = Ax(t) + b(t), \quad x(t_0) = x_0, \tag{3.1.4}$$

where $b(t)$ is a continuous vector valued function. The next theorem provides its solution.

Theorem 3.5

The unique solution of the nonhomogeneous linear differential equation (3.1.4) is given by

$$x(t) = e^{A(t-t_0)}x_0 + e^{At} \int_{t_0}^{t} e^{-A\tau} b(\tau)d\tau. \tag{3.1.5}$$

Proof

Since $b(t)$ is continuous, the function $x(t)$ defined above is differentiable everywhere and

$$\frac{d}{dt}x(t) = Ae^{A(t-t_0)}x_0 + Ae^{At} \int_{t_0}^{t} e^{-A\tau} b(\tau)d\tau + e^{At}e^{-At}b(t)$$

$$= Ax(t) + b(t).$$

So, the solution $x(t)$ as defined by equation (3.1.5) satisfies the nonlinear differential equation (3.1.4). To see that it is the unique solution one can proceed along the lines of the proof of Theorem 3.2. $\qquad\square$

In a similar way one can show that if matrix A in equation (3.1.4) is a continuous time-dependent matrix, that is $A = A(t)$ where each entry is a continuous function of time, equation (3.1.4) has the unique solution (see, for example, Coddington and Levinson (1955))

$$x(t) = \Phi(t, t_0)x_0 + \int_{t_0}^{t} \Phi(t, \tau)b(\tau)d\tau. \qquad (3.1.6)$$

Here $\Phi(t, t_0)$, the **transition matrix** or **fundamental matrix**, is the solution of the matrix differential equation

$$\frac{d}{dt}\Phi(t, t_0) = A(t)\Phi(t, t_0), \quad \Phi(t_0, t_0) = I, \quad \forall t. \qquad (3.1.7)$$

Note

Standard texts on differential equations (see Coddington and Levinson (1955), Section 3.2) show that there always exists what is called a **fundamental set of solutions** to

$$\dot{x}(t) = A(t)x(t), \quad \text{with } A(t) \text{ continuous.} \qquad (3.1.8)$$

That is, this differential equation has n solutions, $\phi_i(t)$, from which all other solutions of equation (3.1.8) can be obtained as a linear combination of $\phi_i(t)$. Moreover, this property does not hold if one considers any set of $n - 1$ functions. So, this fundamental set of solutions is a basis for the set of all solutions of equation (3.1.8). It can be shown that the matrix of fundamental solutions $[\phi_1(t) \cdots \phi_n(t)]$ is invertible for every t. $\qquad\square$

The solution $\Phi(t, t_0)$ can rarely be obtained in terms of standard functions. Consequently, one usually has to resort to numerical integration techniques to calculate the solution.

Corollary 3.6

1. If the vector of control variables $u_i(.)$ chosen by player i is continuous, $i = 1, \cdots, N$, the differential equation (3.1.1) has a unique solution $x(.)$.

2. If the vector of control variables $u_i(.)$ chosen by player i are affine functions of the state variables, that is $u_i(t) = F_i(t)x(t) + v_i(t)$ with both $F_i(.)$ and $v_i(.)$ continuous, $i = 1, \cdots, N$, the differential equation (3.1.1) has a unique solution. $\qquad\square$

3.2 Existence–uniqueness results for differential equations

3.2.1 General case

In the previous section we saw that any linear homogeneous set of differential equations (3.1.4) has a unique solution through the point x_0 provided the vector valued function $b(t)$ is continuous. In this section we will consider a set of nonlinear differential equations

$$\dot{x} = f(t, x), \tag{3.2.1}$$

where $f : \mathbb{R}^n \to \mathbb{R}^n$.

Definition 3.2

A function $g(t)$ is a solution of the differential equation (3.2.1) on an interval I if $g(t)$ is differentiable on I and if for all $t \in I$

$$\dot{g}(t) = f(t, g(t)). \qquad \Box$$

We will present conditions on the function f from which one can conclude that the nonlinear system (3.2.1) has a unique solution through each point $x_0 \in \mathbb{R}^n$ defined on a maximal interval of existence $(\alpha, \beta) \subset \mathbb{R}$. In general, it is not possible to solve the nonlinear set of differential equations (3.2.1). An indication of the limitation of any general existence theorem can be seen by considering the next example.

Example 3.3

Consider the differential equation

$$\dot{x}(t) = x^2(t), \text{ with } x(1) = -1.$$

By straightforward differentiation it is easily verified that $g(t) = -\frac{1}{t}$ is a solution to this differential equation. However, this solution does not exist at $t = 0$, although $f(t, x) = x^2$ is continuous there. $\qquad \Box$

The example shows that any general existence theorem will necessarily have to be of a local nature, and existence on $\mathbb{R}$ can only be asserted under additional conditions on f. Notice that if f is an integrable function, then

$$g(t) := x(0) + \int_0^t f(\tau, x(\tau))d\tau, \ g(0) = x_0 \tag{3.2.2}$$

is a solution of the differential equation

$$\dot{x}(t) = f(t, x), \text{ with } x(0) = x_0.$$

Consequently one can show that, in general, the set of differential equations (3.2.1) will have a solution if the function f is continuous. However, continuity of the function f in (3.2.1) is not sufficient to guarantee uniqueness of the solution as the next example illustrates.

Example 3.4

Consider the differential equation

$$\dot{x} = 3x^{2/3}, \quad x(0) = 0.$$

Then by straightforward substitution we see that $x_1(t) = t^3$ as well as $x_2(t) = 0$ satisfy this differential equation.

Notice that the function $f(x) = x^{2/3}$ is continuous at $x = 0$ but that it is not differentiable there. □

A feature of nonlinear differential equations that differs from linear differential equations is that even when the function $f(t, x)$ in equation (3.2.1) is continuous, the solution $x(t)$ of the differential equation (3.2.1) may become unbounded at some finite time $t = t_1$; i.e. the solution may only exist on some proper interval (t_{-1}, t_1). In that case t_1 is called a **finite escape time** of the differential equation (3.2.1). This point was, in fact, already illustrated in Example 3.3. We will reconsider this example for an arbitrary initial condition.

Example 3.5

Consider the differential equation

$$\dot{x} = x^2, \quad x(0) = 1/x_0 > 0.$$

Then, by straightforward differentiation it is easily verified that $x(t) = \frac{1}{x_0 - t}$ satisfies the differential equation; but, this solution is only defined for $t \in (-\infty, x_0)$ and $\lim_{t \uparrow x_0} = \infty$.

The interval $(-\infty, x_0)$ is called the maximal interval of existence of the solution of this differential equation, and $t = x_0$ is a finite escape time of this differential equation. Notice that the function $x(t) = \frac{1}{x_0 - t}$ has another branch defined on the interval (x_0, ∞); however, this branch is not considered since the initial time $t = 0 \notin (x_0, \infty)$ due to our assumption that $x_0 > 0$. □

Before stating the fundamental existence–uniqueness theorem for sets of linear differential equations we recall from analysis a property (see, for example, Rudin (1964)) which is helpful to verify whether the derivative of a function f is continuous. A function f which has a continuous derivative is called **continuous differentiable**. The set of functions that are continuous differentiable will be denoted by C^1.

Theorem 3.7

Suppose $f : \mathbb{R}^n \to \mathbb{R}^n$. Then $f(x)$ is continuous differentiable if and only if the partial derivatives $\frac{\partial f_i}{\partial x_j}$, $i, j = 1, \cdots, n$, exist and are continuous. □

Theorem 3.8 (the fundamental existence–uniqueness theorem)

Assume that $f(t,x)$ is continuous and the partial derivatives $\frac{\partial f(t,x)}{\partial x_i}$, $i = 1, \cdots, n$, exist and are continuous. Then for each x_0 there exists a $t_{-1} < t_0$ and $t_1 > t_0$ such that the differential equation

$$\dot{x}(t) = f(t,x), \quad x(t_0) = x_0,$$

has a unique solution $x(t)$ on the interval (t_{-1}, t_1). Moreover, this interval of existence is maximal. That is, if a solution also exists on another interval J, then $J \subset (t_{-1}, t_1)$. Notice that t_{-1} may be $-\infty$ and t_1 may be ∞. $\qquad\square$

Note

In the above theorem it is assumed that $f(t,x)$ satisfies a number of conditions with respect to its second argument, x, on $\mathbb{R}^n$. In fact it suffices that these conditions are satisfied on an open subset $E \subset \mathbb{R}^n$ containing x_0. For details on this issue, see Perko (2001). $\qquad\square$

A proof of Theorem 3.8, as well as of the next theorem, can be found in Chapter 2 of Perko (2001) or Chapter 1 of Coddington and Levinson (1955). Theorem 3.9 below states that, under the conditions of Theorem 3.8, solutions with an initial state vector y that are close to x_0 stay close to the solution trajectory through x_0 on a time interval $[t_0 - t, t_0 + t]$. Notice that this time interval is a closed interval. In particular, the theorem does **not** state that the solutions on the whole maximal interval of existence stay close. Figure 3.1 illustrates this point.

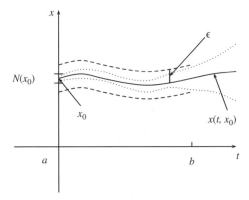

Figure 3.1 Dependence on initial conditions

Theorem 3.9

Assume that $f(t,x)$ is continuous and the partial derivatives $\frac{\partial f(t,x)}{\partial x_i}$, $i = 1, \cdots, n$, exist and are continuous. Suppose that the differential equation (3.2.1) has a solution on a closed interval $[a,b]$. Denote this solution by $x(t,x_0)$. Let ϵ be an arbitrarily chosen positive number. Then there is an open neighborhood $N(x_0)$ of x_0 such that for all solutions $x(t, y)$ of equation (3.2.1), with $x(t_0) = y \in N(x_0)$, $\sup_{t \in [a,b]} |x(t, x_0) - x(t, y)| < \epsilon$. Furthermore, the function $y \rightarrow x(t, y)$ is differentiable and for all h such that $x_0 + h \in N(x_0)$ we have

$$x(t, x_0 + h) = x(t, x_0) + \Phi(t)h + 0(h),$$

where, with $A(t) := f_x(t, x(t, x_0))$, $\Phi(t)$ is the solution of $\dot{\Phi}(t) = A(t)\Phi(t)$, $\Phi(0) = I$. This formula holds uniformly for $a \leq t \leq b$.

Example 3.6

1. Reconsider $\dot{x} = x^2$, $x(0) = 1/x_0 > 0$. The partial derivative of $f(x) = x^2$ is $2x$. Obviously, this derivative is continuous. So, Theorem 3.8 applies and we conclude that for every x_0 there exists a $t_1 > 0$ such that this differential equation has a unique solution on the open interval $(0, t_1)$.

2. Reconsider $\dot{x} = 3x^{2/3}$, $x(0) = 0$. The partial derivative of $f(x) = 3x^{2/3}$ is $\frac{2}{x^{1/3}}$. This derivative is not continuous at $x = 0$. So, Theorem 3.8 does not apply.

   Notice from the Note following Theorem 3.8 that if, e.g., $x(0) = 1$ the differential equation does have a unique solution on some open interval $(0, t_1)$. For, choose for instance $E = (0, 2)$. Then the derivative of f is continuous on E. □

Example 3.7

Consider the set of feedback Nash differential equations

$$\dot{k}_1 = 2ak_1 - s_1 k_1^2 - 2s_2 k_1 k_2 + q_1; \ k_1(t_0) = k_{10}$$
$$\dot{k}_2 = 2ak_2 - s_2 k_2^2 - 2s_1 k_1 k_2 + q_2; \ k_2(t_0) = k_{20}$$

where a, s_i, q_i, $k_{i0} \in \mathbb{R}$. Then with

$$f(k_1, k_2) := \begin{pmatrix} 2ak_1 - s_1 k_1^2 - 2s_2 k_1 k_2 + q_1 \\ 2ak_2 - s_2 k_2^2 - 2s_1 k_1 k_2 + q_2 \end{pmatrix},$$

the matrix of partial derivatives $\frac{\partial f_i}{\partial k_j} = \begin{bmatrix} 2a - 2s_1 k_1 - 2s_2 k_2 & -2s_2 k_1 \\ -2s_1 k_2 & 2a - 2s_2 k_2 - 2s_1 k_1 \end{bmatrix}$.

Obviously, all partial derivatives exist and are continuous. So, according to Theorem 3.7, $f \in C^1$. From Theorem 3.8 we therefore conclude that for each initial condition (k_{10}, k_{20}) there is an interval where this set of differential equations has a unique solution. □

Conditions also exist in literature which guarantee the existence of a solution of equation (3.2.1) on the whole real line. From Perko (2001), Theorem 3.1.3, and Coddington and Levinson (1955) Chapter 1, we recall the next result.

Theorem 3.10

Suppose $f(t, x) \in C$ and there exist a constant L such that

$$|f(t, x) - f(t, y)| \leq L|x - y| \tag{3.2.3}$$

for all $x, y \in \mathbb{R}^n$. Then for all $x_0 \in \mathbb{R}^n$ the initial value problem

$$\dot{x}(t) = f(t, x), \quad x(t_0) = x_0,$$

has a unique solution $x(t) \in C^1$ defined for all $t \in \mathbb{R}$. $\qquad\square$

The condition (3.2.3) is in literature known as the **global Lipschitz[1] condition**. This Lipschitz condition is not a condition which is almost always satisfied. The function $f(k_1, k_2)$ in Example 3.7 does not, for instance, satisfy this condition.

Example 3.8

Consider the differential equation

$$\dot{x} = \frac{x}{1 + x^2}.$$

With $f(x) = \frac{x}{1+x^2}$,

$$|f(x) - f(y)| = \frac{\left| x(1 + y^2) - y(1 + x^2) \right|}{(1 + x^2)(1 + y^2)}$$

$$= |x - y| \frac{|1 - xy|}{(1 + x^2)(1 + y^2)}$$

$$\leq |x - y| \left(1 + \frac{|x|}{1 + x^2} \frac{|y|}{1 + y^2} \right)$$

$$\leq 2|x - y|.$$

So, f satisfies the global Lipschitz condition and consequently the solution to the differential equation exists globally. $\qquad\square$

3.2.2 Control theoretic extensions

In the next chapter we will be dealing with optimal control problems where the goal is to optimize a preference function. This preference function depends both on the state variable and the used amount of control to manipulate this state variable. The dynamics of the state variable over time are described by the differential equation

$$\dot{x}(t) = f(x, u, t), \quad x(t_0) = x_0. \tag{3.2.4}$$

Here $u(t) \in \mathbb{R}^m$ is the control used at time t by the optimizer. As we have seen in the previous subsection the optimization problem is ill-defined if we do not make any further assumptions on the function f. Either equation (3.2.4) may not have any solutions (even in the extended sense, see Definition 3.3 below) or a solution may fail to exist for some control functions on the time interval we are interested in. One way out of this problem is

---

[1]Lipschitz was a famous German mathematician who lived from 1823–1903.

to restrict the optimization problem to those control functions for which equation (3.2.4) does have a unique solution on the considered interval and formulate necessary conditions which the optimal solution has to satisfy under the assumption that such a unique solution exists. This approach then provides a mechanism to discriminate between solutions which definitely are not optimal. Finally if one is left with a candidate solution, one always has to verify whether the solution makes any sense at all by checking whether it satisfies the differential equation (3.2.4) (we will address the issue of how to derive optimal solutions in more detail in Section 4.5). However, notice that this approach does not shed any light on the question of what the corresponding set of admissible control functions looks like. There are two important cases in which we a priori know that equation (3.2.4) has a unique solution. These are the cases where either the function f satisfies a global Lipschitz condition or f is linear (see equation 3.2.6 below).

A problem with the solution concept introduced in Definition 3.2 is that in control theory this concept is too tight. In control theory one would like to allow for discontinuous control functions or even for control functions which are only integrable functions.

The continuity of f guarantees that a solution of equation (3.2.1) is C^1. However, it is clear that the integral in equation (3.2.2) also makes sense for many functions f which are not continuous. If a continuously differentiable solution of equation (3.2.1) is not demanded, the continuity restriction on f can be relaxed by looking for an absolutely continuous function[2] $g(t)$ defined on an interval I such that

$$\dot{g}(t) = f(t, g(t)) \tag{3.2.5}$$

for all $t \in I$, except on a set of Lebesgue-measure[3] zero.

Definition 3.3

If there exists an interval I and absolutely continuous function $g(t)$ satisfying equation (3.2.5) for all $t \in I$ (except on a subset of I that has measure zero), then $g(t)$ is called a **solution of equation (3.2.1) in the extended sense on I**.

The absolute continuity of a solution guarantees that $\dot{g}$ exists on I except on a set of Lebesgue-measure zero, see Riesz and Sz.-Nagy (1972) for example. Therefore, Definition 3.3 makes sense.

---

[2]Formally: a continuous function $g(t)$ is called **absolutely continuous** if, given $\epsilon > 0$, there is a $\delta > 0$ such that $\sum_{k=1}^{m} |g(t_k) - g(t_k')| \leq \epsilon$ for every finite collection $\{(t_k, t_k')\}$ of nonoverlapping intervals with $\sum_{k=1}^{m} |t_k - t_k'| \leq \delta$. From the definition it follows straightforwardly that every absolute continuous function is continuous. A function $G : [a, b] \to \mathbb{R}$ can be written as $G(t) = \int_0^t g(t)dt + G(a)$ for some Lebesgue integrable function $g(t)$ if and only if $G(t)$ is absolute continuous on $[a, b]$. In that case $G(t) = \int_0^t G'(t)dt + G(a)$ and $G'(t) = g(t)$ except possibly on a null set of $[a, b]$ (see, for example, Riesz and Sz. Nagy (1972)).

[3]The Lebesgue-measure μ on $\mathbb{R}$ is a function that assigns to almost all subsets (to be precise all Borel sets) of $\mathbb{R}$ a non-negative number. The Lebesgue-measure is implicitly defined for all Borel sets by the rule to assign to each interval $[a, b)$ as $\mu([a, b))$ its length $b - a$. This definition implies that μ assigns the value zero to any set consisting of a finite or even a countable number of points (like, for example, $\mathbb{N}$ or $\mathbb{Q}$). Sets having a measure zero are called *null sets*. A function f is called measurable on an interval J in case for all α and β, the inverse image set of f defined by $\{x \mid x \in J$ and $\alpha \leq f(x) < \beta\}$, is a Borel set. For more details, see, for example, Halmos (1966).

Theorem 3.11

Consider the linear system

$$\dot{x} = A(t)x(t) + b(t), \quad x(t_0) = x_0 \tag{3.2.6}$$

where $A(t)$ and $b(t)$ are integrable functions over the interval $[a, b]$ such that

$$|A(t)| \le k(t) \text{ and } |b(t)| \le k(t)$$

and $k(t)$ is an integrable function over the interval $[a, b]$. Then equation (3.2.6) has a unique solution in the extended sense on $[a, b]$.

Furthermore, if the above holds for all $b < \infty$, then the solution of equation (3.2.6) exists over $[a, \infty)$.

Moreover, in case a fundamental solution Φ for the corresponding homogeneous differential equation $\dot{x}(t) = A(t)x(t)$, $\Phi(t_0, t_0) = I$, is known the solution of equation (3.2.6) is given by the variation-of-constants formula

$$x(T) := \Phi(T, t_0)x_0 + \int_{t_0}^{T} \Phi(T, \tau)b(\tau)d\tau. \tag{3.2.7}$$

(see Coddington and Levinson (1955) page 97, Problem 1). □

Corollary 3.12

Consider the differential equation

$$\dot{x}(t) = Ax(t) + Bu(t) + v(t), \quad x(0) = x_0. \tag{3.2.8}$$

If $u(t) = Fx(t)$ (with F such that $A + BF$ is stable) and $v(t)$ is a square integrable function (that is $\int_0^\infty v^T(t)v(t)dt < \infty$) then equation (3.2.8) has a unique solution in the extended sense on $[0, \infty)$. □

Apart from this global existence result there are also analogues of the fundamental existence and uniqueness results (section 3.2.1). Furthermore, if the differential equation depends on a parameter $\mu \in \mathbb{R}^m$, i.e. the function $f(x, u, t)$ in equation (3.2.4) is replaced by $f(x, u, t, \mu)$, then the solution $x(x_0, u, t, \mu)$ will also depend on the parameter μ. Theorem 3.14, below, states that such differential equations have a unique solution for all parameters μ in some neighborhood of μ_0 in which case for this nominal $\mu = \mu_0$ a unique solution exists. This result is used later on to draw conclusions concerning the existence of solutions in case the differential equation (3.2.4) is 'perturbed'. This property is used in the derivation of conditions that must be satisfied by optimal solutions.

All these results can be obtained directly from the theory of ordinary differential equations. This is because, for a fixed control sequence $u(.)$, the differential equation (3.2.4) reduces to the ordinary differential equation

$$\dot{x}(t) = f_u(x, t), \quad x(t_0) = x_0, \tag{3.2.9}$$

where $f_u(x, t)$ is defined by $f_u(x, t) := f(x, u(t), t)$.

However, first we introduce a class of functions that plays a fundamental role in the analysis.

Definition 3.4

Let $g(x, t) : \mathbb{R}^{n+1} \to \mathbb{R}$ satisfy the following three properties:

(i) for each fixed t is $g(x, t) \in C^1$;

(ii) for each fixed x is $g(x, t)$ measurable in t;

(iii) given any closed and bounded set K of $\mathbb{R}^n$ and interval $[a, b]$, there exists an integrable function $m(t)$ on $[a, b]$ such that

$$|g(x, t)| \leq m(t) \text{ and } \left| \frac{\partial g}{\partial x}(x, t) \right| \leq m(t) \text{ for all } (x, t) \in K \times [a, b].$$

Then $g(x, t)$ is said to belong to the class of functions CD.   □

Theorem 3.13

Consider the differential equation (3.2.4) where $u(.)$ is a measurable function on $[t_0, \infty)$. Assume that with $f_u(x, t) := f(x, u(t), t)$, $f_u(x, t) \in CD$. Then both Theorem 3.8 and Theorem 3.9 hold with respect to the differential equation (3.2.4) if one interprets in both theorems the phrase solution as solution in the extended sense (see, for example, Lee and Markus (1967) and Coddington and Levinson (1955)).   □

Coddington and Levinson (1955) also show the following theorem.

Theorem 3.14

Consider the differential equation

$$\dot{x}(t) = f(x, u, t, \mu), \ x(0) = x_0, \tag{3.2.10}$$

where $u(.)$ is a measurable function on $[a, b]$ and μ is some parameter taking values in the interval $I := [\mu_0 - c, \mu_0 + c]$ for some positive constants μ_0 and c. Assume that for every fixed $\mu \in I$, $f(x, u, t, \mu) \in CD$; for every fixed $t \in [a, b]$, $f(x, u(t), t, \mu)$ is continuous in (x, μ) at $\mu = \mu_0$; and $|f(x, u(t), t, \mu)| \leq m(t)$ for all $(x, t, \mu) \in K \times [a, b] \times I$, where $m(.)$ is some Lebesgue integrable function over $[a, b]$ and K is a compact set of $\mathbb{R}^n$. For $\mu = \mu_0$ let the differential equation (3.2.10) have a (unique) solution on the interval $[a, b]$.

Then there exists a $\delta < 0$ such that, for any fixed μ satisfying $|\mu - \mu_0| > \delta$, the differential equation (3.2.10) has a unique solution x_μ on $[a, b]$. Furthermore, $x_\mu \to x_{\mu_0}$ uniformly over $[a, b]$ if $\mu \to \mu_0$.   □

Example 3.9

Consider

$$\dot{x}(t) = ax^2(t) + tx(t)u(t), \ x(0) = x_0.$$

Assume that $u(t)$ is chosen from the set of piecewise continuous functions. Then for each fixed t, the partial derivative of $f_u(x, t) = ax^2 + txu(t)$ w.r.t. x is $2ax + tu(t)$. This function is clearly continuous for every fixed t, so condition (**i**) in Definition 3.4 is satisfied. Moreover, for each fixed x the function $f_u(x, t)$ is piecewise continuous in t. So in particular it is measurable. Thus condition (**ii**) in Definition 3.4 holds too. Finally, since $u(t)$ is piecewise continuous, $f_u(x, t)$ will be bounded by the integrable function $|a|m^2 + m|tu(t)|$, whenever x belongs to some compact set K^4, for some $m \in \mathbb{R}$.

Therefore, $f_u(t, x) \in CD$. Consequently, the differential equation has a solution in the extended sense.

Next assume that the differential equation has a solution $x_1(t)$ in the extended sense for $a = 1$ on the interval $[0, 1]$. Let $I := [\frac{1}{2}, 2]$. Since for every $a \in I, f_u(x, t) \in CD$; for every $t \in [0, 1]$ $f_u(x, t)$ is continuous in (x, a) at $a = 1$ and for every $a \in I, |f_u(x, t)|$ is bounded by the integrable function $2m^2 + m|tu(t)|$ the solution of the differential equation converges uniformly over $[0, 1]$ to $x_1(t)$ if $a \to 1$. $\qquad\square$

3.3   Stability theory: general case

In this section we analyze the behavior of trajectories (solutions) $x(t, x_0)$ of the nonlinear system

$$\dot{x} = f(x), \quad x(t_0) = x_0. \tag{3.3.1}$$

A standard assumption throughout this section will be that the function f in equation (3.3.1) is continuous differentiable. In particular this assumption implies that, for an arbitrary choice of x_0, equation (3.3.1) has a unique solution through the initial state x_0 (see Theorem 3.8). A good place to start the analysis of these trajectories is to determine the so-called equilibrium points of equation (3.3.1) and to describe the behavior of trajectories from equation (3.3.1) near its equilibrium points.

An equilibrium point is a solution of the differential equation (3.3.1) that has the property that if one starts at some initial point in time with the initial state x_0 then the state of the differential equation does not change over time. So, one remains for the rest of the time at this point x_0.

Definition 3.5

A point $x_0 \in \mathbb{R}^n$ is called an **equilibrium point** or **critical point** of equation (3.3.1) if $f(x_0) = 0$. $\qquad\square$

If x_0 is an equilibrium point of equation (3.3.1), then $f(x_0) = 0$ and, by Taylor's Theorem,

$$f(x_0 + h) = f'(x_0)h + o(h^T h),$$

---

[4]For such a set a positive number m always exists such that every element of K satisfies $|x| \le m$.

where $o(h^T h)$ denotes higher-order terms in h. From the above approximation it is reasonable to expect that the behavior of the nonlinear system (3.3.1) near the point $x = x_0$ will be approximated by the behavior of the next so-called **linearization** of (3.3.1) at x_0

$$\dot{y} = f'(x_0)y, \; y(t_0) \in N(0), \tag{3.3.2}$$

where $N(0)$ is some small neighborhood of 0. The next theorem states that this is indeed the case provided the matrix $f'(x_0)$ has no eigenvalues on the imaginary axis.

Theorem 3.15 (stable manifold theorem)

Consider the nonlinear differential equation (3.3.1). Suppose that $f \in C^1$, $f(x_0) = 0$ and that $f'(x_0)$ has k eigenvalues with negative real part and $n - k$ eigenvalues with positive real part. Then the local behavior of equation (3.3.1) near the equilibrium point x_0 coincides with the behavior of the linearized system (3.3.2) near the equilibrium point $y = 0$. $\qquad \square$

Example 3.10

Consider the differential equation

$$\dot{x} = x^2 - 1.$$

The equilibrium points of this differential equation are $x_0 = 1$ and $x_0 = -1$. The derivative of $x^2 - 1$ is $2x$. Therefore, the linearized system near the equilibrium point $x_0 = 1$ is

$$\dot{x} = 2x, \tag{3.3.3}$$

and the linearized system near the equilibrium point $x_0 = -1$ is

$$\dot{x} = -2x. \tag{3.3.4}$$

The solutions of the linearized system (3.3.3) are $x(t) = e^{2t}x_0$ and of the linearized system (3.3.4) $x(t) = e^{-2t}x_0$. So, solutions in the linearized system (3.3.3) diverge rapidly from the equilibrium solution $x = 0$ whereas in the linearized system (3.3.4) solutions converge rapidly to the equilibrium solution $x = 0$ when time proceeds. Theorem 3.15 yields then the conclusion that solutions starting at $t = 0$ in a neighborhood of $x_0 = 1$ will diverge from the constant solution $x(t) = 1$ whereas solutions starting in a neighborhood of $x_0 = -1$ will converge to the constant solution $x(t) = -1$. Figure 3.2 illustrates the situation. $\qquad \square$

This is an additional motivation to study the behavior of the linear system

$$\dot{x}(t) = Ax(t) \tag{3.3.5}$$

in more detail.

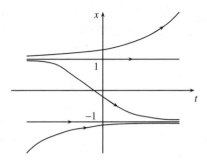

Figure 3.2   Local behavior of solutions of the differential equation $\dot{x} = x^2 - 1$

To that end consider the Jordan canonical form of matrix A again. Let $w_j = u_j + iv_j$ be a generalized eigenvector of the matrix A corresponding to an eigenvalue $\lambda_j = a_j + ib_j$. Notice that if $b_j = 0$ then $v_j = 0$. Assume that matrix A has k real eigenvalues (counting algebraic multiplicities). Consider

$$S := \{u_1, \cdots, u_k, u_{k+1}, v_{k+1}, \cdots, u_m, v_m\}$$

as a basis for $\mathbb{R}^n$ (with $n = 2m - k$).

Definition 3.6

Let $\lambda_j = a_j + ib_j$, $w_j = u_j + iv_j$ and S be as described above. Then we define the *stable, unstable and center subspaces*, E^s, E^u and E^c, respectively, as follows:

$$E^s := \mathrm{Span}\{u_j, v_j | a_j < 0\}$$
$$E^u := \mathrm{Span}\{u_j, v_j | a_j > 0\}$$
$$E^c := \mathrm{Span}\{u_j, v_j | a_j = 0\}.$$

That is, E^s, E^u and E^c are the subspaces of $\mathbb{R}^n$ spanned by the real and imaginary parts of the generalized eigenvectors w_j corresponding to eigenvalues λ_j with negative, positive and zero real parts, respectively.   □

Assume, without loss of generality, that the basis $S = [s_1, \cdots, s_k, s_{k+1}, \cdots, s_l, s_{l+1}, \cdots, s_n]$ is organized as follows. The first k vectors s_i are the vectors that form a basis for E^s, the vectors $s_{k+1}, \cdots, s_l$ form a basis for E^u, and the vectors $s_{l+1}, \cdots, s_n$ form a basis for E^c. The corresponding Jordan canonical form of A is $\mathrm{diag}\{J_s, J_u, J_c\}$. Then the solution of equation (3.3.5) is

$$x(t) = S \, \mathrm{diag}\{e^{J_s t}, e^{J_u t}, e^{J_c t}\} S^{-1} x_0.$$

Now, let $x_0 \in E^s$. Then, x_0 is a linear combination of the first k basis vectors $s_1, \cdots, s_k$.

That is, $x_0 = [s_1 \ s_2 \cdots s_k] \begin{bmatrix} \alpha_1 \\ \vdots \\ \alpha_k \end{bmatrix}$, for some $\alpha_i \in \mathbb{R}$. Introducing $\alpha^T := [\alpha_1, \cdots, \alpha_k]$, and 0_j

as the j-dimensional vector with zero entries, we can rewrite x_0 as $x_0 = S \begin{bmatrix} \alpha \\ 0_{l-k} \\ 0_{n-l} \end{bmatrix}$.
Consequently, if $x_0 \in E^s$

$$x(t) = S \operatorname{diag}\{e^{J_s t}, e^{J_u t}, e^{J_c t}\} S^{-1} S \begin{bmatrix} \alpha \\ 0_{l-k} \\ 0_{n-l} \end{bmatrix} = S \begin{bmatrix} e^{J_s t} \alpha \\ 0_{l-k} \\ 0_{n-l} \end{bmatrix}.$$

By Theorem 3.4 $e^{J_s t} = \operatorname{diag}\{e^{a_i t}\} D$, where D is an invertible matrix that does not contain any exponential terms and $a_i < 0$, $i = 1, \cdots, k$. Since D is invertible and $\alpha \neq 0$, $D\alpha =: \beta \neq 0$ too (since $D = 0$). Similarly it follows that, since matrix S is invertible,
$S \begin{bmatrix} \beta \\ 0_{l-k} \\ 0_{n-l} \end{bmatrix} =: v \neq 0$. So, we conclude that $x(t) = \operatorname{diag}\{e^{a_i t}\} v \neq 0$ and converges exponen-
tially fast to zero (if $t \rightarrow \infty$), whatever $x_0 \in E^s$ is. The same reasoning shows that whenever $x_0 \in E^u$, there exist $a_i > 0$ and $v \neq 0$, containing no exponential terms, such that $x(t) = \operatorname{diag}\{e^{a_i t}\} v$. So in that case $x(t)$ contains entries that become arbitrarily large (if $t \rightarrow \infty$).

Finally, for $x_0 \in E^c$ the above analysis shows that we have to distinguish between two cases. That is, whether the Jordan block has the extended block structure J_{RE} or J_{CE} (with '1s' above the main diagonal) or not. In case the Jordan block has the extended block structure, there exist some initial conditions x_0 for which $x(t)$ has entries that become arbitrarily large whereas for other initial conditions $x(t)$ remains bounded (if $t \rightarrow \infty$). If the Jordan block does not posses this extended block structure the solution $x(t)$ always remains bounded for all initial conditions $x_0 \in E^c$.

Notice that whether A has complex eigenvalues or not determines whether solutions will either oscillate or not. Theorem 3.16 summarizes the above discussion.

Theorem 3.16

Consider the linear system of differential equations

$$\dot{x} = Ax, \quad x(t_0) = x_0 \neq 0. \tag{3.3.6}$$

Then:

1. for all $x_0 \in E^s$, $|x(t)| \rightarrow 0$ if $t \rightarrow \infty$;

2. for all $x_0 \in E^u$, $|x(t)| \rightarrow \infty$ if $t \rightarrow \infty$;

3. there exists an $a > 0$ such that for all $x_0 \in E^c$, $|x(t)| \geq a|x_0|$ if $t \rightarrow \infty$. $\square$

Since the columns of S constitute a basis for $\mathbb{R}^n$ we obtain the next result immediately.

Corollary 3.17

Consider the linear differential equation (3.3.6) with an arbitrarily chosen initial state $x(t_0) = x_0 \in \mathbb{R}^n$. Then $x(t) \rightarrow 0$, if $t \rightarrow \infty$, if and only if $x_0 \in E^s$.

Example 3.11

The matrix

$$A = \begin{bmatrix} -8 & 6 & 5 \\ -8 & 5 & 6 \\ -3 & 4 & 2 \end{bmatrix}$$

has eigenvectors $w_1 = u_1 + iv_1 = \begin{bmatrix} 1 \\ 1 \\ 0 \end{bmatrix} + i \begin{bmatrix} 0 \\ 1 \\ -1 \end{bmatrix}$ corresponding to $\lambda_1 = -2 + i$ and

$u_2 = \begin{bmatrix} 1 \\ 1 \\ 1 \end{bmatrix}$ corresponding to $\lambda_2 = 3$.

The stable subspace, E^s, of equation (3.3.5) is $\text{Span}\left\{ \begin{bmatrix} 1 \\ 1 \\ 0 \end{bmatrix}, \begin{bmatrix} 0 \\ 1 \\ -1 \end{bmatrix} \right\}$, and the unstable

subspace, E^u, of equation (3.3.5) is $\text{Span}\left\{ \begin{bmatrix} 1 \\ 1 \\ 1 \end{bmatrix} \right\}$.   □

Example 3.12

The matrix

$$A = \begin{bmatrix} 0 & 1 & 0 \\ -1 & 0 & 0 \\ 0 & 0 & 2 \end{bmatrix}$$

is in Jordan canonical form and has eigenvalues $\pm i$ and 2.

The center subspace $E^c = \text{Span}\left\{ \begin{bmatrix} 1 \\ 0 \\ 0 \end{bmatrix}, \begin{bmatrix} 0 \\ 1 \\ 0 \end{bmatrix} \right\}$ and the unstable subspace $E^u =$

$\text{Span}\left\{ \begin{bmatrix} 0 \\ 0 \\ 1 \end{bmatrix} \right\}$. Therefore, by Theorem 3.4,

$$e^{At} = \begin{bmatrix} \cos(t) & \sin(t) & 0 \\ -\sin(t) & \cos(t) & 0 \\ 0 & 0 & e^{2t} \end{bmatrix}.$$

So, for an arbitrarily chosen $x_0 \in E^c$ we see a bounded oscillating behavior of $x(t)$, for

$x_0 \in E^u$ we see that $x(t) = \begin{bmatrix} 0 \\ 0 \\ e^{2t}\alpha \end{bmatrix}$ and for x_0 not in either of these subspaces E^c or E^u, a

mixture of both behaviors, that is the first two entries of $x(t)$ are oscillatory and the third entry grows exponentially.   □

In the above Example 3.12 all solutions in E^c behave oscillatory, but the trajectories remain bounded if time passes. The next example illustrates that trajectories in an initial state belonging to E^c may also diverge.

Example 3.13

The next matrix A is in Jordan canonical form.

$$A = \begin{bmatrix} 0 & 1 & 0 \\ 0 & 0 & 0 \\ 0 & 0 & 2 \end{bmatrix}.$$

The eigenvalues of A are 0 and 2. The center subspace $E^c = \text{Span}\left\{ \begin{bmatrix} 1 \\ 0 \\ 0 \end{bmatrix}, \begin{bmatrix} 0 \\ 1 \\ 0 \end{bmatrix} \right\}$. So according to Theorem 3.4

$$e^{At} = \begin{bmatrix} 1 & t & 0 \\ 0 & 1 & 0 \\ 0 & 0 & e^{2t} \end{bmatrix}.$$

Therefore, for $x_0 = \begin{bmatrix} \alpha \\ 0 \\ 0 \end{bmatrix}$, with $\alpha \neq 0$, $x(t) = \begin{bmatrix} \alpha \\ 0 \\ 0 \end{bmatrix}$ whereas for any other

$x_0 = \begin{bmatrix} \alpha \\ \beta \\ 0 \end{bmatrix} \in E^c$ $x(t) = \begin{bmatrix} \alpha + \beta t \\ \beta \\ 0 \end{bmatrix}$. $\qquad\qquad\square$

Notice that if matrix A has only eigenvalues with a negative real part, all trajectories of the linear system (3.3.2) converge to the origin, whereas in the case that all eigenvalues have a positive real part, all trajectories diverge from the origin. In the first case we call matrix A **stable** (or **Hurwitz**) (or the system (3.3.2) stable) and the origin a **sink**. In the second case matrix A is called **anti-stable** and the origin a **source**.

3.4 Stability theory of planar systems

In the previous section the existence and stability of solutions for a set of n differential equations was studied. This section considers the special case $n = 2$, the so-called planar systems. That is,

$$\dot{x}_1 = f_1(x_1, x_2) \tag{3.4.1}$$
$$\dot{x}_2 = f_2(x_1, x_2), \tag{3.4.2}$$

where both $f_1, f_2 \in C^1$. One can think of the system (3.4.1) and (3.4.2) as the equations of motion of a particle moving about in the plane. The system tells us in this case that when the particle is at the point (x_1, x_2), it will be moving so that its velocity vector $[\dot{x}_1, \dot{x}_2]^T$ will be the vector $[f_1(x_1, x_2), f_2(x_1, x_2)]^T$. We can visualize this as a vector with its tail at the point (x_1, x_2) pointing in the direction of the particle's motion. Therefore, one way to make a plot of the dynamics of system (3.4.1) and (3.4.2) is to picture the vector $[f_1(x_1, x_2), f_2(x_1, x_2)]^T$ pointing out from the point (x_1, x_2) for every point in the (x_1, x_2)-plane. To avoid hopelessly cluttering the picture, when we actually draw this graph, we usually ignore the length of the velocity vectors and only draw short vectors that point in the correct direction. We refer to the (x_1, x_2)-plane as the **phase plane**. Since to each point $(x_1, x_2) \in \mathbb{R}^2$ the mapping $f := (f_1, f_2)$ assigns a vector $f(x_1, x_2) \in \mathbb{R}^2$ we say that f defines a **vector field** on $\mathbb{R}^2$. A **solution curve** through the point $(\bar{x}_1, \bar{x}_2)$ is a curve in the phase plane obtained from the unique solution of the initial value problem (3.4.1) and (3.4.2), with $x_1(0) = \bar{x}_1$ and $x_2(0) = \bar{x}_2$, by considering all points $(x_1(t), x_2(t))$ in the phase plane for all $t \in \mathbb{R}$. Usually, we indicate by an arrow in which direction the solutions evolve when time proceeds. The set of all solution curves in the phase plane is called the **phase portrait**, or sometimes **phase diagram**, of system (3.4.1) and (3.4.2). So, the phase diagram of system (3.4.1) and (3.4.2) is a graph where the axes are just the variables x_1 and x_2 and instead of plotting x_1 against time and x_2 against time, we look at the behavior of x_2 against x_1 when time evolves, for the solutions of (3.4.1) and (3.4.2).

Example 3.14

Consider the decoupled system

$$\dot{x}_1 = -x_1; \; x_1(t_0) = c_1$$
$$\dot{x}_2 = 2x_2; \; x_2(t_0) = c_2.$$

At each point (x_1, x_2) we draw a vector in the direction $\begin{bmatrix} -x_1 \\ 2x_2 \end{bmatrix}$. In Figure 3.3(a), we have drawn the vector $[-2 \; 0]^T$ at the point $(2, 0)$, the vector $[-3 \; 4]^T$ at the point $(3, 2)$, and the vector $[-1 \; 4]^T$ at the point $(1, -2)$, along with a sample of other choices. From this figure, we get a feeling that the motion which this system describes is roughly hyperbolic. In fact we are actually able to calculate this motion in this example. Note that $x_1(t) = c_1 e^{-t}$ and $x_2(t) = c_2 e^{2t}$. By elimination of the time parameter t we see that the solutions of the system move along the algebraic curves $x_2 = k/x_1^2$ in the phase plane, where the constant $k = c_1^2 c_2$. Some solution curves are sketched in Figure 3.3(b). □

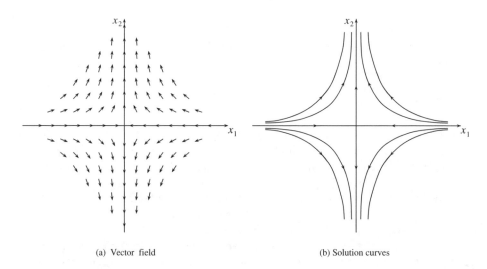

(a) Vector field   (b) Solution curves

Figure 3.3 The set of differential equations $\dot{x}_1 = -x_1$, $\dot{x}_2 = 2x_2$

At each point (x_1, x_2) in the phase plane $\mathbb{R}^2$, the solution curves of (3.4.1) and (3.4.2) are at a tangent to the vectors in the vector field $\begin{bmatrix} f_1(x_1, x_2) \\ f_2(x_1, x_2) \end{bmatrix}$. This follows since, at time $t = t_0$, the velocity vector $v_0 = \dot{x}(t_0)$ is at a tangent to the curve $x := (x_1(t), x_2(t))$ at the point $x_0 = x(t_0)$ and $\dot{x} = [f_1(x_1, x_2), f_2(x_1, x_2)]^T$. Moreover, since for each $[\bar{x}_1, \bar{x}_2]^T \in \mathbb{R}^2$ the initial value problem (3.4.1) and (3.4.2), with $x_1(0) = \bar{x}_1$ and $x_2(0) = \bar{x}_2$, has a unique solution, solution curves in the phase plane do not have any points in common.

Next we discuss the various phase portraits that are possible for the linear system

$$\dot{x} = Ax \tag{3.4.3}$$

when $x \in \mathbb{R}^2$ and A is a 2×2 matrix.

From the Jordan canonical form we recall that there is an invertible matrix S such that the matrix

$$J = SAS^{-1}$$

has either one of the following forms

$$J_{\mathbb{R}} = \begin{bmatrix} \lambda_1 & 0 \\ 0 & \lambda_2 \end{bmatrix}, \quad J_{CE} = \begin{bmatrix} \lambda & 1 \\ 0 & \lambda \end{bmatrix}, \quad \text{or } J = \begin{bmatrix} a & b \\ -b & a \end{bmatrix}.$$

Then, by Theorem 3.4,

$$e^{J_{\mathbb{R}}t} = \begin{bmatrix} e^{\lambda_1 t} & 0 \\ 0 & e^{\lambda_2 t} \end{bmatrix}, \quad e^{J_{RE}} = e^{\lambda t}\begin{bmatrix} 1 & t \\ 0 & 1 \end{bmatrix}, \quad \text{and } e^{J_C} = e^{at}\begin{bmatrix} \cos(bt) & \sin(bt) \\ -\sin(bt) & \cos(bt) \end{bmatrix},$$

respectively.

Consequently $e^{At} = S^{-1}e^{Jt}S$. Now let $y = Sx$. Then $\dot{y} = S\dot{x}$. So y satisfies

$$\dot{y} = Jy. \tag{3.4.4}$$

That is, using a linear transformation, all phase portraits for the linear system (3.4.3) can be derived from one of the following cases. The first three cases are the ones that occur most frequently and correspond with the case that A has no eigenvalues on the imaginary axis.

Case I. (Saddle) $J = J_{\mathbb{R}} = \begin{bmatrix} \lambda_1 & 0 \\ 0 & \lambda_2 \end{bmatrix}$ with $\lambda_2 < 0 < \lambda_1$.

The phase portrait for the linear system (3.4.4) for this case is given in Figure 3.4 (see Example 3.14). If $\lambda_1 < 0 < \lambda_2$, the arrows in Figure 3.4 are reversed. Whenever matrix A has two real eigenvalues of opposite sign, the stable and unstable subspaces of equation (3.4.3) are determined by the eigenvectors of A corresponding to the stable and unstable eigenvalue of A, respectively. The four non-zero solution curves that approach the equilibrium point zero as $t \to \pm\infty$ are called **separatrices** of the system.

Case II. (Stable node) $J = J_{\mathbb{R}} = \begin{bmatrix} \lambda_1 & 0 \\ 0 & \lambda_2 \end{bmatrix}$ with $\lambda_1 \leq \lambda_2 < 0$ or $J = J_{RE} = \begin{bmatrix} \lambda & 1 \\ 0 & \lambda \end{bmatrix}$ with $\lambda < 0$.

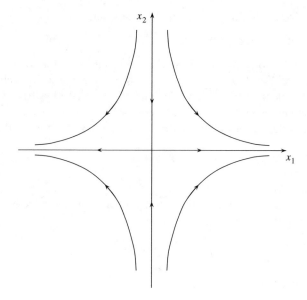

Figure 3.4   A saddle at the origin

The phase portrait for the linear system (3.4.4) in this case is given in Figure 3.5. The origin is referred to as a stable node in each of these cases. In case $\lambda_1 \geq \lambda_2 > 0$ or if $\lambda > 0$, the arrows in Figure 3.5 are reversed and the origin is referred to as an **unstable node**. Notice that the direction in which the solution curves approach the equilibrium

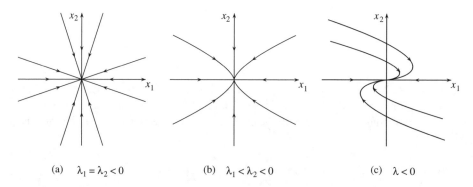

   (a) $\lambda_1 = \lambda_2 < 0$ (b) $\lambda_1 < \lambda_2 < 0$ (c) $\lambda < 0$

Figure 3.5   A stable node at the origin

point is, with only one exception, determined by the eigenvector corresponding to the largest eigenvalue. Only if the initial state is in the direction of the eigenvector corresponding to the smallest eigenvalue, will this state approach the origin along the line in this direction.

    Case III. (Stable focus) $J = J_{\mathbb{C}} = \begin{bmatrix} a & b \\ -b & a \end{bmatrix}$ with $a < 0$.

With $(x_1(0), x_2(0)) = (c_1, c_2)$, we have that the solution of the initial value problem (3.4.3) is

$$(x_1(t), x_2(t)) = (e^{at}(c_1 \cos bt + c_2 \sin bt), e^{at}(-c_1 \sin bt + c_2 \cos bt)).$$

So,

$$x_1^2(t) + x_2^2(t) = e^{2at}(c_1^2 + c_2^2),$$

which shows that the solution curves spiral towards the origin. The phase portrait is given in Figure 3.6. If $a > 0$, the solution curves spiral away from the origin and the origin is called an **unstable focus**.

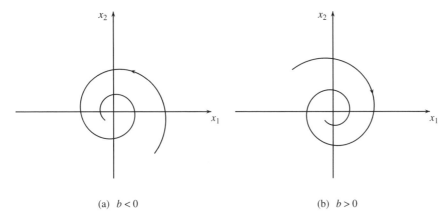

(a) $b < 0$ \qquad\qquad\qquad\qquad (b) $b > 0$

Figure 3.6 A stable focus at the origin

Case IV. (Center) $J = J_\mathbb{C} = \begin{bmatrix} 0 & b \\ -b & 0 \end{bmatrix}$.

This case occurs if A has an eigenvalue on the imaginary axis.

It is easily verified (see Case III) that the solution curves are now circles about the origin. Figure 3.7 illustrates this case. This implies that for the general case solution curves are ellipses.

Finally, for the sake of completeness, we illustrate in Figure 3.8 the phase portraits if matrix A has a zero eigenvalue. In these cases the origin is called a **degenerate equilibrium point**.

The construction of the phase portrait for a general planar system (3.4.1) and (3.4.2) is usually much more involved, since in most cases it is not possible to derive an explicit solution for this set of differential equations. Below we present an algorithm that is usually very helpful in drawing the phase portrait of a planar system; but first another global result on the nonexistence of periodic solutions is stated which is useful in the final construction phase of the algorithm. A solution is called a **periodic solution** if it is a closed solution curve of (3.4.1) and (3.4.2) which is not an equilibrium point of (3.4.1) and (3.4.2). Informally speaking, it is a solution which repeats after a finite time.

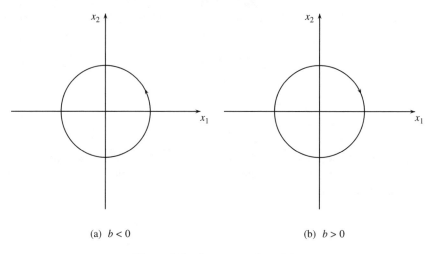

(a) $b < 0$ (b) $b > 0$

Figure 3.7   A center at the origin

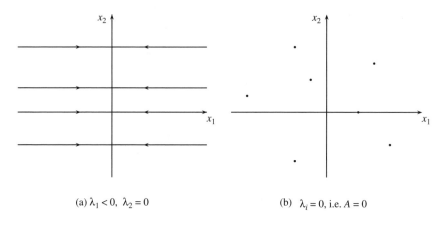

(a) $\lambda_1 < 0,\ \lambda_2 = 0$ (b) $\lambda_i = 0$, i.e. $A = 0$

Figure 3.8   A degenerate equilibrium at the origin

Theorem 3.18 (Bendixson)[5]

Suppose $f = (f_1, f_2) \in C^1(E)$ where E is a simply connected region in $\mathbb{R}^2$ (informally speaking, E is free of holes or cuts in its interior). If

$$\frac{\partial f_1}{\partial x_1} + \frac{\partial f_2}{\partial x_2}$$

is not identically zero and does not change sign in E, then (3.4.1) and (3.4.2) have no periodic solution lying entirely in E.   □

---

[5]Swedish mathematician who lived from 1861–1935.

Example 3.15

Consider the set of differential equations

$$\dot{x}_1 = 2x_1 - x_2^3, \ x_1(0) = x_{10},$$
$$\dot{x}_2 = x_1^2 - x_2, \ x_2(0) = x_{20}.$$

Then the matrix of partial derivatives is

$$\frac{\partial f_i}{\partial x_j} = \begin{bmatrix} 2 & -3x_2^2 \\ 2x_1 & -1 \end{bmatrix}.$$

So, $f \in C^1$. Furthermore, $\frac{\partial f_1}{\partial x_1} + \frac{\partial f_2}{\partial x_2} = 2 - 1 = 1$. So, on whole $\mathbb{R}^2$ this sum of partial derivatives is positive. Therefore, this set of differential equations has no periodic solution.

Algorithm (Algorithm to construct a phase portrait)

Consider the planar system

$$\dot{x}_1 = f_1(x_1, x_2) \tag{3.4.5}$$
$$\dot{x}_2 = f_2(x_1, x_2), \tag{3.4.6}$$

where both $f_1, f_2 \in C^1$. Assume that at every equilibrium point, $x^* = (x_1^*, x_2^*)$, of this system the derivative, $f'(x^*)$, has no eigenvalues on the imaginary axis. Then, the following steps provide essential information for drawing the phase portrait of the planar system (3.4.6) (see also Simon and Blume (1994)).

Step 1 Find the equilibria (x_1^*, x_2^*) of the planar system by solving the set of equations

$$f_1(x_1, x_2) = 0$$
$$f_2(x_1, x_2) = 0.$$

Step 2 Determine the derivative $f'(x^*)$ at every equilibrium point. Calculate the eigenvalues of this derivative. Then, the local behavior of the solution curves near each equilibrium point coincides with one of the three phase portraits discussed above in Case I, II or III (since we assumed that all derivatives have no eigenvalues on the imaginary axis and the assumptions of Theorem 3.15 are satisfied).

Step 3 Find all points where a solution curve either points in a vertical (x_2) or in a horizontal (x_1) direction. A solution curve points in a vertical direction if $\dot{x}_1 = 0$ and $\dot{x}_2 \neq 0$. If $\dot{x}_2 > 0$ the solution evolves in the upward direction when time proceeds; when $\dot{x}_2 < 0$ the solution will evolve downwards. Since $\dot{x}_1 = f_1(x_1, x_2)$, the locus of points (x_1, x_2) where a solution curve points in a vertical direction is found by determining all points (x_1, x_2) that satisfy $f_1(x_1, x_2) = 0$ (and $f_2(x_1, x_2) \neq 0$). A curve along which the vector field always points in the same direction is called an **isocline** of the system. In a similar way determine the locus of points (x_1, x_2) where a solution curve points in a horizontal direction by determining all points (x_1, x_2) that satisfy $f_2(x_1, x_2) = 0$ (and $f_1(x_1, x_2) \neq 0$). Again, by inspection of the sign of $f_1(x_1, x_2)$ at an arbitrary point x^0 of this locus one can determine in which direction the solution curve will evolve for the whole set of points on this locus that is connected to this point x^0.

Step 4 The isoclines which are found in Step 3 divide the phase plane into regions called **sectors**. Within each sector, the sign of the derivatives $\dot{x}_1$ and $\dot{x}_2$ does not change since the functions f_i are continuous. So, the vector field can only point into one quadrant in any given sector and this quadrant can be determined by inspection of the sign of $f_i(x_1, x_2)$ at an arbitrary point x^0 in this sector. Perform this inspection for any sector and indicate the direction of the vector field by an arrow.

Step 5 Based on the information obtained in the previous steps (and maybe other insights obtained from, for example, Theorem 3.18) try to sketch representative solution curves which follow the vector fields' directions. □

Example 3.16

Consider the planar system

$$\dot{x}_1 = x_2 \tag{3.4.7}$$

$$\dot{x}_2 = x_1 - x_1^2 + x_2. \tag{3.4.8}$$

We will follow the algorithm above to obtain an insight into the phase portrait of this system.

Step 1 The equilibrium points are the solutions of the set of equations

$$x_2 = 0$$

$$x_1 - x_1^2 + x_2 = 0.$$

So, $(0, 0)$ and $(0, 1)$ are the equilibrium points.

Step 2 The derivative of f is $\begin{bmatrix} 0 & 1 \\ 1 - 2x_1 & 1 \end{bmatrix}$. So, $f'(0, 0) = \begin{bmatrix} 0 & 1 \\ 1 & 1 \end{bmatrix}$ and the eigenvalues of this matrix are $\lambda_1 = (1 + \sqrt{5})/2 > 0$ and $\lambda_2 = (1 - \sqrt{5})/2 < 0$. Therefore $(0, 0)$ is a saddle-point equilibrium. At the equilibrium point $(0, 1)$ $f'(0, 1) = \begin{bmatrix} 0 & 1 \\ -1 & 1 \end{bmatrix}$, which eigenvalues are $\frac{1}{2} \pm \frac{1}{2}\sqrt{2}i$. So $(0, 1)$ is an unstable focus.

Step 3 The isocline where solution curves point in a vertical direction is $x_2 = 0$, and the isocline where solution curves point in a horizontal direction is $x_2 = x_1^2 - x_1$. So the plane is divided into five sectors.

Step 4 For each sector (see Figure 3.9(a)) we next determine the direction of the vector field by inspection of this direction at an arbitrary point in the sector. For instance,

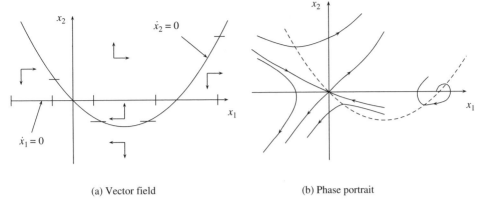

(a) Vector field   (b) Phase portrait

Figure 3.9   Example 3.43

at $(0, 1)$ $\dot{x}_1 = 1 > 0$ and $\dot{x}_2 = 1 > 0$. So both x_1 and x_2 grow if time proceeds. Therefore in that sector the direction of the vector field is northeast. Similarly one can verify that at $(3, 1)$, $\dot{x}_1 > 0$ and $\dot{x}_2 < 0$. So x_1 grows and x_2 declines when time proceeds. Consequently the direction of the vector field is southeast. At $(1, -1)$, $(\frac{1}{2}, -\frac{1}{8})$ and $(-3, 1)$ the direction of the vector field is southwest, northwest and southeast, respectively.

Step 5 To see whether periodic solutions might exist observe that $\frac{\partial f}{\partial x_1} + \frac{\partial f}{\partial x_2} = 0 + 1 = 1$. So, according to Theorem 3.18 periodic solutions to this set of differential equations do not exist. The complete picture, as far as one can deduce from the previous steps, is plotted in Figure 3.9(b). $\qquad\square$

3.5   Geometric concepts

We now turn to some important concepts in linear system theory. To introduce these concepts we consider system (3.1.1) with the number of players equal to 1, that is $N = 1$, and we drop the subscript i. So, consider

$$\dot{x} = Ax + Bu. \tag{3.5.1}$$

First, we consider the notion of controllability. Informally speaking, the system is called controllable if it is possible to regulate the state from any initial position towards any chosen final position, and this within an arbitrary chosen time period.

Definition 3.7

The dynamical system described by equation (3.5.1) or the pair (A, B) is said to be **controllable** if, for any initial state x_0, $t_1 > t_0$ and final state x_1, there exists a piecewise continuous input $u(.)$ such that the solution of (3.5.1) satisfies $x(t_1) = x_1$. Otherwise, the system or the pair (A, B) is said to be uncontrollable. $\qquad\square$

The controllability of a system can be verified through some algebraic or geometric criteria (Zhou, Doyle and Glover, 1996).

Theorem 3.19

The following are equivalent:

1. (A, B) is controllable.

2. The matrix $[A - \lambda I, \ B]$ has full row rank for all $\lambda \in \mathbb{C}$.

3. The eigenvalues of $A + BF$ can be freely assigned (with the restriction that complex eigenvalues are in conjugate pairs) by a suitable choice of F.

4. The so-called controllability matrix

$$C := [B, \ AB, \ A^2B, \ \cdots, \ A^{n-1}B]$$

has full row rank. □

Example 3.17

1. Let $A_1 = \begin{bmatrix} 1 & 0 \\ 2 & -1 \end{bmatrix}$ and $B = \begin{bmatrix} 1 \\ 0 \end{bmatrix}$. Then the controllability matrix $C = \begin{bmatrix} 1 & 1 \\ 0 & 2 \end{bmatrix}$. C has full row rank, therefore (A_1, B) is controllable.

2. Let $A_2 = \begin{bmatrix} 1 & 2 \\ 0 & -1 \end{bmatrix}$ and B as before. Then the controllability matrix $C = \begin{bmatrix} 1 & 1 \\ 0 & 0 \end{bmatrix}$. C has not full row rank, therefore (A_2, B) is not controllable.

Stability plays an important role in robust control theory. From section 3.3 we recall the next definition.

Definition 3.8

The dynamical system $\dot{x} = Ax$ is said to be **stable** if all the (possibly complex) eigenvalues of A are in the open left half of the complex plane, i.e. the real part of every eigenvalue of A is strictly smaller than zero. A matrix A with such a property is said to be stable. □

Definition 3.9

The dynamical system described by equation (3.5.1) or the pair (A, B) is said to be **stabilizable** if, for any initial state x_0, there exists a piecewise continuous input $u(.)$ such that the solution of (3.5.1) converges to zero. □

From this definition it is clear that whenever a system is controllable it is also stabilizable. This observation also follows directly from the next well-known theorem which gives several characterizations for the stabilizability of a system (Zhou, Doyle and Glover, 1996).

Theorem 3.20

The following are equivalent:

1. (A, B) is stabilizable.

2. The matrix $[A - \lambda I, \ B]$ has full row rank for all $\lambda \in \mathbb{C}_0^+$.

3. There exists a matrix F such that the eigenvalues of $A + BF$ are all located in the left half of the complex plane, $\mathbb{C}^-$. □

Example 3.18

Consider $A = \begin{bmatrix} 3 & 0 & 0 \\ 0 & 2 & 0 \\ 0 & 0 & -1 \end{bmatrix}$ and $B = \begin{bmatrix} 1 \\ 1 \\ 0 \end{bmatrix}$. Then

$$[A - \lambda I, \ B] = \begin{bmatrix} 3 - \lambda & 0 & 0 & 1 \\ 0 & 2 - \lambda & 0 & 1 \\ 0 & 0 & -1 - \lambda & 0 \end{bmatrix}.$$

Obviously, for $\lambda \in \mathbb{C}_0^+$ which are not an eigenvalue of A, matrix A is invertible and consequently the above matrix has full row rank. Therefore, only $\lambda = 3$ and $\lambda = 2$ require a further inspection of this matrix. For $\lambda = 3$ this matrix equals $\begin{bmatrix} 0 & 0 & 0 & 1 \\ 0 & -1 & 0 & 1 \\ 0 & 0 & -4 & 0 \end{bmatrix}$,
which clearly has full row rank too. In a similar way one can verify that for $\lambda = 2$ the matrix also has full row rank. So, (A, B) is stabilizable. □

Next assume that not all state variables of the system (3.5.1) can be observed directly. That is, consider the situation that the state variables evolve over time as described in (3.5.1), but that we can only observe some output variables, $y(t)$, of this system. More specifically, assume that we can only observe a set of variables over time which are linearly related to the state variables of the system (3.5.1). That is, consider the system

$$\dot{x}(t) = Ax(t) + Bu(t), \ x(0) = x_0; \tag{3.5.2}$$
$$y(t) = Cx(t).$$

Here, $y(t)$ denotes the observed (or output) variables of the system at time t. In case x_0 is unknown, we will call this system (3.5.2) observable if we can recover x_0 from the knowledge of an input $u(.)$ and output $y(.)$ sequence for some time interval $[0, t_1]$. Denoting the output of system (3.5.2) at time t, induced by the input sequence $u(.)$ and initial state x_0, by $y(t, x_0, u)$, we obtain the following formal definition.

Definition 3.10

The dynamical system (3.5.2), or the pair (C, A), is said to be **observable** if there exists a $t_1 > 0$ such that, whatever the input $u(.)$ has been on $[0, t_1]$, from $y(t, x_0, u) = y(t, x_1, u)$ on $[0, t_1]$ it follows that $x_0 = x_1$. □

It is easily verified that in the above definition the input $u(.)$ does not play any role. For that reason one can introduce the notion of observability, without loss of generality, for the system (3.5.2) with $u(.) = 0$. This clarifies why we call the pair (C, A) observable.

Next, consider for a fixed t_1 the set

$$V(t_1) := \{x_0 \mid y(t, x_0, 0) = 0, \ t \in [0, t_1]\}.$$

$V(t_1)$ is called the **unobservable subspace** of (3.5.2) on $[0, t_1]$. This is because $V(t_1)$ is the linear subspace that contains all initial states that are not observable. It is straightforwardly verified that $x_0 \in V(t_1)$ if and only if $Ce^{At}x_0 = 0$, $t \in [0, t_1]$. Using the Cayley–Hamilton theorem one can show that this condition is equivalent to the condition that $CA^k x_0 = 0$, $k = 0, \cdots, n-1$. In particular we infer from this that $V(t_1)$ does not depend on the time t_1. Furthermore, it now follows directly that

$$V(t_1) = \mathrm{Ker} \begin{bmatrix} C \\ CA \\ \vdots \\ CA^{n-1} \end{bmatrix} =: V. \tag{3.5.3}$$

We will call matrix V the **observability matrix** of system (3.5.2). It is now not difficult to show that (see Exercises) the system (3.5.2) is observable if and only if $V(t_1) = \{0\}$ or, equivalently, matrix V has full column rank n.

By transposition of V we see that the system (3.5.2) is observable if and only if

$$V^T = [C^T, \ A^T C^T, \ A^{T^2} C^T, \cdots, \ A^{T^{(n-1)}} C^T]$$

has full row rank. The following theorem reformulates this result.

Theorem 3.21

(C, A) is observable if and only if (A^T, C^T) is controllable. □

Using this result Corollary 3.53 then yields the next characterizations to verify whether system (3.5.2) is observable.

Corollary 3.22

The following are equivalent:

1. (C, A) is observable.

2. The matrix $\begin{bmatrix} A - \lambda I \\ C \end{bmatrix}$ has full column rank for all $\lambda \in \mathbb{C}$.

3. The eigenvalues of $A + LC$ can be freely assigned (with the restriction that complex eigenvalues are in conjugate pairs) by a suitable choice of L.

4. The observability matrix V has full column rank.   □

Example 3.19

1. Let $A_1 = \begin{bmatrix} 1 & 0 \\ 2 & -1 \end{bmatrix}$ and $C = [1, \ 0]$. Then the observability matrix $V = \begin{bmatrix} 1 & 0 \\ 1 & 0 \end{bmatrix}$. C
   has not full column rank, therefore (C, A_1) is not observable.

2. Let $A_2 = \begin{bmatrix} 1 & 2 \\ 0 & -1 \end{bmatrix}$ and C as before. Then the observability matrix $C = \begin{bmatrix} 1 & 0 \\ 1 & 2 \end{bmatrix}$. C
   has full column rank, therefore (C, A_2) is observable.   □

A similar role to that played by stabilizability in controlling a system ultimately towards the zero state is played by the notion **detectability** in determining all initial states of the system which do not ultimately converge to zero. Stated differently, we will call the system (3.5.2) detectable if all initial states which do not converge to zero are observable. The idea is that all states which do not automatically converge to zero might cause problems in controlling the system. Therefore, we want at least to be able to identify (detect or observe) those states. This idea gives rise to the following definition of detectability

Definition 3.11

The dynamical system described by equation (3.5.2), or the pair (C, A), is called **detectable** if $A + LC$ is stable for some L.   □

This definition then leads to the following relationship between detectability and stabilizability (see Theorem 3.20).

Corollary 3.23

(C, A) is detectable if and only if (A^T, C^T) is stabilizable.   □

From this relationship we obtain, using Theorem 3.20 again, the next characterizations for detectability of a pair (C, A). In particular one can see from these characterizations that an observable system is also always detectable, a property which could also have been derived directly from the definitions.

Theorem 3.24

The following are equivalent:

1. (C, A) is detectable.
2. The matrix $\begin{bmatrix} A - \lambda I \\ C \end{bmatrix}$ has full column rank for all $\lambda \in \mathbb{C}_0^+$.

3. There exists a matrix L such that the eigenvalues of $A + LC$ are all located in the left half of the complex plane, $\mathbb{C}^-$.

Example 3.20

Consider $A = \begin{bmatrix} 3 & 0 & 0 \\ 0 & 2 & 0 \\ 0 & 0 & -1 \end{bmatrix}$ and $C = [1, \ 1, \ 0]$. Then

$$\begin{bmatrix} A - \lambda I \\ C \end{bmatrix} = \begin{bmatrix} 3 - \lambda & 0 & 0 \\ 0 & 2 - \lambda & 0 \\ 0 & 0 & -1 - \lambda \\ 1 & 1 & 0 \end{bmatrix}.$$

Obviously for $\lambda \in \mathbb{C}_0^+$ which are not an eigenvalue of matrix A, matrix A is invertible and, consequently, the above matrix has full column rank. Therefore, only $\lambda = 3$ and $\lambda = 2$ require a further inspection of this matrix. For $\lambda = 3$ this matrix equals $\begin{bmatrix} 0 & 0 & 0 \\ 0 & -1 & 0 \\ 0 & 0 & -4 \\ 1 & 1 & 0 \end{bmatrix}$,

which clearly has full column rank too. In a similar way one verifies that for $\lambda = 2$ the matrix also has full column rank. So, (C, A) is detectable. $\qquad\square$

3.6 Performance specifications

In this book it is assumed that all players have clear preferences, represented by a quadratic utility function. Below, the standard framework that will be used throughout the book will be presented. Then, a number of extensions are discussed and it is shown how they can be reformulated into the standard framework.

Our standard framework is the two-person differential game in which player 1 minimizes the quadratic cost function

$$\int_0^T \{x^T(t)Q_1x(t) + u_1(t)^T R_1 u_1(t)\}dt + x^T(T)Q_{1T}x(T) \tag{3.6.1}$$

and player 2 minimizes

$$\int_0^T \{x^T(t)Q_2x(t) + u_2(t)^T R_2 u_2(t)\}dt + x^T(T)Q_{2T}x(T). \tag{3.6.2}$$

The state equation is given by

$$\dot{x}(t) = Ax(t) + B_1 u_1(t) + B_2 u_2(t), \ x(0) = x_0. \tag{3.6.3}$$

In this framework $x(t)$ is the state of the system, $u_i(t)$ is the control variable of player i, and T is either a finite or infinite planning horizon. Furthermore, all matrices are constant.

Without loss of generality, we assume that the matrices Q_i and R_i in the cost functions are symmetric. This is because since if this were not the case, for instance $x^T Q x$ could be rewritten as

$$\frac{1}{2}x^T Q x + \frac{1}{2}x^T Q x = \frac{1}{2}x^T Q x + (\frac{1}{2}x^T Q x)^T$$

$$= x^T \frac{1}{2} Q x + x^T \frac{1}{2} Q^T x$$

$$= x^T \frac{1}{2}(Q + Q^T)x,$$

where the matrix $\frac{1}{2}(Q + Q^T)$ is clearly symmetric. In this set-up it is assumed that any use of its instruments is disliked by each player. That is, the matrices R_i, $i = 1, 2$ are positive definite.

The expressions $x^T(T)Q_{iT}x(T)$, $i = 1, 2$, denote the cost associated with the terminal state of the system, $x(T)$, and is usually called the **scrap value**. The interpretation of this term depends on the application context. Usually it is interpreted as an estimate for the cost incurred after the planning horizon, e.g. for shutting down a factory, or revenues obtained beyond the planning horizon.

For the infinite planning horizon case we require the closed-loop system to be stable. Therefore this scrap-value term is then dropped. Moreover, if the planning horizon is infinite, we will only consider the case that all matrices do not depend on time (the time-invariant case).

In the above formulation of the standard framework it is assumed that player 1 is not concerned about the control efforts used by the second player, and vice versa. In fact later on we do include such considerations. In the cost function for player 1 an additional quadratic term $u_2^T R_{12} u_2$ is then included where R_{12} is a symmetric matrix (and a similar adaptation is performed for player 2's cost function). By doing so one can, for example, use the obtained theoretical insights to draw conclusions for the class of so-called zero-sum games, where the costs for one player are the benefits for the other player. For the sake of simplicity this extension is, however, ignored in the current section.

Linear quadratic differential games can be used in various fields of applications to analyze conflict situations. Therefore, the specific interpretation of the state and control variables can differ enormously. In the next section a number of examples are given with a different background. To fix the idea it is enough for the moment to consider the interpretation that is often used in economics. There the state and control variables are assumed to describe deviations of certain economic variables from their target values or their long-run equilibrium levels. The cost functions then describe the goal of the decision makers to minimize deviations from these targets, where every deviation is quadratically penalized.

Next we comment on a number of straightforward extensions to the standard framework sketched above.

I. The N-player case
The generalization to N players reads as follows. Now, all N players want to minimize their cost function

$$J_i := \int_0^T \{x^T(t)Q_i(t)x(t) + u_i(t)^T R_i(t)u_i(t)\}dt + x^T(T)Q_i(T)x(T) \tag{3.6.4}$$

subject to the dynamics of the system described by

$$\dot{x}(t) = A(t)x(t) + \sum_{i=1}^{N} B_i(t)u_i(t), \quad x(0) = x_0. \tag{3.6.5}$$

The matrices $Q_i(t)$ and $R_i(t)$ are again symmetric in this formulation and R_i are again assumed to be positive definite.

The corresponding equilibrium strategies will usually be stated without proof. This, since the proofs are straightforward generalizations of the two-player case.

II. Discounting
Particularly in problems with an economic background it is often natural to discount future losses to obtain an accurate description of the present value of total future losses. In those cases player i, $i = 1, 2$, is assumed to minimize the cost function

$$J_i := \int_0^T e^{-rt}\{y^T(t)Q_i y(t) + v_i^T(t)R_i v_i(t)\}dt \tag{3.6.6}$$

where r is the discount factor, subject to the dynamical state equation

$$\dot{y}(t) = Ay(t) + B_1 v_1(t) + B_2 v_2(t), \quad y(0) = x_0. \tag{3.6.7}$$

Introducing $x(t) := e^{-\frac{1}{2}rt}y(t)$ and $u_i(t) := e^{-\frac{1}{2}rt}v_i(t)$ this problem can be rewritten as

$$J_i = \int_0^T \left\{ e^{-\frac{1}{2}rt}y^T(t)Q_i e^{-\frac{1}{2}rt}y(t) + e^{-\frac{1}{2}rt}v_i^T(t)R_i e^{-\frac{1}{2}rt}v_i(t) \right\}dt$$

$$= \int_0^T \{x^T(t)Q_i x(t) + u_i^T(t)R_i u_i(t)\}dt,$$

where

$$\dot{x}(t) = \frac{d}{dt}\left\{ e^{-\frac{1}{2}rt}y(t) \right\}$$

$$= -\frac{1}{2}re^{-\frac{1}{2}rt}y(t) + e^{-\frac{1}{2}rt}\dot{y}(t)$$

$$= -\frac{1}{2}rx(t) + e^{-\frac{1}{2}rt}(Ay(t) + B_1 v_1(t) + B_2 v_2(t))$$

$$= (A - \frac{1}{2}rI)x(t) + B_1 u_1(t) + B_2 u_2(t).$$

So discounting future losses by a factor r can be included into our standard framework by just replacing matrix A by $A - \frac{1}{2}rI$.

In case the two players have different opinions about the discounting rate of future cost, the same approach can be used to reformulate the problem into the standard framework. Assume that player i discounts his future losses by a factor r_i. Introducing then

$x_i(t) := e^{-\frac{1}{2}r_i t}y(t)$ and $u_i(t) := e^{-\frac{1}{2}r t}v_i(t)$, the cost functions can be rewritten along the lines above as

$$J_i = \int_0^T \{x_i^T(t)Q_i x_i(t) + u_i^T(t)R_i u_i(t)\}dt,$$

where $x_i(t)$, $i = 1, 2$ satisfy the differential equation

$$\dot{x}_i(t) = (A - \frac{1}{2}r_i I)x_i(t) + B_1 u_1(t) + B_2 u_2(t); \quad x_i(0) = x_0.$$

Next, introduce as the state variable $x^T(t) := [x_1^T(t), x_2^T(t)]$. Then the problem is equivalent to finding the equilibrium strategies for the standard problem with cost function for player 1

$$J_1 = \int_0^T \left\{x^T(t)\begin{bmatrix} Q_1 & 0 \\ 0 & 0 \end{bmatrix}x(t) + u_1^T(t)R_1 u_1(t)\right\}dt;$$

and cost function for player 2

$$J_2 = \int_0^T \left\{x^T(t)\begin{bmatrix} 0 & 0 \\ 0 & Q_2 \end{bmatrix}x(t) + u_2^T(t)R_2 u_2(t)\right\}dt.$$

The dynamics of the system are described by

$$\dot{x}(t) = \begin{bmatrix} A - \frac{1}{2}r_1 I & 0 \\ 0 & A - \frac{1}{2}r_2 I \end{bmatrix}x(t) + \begin{bmatrix} B_1 \\ B_1 \end{bmatrix}u_1(t) + \begin{bmatrix} B_2 \\ B_2 \end{bmatrix}u_2(t), \quad x^T(0) = [x_0^T, x_0^T].$$

$$(3.6.8)$$

If one considers discounting where both players have their own specification r_i of the discount factor which differs for both players, then there is a subtle point which one should keep in mind if one considers an infinite planning horizon. This is best illustrated by first considering the case that both players choose the same discount factor. We will see later on that the notion of stabilizability of the matrix pairs (A, B_i), $i = 1, 2$, is a requirement that has to be satisfied by the system (which we will call individual stabilizability for the moment). Now, if the matrix pairs (A, B_i) are stabilizable by Theorem 3.20 the matrices $[A - \lambda I, B_i]$ have full row rank for all complex numbers $\lambda \in \mathbb{C}_0^+$. If matrix A is substituted by matrix $A - \frac{1}{2}I$, with $r > 0$, it is clear that the matrix $[A - \frac{1}{2}rI, B_i]$ will have full row rank too for all λ in the complex plane that have a real part equal or larger than $\frac{1}{2}r$. So, the discounted system is stabilizable too.

This, unfortunately, does not hold in the case where both players choose a different discount factor. That is, individual stabilizability of the original system $\dot{y} = Ay + B_1 v_1 + B_2 v_2$ does not imply that the transformed system (3.6.8) has this property too. So, individual stabilizability of this system (3.6.8) has to be imposed instead of individual stabilizabilty of the original system if, in the theory dealing with the standard framework, individual stabilizability of the system is required. Example 3.21 illustrates this point.

Example 3.21

Consider $A = \begin{bmatrix} 3 & 0 & 0 \\ 0 & 2 & 0 \\ 0 & 0 & -1 \end{bmatrix}$ and $B_1 = B_2 = \begin{bmatrix} 1 \\ 1 \\ 0 \end{bmatrix}$. Then both (A, B_1) and (A, B_2) are stabilizable (see Example 3.18). In particular this also implies that with $B := [B_1, B_2]$ (A, B) is also stabilizable. Next consider a discount factor $r_1 = 4$ by player 1 and a discount factor $r_2 = 2$ by player 2. Then matrix

$$\begin{bmatrix} A - \frac{1}{2} r_1 I - \lambda I & 0 & B_1 & B_2 \\ 0 & A - \frac{1}{2} r_2 I - \lambda I & B_1 & B_2 \end{bmatrix}$$

$$= \begin{bmatrix} 1 - \lambda & 0 & 0 & 0 & 0 & 0 & 1 & 1 \\ 0 & -\lambda & 0 & 0 & 0 & 0 & 1 & 1 \\ 0 & 0 & -3 - \lambda & 0 & 0 & 0 & 0 & 0 \\ 0 & 0 & 0 & 2 - \lambda & 0 & 0 & 1 & 1 \\ 0 & 0 & 0 & 0 & 1 - \lambda & 0 & 1 & 1 \\ 0 & 0 & 0 & 0 & 0 & -2 - \lambda & 0 & 0 \end{bmatrix}.$$

It is easily verified that for $\lambda = 1$ this matrix does not have full row rank. So, the transformed system is not stabilizable and, consequently, not individually stabilizable. $\square$

For completeness, we state below the conditions under which stabilizability of the standard system (A, B) implies stabilizability of the transformed system.

Proposition 3.25

Assume (A, B) is stabilizable and $r_i > 0$. Then $\left(\begin{bmatrix} A - r_1 I & 0 \\ 0 & A - r_2 I \end{bmatrix}, \begin{bmatrix} B \\ B \end{bmatrix} \right)$ is stabilizable if and only if for all $\mu \in \sigma(A - r_1 I) \cap \sigma(A - r_2 I) \cap \mathbb{C}_0^+$ the following property holds

$$\text{Span}\{N((A - (r_1 + \mu)I)^T), N((A - (r_2 + \mu)I)^T)\} \cap N(B^T) = \{0\}. \qquad \square$$

A proof of this result is provided in the Appendix to this chapter. Notice that, in case the matrices $A - r_1 I$ and $A - r_2 I$ have no unstable eigenvalues in common, the conditions stated in Proposition 3.25 are trivially satisfied. So, under these conditions the transformed system is stabilizable provided the standard system (A, B) is stabilizable.

III. Cost functions containing cross products
Consider the case that the cost functions contain cross-product terms. That is, player i minimizes the quadratic cost function

$$\int_0^T \{x^T(t) Q_i x(t) + 2x^T(t) S_i v_i(t) + v_i(t)^T R_i v_i(t)\} dt, \ i = 1, 2, \qquad (3.6.9)$$

subject to the standard state equation

$$\dot{x}(t) = Ax(t) + B_1 v_1(t) + B_2 v_2(t), \ x(0) = x_0, \qquad (3.6.10)$$

where both Q_i and R_i are symmetric; R_i is additionally assumed to be positive definite, and S_i is an arbitrary matrix.

To reduce this problem to the one covered by the standard case consider the following identity, obtained by completing the square:

$$x^T Q_i x + 2x^T S_i v_i + v_i^T R_i v_i = (v_i + R_i^{-1} S_i^T x)^T R_i (v_i + R_i^{-1} S_i^T x) + x^T (Q_i - S_i R_i^{-1} S_i^T) x.$$

Introducing

$$u_i = v_i + R_i^{-1} S_i^T x$$

the standard system (3.6.10) becomes equivalent to

$$\dot{x}(t) = (A - B_1 R_1^{-1} S_1^T - B_2 R_2^{-1} S_2^T) x(t) + B_1 u_1(t) + B_2 u_2(t), \ x(0) = x_0, \qquad (3.6.11)$$

and the cost functions are equivalent to

$$\int_0^T \{x^T(t)(Q_i - S_i R_i^{-1} S_i^T) x(t) + u_i(t)^T R_i u_i(t)\} dt, \ i = 1, 2. \qquad (3.6.12)$$

So, the equilibrium strategies for problem (3.6.9) and (3.6.10) can be found by determining the equilibrium strategies for the standard problem (3.6.11) and (3.6.12). The following statements hold.

1. The equilibrium strategies u_i^* and v_i^* of both problems are related by $u_i^* = v_i^* + R_i^{-1} S_i^T x$.

2. The optimal costs for the two problems are the same.

3. The closed-loop trajectories (when the equilibrium strategies are implemented) are the same.

Finally notice that concerning the individual stabilizability issue there are again problems here. That is, if the pairs (A, B_i), $i = 1, 2$, are stabilizable the pairs $(A - B_1 R_1^{-1} S_1^T - B_2 R_2^{-1} S_2^T, B_i)$ may not be stabilizable.

IV. Affine systems and cost functions
The next extension we consider is when the system dynamics are subject to exogenous terms and/or the cost functions of the players contain linear terms. As we will see in this case, by making some redefinitions the problem can be reformulated as a standard problem. Assume that player i likes to minimize the quadratic cost function

$$\int_0^T \{y^T(t) Q_i y(t) + 2s_i^T y(t) + 2y^T V_i v_i + v_i(t)^T R_i v_i(t) + 2w_i^T v_i(t)\} dt, \ i = 1, 2, \quad (3.6.13)$$

subject to the state equation

$$\dot{y}(t) = Ay(t) + B_1 v_1(t) + B_2 v_2(t) + c(t), \ y(0) = x_0. \qquad (3.6.14)$$

Here Q_i and R_i are again symmetric matrices, V_i are arbitrary matrices, R_i are positive definite, and s_i, w_i and $c(t)$ are known vectors.

This problem can be rephrased into the standard framework by introducing the artificial state $x_2(t) = e$, where e is the constant vector whose entries are all 1, that is: $e^T = [1, \cdots, 1]$. With this definition, x_2 is the unique solution of the differential equation

$$\dot{x}_2(t) = 0, \ x_2(0) = e.$$

Next introduce as a new state $x^T(t) := \left[y^T(t), \ x_2^T(t) \right]$ and $C(t) := \text{diag}\{c_i(t)\}$, where $c_i(t)$ is the ith entry of $c(t)$. Using this new state $x(t)$, system (3.6.14) is equivalent to

$$\dot{x}(t) = \begin{bmatrix} A & C(t) \\ 0 & 0 \end{bmatrix} x(t) + \begin{bmatrix} B_1 \\ 0 \end{bmatrix} v_1(t) + \begin{bmatrix} B_2 \\ 0 \end{bmatrix} v_2(t), \ x^T(0) = [x_0^T, \ e^T]. \tag{3.6.15}$$

Introducing $S_i := \text{diag}\{s_{ij}\}$ and $W_i := \text{diag}\{w_{ij}\}$, where s_{ij} and w_{ij} are the jth entry of s_i and w_i, respectively, the cost functions are equivalent to

$$\int_0^T \left\{ [x^T(t), \ v_i^T(t)] \begin{bmatrix} Q_i & S_i & V_i \\ S_i^T & 0 & W_i \\ V_i^T & W_i^T & R_i \end{bmatrix} \begin{bmatrix} x(t) \\ v_i(t) \end{bmatrix} \right\} dt, \ i = 1, 2. \tag{3.6.16}$$

This fits into the framework of Case III, considered above. According to Case III, the problem can be rewritten into the standard form by using the input transformation $u_i = v_i + R_i^{-1} [V_i^T, \ W_i^T] x(t)$. This yields the equivalent system

$$\dot{x}(t) = \begin{bmatrix} A - B_1 R_1^{-1} V_1^T - B_2 R_2^{-1} V_2^T & C(t) - B_1 R_1^{-1} W_1^T - B_2 R_2^{-1} W_2^T \\ 0 & 0 \end{bmatrix} x(t)$$
$$+ \begin{bmatrix} B_1 \\ 0 \end{bmatrix} u_1(t) + \begin{bmatrix} B_2 \\ 0 \end{bmatrix} u_2(t), \ x^T(0) = [x_0^T, \ e^T]. \tag{3.6.17}$$

Whereas the equivalent cost functions are

$$\int_0^T \left\{ x^T(t) \begin{bmatrix} Q_i - V_i R_i^{-1} V_i^T & S_i - V_i R_i^{-1} W_i^T \\ S_i^T - W_i R_i^{-1} V_i^T & -W_i R_i^{-1} W_i^T \end{bmatrix} x(t) + u_i(t)^T R_i u_i(t) \right\} dt, \ i = 1, 2. \tag{3.6.18}$$

Notice that this approach does not work on an infinite horizon since stabilizability of the system is required. In this case, without making further assumptions on the exogenous term $c(t)$ in system (3.6.14) and considering discounted cost functions, the state does not converge to zero. Although the reader might derive this case by combining the results here and those from Case II, we will deal with this case in the next seperate Case V. This is because one frequently encounters this model in literature.

V. Infinite horizon affine systems and discounted affine cost functions
Here we consider the case that player i minimizes the quadratic cost function

$$\int_0^\infty e^{-r_i t} \{ y^T(t) Q_i y(t) + 2 s_i^T y(t) + 2 y^T V_i \tilde{v}_i + \tilde{v}_i(t)^T R_i \tilde{v}_i(t) + 2 w_i^T \tilde{v}_i(t) \} dt, \ i = 1, 2,$$

$$\tag{3.6.19}$$

subject to the state equation

$$\dot{y}(t) = Ay(t) + B_1\tilde{v}_1(t) + B_2\tilde{v}_2(t) + c(t), \ y(0) = x_0, \tag{3.6.20}$$

where the notation is as in Case IV. r_i denotes the discount factor of player i. We will only deal with the case where $r_1 = r_2$. The case where both discount factors differ can be solved in a similar way to Case II and is left as an exercise to the reader. An additional assumption is that $c(t)$ satisfies the differential equation

$$\dot{c}(t) = C_1c(t) + C_2e, \ c(0) = c_0,$$

where the eigenvalues of matrix C_1 all have a real part strictly smaller than $\frac{1}{2}r$. Notice that this formulation includes the particular case that $c(t)$ is constant for all t.

Introducing as the new state variable

$$x^T(t) := e^{-\frac{1}{2}t}\left[y^T(t), \ c^T(t), \ e^T\right]$$

and controls $v_i = e^{-\frac{1}{2}t}\tilde{v}(t)$ the cost functions (3.6.19) and system (3.6.20) can be rewritten, respectively, as

$$\int_0^\infty \left\{ \left[x^T(t), \ v_i^T(t)\right] \begin{bmatrix} Q_i & 0 & S_i & V_i \\ 0 & 0 & 0 & 0 \\ S_i^T & 0 & 0 & W_i \\ V_i^T & 0 & W_i^T & R_i \end{bmatrix} \begin{bmatrix} x(t) \\ v_i(t) \end{bmatrix} \right\} dt \tag{3.6.21}$$

and

$$\dot{x}(t) = \begin{bmatrix} A - \frac{1}{2}rI & I & 0 \\ 0 & C_1 - \frac{1}{2}rI & C_2 \\ 0 & 0 & -\frac{1}{2}rI \end{bmatrix} x(t) + \begin{bmatrix} B_1 \\ 0 \\ 0 \end{bmatrix} v_1(t) + \begin{bmatrix} B_2 \\ 0 \\ 0 \end{bmatrix} v_2(t). \tag{3.6.22}$$

Using the approach from Case III again, this problem can be rewritten in our standard form by using the input transformation $u_i = v_i + R_i^{-1}\left[V_i^T, \ 0, \ W_i^T\right]x(t)$. This yields the equivalent system

$$\dot{x}(t) = \begin{bmatrix} A - \frac{1}{2}I - B_1R_1^{-1}V_1^T - B_2R_2^{-1}V_2^T & I & -B_1R_1^{-1}W_1^T - B_2R_2^{-1}W_2^T \\ 0 & C_1 - \frac{1}{2}rI & C_2 \\ 0 & 0 & -\frac{1}{2}rI \end{bmatrix} x(t)$$

$$+ \begin{bmatrix} B_1 \\ 0 \\ 0 \end{bmatrix} u_1(t) + \begin{bmatrix} B_2 \\ 0 \\ 0 \end{bmatrix} u_2(t), \ x^T(0) = \left[x_0^T, \ c_0^T, \ e^T\right], \tag{3.6.23}$$

whereas the equivalent cost functions are

$$\int_0^\infty \left\{ x^T(t) \begin{bmatrix} Q_i - V_iR_i^{-1}V_i^T & 0 & S_i - V_iR_i^{-1}W_i^T \\ 0 & 0 & 0 \\ S_i^T - W_iR_i^{-1}V_i^T & 0 & -W_iR_i^{-1}W_i^T \end{bmatrix} x(t) + u_i(t)^T R_i u_i(t) \right\} dt, \ i = 1, 2. \tag{3.6.24}$$

Notice that system (3.6.23) is in general not individually stabilizable if the original system (3.6.19) is individually stabilizable.

VI. Tracking systems
In fact a special case of Case IV are the systems in which each of the players likes to track a prespecified state trajectory using a prespecified ideal control path. That is, player i likes to minimize the quadratic cost function

$$\int_0^T \{(y(t) - \bar{y}_i(t))^T Q_i(y(t) - \bar{y}_i(t)) + (v_i(t) - \bar{v}_i(t))^T R_i(v_i(t) - \bar{v}_i(t))\} dt, \ i = 1, 2,$$

$$(3.6.25)$$

subject to the state equation

$$\dot{y}(t) = Ay(t) + B_1 v_1(t) + B_2 v_2(t), \ y(0) = x_0, \tag{3.6.26}$$

where Q_i, R_i are again symmetric matrices, R_i are positive definite and $\bar{y}_i$ and $\bar{v}_i$ are prespecified ideal state and control paths, respectively. Introducing $u_i := v_i - \bar{v}_i$ this problem can be rewritten into the framework studied in Case IV. The cost functions become

$$\int_0^T \{y^T(t) Q_i y(t) - 2\bar{y}_i^T(t) Q_i y(t) + u_i^T(t) R_i u_i(t)\} dt + \int_0^T \bar{v}_i^T(t) Q_i \bar{v}_i(t) dt, \ i = 1, 2, \ (3.6.27)$$

whereas the state equation is equivalent to

$$\dot{y}(t) = Ay(t) + B_1 u_1(t) + B_2 u_2(t) + c(t), \ y(0) = x_0, \tag{3.6.28}$$

where $c(t) := B_1 \bar{v}_1(t) + B_2 \bar{v}_2(t)$. Since the second part of equation (3.6.27) does not depend on the controls used by the different players, this cost and state description completely fits with (3.6.13) and (3.6.14).

In a similar way one can deal with the infinite horizon case, but notice that the approach taken in Case V requires that a combination of the ideal paths should satisfy a differential equation analogous to the one we formulated for the exogenous term $c(.)$.

If we assume a priori that the prespecified state and control paths are formulated by both players consistent with the system, then we can pursue a more direct approach invoking a different transformation. That is, assume that $\bar{y}_i$ are both differentiable and generated according the differential equation

$$\dot{\bar{y}}_i(t) = A\bar{y}_i(t) + B_1\bar{v}_1(t) + B_2\bar{v}_2(t); \ \bar{y}_i(0) = \bar{y}_{i0}, \ i = 1, 2.$$

Then, with $y_i(t) := y(t) - \bar{y}_i(t)$ and u_i as above, equations (3.6.25) and (3.6.26) (with $T = \infty$) can be rewritten as

$$\int_0^\infty \{y_i^T Q_i y_i(t) + u_i^T(t) R_i u_i(t)\} dt, \ i = 1, 2, \tag{3.6.29}$$

and

$$\dot{y}_i(t) = Ay_i(t) + B_1v_1(t) + B_2v_2(t), \ y_i(0) = x_0 - \bar{y}_{i0}. \qquad (3.6.30)$$

Introducing the state variable $x^T(t) := \begin{bmatrix} y_1^T, & y_2^T \end{bmatrix}$ (3.6.29,3.6.30) can then be rewritten into the standard framework.

Notice that if we consider discounted cost functions, the class of prescribed ideal control and state trajectories can be chosen to be much larger by using a similar approach to Case V. This already leads us to the final remark we want to make here, which is, that by considering any combination of the above cases we obtain a large class of problems which can all be analyzed using our standard framework.

3.7 Examples of differential games

In this section we present a number of examples which fit into the standard framework presented above.

Example 3.22   Monetary and fiscal policy interaction in the EMU

This model was presented by van Aarle *et al.* (2001) and extends the policy modelling approach taken by Neck and Dockner (1995).

To study macroeconomic policy design in the European Monetary Union (EMU) the next comprehensive model is considered. It is assumed that the EMU consists of two symmetric, equal sized (blocks of) countries that share a common central bank, the ECB. External interaction of the EMU countries with the non-EMU countries and also the dynamic implications of government debt and net foreign asset accumulation are ignored. The model consists of the following equations:

$$y_1(t) = \delta_1 s(t) - \gamma_1 r_1(t) + \rho_1 y_2(t) + \eta_1 f_1(t) \qquad (3.7.1)$$

$$y_2(t) = -\delta_2 s(t) - \gamma_2 r_2(t) + \rho_2 y_1(t) + \eta_2 f_2(t) \qquad (3.7.2)$$

$$s(t) = p_2(t) - p_1(t) \qquad (3.7.3)$$

$$r_i(t) = i_E(t) - \dot{p}_i(t), \ i = 1, 2, \qquad (3.7.4)$$

$$m_i(t) - p_i(t) = \kappa_i y_i(t) - \lambda_i i_E(t), \ i = 1, 2, \qquad (3.7.5)$$

$$\dot{p}_i(t) = \xi_i y_i(t), \ i = 1, 2, \qquad (3.7.6)$$

in which y_i denotes the real output, s competitiveness of country 2 vis-à-vis country 1, r_i the real interest rate, p_i the price level, f_i the real fiscal deficit, i_E the nominal interest rate and m_i the nominal money balances of country (block) $i \in \{1, 2\}$. All variables are in logarithms, except for the interest rate which is in perunages. The variables denote deviations from their long-term equilibrium (balanced growth path) that has been normalized to zero, for simplicity.

Equations (3.7.1) and (3.7.2) represent output in the EMU countries as a function of competitiveness in intra-EMU trade, the real interest rate, the foreign output and the domestic fiscal deficit. Competitiveness is defined in equation (3.7.3) as the output price differential. Real interest rates are defined in equation (3.7.4) as the difference between the EMU-wide nominal interest rate, i_E, and domestic inflation. Notice that equation (3.7.4) implies that, temporarily, real interest rates may diverge among countries if inflation rates are different. Equation (3.7.5) provides the demand for the common currency where it is assumed that the money market is in equilibrium. The structural model (3.7.1–3.7.6) models an integrated economy with several kinds of cross-country effects. Besides the common nominal interest rate there are two other important direct cross-country spillovers that affect domestic output: (i) the intra-EMU competitiveness channel (as measured by the elasticity δ), and (ii) the foreign output channel (as measured by the elasticity ρ).

It is assumed that the common nominal interest rate is set by the ECB (that is, an interest rate targeting approach is proposed here). Domestic output and inflation are related through a Phillips curve type relation in equation (3.7.6).

The model (3.7.1)–(3.7.6) can be reduced to two output equations:

$$y_1(t) = b_1 s(t) - c_1 i_E(t) + a_1 f_1(t) + \frac{\rho_1}{k_1} a_2 f_2(t) \tag{3.7.7}$$

$$y_2(t) = -b_2 s(t) - c_2 i_E(t) + \frac{\rho_2}{k_2} a_1 f_1(t) + a_2 f_2(t) \tag{3.7.8}$$

in which a_i, b_i, c_i, k_i and ϕ_i (below) are parameters related to the original model parameters (see van Aarle, Engwerda and Plasmans (2002) for details). The dynamics of the model are then represented by the following first-order linear differential equation with competitiveness, $s(t)$, as the scalar state variable and the national fiscal deficits, $f_i(t)$ $i = 1, 2$, and the common interest rate, $i_E(t)$, as control variables:

$$\dot{s}(t) = \phi_4 s(t) - \phi_1 f_1(t) + \phi_2 f_2(t) + \phi_3 i_E(t), \quad s(0) = s_0. \tag{3.7.9}$$

The initial value of the state variable, s_0, measures any initial disequilibrium in intra-EMU competitiveness. Such an initial disequilibrium in competitiveness could be the result of, for example, differences in fiscal policies in the past or some initial disturbance in one country.

The aim of the fiscal authorities is to use their fiscal policy instrument such that the following quadratic loss functions are minimized. The loss functions express the countries' concern towards domestic nominal inflation, domestic real output and domestic real fiscal deficit[6].

$$\min_{f_i} J_i = \min_{f_i} \int_0^\infty e^{-\theta t} \{ \alpha_i p_i^2(t) + \beta_i y_i^2(t) + \chi_i f_i^2(t) \} dt, \quad i = 1, 2. \tag{3.7.10}$$

Here θ denotes the rate of time preference and α_i, β_i and χ_i ($i \in \{1, 2\}$) represent preference weights that are attached to the stabilization of inflation, output and fiscal

---

[6]Note that in a monetary union the fiscal players are assumed not to have any direct control over the nominal interest rate since this control is generally left to the common central bank.

deficits, respectively. Preference for a low fiscal deficit reflects the goal to prevent excessive deficits. This aim is on the one hand a reflection of the rules that were agreed in the Stability and Growth Pact (SGP) that sanctions such excessive deficits in the EMU. On the other hand, costs could also result from undesirable debt accumulation and intergenerational redistribution that high deficits imply and, in that interpretation, χ_i could also reflect the priority attached to fiscal retrenchment and consolidation.

As stipulated in the Maastricht Treaty, the ECB directs the common monetary policy towards stabilizing inflation and, as long as this is not in contradiction to inflation stabilization, stabilizing output in the aggregate EMU economy. Moreover, we will assume that the active use of monetary policy implies costs for the monetary policy-maker: all other things being equal it would like to keep its policy instrument constant, avoiding large swings. Consequently, we assume that the ECB is confronted with the following optimization problem:

$$\min_{i_E} J_E^A = \min_{i_E} \int_0^\infty e^{-\theta t} \left\{ (\alpha_{1E}\dot{p}_1(t) + \alpha_{2E}\dot{p}_2(t))^2 + (\beta_{1E}y_1(t) + \beta_{2E}y_2(t))^2 + \chi_E i_E^2(t) \right\} dt.$$

$$(3.7.11)$$

Alternatively, one could consider a case where the ECB is governed by national interests rather than by EMU-wide objectives. In that scenario, the ECB would be a coalition of the former national central banks that decide cooperatively on the common monetary policy that is based on individual, national interests rather than on EMU-wide objectives. In this scenario the monetary policy of the ECB will typically be more sensitive to individual country variables. Then the ECB seeks to minimize a loss function, which is assumed to be quadratic in the individual countries' inflation rates and outputs – rather than in EMU-wide inflation and output as in equation (3.7.11) – and the common interest rate. That is, it considers

$$\min_{i_E} J_E^N = \min_{i_E} \int_0^\infty e^{-\theta t} \left\{ \alpha_{1E}\dot{p}_1^2(t) + \alpha_{2E}\dot{p}_2^2(t) + \beta_{1E}y_1^2(t) + \beta_{2E}y_2^2(t) + \chi_E i_E^2(t) \right\} dt.$$

$$(3.7.12)$$

The loss function in equation (3.7.12) can also be interpreted as a loss function in which the ECB is a coalition of national central bankers who all have a share in the decision making proportional to the size of their economies.

Below, we will only further elaborate the model if the ECB objective is represented by equation (3.7.11). In a similar way the appropriate formulae can be obtained if equation (3.7.12) is used as the ECB's performance criterion.

Using equation (3.7.6), equation (3.7.10) and (3.7.11) can be rewritten as:

$$J_i = d_i \int_0^\infty e^{-\theta t} \left\{ y_i^2(t) + \frac{\chi_i}{d_i} f_i^2(t) \right\} dt, \quad i = 1, 2,$$

$$(3.7.13)$$

$$J_E^A = \int_0^\infty e^{-\theta t} \left\{ d_{1E}y_1^2(t) + d_{2E}y_2^2(t) + 2d_{3E}y_1(t)y_2(t) + \chi_E i_E^2(t) \right\} dt,$$

$$(3.7.14)$$

where $d_i := \alpha_i \xi_i^2 + \beta_i$, $d_{iE} := \alpha_{iE}^2 \xi_i^2 + \beta_{iE}^2$ with $i = 1, 2$, and $d_{3E} := \alpha_{1E} \alpha_{2E} \xi_1 \xi_2 + \beta_{1E} \beta_{2E}$.[7]

Defining $x^T(t) := [s(t), f_1(t), f_2(t), i_E(t)]$, equation (3.7.7) and (3.7.8) can be rewritten as

$$y_1(t) = \left[b_1, a_1, \frac{\rho_1}{k_1} a_2, -c_1 \right] x(t) =: m_1 x(t)$$

$$y_2(t) = \left[-b_2, \frac{\rho_2}{k_2} a_1, a_2, -c_2 \right] x(t) =: m_2 x(t).$$

Introducing e_j as the jth standard basis vector of $\mathbb{R}^4$ (i.e. $e_1 := [1\ 0\ 0\ 0]^T$, etc.), $M_i := m_i^T m_i + \frac{\chi_i}{d_i} e_{i+1}^T e_{i+1}$, $i = 1, 2$, and $M_E^A := d_{1E} m_1^T m_1 + d_{2E} m_2^T m_2 + 2 d_{3E} m_1^T m_2 + \chi_E e_4^T e_4$, the policy makers' loss functions (3.7.13)–(3.7.14) can be written as:

$$J_i = d_i \int_0^\infty e^{-\theta t} \{ x^T(t) M_i x(t) \} dt, \quad i = 1, 2, \tag{3.7.15}$$

$$J_E^A = \int_0^\infty e^{-\theta t} \{ x^T(t) M_E^A x(t) \} dt. \tag{3.7.16}$$

This formulation of the problem fits into the framework we considered in Cases II and III. Using the transformations outlined there, the model can then be rewritten into the standard framework. ☐

Example 3.23   The transboundary acid rain problem

This problem was presented by Mäler and de Zeeuw (1998). Suppose there is a group of n countries emitting e_i, $i = 1, \cdots, n$, tons of sulfur or nitrogen oxides (for short: sulfur). This sulfur is partly transferred to other countries by winds. Let matrix B denote the transport matrix. So, entry b_{ij} is the fraction of country j's emissions e_j that is deposited in country i. Then Be is the vector of depositions in the n countries as a consequence of the emissions by the same countries. In addition to this each of the countries receives the so-called 'background' depositions from the countries outside the group as well as from the sea, which are assumed to be a given.

At the beginning of the acidification process no damage is done to the soil and even more nutrients become available for plants, but for simplicity we will ignore this aspect here. Above a certain critical load the acid buffer stock decreases and the depletion d_i of this buffer stock indicates how much damage is done to the soil in country i, $i = 1, \cdots, n$. In fact there are different acid buffers which are depleted one by one. As long as the acidification process is in the first acid buffer, the soil becomes less productive but can recover when the depositions are at, or below, the critical load again, so that the damage is reversible. However, when this buffer is depleted, the soil cannot recover and will loose some of its productivity for ever. Moreover, when the last acid buffer is depleted, the soil

---

[7]In the case that national variables feature in the ECB objective function, as in equation (3.7.12), $d_{iE} = \alpha_{iE} \xi_i^2 + \beta_{iE}$ with $i = \{1, 2\}$ and $d_{3E} = 0$.

will even become non-productive. We will model this acidification process for the first buffer stock.

Critical loads are given as grams of sulfur per year per square meter and can differ substantially from one region to the other, depending on the characteristics of the soil, the bedrock, the vegetation cover, the precipitation, etc. Since the analysis is performed at the aggregation level of countries, a measure is needed for a country's critical load. In Mäler and de Zeeuw (1998) it is indicated (and actually computed) how a rough estimate of these critical loads can be obtained. If the background depositions were to exceed the critical load in one of the countries, the group cannot control the acidification process in that country. Therefore it is assumed that this is not the case. In fact, the difference between the total depositions and the critical loads does matter. Therefore, the background depositions can be subtracted on both sides, so that the starting point of the analysis consists of the internal depositions Be, on the one hand, and the critical loads minus the background depositions, denoted by the vector c, on the other hand. Each country faces the trade-off between the costs of reducing emissions and the benefits of lower damage to the environment. Each country chooses a time path for its emissions with the objective to minimize a discounted stream of costs and damages subject to the depletion of the acid buffer stocks. Therefore, the problem of country i can be formulated as the optimal control problem to minimize w.r.t. e_i the cost function

$$\int_0^\infty e^{-rt}\{C_i(e_i(t)) + D_i(d_i(t))\}dt, \ i = 1, \cdots, n,$$

subject to the state equation

$$\dot{d}(t) = Be(t) - c, \ d(0) = d_0,$$

where r denotes the interest rate, C the cost function of the reduction of the emissions and D the damage function of the depletion of the acid buffer stocks. In the analysis, the steady-state levels of the depletion are simply used as an indicator of how serious the situation is. Next assume that the cost and damage functions are given by

$$C_i(e_i) = \gamma_i(e_i - \bar{e}_i)^2; \ D_i(d_i) = \delta_i d_i^2, \ \gamma_i > 0, \delta_i > 0$$

where $\bar{e}_i$ are a set of prespecified values which are chosen such that countries with lower per capita sulfur emissions have higher marginal emission reduction costs. It is clear again that by a combination of Cases II and V this problem can be written into our standard framework. □

Example 3.24   Dynamic duopoly with 'sticky' prices

This model is based on the model presented by Fershtman and Kamien (1987) (see also Dockner *et al.*, 2000).

In this example we consider dynamic duopolistic competition in a market for a homogeneous good. That is, we consider a market with only two sellers offering an identical product, e.g. two companies that produce water, safe for drinking. It is assumed that the market price does not adjust instantaneously to the price indicated by the demand function. There is a lag in the market price adjustment so the price is said to be 'sticky'. This scenario is modelled as a differential game where the dynamics describe the evolution of market price over time. The dynamics include the reasonable feature that, when the parameter measuring the speed of price adjustment tends to infinity, price converges to its value on the demand function.

The model assumptions are as follows. Demand is linear in price, and production costs are quadratic in output. Denote by $u_i(t) \geq 0$ the output rate of company i, $i = 1, 2$. The relationship between the output of both companies and the price (the linear instantaneous inverse demand function) is given by $p(t) = a - (u_1(t) + u_2(t))$, where $p(t)$ is the market price at time t and $a > 0$ is a constant. To model that the market price does not adjust instantaneously to the price indicated by the demand function we let the rate of change of the market price be a function of the difference between the current market price and the price indicated by the linear demand function (for any particular aggregate output). That is

$$\dot{p}(t) = s\{a - (u_1(t) + u_2(t)) - p(t)\}, \quad p(0) = p_0, \tag{3.7.17}$$

where $s \in (0, \infty)$ is the adjustment speed parameter. Notice that for larger values of s the market price adjusts quicker along the demand function. For simplicity, the companies are assumed to have a quadratic production cost function $C(u_i) := c_i u_i + u_i^2$, where $c_i \in (0, a)$, $i = 1, 2$, are fixed parameters. Consequently, the discounted profit of company i is given by

$$J_i(u_1, u_2) = \int_0^\infty e^{-rt}\{p(t)u_i(t) - c_i u_i(t) - \frac{1}{2}u_i^2(t)\}dt, \tag{3.7.18}$$

in which $r > 0$ is the discount rate. This model formulation (3.7.17) and (3.7.18) fits into the affine discounted linear quadratic framework (if we ignore for the moment the fact that $u_i \geq 0$). $\square$

Example 3.25 Robust control: the H_∞ disturbance attenuation problem

Consider the next Philips multiplier-accelerator model, to represent the dynamics of an economic system for which an optimal stabilization policy is to be designed. This model is represented by the pair of equations

$$y(t) = (1 - \alpha)x(t) + \beta\dot{x}(t) + g(t), \quad \alpha \in (0, 1), \quad \beta > 0, \tag{3.7.19}$$

$$\dot{x}(t) = \gamma(y(t) - x(t)), \quad x(0) = x_0, \quad \gamma > 0. \tag{3.7.20}$$

Equation (3.7.19) is the Harrodian demand relation specifying current demand y as the sum of private sector demand $(1 - \alpha)x + \beta\dot{x}$ and public sector demand g. Equation (3.7.20) represents the dynamic adjustment mechanism of supply x to demand y. The variables x, y and g are assumed to be measured as deviations from levels defining a desired equilibrium position. Substitution of equation (3.7.19) into equation (3.7.20) after some rewriting yields the following reduced form model

$$\dot{x}(t) = -\alpha\delta x(t) + \delta g(t), \ x(0) = x_0, \ \delta = \frac{\gamma}{1 - \beta\gamma}. \tag{3.7.21}$$

The objective is to design for this system an optimal public sector demand policy that keeps disequilibrium income near its desired value of zero without excessive deviation of public sector demand from its desired value, also zero. This objective can be formalized by assuming a quadratic preference functional of the form

$$J_0 = \int_0^\infty \{x^2(t) + \phi g^2(t)\}dt, \ \phi > 0, \tag{3.7.22}$$

where ϕ measures the relative importance of the two performance cost elements: the larger ϕ, the larger the cost of using the control g to force x to equilibrium.

With this, we constructed a mathematical model describing the behavior of the economic system and formalized an objective which implicitly defines our control policy. However, we apply this control policy to our system and not to our model. Obviously, since we never have complete information regarding the system, the model will not describe the real world exactly. Because we do not know how sensitive our objectives are with respect to the differences between our model and the real system, the behavior obtained might differ significantly from the mathematically predicted behavior. Hence our control policy will not in general be suitable for our system and the behavior we obtain can be completely surprising.

Therefore it is extremely important that, when we look for a control policy for our system, we keep in mind that our mathematical model is not a perfect description of reality. This leads to the study of robust control policies for our system. In this context robustness means that the stability property is preserved when the control policy is used to control a more complex model as long as the new model is close to our original model.

One approach, which stems from the 1960s, is the linear quadratic gaussian (LQG) or H_2 approach. In this approach the uncertainty is modelled as a white noise gaussian process added as an extra input to the system (3.7.21). A disadvantage of this approach is that uncertainty cannot always be modelled as white noise. While measurement noise can be described quite well by a random process, this is not the case with, for example, parameter uncertainty and external effects entering the system.

So our goal is to obtain stability where, instead of trying to obtain this for one model, we seek one control policy which will stabilize any element from a certain class of models. It is then hoped that a control policy which stabilizes all elements of this class of models also stabilizes the actual system.

To that end, we consider the following adapted model of system (3.7.21)

$$\dot{x}(t) = -\alpha\delta x(t) + \delta g(t) + w(t), \ x(0) = x_0, \tag{3.7.23}$$

where the input $w(t)$ is an extraneous input representing the disturbance acting on the system. We assume that this unknown disturbance is finite in the sense that $\int_0^\infty w^2(t)dt$ exists as a finite number (i.e. $w(t)$ is **square integrable** or, stated differently, $w(t) \in L_2$).

With this formulation, the preference function (3.7.22) is a function of the extraneous disturbance $w(t)$ and consequently the induced optimal policy too. Since $w(t)$ is unknown this does not make any sense. So we have to adapt our preference function too. Now, assume that our system is at time $t = 0$ in equilibrium, i.e. $x_0 = 0$. Then, as our new goal, we like to find that control policy which minimizes the worst possible outcome with respect to $w(.)$ of the objective function

$$J(g, w) = \frac{\int_0^\infty \{x^2(t) + \phi g^2(t)\} dt}{\int_0^\infty w^2(t) dt}. \tag{3.7.24}$$

Or, mathematically more precise, we need to find

$$\bar{J} := \inf_g \sup_w J(g, w). \tag{3.7.25}$$

This should be read so that we first fix the trajectory of $g(.)$ and look for that trajectory $w(.)$ for which the value of J is as large as possible, say this value equals $\bar{J}(g)$. Next we determine among all possible control policies $g(.)$ the control policy for which the value $\bar{J}(g)$ is as small as possible. Within this framework it is usually assumed that at any point in time the current state of the system can be observed, so that the state can be used in the design of the optimal control policy. Obviously, under these operating conditions one should be able to react immediately to unexpected state deviations. Therefore, one may hope that under these conditions a robust controller will be an effective instrument in achieving stability in practical applications.

The value $\bar{J}$ in equation (3.7.25) is called the **upper value** of the game. Conversely, the **lower value**, $\underline{J}$, of the game is defined as

$$\underline{J} := \sup_w \inf_g J(g, w). \tag{3.7.26}$$

To determine the lower value, we first consider for a fixed w the control policy which minimizes $J(w)$ and next we look for that w for which $J(w)$ becomes as large as possible. Since

$$\sup_w J(g, w) \geq \sup_w \inf_g J(g, w)$$

we see that by taking on both sides of this inequality the infimum with respect to g that the upper value of the game is always larger than or equal to its lower value, that is

$$\bar{J} \geq \underline{J}, \tag{3.7.27}$$

which clarifies the terminology.

Now assume that the game has a **saddle-point solution**, i.e. there exists a pair of strategies $(\bar{g}, \bar{w})$ such that

$$J(\bar{g}, w) \leq J(\bar{g}, \bar{w}) \leq J(g, \bar{w}), \quad \text{for all admissible } g, w. \tag{3.7.28}$$

Then, the upper and lower value of the game coincide and the **value of the game** equals $J(\bar{g}, \bar{w})$. This result is outlined in the next theorem and a proof of it can be found in the Appendix to this chapter.

Theorem 3.26

Assume that the game $J(g, w)$ has a saddle-point solution $(\bar{g}, \bar{w})$. Then

$$\bar{J} = \underline{J} = J(\bar{g}, \bar{w}).$$ □

Next, following the lines of Başar and Olsder (1999), assume for the moment that there exists a control policy g^* that achieves the upper value in equation (3.7.25). Denote this upper value by $\hat{\lambda}$. Then equation (3.7.25) can be rewritten as

$$\int_0^\infty \{x^2(t) + \phi g^2(t)\}dt \leq \hat{\lambda} \int_0^\infty w^2(t)dt, \ \forall w \in L_2 \tag{3.7.29}$$

and there exists no other control policy g and corresponding $\lambda < \hat{\lambda}$ such that

$$\int_0^\infty \{x^2(t) + \phi g^2(t)\}dt \leq \lambda \int_0^\infty w^2(t)dt, \ \forall w \in L_2. \tag{3.7.30}$$

Introducing the parameterized (in $\lambda \geq 0$) family of cost functions

$$L_\lambda(g, w) := \int_0^\infty \{x^2(t) + \phi g^2(t)\}dt - \lambda \int_0^\infty w^2(t)dt, \tag{3.7.31}$$

equations (3.7.29) and (3.7.30) are equivalent to the problem of finding the 'smallest' value of $\lambda \geq 0$ under which the upper value of the game defined by the differential equation (3.7.23) (with $x_0 = 0$) and objective function (3.7.31) is nonpositive, and finding the corresponding control policy that achieves this upper value.

Now, for a fixed $\lambda \geq 0$, consider the two-person differential game in which player 1 (the government) minimizes w.r.t. g the quadratic cost function

$$J_1 := L_\lambda(g, w) \tag{3.7.32}$$

and player 2 (nature) minimizes w.r.t. w the quadratic cost function

$$J_2 := -L_\lambda(g, w), \tag{3.7.33}$$

subject to the state equation (3.7.23). Then, if this zero-sum differential game has a Nash solution $(\bar{g}, \bar{w})$ (that is $J_1(\bar{g}, \bar{w}) \leq J_1(g, \bar{w})$ and $J_2(\bar{g}, \bar{w}) \leq J_2(\bar{g}, w)$),

$$L_\lambda(\bar{g}, w) \leq L_\lambda(\bar{g}, \bar{w}) \leq L_\lambda(g, \bar{w}).$$

So $(\bar{g}, \bar{w})$ is a saddle-point solution for $L_\lambda(g, w)$ which implies, according to Theorem 3.65, that it also yields an upper value for the corresponding game. In particular (with $x_0 = 0$) $\bar{g}$ delivers a performance level of at least λ, that is, for all $w \in L_2$

$$\int_0^\infty \{x^2(\bar{g}) + \phi \bar{g}^2\}dt \leq \lambda \int_0^\infty w^2(t)dt.$$

Notice that in practice it may be better to consider a suboptimal controller instead of the optimal controller that achieves the lower bound for the performance level. A suboptimal controller achieving almost the lower bound may have additional beneficial properties compared with optimal ones, such as ease of implementation and better performance for a large set of disturbances that usually occur. □

3.8 Information, commitment and strategies

What information players have on the game, whether or not they can commit themself to the proposed decisions and which control strategies are used by the different players are essential elements in determining the outcome of a game. Without a proper specification of these three elements, a game is ill-defined and usually leads to ambiguous statements.

Concerning the information aspect of the game we assume in this book that the players know the system and each other's preferences. As far as the commitment issue is concerned we assume that all players are able to implement their decisions. With respect to the choice of control strategies we make different assumptions. Basically, we consider two types of strategies: strategies where players base their actions purely on the initial state of the system and time (open-loop strategies) and strategies where players base their actions on the current state of the system (feedback strategies). Note that the implementation of the second type of strategies requires a full monitoring of the system. To implement this strategy each player has to know at each point in time the exact state of the system. On the other hand, an advantage of this strategy is that as far as the commitment issue is concerned it is much less demanding. If, due to some external cause, the state of the system changes during the game this has no consequences for the actions taken by the players. They are able to respond to this disturbance in an optimal way. This, in contrast to the open-loop strategy which implies that the players cannot adapt their actions during the game in order to account for the unforeseen disturbance without breaking their commitment. Since all players are confronted with this commitment promise, one might expect that under such conditions the players will try to renegotiate on the agreed decisions. So open-loop strategies make sense particularly for those situations where the model is quite robust or the players can commit themself strongly. A practical advantage of the open-loop strategy is that it is, usually, numerically and analytically more tractable than the feedback strategy.

3.9 Notes and references

There is a vast amount of literature on linear dynamical systems. Two classical works are Kwakernaak and Sivan (1972) and Kailath (1980). Some of the more recent work which has been done in this area is reported in, for example, Zhou, Doyle and Glover (1996) and Polderman and Willems (1998). The last-mentioned work approaches the set of linear systems from a so-called behavioral point of view. Models are viewed as relationships between certain variables. The collection of all time trajectories, which the dynamical model allows, is called the behavior of the system. The main difference with the more conventional approach is that it does not start with an input/output representation.

Whether such an approach may also yield additional insights in a game setting is not yet clear.

Books which have been extensively consulted for the sections on differential equations are Perko (2001) and Coddington and Levinson (1955). These books are very suitable for those interested in either more or mathematical details on this topic. Finally, readers who are interested in more elaborated examples on differential games in economics are referred to Dockner *et al.* (2000).

3.10 Exercises

1. Calculate e^{At} for the matrices in Exercise 2.21 and Exercise 2.22.

2. Calculate the solution $x(t)$ at $t = 2$, with $x(0) = [1, \ 1]^T$, of the nonhomogeneous differential equations

   (a) $\dot{x}(t) = \begin{bmatrix} 3 & -4 \\ 2 & -3 \end{bmatrix} x(t) + \begin{bmatrix} 1 \\ 0 \end{bmatrix}$;

   (b) $\dot{x}(t) = \begin{bmatrix} -1 & 1 \\ 0 & -1 \end{bmatrix} x(t) + \begin{bmatrix} 1 \\ e^{-t} \end{bmatrix}$;

   (c) $\dot{x}(t) = \begin{bmatrix} -8 & 10 \\ -5 & 6 \end{bmatrix} x(t) + \begin{bmatrix} 0 \\ 1 \end{bmatrix}$.

3. Show that the following differential equations have for every value x_0 at $t = 0$ a solution on some interval $t_1 < 0 < t_2$.

   (a) $\dot{x}(t) = 4x(t) + x^3(t) + e^{-x(t)}$;

   (b) $\dot{x}(t) = \dfrac{3x(t) - x^3(t)}{e^{-x(t)} + x^2(t)}$;

   (c) $\dot{x}(t) = 2tx(t) + e^t x^3(t) + \dfrac{t+1}{(t^2+1)e^{-x(t)}}$.

4. Assume that $a(t)$, $b(t)$, $q(t)$ and $r(t)$ are continuous functions and $r(t) > 0$ for all t. Show that the following differential equation has, for every value k_0 at $t = 0$, a solution on some interval $[0, t_1)$.

$$k(t) = -2a(t)k(t) + \frac{b^2(t)}{r(t)} k^2(t) - q(t).$$

5. Show that the following differential equations have a solution on $\mathbb{R}$ for every initial value x_0 at $t = 0$.

   (a) $\dot{x}(t) = \dfrac{x(t)}{1 + e^{x^2(t)}}$;

   (b) $\dot{x}(t) = \dfrac{t}{1 + t^4} x(t)$.

6. Determine the equilibrium points and the local behavior of solutions near this point for the following differential equations.

   (a) $\dot{x}(t) = x^3(t) - x(t)$;

   (b) $\dot{x}(t) = x(t)e^{-x(t)}$.

7. Determine the stable, unstable and center subspaces, respectively, for the differential equation $\dot{x}(t) = Ax(t)$, $x(0) = x_0$, if matrix A equals

$$\begin{bmatrix} -1 & 1 \\ 0 & 1 \end{bmatrix}; \begin{bmatrix} -1 & 1 \\ 0 & -1 \end{bmatrix}; \begin{bmatrix} 3 & 2 \\ -3 & -2 \end{bmatrix} \text{ and } \begin{bmatrix} -3 & 2 \\ -5 & 3 \end{bmatrix}.$$

8. Determine the equilibrium points and the local behavior of solutions near this point for the following set of differential equations.

   (a) $\dot{x}_1(t) = x_1^3(t) - x_1(t)$, $\dot{x}_2(t) = x_2(t) - x_1(t)$;
   (b) $\dot{x}_1(t) = x_1(t)e^{-x_2(t)}$, $\dot{x}_2(t) = (x_1(t) + 1)x_2(t)$.

9. Make a phase portrait for the following planar systems.

   (a) $\dot{x}_1(t) = x_1(t) + x_2(t) - 1$, $\dot{x}_2(t) = (x_1^2(t) + 1)x_2(t)$;
   (b) $\dot{x}_1(t) = x_1(t) + x_2(t) - 2$, $\dot{x}_2(t) = -x_2(t) + x_1(t)$;
   (c) $\dot{x}_1(t) = x_1^2(t) - x_2(t)$, $\dot{x}_2(t) = x_2(t) - x_1(t)$;
   (d) $\dot{x}_1(t) = x_2^2(t) - 1$, $\dot{x}_2(t) = x_1(t) + x_2(t)$;
   (e) $\dot{x}_1(t) = x_1(t) - x_2^2(t)$, $\dot{x}_2(t) = -x_2(t)(x_1(t) - 1)$;
   (f) $\dot{x}_1(t) = -x_1(t) + x_2^2(t)$, $\dot{x}_2(t) = -x_2(t)(x_1(t) - 1)$.

10. Consider the system

$$\dot{x}(t) = Ax(t) + Bu(t); \quad y(t) = Cx(t),$$

where A, B and C are as specified below. Verify whether this system is controllable, stabilizable, observable and/or detectable if

   (a) $A = \begin{bmatrix} 0 & 0 \\ 0 & 0 \end{bmatrix}$, $B = \begin{bmatrix} 1 & 2 \\ 0 & 3 \end{bmatrix}$, $C = \begin{bmatrix} 1 & 0 \\ 1 & 1 \end{bmatrix}$;

   (b) $A = \begin{bmatrix} 0 & 0 \\ 0 & -1 \end{bmatrix}$, $B = \begin{bmatrix} 1 \\ 1 \end{bmatrix}$, $C = \begin{bmatrix} 2 & -1 \end{bmatrix}$;

   (c) $A = \begin{bmatrix} 1 & 1 \\ 0 & -1 \end{bmatrix}$, $B = \begin{bmatrix} 1 \\ 0 \end{bmatrix}$, $C = \begin{bmatrix} 1 & 0 \end{bmatrix}$;

   (d) $A = \begin{bmatrix} 1 & 0 \\ 4 & -1 \end{bmatrix}$, $B = \begin{bmatrix} 0 \\ 1 \end{bmatrix}$, $C = \begin{bmatrix} 1 & 0 \end{bmatrix}$;

   (e) $A = \begin{bmatrix} 1 & 0 \\ 1 & 2 \end{bmatrix}$, $B = \begin{bmatrix} 1 \\ -1 \end{bmatrix}$, $C = \begin{bmatrix} 1 & 0 \end{bmatrix}$.

11. Show that equation (3.5.2) is observable if and only if $V(t_1) = \{0\}$.

12. Show that if (A, B_i), $i = 1, 2$, is stabilizable $(A + [B_1 \ B_2]\begin{bmatrix} F_1 \\ F_2 \end{bmatrix}, B_i)$, $i = 1, 2$, is not necessarily stabilizable.

13. Rewrite the following models into the standard framework

$$\min \int_0^T \{x^T(t)Q_i(t)x(t) + u^T(t)R_i(t)u(t)\}dt + x^T(T)Q_Tx(T);$$
$$\text{subject to } \dot{x}(t) = A(t)x(t) + B_1(t)u_1(t) + B_2(t)u_2(t), \quad x(0) = x_0.$$

Indicate explicitly the values of the different matrices.

(a) $\min_v J_1 = \int_0^6 s^2(t) + v^2(t)dt$; $\min_w J_2 = \int_0^6 k^2(t) + 2w^2(t)dt$; where
$\dot{s}(t) = -s(t) + k(t) - 2v(t) + w(t)$, $s(0) = 1$;
$\dot{k}(t) = -2k(t) + s(t) + v(t) - 3w(t)$, $k(0) = 2$.

(b) $\min_v J_1 = \int_0^5 \{2s^2(t) + v^2(t)\}dt + s^2(5)$; $\min_w J_2 = \int_0^2 10k^2(t) + 4w^2(t)dt$; where
$\dot{s}(t) = -s(t) + k(t) - 2v(t) + 2w(t)$, $s(0) = 1$;
$\dot{k}(t) = -4k(t) + 2s(t) + v(t) - 4w(t)$, $k(0) = 1$.

(c) $\min_v J_1 = \int_0^4 e^{-2t}\{2s^2(t) + v^2(t)\}dt$; $\min_w J_2 = \int_0^4 e^{-t}\{10k^2(t) + 4w^2(t)\}dt$; where
$\dot{s}(t) = -2s(t) + 2k(t) - 2v(t) + w(t)$, $s(0) = 1$;
$\dot{k}(t) = -2k(t) + s(t) + v(t) - w(t)$, $k(0) = 1$.

(d) $\min_v J_1 = \int_0^3 \{s^2(t) + v^2(t)\}dt + 2s^2(3)$; $\min_w J_2 = \int_0^3 k^2(t) + 2w^2(t)dt$; where
$\dot{s}(t) = -s(t) + k(t) - 2v(t) + w(t) + 1$, $s(0) = 1$;
$\dot{k}(t) = -2k(t) + v(t) - 3w(t) + 2$, $k(0) = 1$.

(e) $\min_v J_1 = \int_0^2 s^2(t) - 2s(t)v(t) + 2v^2(t)dt + s^2(2)$;

$\min_w J_2 = \int_0^2 2k^2(t) + 2k(t)w(t) + w^2(t)dt$; where
$\dot{s}(t) = -s(t) + k(t) - 2v(t) + w(t)$, $s(0) = 1$;
$\dot{k}(t) = -2k(t) + s(t) + v(t) - 3w(t)$, $k(0) = 1$.

(f) $\min_v J_1 = \int_0^1 e^{-t}\{s^2(t) + 2s(t) + 2v^2(t)\}dt + s^2(1)$;

$\min_w J_2 = \int_0^1 e^{-t}\{2k^2(t) - k(t)w(t) + w^2(t)\}dt + 2k^2(1)$; where
$\dot{s}(t) = -s(t) - 2v(t) + w(t) + 1$, $s(0) = 1$;
$\dot{k}(t) = -2k(t) + s(t) + v(t) - 3w(t) + 1$, $k(0) = 1$.

14. Consider the problem that player i minimizes the quadratic cost function

$$\int_0^\infty e^{-r_i t}\{y^T(t)Q_i y(t) + 2s_i^T y(t) + 2y^T V_i \tilde{v}_i + \tilde{v}_i(t)^T R_i \tilde{v}_i(t) + 2w_i^T \tilde{v}_i(t)\}dt, \quad i = 1, 2,$$

subject to the state equation

$$\dot{y}(t) = Ay(t) + B_1 \tilde{v}_1(t) + B_2 \tilde{v}_2(t) + c(t), \quad y(0) = y_0, \qquad (3.10.1)$$

where r_i denotes the discount factor of player i. Also assume that $c(t)$ satisfies the differential equation

$$\dot{c}(t) = C_1 c(t) + C_2 e, \quad c(0) = c_0,$$

where the eigenvalues of matrix C_1 all have a real part strictly smaller than $\frac{1}{2}r$ and $e^T = [1 \cdots 1]$. Reformulate this game into the standard framework.

15. Assume that player i likes to minimize the quadratic cost function

$$\int_0^\infty e^{-rt}\{(y(t) - \bar{y}_i(t))^T Q_i(y(t) - \bar{y}_i(t)) + (v_i(t) - \bar{v}_i(t))^T R_i(v_i(t) - \bar{v}_i(t))\}dt, \quad i = 1, 2,$$

subject to the state equation

$$\dot{y}(t) = Ay(t) + B_1 v_1(t) + B_2 v_2(t), \quad y(0) = y_0,$$

where Q_i and R_i are positive definite, and $\bar{y}_i$ and $\bar{v}_i$ are prespecified ideal state and control paths, respectively. Assume that $\bar{y}_i$ are both differentiable and generated according the differential equation

$$\dot{\bar{y}}_i(t) = A\bar{y}_i + B_1\bar{v}_1(t) + B_2\bar{v}_2(t) + c(t); \ \bar{y}_i(0) = \bar{y}_{i0}, \ i = 1, 2,$$

where $c(t)$ satisfies the differential equation

$$\dot{c}(t) = C_1 c(t) + C_2 e, \ c(0) = c_0, \ e^T = [1 \cdots 1].$$

Reformulate this problem into the standard framework, assuming that all eigenvalues of matrix C_1 have a real part smaller than $\frac{1}{2}r$.

3.11 Appendix

Proof of Proposition 3.25

For notational convenience introduce matrix

$$C(\lambda) := \begin{bmatrix} A - (r_1 + \lambda)I & 0 & B \\ 0 & A - (r_2 + \lambda)I & B \end{bmatrix}.$$

Then $\left(\begin{bmatrix} A - r_1 I & 0 \\ 0 & A - r_2 I \end{bmatrix}, \begin{bmatrix} B \\ B \end{bmatrix} \right)$ is not stabilizable if and only if there exists a $\mu \in \mathbb{C}_0^+$ and a vector $\xi^T = [\xi_1^T, \ \xi_2^T] \neq 0$ such that

$$\xi^T C(\mu) = 0.$$

This equality is satisfied if and only if the next three equalities hold simultaneously

$$(i) \ (A - (r_1 + \mu)I)^T \xi_1 = 0; \ (ii) \ (A - (r_2 + \mu)I)^T \xi_2 = 0; \ (iii) \ B^T(\xi_1 + \xi_2) = 0.$$

First assume that $\xi_1 = 0$. Then, according to (ii) and (iii), there exists a vector $\xi_2 \neq 0$ such that both $(A - (r_2 + \mu)I)^T \xi_2 = 0$ and $B^T(\xi_2) = 0$, or stated differently, $\xi^T[A - r_2 I - \mu I, B] \neq 0$. This implies (see Theorem 3.20) that $(A - r_2 I, B)$ is not stabilizable. However, as we argued before, stabilizability of (A, B) implies that for any $r > 0$ $(A - rI, B)$ is also stabilizable. Therefore our assumption on ξ_1 must be wrong and, thus, $\xi_1 \neq 0$. In a similar way one can show that also $\xi_2 \neq 0$. Since both ξ_1 and ξ_2 differ from zero, we conclude from the equalities (i) and (ii) that μ must be an eigenvalue of both $A - r_1 I$ and $A - r_2 I$. Furthermore equality (iii) states that with these choices of $\xi_j \in N((A - (r_j + \mu)I)^T)$, $j = 1, 2$, the sum of these vectors must belong to $N(B^T)$. Which completes the proof. $\square$

Proof of Theorem 3.26

According to equation (3.7.27), $\bar{J} \geq \underline{J}$. So, what is left to be shown is that under the stated assumption $\bar{J} \leq \underline{J}$. Obviously,

$$\bar{J} = \inf_{g} \sup_{w} J(g, w) \leq \sup_{w} J(\bar{g}, w). \tag{3.11.1}$$

Since $J(\bar{g}, w) \le J(\bar{g}, \bar{w})$ we next conclude that

$$\sup_{w} J(\bar{g}, w) \le \sup_{w} J(\bar{g}, \bar{w}) = J(\bar{g}, \bar{w}). \tag{3.11.2}$$

Similarly, since $J(\bar{g}, \bar{w}) \le J(g, \bar{w})$ for all g,

$$J(\bar{g}, \bar{w}) = \inf_{g} J(\bar{g}, \bar{w}) \le \inf_{g} J(g, \bar{w}). \tag{3.11.3}$$

Notice that

$$\inf_{g} J(g, \bar{w}) \le \sup_{w} \inf_{g} J(g, w) = \underline{J}. \tag{3.11.4}$$

The conclusion now follows directly by lining up the results (3.11.1)–(3.11.4). That is,

$$\bar{J} \le \sup_{w} J(\bar{g}, w) \le J(\bar{g}, \bar{w}) \le \inf_{g} J(g, \bar{w}) \le \underline{J}. \qquad \square$$

4

Optimization techniques

In an optimal control problem an optimality criterion is given which assigns a certain number to each evolution of the underlying dynamic system. The problem is then to find an admissible control function which maximizes/minimizes the optimality criterion in the class of all admissible control functions. In this chapter we present on the one hand some less well-known optimization techniques that will be used later on to analyze optimal control problems. On the other hand, we present the basic theorems that are available as tools for solving optimal control problems.

Section 4.1 considers some non-standard theory concerning differentiation of functions. The Gateaux and Fréchet differentials are introduced and it is shown how these differentials can be used to calculate the matrix derivative of a matrix valued function. In section 4.2 the unconstrained optimal control problem is introduced and the first-order necessary condition, known as the Euler–Lagrange equation, is derived. We use the so-called variational approach to establish this equation. Furthermore both a second-order neccessary condition for the existence of an optimal solution and a second-order sufficient conditon for the existence of an optimal solution are provided.

Section 4.3 deals with the well-known maximum principle of Pontryagin. This is followed in section 4.4 by a discussion of the dynamic programming principle.

4.1 Optimization of functions

We assume that the reader is familiar with the notion of a derivative – a basic notion which is very helpful in the analysis of functions $f : \mathbb{R}^n \to \mathbb{R}$. The next section considers functions of matrices $f : \mathbb{R}^{n \times m} \to \mathbb{R}$ and in particular analyzes its extremal locations. Although one can consider these functions as functions from $\mathbb{R}^{nm} \to \mathbb{R}$ and therefore use the classical notion of a derivative to determine extremum locations, it turns out that the extremum behavior of these functions can be analyzed much more simply using the notion of the Gateaux differential. This notion has been developed to analyze functions

$f : X \rightarrow Y$, where X is a vector space and Y a normed space[1]. So X may represent, for instance, the set of all $n \times m$ matrices, or the set of continuous functions with the usual definition of addition and scalar multiplication of functions. We will recall in this section some basic facts on the Gateaux differential which will be used in the next chapter. More details and references on this subject can be found in Luenberger (1969). The Gateaux differential generalizes the concept of a directional derivative. Its formal definition is as follows.

Definition 4.1

Assume $f : X \rightarrow Y$. Let $x \in X$ and let h be arbitrary in X. If the limit

$$\lim_{\alpha \to 0} \frac{1}{\alpha} [f(x + \alpha h) - f(x)] \tag{4.1.1}$$

exists, this limit is called the **Gateaux differential** of f at x with increment h and is denoted by $\delta f(x, h)$. If this limit (4.1.1) exists for each $h \in X$, the mapping f is called **Gateaux differentiable** at x. □

If $Y = \mathbb{R}$ the Gateaux differential $\delta f(x, h)$, if it exists, coincides with the derivative of $f(x + \alpha h)$ with respect to α at $\alpha = 0$, that is

$$\delta f(x, h) = \frac{df(x + \alpha h)}{d\alpha} \text{ at } \alpha = 0. \tag{4.1.2}$$

Example 4.1

Consider $f(x) = x^T P x$. Then the Gateaux differential of f at x with increment Δx is

$$\lim_{\alpha \to 0} \frac{1}{\alpha} [f(x + \alpha \Delta x) - f(x)] = \lim_{\alpha \to 0} \frac{1}{\alpha} [(x + \alpha \Delta x)^T P(x + \alpha \Delta x) - x^T P x]$$

$$= \lim_{\alpha \to 0} \frac{1}{\alpha} [\alpha (\Delta x)^T P x + \alpha x^T P \Delta x + \alpha^2 (\Delta x)^T P \Delta x]$$

$$= (\Delta x)^T P x + x^T P \Delta x$$

$$= 2x^T P \Delta x.$$

By taking Δx equal to the ith standard basis vector in $\mathbb{R}^n$ we obtain that the Gateaux differential of f is $2x^T P$. □

An immediate consequence is the next first-order necessary condition for a function that has an extremum at some 'point' x_0 where the function is Gateaux differentiable.

---

[1]A normed space is a vector space where the notion of the length of a vector has been defined. $\mathbb{R}^n$, where the length of each vector is defined by its Euclidean length, is an example of a normed space. For a formal definition see Lancaster and Tismenetsky (1985).

Theorem 4.1

Assume $f : X \to \mathbb{R}$ has a Gateaux differential for all $x \in X$. Then, if f has an extremum at x_0,

$$\delta f(x_0, h) = 0, \text{ for all } h \in X.$$

Proof

For every $h \in X$, the real valued function $F(\alpha) := f(x_0 + \alpha h)$ is a function of the scalar variable α. Since F has an extremum at $\alpha = 0$, its derivative $F'(0)$ must be zero. However, according to equation (4.1.2)

$$F'(0) = \frac{df(x + \alpha h)}{d\alpha}\big|_{\alpha=0} = \delta f(x, h). \qquad \square$$

The existence of the Gateaux differential is a rather weak requirement. Its existence does not, in general, imply continuity of the map f. On the other hand, if X is also a normed space we can introduce the notion of Fréchet differentiability, which coincides in $\mathbb{R}^n$ with the usual definition of differentiability. In Theorem 4.2 we will show that if f is Fréchet differentiable, then this derivative coincides with the Gateaux differential. Therefore, in case f is Fréchet differentiable, one can use the Gateaux differential to calculate this derivative. Anticipating this result we use the same notation for the Fréchet derivative in the next definition.

Definition 4.2

Let $f : X \to Y$, where both X and Y are normed spaces. Consider $x_0 \in X$. Then, if there is a linear and continuous map $\delta f(x_0, h)$ in h such that

$$\lim_{|h| \to 0} \frac{|f(x_0 + h) - f(x_0) - \delta f(x_0, h)|}{|h|} = 0, \qquad (4.1.3)$$

for all $h \in X$, f is said to be **Fréchet differentiable** at x_0. $\delta f(x_0, h)$ is called the **Fréchet differential** of f at x_0 with increment h. Since, by definition, the Fréchet differential $\delta f(x_0, h)$ is of the form $A_{x_0} h$, the correspondence $x \to A_x$ defines a transformation from X into the normed linear space of all bounded linear operators from X to Y. This transformation is called the **Fréchet derivative** of f, and is denoted by ∂f. Partial derivatives and differentials are denoted by ∂_i and δ_i, respectively, where the index refers to the corresponding argument. $\qquad \square$

Theorem 4.2

If the Fréchet differential of f exists at x_0, then the Gateaux differential exists at x_0 and they coincide.

Proof

By definition (see equation (4.1.3)) the Fréchet differential $\delta f(x_0, h)$ satisfies for all $h \in X$

$$\lim_{|h| \to 0} \frac{|f(x_0 + h) - f(x_0) - \delta f(x_0, h)|}{|h|} = 0.$$

In particular the above limit holds for all $h = \alpha v$, where the scalar $\alpha \to 0$ and $0 \neq v \in X$. That is,

$$\lim_{|\alpha| \to 0} \frac{|f(x_0 + \alpha v) - f(x_0) - \delta f(x_0, \alpha v)|}{|\alpha v|} = 0.$$

Or, using the linearity of $\delta f(x_0, \alpha v)$,

$$\lim_{|\alpha| \to 0} \frac{|f(x_0 + \alpha v) - f(x_0) - \alpha \delta A_{x_0} v|}{|\alpha v|} = 0.$$

So $\lim_{\alpha \to 0} \frac{1}{\alpha} [f(x_0 + \alpha v) - f(x_0)]$ exists and coincides with $\delta f(x_0, v)$. $\qquad\square$

Much of the theory of ordinary derivatives can be generalized to Fréchet derivatives. In particular we have the following generalization. Its proof is similar to the ordinary case and is therefore omitted.

Theorem 4.3

Assume f and g are Fréchet differentiable on X. Then for any scalars $\lambda, \mu \in \mathbb{R}$ the functions $\lambda f(x) + \mu g(x)$ and $f(g(x))$ are also Fréchet differentiable. Moreover,

$$\partial(\lambda f + \mu g)(x) = \lambda \partial f(x) + \mu \partial g(x)$$
$$\partial f(g(x)) = \partial f(g(x)) \partial g(x).$$

$\qquad\square$

Example 4.2

Consider the matrix function $f : \mathbb{R}^{n \times m} \to \mathbb{R}^{n \times n}$ defined by

$$f(F, P(F)) = (A + BF)^T P + P(A + BF) + Q + F^T RF,$$

where $P(F)$ is a Fréchet differentiable function of F, and all matrices A, B, Q, R and F have appropriate dimensions. Then according the chain rule in Theorem 4.3

$$\partial f(F, P(F)) = \partial_1 f(F, P(F)) + \partial_2 f(F, P(F)) \partial P(F).$$

To determine the partial derivatives $\partial_i f(F, P(F))$ we consider the corresponding Gateaux differentials (see Theorem 4.2). To that end let ΔF and ΔP be arbitrary perturbation matrices of F and P, respectively. Then,

$$
\begin{aligned}
\delta_1 f(F, P(F), \Delta F) &= \lim_{\alpha \to 0} \frac{f(F + \alpha \Delta F, P) - f(F, P)}{\alpha} \\
&= \lim_{\alpha \to 0} \frac{\alpha \Delta F^T (RF + B^T P) + \alpha (F^T R + PB) \Delta F + \alpha^2 \Delta F^T R \Delta F}{\alpha} \\
&= \Delta F^T (B^T P + RF) + (PB + F^T R) \Delta F.
\end{aligned}
$$

Whereas

$$
\begin{aligned}
\delta_2 f(F, P(F), \Delta P) &= \lim_{\alpha \to 0} \frac{f(F, P + \alpha \Delta P) - f(F, P)}{\alpha} \\
&= (A + BF)^T \Delta P + \Delta P (A + BF). \qquad \square
\end{aligned}
$$

4.2 The Euler–Lagrange equation

In this section we derive the first-order necessary conditions for the basic dynamic optimization problem to find a control function $u(.)$ that minimizes the cost functional

$$
J(x_0, u) := \int_0^T g(t, x(t), u(t)) dt + h(x(T)) \tag{4.2.1}
$$

where the state variable $x(t)$ satisfies the differential equation:

$$
\dot{x}(t) = f(t, x(t), u(t)), \quad x(0) = x_0. \tag{4.2.2}
$$

Here $x(.) \in \mathbb{R}^n$, $u(.) \in \mathbb{R}^m$ and x_0 is a given vector in $\mathbb{R}^n$. It is assumed that the dynamic process starts from $t = 0$ and ends at the fixed terminal time $T > 0$. Throughout this chapter the following assumptions on the functions f, g and h are made.

Basic Assumptions A

(i) $f(t, x, u)$ and $g(t, x, u)$ are continuous functions on $\mathbb{R}^{1+n+m}$. Moreover, for both f and g all partial derivatives w.r.t. x and u exist and are continuous.

(ii) $h(x) \in C^1$. $\qquad \square$

Moreover we assume in this section that the set of admissible control functions, $\mathcal{U}$, consists of the set of functions that are continuous on $[0, T]$.

Note

1. Since f and g are continuous in u and $u(.)$ is a continuous function, $f_u(t, x) := f(t, x, u(t))$ and $g_u(t, x) := g(t, x, u(t))$ are continuous functions too. So, according to

Theorem 3.8 the differential equation (4.2.2) has for every $u \in \mathcal{U}$ locally a unique solution.

2. If the solution $x(t)$ exists on the whole interval $[0, T]$, $g_u(t, x(t))$ is continuous too (since $g_u(t, x)$ and $x(.)$ are continuous). So, in that case the cost functional equation (4.2.1) is well-defined too. □

Inspired by the theory of static optimization we introduce for each t in the interval $[0, T]$ the quantity $\lambda(t)[f(t, x(t), u(t)) - \dot{x}(t)]$, where the Lagrange multiplier $\lambda(t)$ (also called **costate** variable) is an arbitrarily chosen row vector. Since $f(t, x(t), u(t)) - \dot{x}(t) = 0$ for every t, in particular

$$\int_0^T \lambda(t)[f(t, x(t), u(t)) - \dot{x}(t)]dt = 0. \tag{4.2.3}$$

We can append this quantity (4.2.3) to the cost function (4.2.1) to obtain a cost function $\bar{J}$ which coincides with the original cost function J if the dynamical constraint of equation (4.2.2) is satisfied. That is,

$$\bar{J} := \int_0^T \{g(t, x, u) + \lambda(t)f(t, x, u) - \lambda(t)\dot{x}(t)\}dt + h(x(T)).$$

Introducing the **Hamiltonian function** H as

$$H(t, x, u, \lambda) := g(t, x, u) + \lambda(t)f(t, x, u), \tag{4.2.4}$$

we can rewrite $\bar{J}$ as

$$\bar{J} = \int_0^T \{H(t, x, u, \lambda) - \lambda(t)\dot{x}(t)\}dt + h(x(T)). \tag{4.2.5}$$

Integration by parts shows that

$$-\int_0^T \lambda(t)\dot{x}(t)dt = -\lambda(T)x(T) + \lambda(0)x_0 + \int_0^T \dot{\lambda}(t)x(t)dt. \tag{4.2.6}$$

Hence, substituting this result into equation (4.2.5), $\bar{J}$ can be further rewritten as

$$\bar{J} = \underbrace{\int_0^T \{H(t, x, u, \lambda) + \dot{\lambda}(t)x(t)\}dt}_{\bar{J}_1} + \underbrace{h(x(T)) - \lambda(T)x(T)}_{\bar{J}_2} + \underbrace{\lambda(0)x_0}_{\bar{J}_3}. \tag{4.2.7}$$

This expression (4.2.7) of $\bar{J}$ has three additive terms, $\bar{J}_1$, $\bar{J}_2$ and $\bar{J}_3$. The first term, $\bar{J}_1$, concerns the whole planning period $[0, T]$, the $\bar{J}_2$ term deals exclusively with the final time, and the third term, $\bar{J}_3$, deals only with the initial time.

Once again, we stress the fact that the choice of the $\lambda(t)$ path will have no effect on the value of $\bar{J}$, as long as the state $x(t)$ satisfies the differential equation $\dot{x}(t) = f(t, x(t), u(t))$, or stated differently,

$$\dot{x}(t) = \frac{\partial H}{\partial \lambda} \text{ for all } t \in [0, T]. \tag{4.2.8}$$

To relieve us from further worries about the effect of $\lambda(t)$ on $\bar{J}$, we simply impose equation (4.2.8) as a necessary condition for the maximization of $\bar{J}$.

Next assume that $u^*(t) \in \mathcal{U}$ is an optimal control path generating the minimum value of J and $x^*(t)$ is the corresponding optimal state trajectory. If we perturb this optimal $u^*(t)$ path with a continuous perturbing curve $p(t)$, we can generate 'neighboring' control paths

$$u(t) = u^*(t) + \epsilon p(t) \tag{4.2.9}$$

where ϵ is a 'small' scalar. From Theorem 3.14 we know that for a 'small enough' ϵ, each corresponding neighboring control path induces a corresponding 'neighboring' state trajectory $x(t, \epsilon)$ which is also defined on the whole interval $[0, T]$. The cost induced by this control function now also becomes a function of the scalar ϵ too. From equation (4.2.7) we have

$$\bar{J}(\epsilon) = \int_0^T \{ H(t, x(t, \epsilon, p), u^* + \epsilon p, \lambda) + \dot{\lambda}(t) x(t, \epsilon, p) \} dt + h(x(T, \epsilon, p)) - \lambda(T) x(T, \epsilon, p)$$
$$+ \lambda(0) x_0. \tag{4.2.10}$$

By assumption $\bar{J}(\epsilon)$ has a minimum at $\epsilon = 0$. Notice that due to our differentiability assumptions on f, g and h, $\bar{J}(\epsilon)$ is a differentiable function w.r.t. ϵ. So the first-order condition implies that $\frac{d\bar{J}(\epsilon)}{d\epsilon} = 0$ at $\epsilon = 0$. Evaluating the derivative of $\bar{J}(\epsilon)$ yields

$$\frac{d\bar{J}(\epsilon)}{d\epsilon} = \int_0^T \left\{ \frac{\partial H}{\partial x} \frac{dx(t, \epsilon, p)}{d\epsilon} + \frac{\partial H}{\partial u} p(t) + \dot{\lambda}(t) \frac{dx(t, \epsilon, p)}{d\epsilon} \right\} dt \tag{4.2.11}$$
$$+ \frac{\partial h(x(T))}{\partial x} \frac{dx(T, \epsilon, p)}{d\epsilon} - \lambda(T) \frac{dx(T, \epsilon, p)}{d\epsilon}.$$

Notice that $\lambda(t)$ is arbitrary so far. Therefore the equality $\frac{d\bar{J}(\epsilon)}{d\epsilon} = 0$ at $\epsilon = 0$ should in particular hold if we choose $\lambda(t)$ as the solution of the following boundary-value problem

$$\frac{\partial H(t, x^*, u^*, \lambda)}{\partial x} + \dot{\lambda} = 0, \text{ with } \frac{\partial h(x^*(T))}{\partial x} - \lambda(T) = 0.$$

Since both the partial derivatives f_x and g_x are continuous they are in particular integrable on the interval $[0, T]$. So, according to Theorem 3.5, this linear differential equation has a solution. Choosing $\lambda(t)$ in this way, from equation (4.2.11) $\frac{d\bar{J}(\epsilon)}{d\epsilon} = 0$ at $\epsilon = 0$ if and only if

$$\int_0^T \frac{\partial H(t, x^*, u^*, \lambda)}{\partial u} p(t) dt = 0.$$

This equality should hold for any continuous function $p(t)$ on $[0, T]$. Choosing $p(t) = \frac{\partial H^T(t, x^*, u^*, \lambda)}{\partial u}$ shows that another necessary condition is that

$$\frac{\partial H(t, x^*, u^*, \lambda)}{\partial u} = 0.$$

So, we reach the next theorem.

Theorem 4.4 (Euler–Lagrange)

Consider the optimization problem given by equations (4.2.1) and (4.2.2). Let $H(t, x, u, \lambda) := g(t, x, u) + \lambda f(t, x, u)$. Assume that the functions f, g and h satisfy the Basic Assumptions A. If $u^*(t) \in \mathcal{U}$ is a control that yields a local minimum for the cost functional (4.2.1), and $x^*(t)$ and $\lambda^*(t)$ are the corresponding state and costate, then it is necessary that

$$\dot{x}^*(t) = f(t, x^*, u^*) \left(= \frac{\partial H(t, x^*, u^*, \lambda)}{\partial \lambda} \right), \quad x^*(0) = x_0; \quad (4.2.12)$$

$$\dot{\lambda}^*(t) = -\frac{\partial H(t, x^*, u^*, \lambda^*)}{\partial x}; \quad \lambda^*(T) = \frac{\partial h(x^*(T))}{\partial x} \quad (4.2.13)$$

and, for all $t \in [0, T]$,

$$\frac{\partial H(t, x^*, u^*, \lambda^*)}{\partial u} = 0. \quad (4.2.14)$$

□

Notice that the dynamic optimization problem has, in equation (4.2.14), been reduced to a static optimization problem which should hold at every single instant of time. The optimal solutions of the instantaneous problems give the optimal solution to the overall problem.

Equations (4.2.12) and (4.2.13) constitute $2n$ differential equations with n boundary conditions for x^* specified at $t = 0$ and n boundary conditions for λ^* specified at $t = T$. This is referred to as a **two-point boundary-value problem**. In principle, the dependence on u^* can be removed by solving u^* as a function of x^* and λ^* from the m algebraic equations (4.2.14). In practice, however, equation (4.2.14) can rarely be solved analytically. Even if it is possible, the resulting two-point boundary-value problem is likely to be difficult to solve. The next examples demonstrate the use of Theorem 4.4, together with its pitfalls.

Example 4.3

Consider the minimization of

$$\int_0^5 \left\{ x(t) - \frac{1}{2} u^2(t) \right\} dt \quad (4.2.15)$$

subject to

$$\dot{x}(t) = u(t), \quad x(0) = 1. \quad (4.2.16)$$

The Hamiltonian function for this problem is

$$H(t, x, u, \lambda) := x - \frac{1}{2}u^2 + \lambda u. \tag{4.2.17}$$

If there exists a continuous optimal control path $u^*(.)$ and a corresponding state trajectory $x^*(.)$ then, by Theorem 4.4, necessarily there exists a costate trajectory $\lambda^*(.)$ satisfying the differential equation

$$\dot{\lambda} = -1, \ \lambda(T) = 0. \tag{4.2.18}$$

It is easily verified that the solution of this differential equation (4.2.18) is

$$\lambda(t) = -t + 5. \tag{4.2.19}$$

Moreover, by straightforward differentiation of the Hamiltonian w.r.t. u, it is seen that Theorem 4.4 dictates that the candidate optimal control path $u^*(.)$ satisfies

$$-u + \lambda = 0. \tag{4.2.20}$$

Combining equations (4.2.20) and (4.2.19) one obtains that the candidate optimal control path satisfies

$$u^*(t) = \lambda^*(t) = -t + 5. \tag{4.2.21}$$

Moreover, elementary calculations show that, after substitution of this result into equation (4.2.16) the corresponding candidate optimal state trajectory is given by

$$x^*(t) = -\frac{1}{2}t^2 + 5t + 1.$$

So, if our minimization problem has a solution, then the optimal control trajectory is given by equation (4.2.21). Therefore, all we need to verify is whether this solution solves the problem (4.2.15). However, taking a closer look at our problem statement, we see that if we take u arbitrarily negative, the state also becomes negative, yielding a cost which is definitely smaller than the cost resulting from our candidate optimal solution (4.2.21). That is, the problem has no solution. This shows that one has to be careful in using theorems like Theorem 4.4. $\qquad \square$

Example 4.4

Consider the problem to minimize

$$\int_0^T \{x^2(t) + u^2(t)\}dt + ex^2(T) \tag{4.2.22}$$

subject to

$$\dot{x}(t) = ax(t) + bu(t), \ x(0) = 1 \ (b \neq 0). \tag{4.2.23}$$

In this problem, the state variable $x(t)$ might be interpreted as the deviation of some quantity from a desired level. If no effort is made, $u(t) = 0$, the variable $x(t)$ will grow (or decline) at a rate a if $a > 0$ (or $a < 0$). The variable $x(t)$ can be brought closer to zero by choosing a nonzero control $u(t)$, but only at some cost. The criterion (4.2.22) expresses a balance between the cost of having a nonzero deviation $x(t)$ and the cost of the effort required to make the deviation smaller. Finally, there is some additional cost involved in case the quantity at the end of the planning horizon still differs from its desired level.

In view of the positive initial condition $x(0) = 1$, an optimal control function $u(t)$ is likely to be negative. Moreover, when the value of $x(t)$ gets close to zero, one expects that the control effort will be decreased as well. When the state $x(t)$ reaches zero it is clearly optimal to let $u(t) = 0$ from that time on, since this policy will keep the deviation at zero while no control effort needs to be delivered.

Now we use Theorem 4.4 to determine the optimal control path. With

$$f = ax + bu, \ g = x^2 + u^2 \text{ and } h = ex^2$$

the Hamiltonian for this problem is

$$H = x^2 + u^2 + \lambda(ax + bu).$$

So, by Theorem 4.4, if $u^*(.)$ is the optimal control path, and $x^*(.)$ the corresponding optimal state trajectory then it is necessary that there exists a costate trajectory $\lambda^*(.)$ such that

$$\dot{x}^*(t) = ax^*(t) + bu^*(t), \ x^*(0) = 1 \tag{4.2.24}$$

$$\dot{\lambda}^*(t) = -(2x^*(t) + a\lambda^*(t)), \ \lambda^*(T) = 2ex^*(T) \tag{4.2.25}$$

$$2u^*(t) + b\lambda^*(t) = 0. \tag{4.2.26}$$

From equation (4.2.26) we obtain

$$u^*(t) = -\frac{1}{2}b\lambda^*(t). \tag{4.2.27}$$

Substitution of equation (4.2.27) into equation (4.2.24) shows that x^* and λ^* should satisfy the following two-point boundary-value problem

$$\begin{bmatrix} \dot{x}^*(t) \\ \dot{\lambda}^*(t) \end{bmatrix} = \begin{bmatrix} a & -\frac{1}{2}b^2 \\ -2 & -a \end{bmatrix} \begin{bmatrix} x^*(t) \\ \lambda^*(t) \end{bmatrix}, \ x^*(0) = 1; \ \lambda^*(T) - 2ex^*(T) = 0. \tag{4.2.28}$$

By solving this two-point boundary-value problem and next substituting the resulting $\lambda^*(t)$ into equation (4.2.27) one can then calculate the candidate optimal control trajectory. In fact one can solve this two-point boundary-value problem analytically. However, this requires going through a number of analytical calculations which do not provide much additional insight into the problem (see for example Kwakernaak and Sivan, 1972). Since we will solve this problem later on using the principle of dynamic programming, which yields more insight into the problem, a more detailed analysis of the solution is postponed until later.   □

Example 4.5

Consider a monopolistic firm that produces a single commodity with a total cost function

$$C(t) = C(q(t)),$$

where the output $q(t)$ at time t is always set equal to the quantity demanded. The quantity demanded is assumed to depend not only on its price $p(t)$, but also on the rate of its change $\dot{p}(t)$, that is $q(t) = q(p(t), \dot{p}(t))$. This price change, $\dot{p}(t)$, is completely determined by the monopolist and, therefore, can be viewed as the control instrument of the monopolist. Since, obviously, $\dot{p}(t)$ completely determines the future price, given some initial price p_0 at time $t = 0$, the dynamics of the price process are described by the differential equation

$$\dot{p}(t) = u(t), \ p(0) = p_0. \tag{4.2.29}$$

The firm's profit at time t, $\pi(t)$, is its revenues minus its cost at that time, i.e.

$$\pi(t) = p(t)q(t) - C(q(t)).$$

Now assume that the objective of the firm is to find an optimal price path, $p(.)$, that maximizes the total profit over a finite time period $[0, T]$. This period is assumed to be not too long in order to justify the assumptions of fixed demand and cost functions, as well as the omission of a discount factor. The objective of the monopolist is therefore to maximize

$$\Pi = \int_0^T \pi(t)dt.$$

Or, equivalently, introducing $J(p_0, u) = -\Pi$, the problem is to find a control function $u(.)$ that minimizes the cost functional

$$J(p_0, u) = \int_0^T C(q(p(t), u(t))) - p(t)q(p(t), u(t))dt \tag{4.2.30}$$

where the state variable $p(t)$ satisfies the differential equation (4.2.29). Assuming that both the cost and demand functions are sufficiently smooth we can now use Theorem 4.4 to determine the optimal price path. With

$$f = u, \ g = \pi(p, u), \ \text{and } h = 0$$

we conclude from this theorem that if $u^*(.)$ is the optimal control path, and $p^*(.)$ the corresponding optimal price path then it is necessary that there exists a costate path $\lambda^*(.)$ such that

$$\dot{p}^*(t) = u^*(t), \ p^*(0) = p_0 \tag{4.2.31}$$

$$\dot{\lambda}^*(t) = -\frac{\partial \pi}{\partial p}(p^*(t), u^*(t)), \ \lambda^*(T) = 0 \tag{4.2.32}$$

$$\frac{\partial \pi}{\partial u}(p^*(t), u^*(t)) + \lambda^*(t) = 0. \tag{4.2.33}$$

According to equation (4.2.33)

$$\lambda^*(t) = -\frac{\partial \pi}{\partial u}(p^*(t), u^*(t)).$$

Substitution of this result into equation (4.2.32) yields

$$\frac{d}{dt}\left\{\frac{\partial \pi}{\partial u}(p^*(t), u^*(t))\right\} = \frac{\partial \pi}{\partial p}(p^*(t), u^*(t)), \quad \frac{\partial \pi}{\partial u}(p^*(T), u^*(T)) = 0. \tag{4.2.34}$$

Equations (4.2.31)–(4.2.34) constitute a two-point boundary-value problem that can be solved by substitution of u^* from equation (4.2.31) into equation (4.2.34). This yields then an ordinary second-order differential equation in the price p which has to be solved under the conditions that $p(0) = p_0$ and $p(T), \dot{p}(T)$ satisfy the boundary condition implied by equation (4.2.34).

In this case Theorem 4.4 is also useful to develop some economic theoretical insights into the problem. To that end we proceed by elaborating the left-hand side of equation (4.2.34). Note that

$$\frac{d}{dt}\left\{\frac{\partial \pi}{\partial u}(p^*(t), u^*(t))\right\} = \frac{\partial^2 \pi}{\partial u \partial p}(p^*(t), u^*(t))\dot{p}^*(t) + \frac{\partial^2 \pi}{\partial u^2}(p^*(t), u^*(t))\dot{u}^*(t).$$

So, equation (4.2.34) can be rewritten as

$$\frac{\partial^2 \pi}{\partial u \partial p}(p^*(t), u^*(t))\dot{p}^*(t) + \frac{\partial^2 \pi}{\partial u^2}(p^*(t), u^*(t))\dot{u}^*(t) - \frac{\partial \pi}{\partial p}(p^*(t), u^*(t)) = 0.$$

Multiplying this equation through by $u^*(t)$ we get

$$u^*(t)\left[\frac{\partial^2 \pi}{\partial u \partial p}(p^*(t), u^*(t))\dot{p}^*(t) + \frac{\partial^2 \pi}{\partial u^2}(p^*(t), u^*(t))\dot{u}^*(t) - \frac{\partial \pi}{\partial p}(p^*(t), u^*(t))\right] = 0.$$

$$\tag{4.2.35}$$

Next consider the expression $y(t) := u^*(t)\frac{\partial \pi}{\partial u}(p^*(t), u^*(t)) - \pi(p^*(t), u^*(t))$. By straightforward differentiation and using equation (4.2.31) we see that its derivative, $\frac{dy}{dt}(t)$, coincides with the left-hand side of equation (4.2.35). Consequently, the solution of equation (4.2.35) is $y(t) = $ constant. So a necessary optimality condition is that

$$\pi(p^*(t), u^*(t)) - u^*(t)\frac{\partial \pi}{\partial u}(p^*(t), u^*(t)) = c,$$

where c is some constant. Or, after some rewriting,

$$\frac{\partial \pi}{\partial u}(p^*(t), u^*(t))\frac{u^*(t)}{\pi(p^*(t), u^*(t))} = 1 - \frac{c}{\pi(p^*(t), u^*(t))}. \tag{4.2.36}$$

Economists will recognize on the left-hand side of equation (4.2.36) the partial elasticity of the profit with respect to the rate of change of the price. In fact the constant c also has an economic interpretation. It represents the monopoly profit in case the profit function does not depend on the rate of change of the price. For a further discussion of this issue we refer to Chiang (1992). □

We conclude this section by considering the second-order derivative of $\bar{J}(\epsilon)$. Notice that a solution of equations (4.2.12)–(4.2.14) will only be a minimum if $\frac{d^2\bar{J}}{d\epsilon^2} \geq 0$ at $\epsilon = 0$. On the other hand, if $\frac{d^2\bar{J}}{d\epsilon^2} > 0$ at $\epsilon = 0$ equations (4.2.12)–(4.2.14) will certainly provide a minimum. Assuming that $f, g \in C^2$, differentiation of equation (4.2.11) shows that the second-order derivative of $\bar{J}(\epsilon)$ equals

$$
\frac{d^2\bar{J}(\epsilon)}{d\epsilon^2}(\epsilon) = \int_0^T \left\{ \frac{dx^T(t,\epsilon,p)}{d\epsilon} \frac{\partial^2 H}{\partial x^2} \frac{dx(t,\epsilon,p)}{d\epsilon} + 2 \frac{dx^T(t,\epsilon,p)}{d\epsilon} \frac{\partial^2 H}{\partial u \partial x} p(t) + p^T(t) \frac{\partial^2 H}{\partial u^2} p(t) \right.
$$
$$
\left. + \left(\frac{\partial H}{\partial x} + \dot{\lambda}(t) \right) \frac{d^2 x(t,\epsilon,p)}{d\epsilon^2} \right\} dt
$$
$$
+ \frac{dx^T(T,\epsilon,p)}{d\epsilon} \frac{\partial^2 h(x(T))}{\partial x^2} \frac{dx(T,\epsilon,p)}{d\epsilon} + \frac{\partial h(x(T))}{\partial x} \frac{d^2 x(T,\epsilon,p)}{d\epsilon^2} - \lambda(T) \frac{d^2 x(T,\epsilon,p)}{d\epsilon^2}.
$$

Using the first-order conditions of equations (4.2.13) and (4.2.14) and introducing the notation $q(t,p) := \frac{dx(t,\epsilon,p)}{d\epsilon}$ at $\epsilon = 0$, $\frac{d^2\bar{J}(0)}{d\epsilon^2}$ can be rewritten as

$$
V(x^*,p) := \int_0^T [q(t,p), p^T(t)] H'' \begin{bmatrix} q(t,p) \\ p(t) \end{bmatrix} dt + q^T(T,p) \frac{\partial^2 h(x^*(T))}{\partial x^2} q(T,p), \quad (4.2.37)
$$

where H'' is the matrix obtained from the second-order derivative of the hamiltonian $H(t,x^*,u^*,\lambda^*)$ as $H'' = \begin{bmatrix} H_{xx} & H_{xu} \\ H_{ux} & H_{uu} \end{bmatrix}$. So, we have the following result.

Lemma 4.5

Consider $V(x^*,p)$ in equation (4.2.37). Then,

(i) if $u^*(t)$ provides a minimum for the optimal control problem (4.2.1) and (4.2.2) then $V(x^*,p) \geq 0$ for all functions $p(t) \in \mathcal{U}$;

(ii) if $V(x^*,p) > 0$ for all functions $p(t) \in \mathcal{U}$ and the conditions of Theorem 4.4 are satisfied, $u^*(t)$ is a control path that yields a local minimum for the cost function (4.2.1). □

It will be clear that usually the conditions of this lemma are hard to verify. However, for instance in the linear quadratic case, one can derive some additional insights from it.

4.3 Pontryagin's maximum principle

In the previous section it was assumed that the control functions are continuous. In applications, this assumption is, however, often too restrictive. Therefore, a more general problem setting is called for. In section 4.2, at each point in time t $u(t)$ could (in principle) be chosen arbitrarily in $\mathbb{R}^m$ and, moreover, $u(.)$ was assumed to be continuous. We now assume that there is a subset $U \subset \mathbb{R}^m$ such that for all $t \in [0, T]$, $u(t) \in U$. So, $u(t)$ may not be chosen arbitrarily anymore but is restricted to the set U. We do not make any further assumptions here on this set U. So, in particular it may be $\mathbb{R}^m$. Additional to our Basic Assumptions A (see section 4.2), we assume that the set of admissible control functions $\mathcal{U}$ now consists of the set of measurable functions from $[0, T]$ into U for which the differential equation (4.2.2) has a unique solution (in the extended sense) on $[0, T]$ and the cost function (4.2.1) exists. That is,

$\mathcal{U} := \{u(.) \mid u(t) \in U, \forall t \in [0, T]; u(.)$ is measurable on $[0, T]$; $x(t, u(t))$ is an abso-lutely continuous function which satisfies equation (4.2.2) on $[0, T]$ except on a set S which has a Lebesgue measure zero; and equation (4.2.1) exists. $\}$

The price we have to pay for considering this more general problem is that its solution requires a more subtle analysis. This is because the problem easily gets ill-defined due to the fact that either the differential equation (4.2.2) does not have a unique solution or the cost function (4.2.1) does not exist. By restricting the set of admissible control functions from the outset to those control functions for which the problem setting does make sense, we partly replace the analytical difficulties by the problem to verify whether some pre-specified control functions are admissible. Below we present some additional assumptions which make it possible to avoid some of the above mentioned existence problems.

Note

1. Since f and g are continuous in u, $f_u(t, x) := f(t, x, u(t))$ and $g_u(t, x) := g(t, x, u(t))$ are also measurable functions. As a consequence the cost function (4.2.1) exists if one also assumes that U is bounded.

2. By Theorem 3.13 the differential equation (4.2.2) has a local unique solution for all measurable $u(.)$ for which $f_u(t, x) \in CD$.

3. If f satisfies the inequality

$$|f(t, x, u)| \le L|x| + N, \ t \in [0, T], \ x \in \mathbb{R}^n, \ u \in U,$$

for some positive constants L and N, the differential equation (4.2.2) has a unique solution for all measurable functions u (see for example Lee and Markus (1967)). $\qquad\square$

Then we have the next generalization of the Euler–Lagrange Theorem 4.4, the so-called maximum principle of Pontryagin.

Theorem 4.6 (Pontryagin)

Consider the optimization problem given by equations (4.2.1) and (4.2.2). Assume the functions f, g and h satisfy the Basic Assumptions A and introduce $H := g + \lambda f$. Let $u^*(t) \in \mathcal{U}$ be a control that yields a local minimum for the cost functional (4.2.1), and let $x^*(t)$ be the corresponding state trajectory. Then there exists a costate function $[\lambda^*]^T : [0, T] \to \mathbb{R}^n$ satisfying

$$\dot{x}^*(t) = f(t, x^*, u^*) \left(= \frac{\partial H(t, x^*, u^*, \lambda)}{\partial \lambda} \right), \ x^*(0) = x_0; \qquad (4.3.1)$$

$$\dot{\lambda}^*(t) = -\frac{\partial H(t, x^*, u^*, \lambda^*)}{\partial x}; \ \lambda^*(T) = \frac{\partial h(x^*(T))}{\partial x} \qquad (4.3.2)$$

and, for all $t \in (0, T)$ at which u^* is continuous,

$$H(t, x^*, u^*, \lambda^*) = \min_{u \in U} H(t, x^*, u, \lambda^*), \qquad (4.3.3)$$

that is

$$u^*(t) = \arg\min_{u \in U} H(t, x^*, u, \lambda^*). \qquad \square$$

Notice that, apart from the fact that we consider a larger set of admissible control functions, the main difference between Pontryagin's theorem and the Euler–Lagrange theorem is that the maximum principle tells us that the Hamiltonian is minimized at the optimal trajectory and that it is also applicable when the minimum is attained at the boundary of U.

The proof of Theorem 4.6 is given in the Appendix to this chapter. Below a number of examples are provided to illustrate the principle.

Example 4.6

The first example is rather elementary and illustrates the fact that the optimal control may occur at the boundary of the control set U. Consider the problem to minimize

$$\int_0^T x(t)dt - 2x(T),$$

subject to the state equation

$$\dot{x}(t) = u(t), \quad x(0) = 1 \text{ and } u(t) \in [-1, 1].$$

Notice that with $f := u$, $g := x$ and $h := -2x$, the Basic Assumptions A are satisfied. The Hamiltonian for this system is

$$H(t, x, u, \lambda) := x + \lambda u. \qquad (4.3.4)$$

If there exists an optimal control function $u^*(.)$ for this problem then, by Theorem 4.6, there necessarily exists a costate function $\lambda^*(.)$ satisfying

$$\dot{\lambda}(t) = -1, \quad \lambda(T) = -2. \qquad (4.3.5)$$

Furthermore,

$$u^*(t) = \arg\min_{u(t) \in [-1, 1]} \{x^*(t) + \lambda^*(t)u(t)\}.$$

Since the right-hand side of this equation is linear in $u(t)$, it is easily verified that the optimal control is

$$u^*(t) = \begin{cases} 1 & \text{if } \lambda^*(t) < 0 \\ \text{undefined} & \text{if } \lambda^*(t) = 0 \\ -1 & \text{if } \lambda^*(t) > 0. \end{cases} \qquad (4.3.6)$$

From equations (4.3.5) it is easily verified that $\lambda^*(t) = -t + T - 2$. So, $\lambda^*(t) < 0$ if $T - 2 < t \leq T$ and $\lambda^*(t) > 0$ if $0 \leq t < T - 2$. In particular notice that $\lambda^*(t) < 0$ for all $t \in [0, T]$ if $T < 2$. From equation (4.3.6) we conclude then that

$$\text{if } T \leq 2, u^*(t) = 1$$

$$\text{if } T > 2, u^*(t) = \begin{cases} 1 \text{ if } T - 2 < t \leq T \\ -1 \text{ if } 0 \leq t < T - 2. \end{cases}$$

Notice that if $T > 2$ the optimal control is uniquely defined except at one point in time, $t = T - 2$. Since the specific choice of u at $t = T - 2$ does not influence the cost function, we may choose the control at this point as we like. Since the optimal control switches in this case from one bound of the control set to the other bound of the control set, the control is called a **bang-bang** control. □

Example 4.7

Consider the problem to minimize

$$\int_0^2 (u(t) - 1)x(t)dt, \tag{4.3.7}$$

subject to

$$\dot{x}(t) = u(t); \ x(0) = 0 \text{ and } u(t) \in [0, 1]. \tag{4.3.8}$$

It is easily verified that, with $f := u$, $g := (u - 1)x$ and $h = 0$, the Basic Assumptions A are satisfied. The Hamiltonian for this system is

$$H(t, x, u, \lambda) := (u - 1)x + \lambda u. \tag{4.3.9}$$

So, from Theorem 4.6 we have that if there exists an optimal control function for this problem, then necessarily there exists a costate function $\lambda^*(.)$ satisfying

$$\dot{\lambda}(t) = 1 - u(t); \ \lambda(2) = 0. \tag{4.3.10}$$

Furthermore,

$$u^*(t) = \arg\min_{u(t) \in [0,1]} \{(u(t) - 1)x^*(t) + \lambda^*(t)u(t)\}. \tag{4.3.11}$$

To solve this problem, we first rewrite equation (4.3.11) as

$$u^*(t) = \arg\min_{u(t) \in [0,1]} \{(x^*(t) + \lambda^*(t))u(t) - x^*(t)\}. \tag{4.3.12}$$

From equation (4.3.12) we see that the function which has to be minimized w.r.t. $u(t)$, $(x^*(t) + \lambda^*(t))u(t) - x^*(t)$, is linear in $u(t)$. Since $u(t) \in [0, 1]$ this minimum is attained by either taking $u(t) = 1$ or $u(t) = 0$. This is dependent on which sign the term $x^*(t) + \lambda^*(t)$ has. That is,

$$u^*(t) = \begin{cases} 1 \text{ if } x^*(t) + \lambda^*(t) < 0 \\ 0 \text{ if } x^*(t) + \lambda^*(t) > 0 \end{cases} \qquad (4.3.13)$$

Equations (4.3.8) and (4.3.10) constitute a two-point boundary-value problem. Due to the simple structure of the problem we can solve this problem analytically. To that end consider the function $\lambda^*(t)$ in some more detail. Since $u(t) \in [0, 1]$ it is clear from equation (4.3.10) that $\dot{\lambda}(t) \geq 0$ on $[0, 2]$. Consequently, since $\lambda^*(2) = 0$, $\lambda^*(t) \leq 0$ on $[0, 2]$. Now assume that $\lambda^*(0) = 0$. This implies that $\lambda^*(t) = 0$ for all $t \in [0, 2]$, which is only possible (see equation (4.3.10)) if $u(t) = 1$ for all $t \in [0, 2]$. However, this implies (see equation (4.3.8)) that $x(t) > 0$ on $(0, 2]$ and, thus, also $x(t) + \lambda(t) > 0$ on $(0, 2]$. According to equation (4.3.13) this implies, however, that $u^*(t) = 0$ on $(0, 2]$, which contradicts our previous conclusion that $u^*(t) = 1$ on $(0, 2]$. Therefore, we conclude that $\lambda^*(0) < 0$. Since $x(0) = 0$, consequently, $u^*(t) = 1$ on some interval $t \in [0, t_1]$. Substitution of this result into equations (4.3.8) and (4.3.10) shows that $x^*(t)$ and $\lambda^*(t)$ satisfy

$$x^*(t) = t \text{ and } \lambda^*(t) = \lambda^*(0), \ \forall t \in [0, t_1].$$

The length of this interval, t_1, is determined by the point in time where $x^*(t_1) + \lambda^*(t_1) = 0$. That is, the point in time t_1 where $t_1 + \lambda^*(0) = 0$. From that point in time t_1 onwards, the optimal control is $u^*(t) = 0$. The corresponding state and costate trajectory are (see equations (4.3.8) and (4.3.10) again with $u(.) = 0$)

$$x^*(t) = -\lambda^*(0) \text{ and } \lambda^*(t) = 2\lambda^*(0) + t, \ \forall t \in [t_1, t_2].$$

Notice that, since $\lambda^*(.)$ is increasing and $x^*(.)$ remains constant, $\lambda^*(t) + x^*(t)$ will increase from $t = t_1$ on. In particular this implies that $\lambda^*(t) + x^*(t) > 0$ for all $t > t_1$ and, thus, $u^*(t) = 0$ on $(t_1, 2]$. Finally, due to the fact that $\lambda^*(2) = 0$, we infer that $0 = 2\lambda^*(0) + 2$. So, $\lambda^*(0) = -1$ and, consequently, the candidate optimal control trajectory is

$$u^*(t) = \begin{cases} 1 \text{ for all } t \in [0, 1] \\ 0 \text{ for all } t \in (1, 2] \end{cases}$$

The control and corresponding state trajectory are illustrated in Figure 4.1. $\square$

Example 4.8

Consider the minimization of

$$\int_0^2 2x^2(t) + u^2(t) dt,$$

subject to

$$\dot{x}(t) = x(t) + 2u(t), \ x(0) = 1 \text{ and } u \in [-1, 1].$$

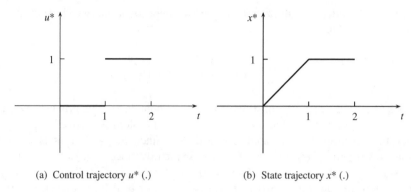

(a) Control trajectory u* (.)   (b) State trajectory x* (.)

Figure 4.1 Optimal control and state trajectory for Example 4.7

With $f := x + 2u$, $g := 2x^2 + u^2$ and $h := 0$, the Hamiltonian for this problem is

$$H(t, x, u, \lambda) := 2x^2 + u^2 + \lambda(x + 2u). \quad (4.3.14)$$

Differentiation of $H(t, x^*, u, \lambda^*)$ w.r.t. u shows that $u^* = -\lambda^*$ yields the minimum value of H provided that $-1 \leq \lambda^* \leq 1$. In case $|\lambda^*| > 1$, the minimum value is attained at the boundary points of the interval $[-1, 1]$. Comparing the value of $H(t, x^*, -1, \lambda^*)$ and $H(t, x^*, 1, \lambda^*)$ shows that $H(t, x^*, -1, \lambda^*) \leq H(t, x^*, 1, \lambda^*)$ if and only if $-2\lambda^* < 2\lambda^*$. Therefore, $u^* = -1$ if $\lambda^* > 1$ and $u^* = 1$ if $\lambda^* < -1$. Summarizing, we have

$$u^* = \begin{cases} -\lambda^* & \text{if} & \lambda^* \in [-1, 1] \\ -1 & \text{if} & \lambda^* > 1 \\ 1 & \text{if} & \lambda^* < -1 \end{cases}$$

where (x^*, u^*, λ^*) satisfy the dynamical constraints

$$\dot{x}^*(t) = x^*(t) + 2u^*(t), \ x(0) = 1 \quad (4.3.15)$$
$$\dot{\lambda}^*(t) = -4x^*(t) - \lambda^*(t), \ \lambda^*(2) = 0. \quad (4.3.16)$$

From these equations we have to determine the optimal control trajectory. Since the equations are rather simple it is possible to find an analytic solution. To that purpose, we proceed in three steps.

Step 1: Since $\lambda^*(2) = 0$ and, thus, belongs to the interval $[-1, 1]$ we conclude that there is a point in time $t_1 < 2$ such that

$$u^*(t) = -\lambda^*(t), \text{ for all } t \in [t_1, 2].$$

Substitution of this result into equation (4.3.15) shows that (using equation (4.3.16)) $x^*(t)$ satisfies the second-order differential equation:

$$\begin{aligned} \ddot{x}^*(t) &= \dot{x}^*(t) - 2\dot{\lambda}^*(t) \\ &= \dot{x}^*(t) + 8x^*(t) + 2\lambda^*(t) \\ &= \dot{x}^*(t) + 8x^*(t) + x^*(t) - \dot{x}^*(t) \\ &= 9x^*(t). \end{aligned}$$

It is easily verified that the solution of this differential equation is

$$x^*(t) = c_1 e^{3t} + c_2 e^{-3t}, \tag{4.3.17}$$

where c_1 and c_2 are some constants which are indirectly determined by the boundary conditions of equations (4.3.15) and (4.3.16).

Since $\dot{x}^*(t) = x^*(t) - 2\lambda^*(t)$ we obtain from this equation, using equation (4.3.17), that

$$
\begin{aligned}
\lambda^*(t) &= \frac{1}{2}(x^*(t) - \dot{x}^*(t)) \\
&= \frac{1}{2}(c_1 e^{3t} + c_2 e^{-3t} - 3c_1 e^{3t} + 3c_2 e^{-3t}) \\
&= -c_1 e^{3t} + 2c_2 e^{-3t}.
\end{aligned}
$$

From the fact that $\lambda^*(2) = 0$ it follows then that the constants c_1 and c_2 satisfy $0 = -c_1 e^6 + 2c_2 e^{-6}$. So, we conclude that on the interval $[t_1, 2]$ the optimal state and costate trajectories are

$$x^*(t) = c_2(2e^{-12}e^{3t} + e^{-3t}), \tag{4.3.18}$$
$$\lambda^*(t) = 2c_2(-e^{-12}e^{3t} + e^{-3t}), \tag{4.3.19}$$

respectively. Notice that the function $m(t) := -e^{-12}e^{3t} + e^{-3t}$ is a positive, monotonically decreasing function on $[0, 2]$. Consequently, $\lambda^* \geq 0$ if and only if $c_2 \geq 0$.

Now, assume that $t_1 = 0$. Then it follows from the boundary condition $x(0) = 1$ and equation (4.3.18) that $c_2 = \frac{1}{1+2e^{-12}}$. Substitution of c_2 into equation (4.3.19) shows then that $\lambda^*(0) = \frac{2}{1+2e^{-12}}(1 - e^{-12}) > 1$. This violates our assumption that $|\lambda^*(t)| \leq 1$. So, the optimal control switches at some point in time $t_1 > 0$ from $u^* = \lambda^*(t)$ to either $u^*(t) = 1$ or $u^*(t) = -1$. In the next two steps we analyze these two cases separately.

Step 2: Assume that $c_2 > 0$. According to Step 1, this case corresponds to an optimal control $u^* = -1$ that is used during some time interval $[t_2, t_1]$, where t_2 is some point in the time interval $[0, t_1)$ and the switching point t_1 is determined by the equation

$$2c_2(-e^{-12}e^{3t_1} + e^{-3t_1}) = 1$$

(see equation (4.3.19)). On this interval $[t_2, t_1]$ the candidate optimal state and costate equations (x^*, λ^*) are determined by

$$\dot{x}^*(t) = x^*(t) - 2, \quad x^*(t_1) = c_2(2e^{-12}e^{3t_1} + e^{-3t_1}) \tag{4.3.20}$$
$$\dot{\lambda}^*(t) = -4x^*(t) - \lambda^*(t), \quad \lambda^*(t_1) = 1 \tag{4.3.21}$$

(see equations (4.3.15), (4.3.16) and (4.3.18)).
The solution of equation (4.3.20) is

$$x^*(t) = c_3 e^t + 2, \tag{4.3.22}$$

where c_3 is some constant (ignoring for the moment the boundary condition in equation (4.3.20)).

Next, assume that $t_2 = 0$. Then, substitution of the boundary condition $x(0) = 1$ into equation (4.3.22) yields $x^*(t) = -e^t + 2$. Notice that $x^*(t)$ is a monotonically decreasing function. Since, by assumption on c_2, $x^*(t_1) > 0$ we conclude that $x^*(t) > 0$ on $[0, t_1]$. From equation (4.3.21) it is easily seen then that, since the derivative of $\lambda^*(t)$ is positive on $[0, t_1]$ and $\lambda^*(t_1) = 1$, $\lambda^*(t) > 1$ for all $t \in [0, t_1]$. So, the trajectory consisting of

$$u^*(t) = -1 \text{ on } [0, t_1] \text{ together with } u^*(t) = -\lambda^*(t) \text{ on } [t_1, 2] \qquad (4.3.23)$$

is an appropriate candidate optimal control trajectory if we can find a positive c_2 and $t_1 \in (0, 2)$ satisfying

$$-e^{t_1} + 2 = c_2(2e^{-12}e^{3t_1} + e^{-3t_1}) \qquad (4.3.24)$$

$$1 = 2c_2(-e^{-12}e^{3t_1} + e^{-3t_1}). \qquad (4.3.25)$$

From equation (4.3.25) we have that $2c_2 = \frac{1}{-e^{-12}e^{3t_1} + e^{-3t_1}} > 0$, if t_1 exists. So, what is left to be shown is that the equation

$$2(-e^{-12}e^{3t_1} + e^{-3t_1})(-e^{t_1} + 2) = 2e^{-12}e^{3t_1} + e^{-3t_1}$$

has a solution $t_1 \in (0, 2)$. Some elementary calculation shows that we can rewrite this equation as

$$n(t_1) := 2e^{-12+4t_1} - 2e^{-2t_1} - 6e^{-12+3t_1} + 3e^{-3t_1} = 0. \qquad (4.3.26)$$

Since $n(0) > 0$ and $n(2) < 0$, it follows from the continuity of $n(t)$ that the equation (4.3.26) has a solution $t_1 \in (0, 2)$. Which proves that equation (4.3.23) is a control trajectory satisfying all conditions of Theorem 4.6.

Step 3: Next we consider the case $c_2 < 0$. This case corresponds to an optimal control $u^* = 1$ that is used during some time interval $[t_2, t_1]$ (see Step 1). The switching point t_1 is determined by the equation

$$2c_2(-e^{-12}e^{3t_1} + e^{-3t_1}) = -1$$

(see equation (4.3.19)).

During the time interval $[t_2, t_1]$, the candidate optimal state trajectory satisfies

$$\dot{x}(t) = x(t) + 2, \quad x(t_1) = c_2(2e^{-12}e^{3t_1} + e^{-3t_1}). \qquad (4.3.27)$$

(see equations (4.3.15) and (4.3.18)).

The solution $x^*(t)$ of this differential equation (4.3.27) on $[t_2, t_1]$ is

$$x^*(t) = c_2(2e^{-12}e^{2t_1} + e^{-4t_1})e^t - 2 + 2e^{t-t_1}. \qquad (4.3.28)$$

Since $c_2 < 0$ and $t - t_1 \leq 0$ we immediately deduce from this equation (4.3.28) that $x^*(t) < 0$ for all $t \in [t_2, t_1]$. However, this implies (see equation (4.3.16)) that $\lambda^*(t) < -1$ for all $t \in [t_2, t_1]$. So, no switching point will occur anymore and, thus, $t_2 = 0$. However, by assumption $x^*(0)$ should equal 1. This is obviously impossible, so we conclude that the case $c_2 < 0$ does not provide any additional candidate optimal solutions.

So, in conclusion, the only candidate solution we have found is given by equation (4.3.23). It can be shown in a similar way to the ensueing Example 4.10, that this solution indeed solves our optimization problem. □

The previous examples clearly demonstrate the need for sufficient conditions to conclude that a set of trajectories (x^*, u^*, λ^*) satisfying equations (4.3.1)–(4.3.3) are optimal. To that end we consider the so-called minimized Hamiltonian.

Definition 4.3

Consider the Hamiltonian $H(t, x, u, \lambda)$ defined in equation (4.2.4). Then

$$H^0(t, x, \lambda) := \min_{u \in \mathcal{U}} H(t, x, u, \lambda) \qquad (4.3.29)$$

is called the minimized Hamiltonian. □

Notice that the minimized Hamiltonian is a function of three arguments: t, x and λ. From the next lemma we infer that this minimized Hamiltonian is differentiable. This property is used to obtain a sufficient condition for optimality.

Lemma 4.7

Let $g : \mathbb{R}^{n+m} \to \mathbb{R}$ be a function of two variables x and u, with $x \in \mathbb{R}^n$ and $u \in U \subset \mathbb{R}^m$. Suppose that g is continuously differentiable in x, and that the function $g(x, u)$ (seen as a function of u) has a minimum on U for every x. Define

$$g^0(x) = \min_{u \in U} g(x, u).$$

Finally, assume that g^0 is convex. Then, the function g^0 is differentiable, and for any given x_0

$$\frac{\partial g^0}{\partial x}(x_0) = \frac{\partial g}{\partial x}(x_0, u^*)$$

where u^* is any point in U that satisfies

$$g(x_0, u^*) = g^0(x_0).$$

Proof

Take an $x_0 \in \mathbb{R}^n$. As a consequence of the assumed convexity of g^0, there exists a vector $a \in \mathbb{R}^n$ such that

$$g^0(x) \geq g^0(x_0) + a^T(x - x_0) \qquad (4.3.30)$$

for all x. Let $u^* \in U$ be such that $g(x_0, u^*) = g^0(x_0)$. Then

$$g(x, u^*) \geq g^0(x_0) \geq g^0(x_0) + a^T(x - x_0) = g(x_0, u^*) + a^T(x - x_0). \qquad (4.3.31)$$

Define a function $G : \mathbb{R}^n \to \mathbb{R}$ by

$$G(x) := g(x, u^*) - a^T(x - x_0).$$

Because, by assumption, $g(x, u^*)$ is continuously differentiable, $G(x) \in C^1$. Moreover, it follows from equation (4.3.31) that

$$G(x) \geq g(x_0, u^*)$$

for all x, whereas we see from the definition of G that equality holds if $x = x_0$. In other words, $G(x)$ has a minimum at x_0 and hence must satisfy the necessary condition

$$\frac{\partial G}{\partial x}(x_0) = 0.$$

From the definition of G, this means that

$$a = \left[\frac{\partial g}{\partial x}(x_0, u^*)\right]^T.$$

This proves that the vector a which determines the supporting hyperplane at $(x_0, g^0(x_0))$ at the right-hand side of equation (4.3.30) is uniquely determined. Consequently (for example Roberto and Verberg, 1973), g^0 must be differentiable at x_0, and

$$\frac{\partial g^0}{\partial x}(x_0) = a^T = \frac{\partial g}{\partial x}(x_0, u^*). \qquad \square$$

The next theorem is due to Arrow (1968).

Theorem 4.8 (Arrow)

Consider the optimization problem given in equations (4.2.1) and (4.2.2). Assume that f, g and h satisfy the Basic Assumptions A and the scrap-value function h is, moreover, convex. Let the Hamiltonian $H := g + \lambda f$. Suppose (x^*, u^*, λ^*) is a solution to equations (4.3.1)–(4.3.3). If the minimized Hamiltonian H^0 in equation (4.3.29) is convex in x when $u = u^*$ and $\lambda = \lambda^*$, then (x^*, u^*, λ^*) is an optimal solution to the problem (4.2.1) and (4.2.2). Moreover, if H^0 is strictly convex in x when $u = u^*$ and $\lambda = \lambda^*$, the optimal state trajectory $x^*(.)$ is unique (but u^* is not necessarily unique).

Proof

Let J^* denote the cost corresponding to (x^*, u^*, λ^*). Then, from equation (4.2.5),

$$
\begin{aligned}
J^* - J &= \int_0^T \{H(t, x^*, u^*, \lambda^*) - \lambda^*(t)\dot{x}^*(t) - (H(t, x, u, \lambda^*) - \lambda^*(t)\dot{x}(t))\}dt + h(x^*(T)) \\
&\quad - h(x(T)) \\
&= \int_0^T \{H(t, x^*, u^*, \lambda^*) + \dot{\lambda}^*(t)x^*(t) - (H(t, x, u, \lambda^*) + \dot{\lambda}^*(t)x(t))\}dt + h(x^*(T)) \\
&\quad - h(x(T)) + \lambda^*(T)(x(T) - x^*(T)),
\end{aligned}
\qquad \text{(i)}
$$

where the last equality follows again by integration by parts. From the convexity assumption on H^0 and the fact that H^0 is differentiable (see Lemma 4.7) we obtain

$$H^0(t,x,\lambda^*) - H^0(t,x^*,\lambda^*) \geq \frac{\partial H^0}{\partial x}(t,x^*,\lambda^*)(x - x^*).$$

Since $H^0(t,x^*,\lambda^*) = H(t,x^*,u^*,\lambda^*)$, using equation (4.3.2), we can rewrite the above inequality as

$$0 \leq H^0(t,x,\lambda^*) - H(t,x^*,u^*,\lambda^*) + \dot{\lambda}^*(x - x^*)$$

$$\leq H(t,x,u,\lambda^*) - H(t,x^*,u^*,\lambda^*) + \dot{\lambda}^*(x - x^*). \tag{ii}$$

The last inequality is due to the fact that by definition of H^0, $H^0(t,x,\lambda^*) \leq H(t,x,u,\lambda^*)$ for every choice of u. On the other hand it follows from equation (4.3.2) and the convexity of h

$$h(x^*(T)) - h(x(T)) + \lambda^*(T)(x(T) - x^*(T)) = h(x^*(T)) - h(x(T)) + \frac{\partial h}{\partial x}(x^*(T))(x(T)$$

$$- x^*(T)) \leq 0. \tag{iii}$$

From the inequalities (ii) and (iii) it is then obvious that $J^* - J \leq 0$ in (i). Obviously, if the inequality in (ii) is strict, we also have a strict inequality in (i), which proves the uniqueness claim. □

In practice, this theorem is applied as follows. Usually, the process of finding a candidate solution from the necessary conditions already involves the determination of the function $u^*(t,x,\lambda)$, and this allows one to write down the minimized Hamiltonian $H^0(t,x,\lambda)$. If the trajectory $\lambda^*(.)$ corresponding to the candidate solution is also found, one can compute $H^0(t,x,\lambda^*(t))$ for each t and see whether the resulting function in x is convex. One can spare oneself the trouble of computing the trajectory $\lambda^*(.)$ if it turns out that $H^0(t,x,\lambda)$ is convex in x for each t and each occurring value of λ. For, obviously, $H^0(t,x,\lambda^*(t))$ is then convex in x for each t as well and Arrow's theorem applies.

If the Hamiltonian $H(t,x,u,\lambda^*(t))$ is simultaneously convex in both arguments (x,u) for all t it can then be shown that $H^0(t,x,\lambda^*(t))$ is also convex in x for all t (see Exercises). So, if $H(t,x,u,\lambda^*(t))$ is simultaneously convex in (x,u) for all t we can also conclude from Arrow's theorem that $u^*(.)$ will be optimal. If the Hamiltonian $H(t,x,u,\lambda^*(t))$ is simultaneously strictly convex in (x,u) for all t we have an even stronger result. That is, in this case we can also conclude that u^* is unique, a result which is due to Mangasarian (1966).

Finally, we stress the point that it is **not** in general sufficient to verify that the Hamiltonian $H(t,x,u,\lambda)$ is convex in x. This is because $H^0(t,x,\lambda)$ equals $H(t,x,u^*(t,x,\lambda),\lambda)$ and this expression depends on x not only via the second argument but also via the third argument, u^*, of H.

Example 4.9

Reconsider Example 4.7. The minimized Hamiltonian for this example is

$$H^0(t,x,\lambda) = \begin{cases} \lambda & \text{if } x + \lambda < 0, \\ -x & \text{if } x + \lambda > 0. \end{cases}$$

Obviously, for each t and each λ, $H^0(t, x, \lambda)$ is convex in x. So, according to Theorem 4.8, the candidate control function $u^*(.)$ is indeed optimal in this example.　□

Example 4.10

Reconsider Example 4.4. The minimized Hamiltonian for this example is

$$H^0(t, x, \lambda) = x^2 - \frac{1}{4}b^2\lambda^2 + a\lambda x.$$

Obviously, for each t and each λ, $H^0(t, x, \lambda)$ is convex in x. So, according to Theorem 4.8, the candidate control function $u^*(.)$ is optimal in this example. In fact the Hamiltonian is already simultaneously convex in (x, u). So, from Mangasarian's result we conclude that the optimal control is unique.　□

In most applications it is difficult to find an analytical solution of the dynamic optimization problem. In general, for scalar problems, one can still provide a qualitative analysis of the problem. That is, by considering the phase plane formed by the state and costate variable one can still get a rough idea how the optimal state and control trajectory will evolve. As a first example reconsider Example 4.7.

Example 4.11

Reconsider the minimization problem

$$\int_0^2 (u(t) - 1)x(t)dt, \tag{4.3.32}$$

subject to

$$\dot{x}(t) = u(t); \ x(0) = 0 \text{ and } u(t) \in [0, 1]. \tag{4.3.33}$$

The Hamiltonian for this system is

$$H(t, x, u, \lambda) := (u - 1)x + \lambda u. \tag{4.3.34}$$

So if there exists an optimal control function for this problem then, by Theorem 4.6, necessarily there exists a costate function $\lambda^*(.)$ satisfying

$$\dot{\lambda}(t) = 1 - u(t); \ \lambda(2) = 0. \tag{4.3.35}$$

Furthermore,

$$u^*(t) = \arg \min_{u(t) \in [0,1]} \{(u(t) - 1)x^*(t) + \lambda^*(t)u(t)\}. \tag{4.3.36}$$

From equation (4.3.13) we recall that the optimal control is

$$u^*(t) = \begin{cases} 1 \text{ if } x^*(t) + \lambda^*(t) < 0, \\ 0 \text{ if } x^*(t) + \lambda^*(t) > 0. \end{cases} \tag{4.3.37}$$

Combining the expression for u^* in equation (4.3.37) with the set of differential equations (4.3.33) and (4.3.35) yields

$$\begin{bmatrix} \dot{x}(t) \\ \dot{\lambda}(t) \end{bmatrix} = \begin{bmatrix} 0 \\ 0 \end{bmatrix} + \begin{bmatrix} u \\ 1 - u \end{bmatrix}, \quad x(0) = 0; \ \lambda(2) = 0.$$

In the (x, λ) phase plane one gets, depending on the value of u, two different dynamics. One corresponding to $u = 0$ and one corresponding to $u = 1$. At those points in the (x, λ) phase plane where $x + \lambda = 0$, switches take place between the two control regimes. Therefore, the set of points in the (x, λ) plane where $x + \lambda = 0$ is called the **switching** curve.

To understand the complete dynamics of the candidate optimal state and costate trajectory we proceed as follows. First, we determine the dynamics of both control regimes separately. Then, both pictures are combined into one graph. From this graph we then deduce what the optimal state and costate trajectory will look like. From this, we can then infer the optimal control trajectory.

In case $u = 0$, the optimal dynamics satisfy the set of differential equations

$$\begin{bmatrix} \dot{x}(t) \\ \dot{\lambda}(t) \end{bmatrix} = \begin{bmatrix} 0 \\ 1 \end{bmatrix}.$$

The trajectories of this set of differential equations are vertical lines which are passed in an upward direction over time (see Figure 4.2(a)).

In case $u = 1$, the optimal dynamics satisfy the set of differential equations

$$\begin{bmatrix} \dot{x}(t) \\ \dot{\lambda}(t) \end{bmatrix} = \begin{bmatrix} 1 \\ 0 \end{bmatrix}.$$

The trajectories of this set of differential equations are horizontal lines which are passed in a forward direction over time (see Figure 4.2(b)).

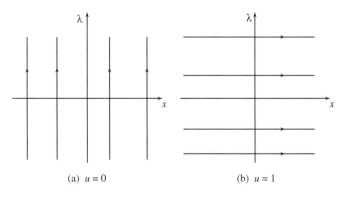

(a) $u = 0$ (b) $u = 1$

Figure 4.2 Phase portrait for Example 4.11 for $u = 0$ and $u = 1$

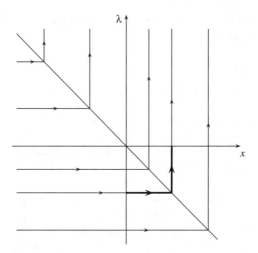

Figure 4.3 Combined phase portrait for Example 4.11

In Figure 4.3 we have sketched the combined phase portrait by copying the part of Figure 4.2(a) that is above the line $x + \lambda = 0$ (where $u^* = 0$ is optimal) and the part of Figure 4.2(b) that is below the line $x + \lambda = 0$ (where $u^* = 1$ is optimal).

Within this class of potential candidate optimal solutions we are looking for that trajectory in Figure 4.3 which starts on the line $x = 0$ and ends on the line $\lambda = 0$ and traverses the trajectory in two time units. From this graph we infer that the optimal control is provided by first choosing $u = 1$ for some time interval followed by a switch towards $u = 0$ for the rest of the time period. Given the symmetry of the graph one might expect that the switch takes place half-way through the planning interval, i.e. at $t = 1$. □

The next example illustrates a case where only a qualitative analysis is possible.

Example 4.12

Advertising by a firm affects its present and future sales. The following example developed by Nerlove and Arrow (1962) treats advertising as an investment in building up some sort of advertising capital usually called goodwill, g. It is assumed that this goodwill depreciates over time at a constant proportional rate δ, that is

$$\dot{g}(t) = -\delta g(t) + u(t), \ g(0) = g_0, \qquad (4.3.38)$$

where $u(t) \geq 0$ is the advertising effort at time t measured, for example, in dollars per unit time. We assume that the sales of the firm at time t, $s(t)$, depend on the goodwill and the price $p(t)$ of the product. Moreover, we assume that the firm is a monopolist and that the price it sets is (apart from a fixed cost term) a function of its goodwill and its advertising efforts. The profits of the firm at time t are then

$$\pi(t) := p(t)s(t) - C(s(t), u(t)).$$

So, the problem the firm is confronted with is to choose its advertising strategy $u(.)$ such that its discounted profits are maximized, that is, to minimize w.r.t. $u(.) \geq 0$

$$\int_0^T -e^{-rt}\pi(g(t), u(t))dt, \qquad (4.3.39)$$

subject to equation (4.3.38) where $r > 0$ is the discount factor.

For simplicity assume that the profit function $\pi(g, u)$ can be written in the form

$$\pi(g, u) = -p_1(g) - p_2(u) + cu$$

where p_1 and p_2 are increasing concave functions and c is a positive constant expressing the cost of advertising. Then, with

$$f = -\delta g + u, \ g = e^{-rt}(p_1(g) + p_2(u) - cu), \text{ and } h = 0,$$

we conclude from Theorem 4.16 that if $u^*(.)$ is the optimal control path, and $g^*(.)$ the corresponding optimal goodwill path, then it is necessary that there exists a costate path $\lambda^*(.)$ such that

$$\dot{g}^*(t) = -\delta g^*(t) + u^*(t), \ g^*(0) = g_0, \qquad (4.3.40)$$

$$\dot{\lambda}^*(t) = e^{-rt}\frac{dp_1}{dg}(g^*(t)) + \delta\lambda^*(t), \ \lambda^*(T) = 0, \qquad (4.3.41)$$

$$H(t, g^*, u^*, \lambda^*) = \min_{u \geq 0} H(t, g^*, u, \lambda^*), \qquad (4.3.42)$$

where $H(t, g, u, \lambda) = -e^{-rt}(p_1(g) + p_2(u) - cu) + \lambda(-\delta g + u)$. To enhance a further explicit solution of the above set of necessary conditions we assume that $p_2(u) = u - \frac{1}{2}u^2$. Then, since H is a convex function of u, the minimum value is attained at the location where H has its global minimum if this location is positive. Otherwise, due to the fact that the location has to be larger than zero, this minimum value is attained at zero. Consequently condition (4.3.42) can be rewritten in shorthand as

$$u^*(t) = \max(1 - c - e^{rt}\lambda^*(t), 0). \qquad (4.3.43)$$

By introducing the variable $\mu(t) = e^{rt}\lambda(t)$ (and consequently $\dot{\mu}(t) = re^{rt}\lambda(t) + e^{rt}\dot{\lambda}(t) = r\mu(t) + e^{rt}\dot{\lambda}(t)$), we see that the set of equations (4.3.40)–(4.3.42) can be rewritten as

$$\dot{g}^*(t) = -\delta g^*(t) + u^*(t), \ g^*(0) = g_0 \qquad (4.3.44)$$

$$\dot{\mu}^*(t) = \frac{dp_1}{dg}(g^*(t)) + (\delta + r)\mu^*(t), \ \mu^*(T) = 0 \qquad (4.3.45)$$

$$u^*(t) = \max(1 - c - \mu^*(t), 0). \qquad (4.3.46)$$

The expression that we found for u^* in equation (4.3.46) now has to be combined with the set of differential equations

$$\begin{bmatrix} \dot{g}(t) \\ \dot{\mu}(t) \end{bmatrix} = \begin{bmatrix} -\delta g \\ \frac{dp_1}{dg}(g(t)) + (\delta + r)\mu(t) \end{bmatrix} + \begin{bmatrix} u \\ 0 \end{bmatrix}, \ g(0) = g_0; \ \mu(T) = 0.$$

In the (g, μ) phase plane we get, depending on the value of u, two different dynamics. One, corresponding with $u = 0$ and one corresponding with $u = 1 - c - \mu$. The switching curve, where switches between the two control regimes take place, is in this case given by the set of points in the (g, μ) plane where $0 = 1 - c - \mu$.

To understand the complete dynamics of the optimal investment process we proceed as follows. First, we determine the dynamics of both control regimes separately. Then, we combine both pictures into one graph and conclude from this graph how the optimal investment path will look.

If $u = 0$, the optimal dynamics satisfy the set of differential equations

$$\begin{bmatrix} \dot{g}(t) \\ \dot{\mu}(t) \end{bmatrix} = \begin{bmatrix} -\delta g \\ \frac{dp_1}{dg}(g(t)) + (\delta + r)\mu(t) \end{bmatrix}.$$

Using the algorithm given in section 3.4 we construct the phase portrait of this system. To that end first recall that, since p_1 is an increasing function, its derivative is positive. Consequently, the planar system has a unique equilibrium point $(g, \mu) = \left(0, -\frac{1}{\delta + r}\frac{dp_1}{dg}(0)\right)$. The matrix of the derivatives at the equilibrium point is

$$\begin{bmatrix} -\delta & 0 \\ \frac{d^2 p_1}{dg^2}(0) & \delta + r \end{bmatrix}.$$

This matrix has one positive and one negative eigenvalue, so the equilibrium point is a saddle. The horizontal isocline is given by $\mu = -\frac{1}{\delta + r}\frac{dp_1(g)}{dg}$, whereas the vertical isocline is $g = 0$. By inspection of the sign of the derivatives $\dot{g}$ and $\dot{\mu}$ in the different sectors the solution curves can then be easily determined and are sketched in Figure 4.4(a).

In a similar way next the dynamical system is analyzed in case $u = 1 - c - \mu$. Then, the planar system again has a unique equilibrium (g^e, μ^e) that is attained at the intersection point of the line $\mu = -\delta g + 1 - c$ and the curve $\mu = -\frac{1}{\delta + r}\frac{dp_1}{dg}(g)$.

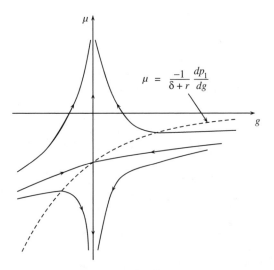

Figure 4.4 (a) Phase portrait for Example 4.12 for $u = 0$

The matrix of the derivatives at the equilibrium point is

$$\begin{bmatrix} -\delta & -1 \\ \frac{d^2 p_1}{dg^2}(g^e) & \delta + r \end{bmatrix}.$$

Since p_1 is concave (and therefore $\frac{d^2 p_1(g)}{dg^2} < 0$) it is easily verified that this matrix has one positive and one negative eigenvalue. So the equilibrium point is again a saddle. The horizontal isocline is given again by $\mu = -\frac{1}{\delta + r} \frac{dp_1(g)}{dg}$, whereas the vertical isocline is $\mu = -\delta g + 1 - c$.

By inspection of the sign of the derivatives $\dot{g}$ and $\dot{\mu}$ in the different sectors we then obtain the phase portrait as sketched in Figure 4.4(b). In Figures 4.4(c) and (d) the

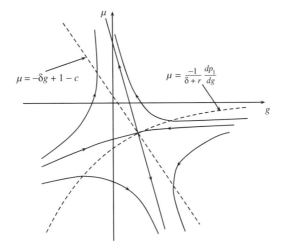

Figure 4.4 (b) Phase portrait for Example 4.12 for $u = 1 - c - \mu$

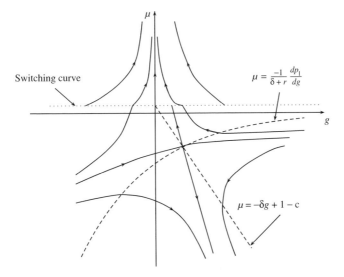

Figure 4.4 (c) Phase portrait for Example 4.12, $c < 1$

combined phase portrait is sketched by copying the part of Figure 4.4(a) that is above the line $\mu = 1 - c$ (where $u^* = 0$ is optimal) and the part of Figure 4.4(b) that is below the line $\mu = 1 - c$ (where $u^* = 1 - c - \mu$ is optimal). From this combined phase portrait it is clear that if $c < 1$ (see Figure 4.4(c)), irrespective of the exact choice of p_1 and other model parameters, the policy $u^* = 1 - c - \mu^*$ is always optimal. On the other hand, the phase portrait sketched in Figure 4.4(d) suggests that, if $c > 1$, different optimal strategies

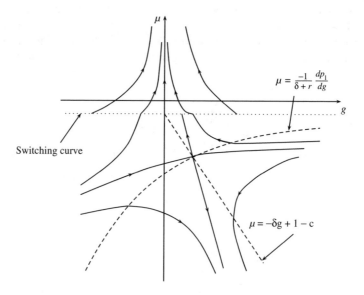

Figure 4.4   (d) Phase portrait for Example 4.12, $c > 1$

might occur. Depending on the exact values of c and g_0 the strategy $u(.) = 0$ is sometimes optimal, whereas in other cases a strategy $u(t) = 1 - c - \mu^*$, $t \in [0, t_0]$ followed by $u(t) = 0, t \in [t_0, T]$ might be optimal.

Finally notice that $H_{xx} = -p_{1,xx}$, $H_{xu} = 0$ and $H_{uu} = -p_{2,uu}$. Since we assumed that p_i, $i = 1, 2$, are concave we conclude that H'' is simultaneously convex in (g, u). Therefore, according to the discussion following on from Theorem 4.8, any solution satisfying the equations (4.3.44)–(4.3.46) will be optimal.

We hope this will give the reader a flavor of how this kind of optimization problem can be analyzed.   □

In many planning problems the choice of a planning period, T, is somewhat artificial. If plans are made for an indefinite period, it may be reasonable to take $T = $ '∞'. There are some difficulties with this choice, however, which must be tackled. One problem is that the value of the criterion might become infinite since integration takes place over an infinitely long interval. Like before, we cope with this problem by restricting the set of all admissible controllers. We just consider those control functions for which the cost

$$J(x_0, u) := \int_0^\infty g(t, x(t), u(t))dt \tag{4.3.47}$$

exists (as a finite number) and the state $x(t)$ converges. The scrap value $h(x(T))$ has been dropped in this expression, because there is no final terminal time. Therefore, the set of admissible control functions will be restriced to

$$\mathcal{U}_{re} := \{u \in \mathcal{U} \mid J(x_0, u) \text{ exists as a finite number and } \lim_{t \to \infty} x(t) \text{ exists.}\} \quad (4.3.48)$$

Another problem is that if for some initial state x_0 the differential equation has a solution on $[0, \infty)$ this does not imply that there is an open neighborhood of x_0 such that for all initial states in this neighborhood a solution also exists on $[0, \infty)$. Reconsider, for example, Example 3.5.

Example 4.13

Consider the differential equation

$$\dot{x} = x^2, \quad x(0) = 0.$$

Then, obviously, $x(.) = 0$ is the solution of this initial-value problem. This solution is defined on $[0, \infty)$ (assuming for the moment that we are only interested in solutions on the positive real half line). However, for every positive initial state $x_0 = 1/p > 0$, the solution of this differential equation is $x(t) = \frac{1}{p-t}$. So, for any positive initial state $x_0 = 1/p$ the maximal interval of existence is $[0, p)$. So, there does not exist an open interval around $x(0) = 0$ such that for every initial $x(0)$ in this interval the differential equation has a solution on $[0, \infty)$. □

In the derivation of the maximum principle on a finite planning horizon we used the property that if the problem (4.2.1) and (4.2.2) has an optimal solution, then there is also a 'neighborhood' of this optimal solution for which the differential equation has a solution. Example 4.13 above illustrates that without further assumptions this property cannot be guaranteed in general. Theorem 3.10 showed that a condition which guarantees that solutions will exist for all initial states on the whole interval $[0, \infty)$ is that $f(t, x(t), u(t))$ is globally Lipschitz. Theorem 4.9, below, does not make this assumption but restricts the analysis to the set of all control sequences for which the solution of the differential equation (4.2.2) exists. It provides the following analogue of Theorem 4.6 (see the Appendix at the end of this chapter for an outline of the proof).

Theorem 4.9

Consider the optimization problem given by equation (4.3.47) and (4.2.2). Assume that f, g and h satisfy the Basic Assumptions A and let $H := g + \lambda f$. Moreover assume that $u \in \mathcal{U}_{re}$.

Let $u^*(t) \in \mathcal{U}_{re}$ be a control that yields a local minimum for the cost function (4.3.47), and let $x^*(t)$ be the corresponding state trajectory.

Let $A(t) := f_x(t, x^*(t), u^*(t))$ and $\Phi(t, t_0)$ be the fundamental matrix of the corresponding linear system

$$\dot{y}(t) = A(t)y(t) \text{ with } \Phi(t_0, t_0) = I. \quad (4.3.49)$$

Assume that for every t_0 (i) there is a bound M such that for all $t \geq t_0$ $|\Phi(t,t_0)| < M$, and (ii) $\lim_{t\to\infty} |\Phi(t,t_0)| = 0$. Furthermore, let the costate function $[\lambda^*]^T : [0,\infty) \to \mathbb{R}^n$ be a solution of the linear differential equation

$$\dot{\lambda}^*(t) = -\frac{\partial H(t,x^*,u^*,\lambda^*)}{\partial x}. \tag{4.3.50}$$

Finally, assume that $\lim_{t\to\infty} \lambda^*(t)x^*(t)$ exists. Then, for all $t \in (0,\infty)$ at which u^* is continuous,

$$H(t,x^*,u^*,\lambda^*) = \min_{u\in U} H(t,x^*,u,\lambda^*). \tag{4.3.51}$$

$\square$

An important distinction between the above theorem and Theorem 4.6 is that the constraints on the costate variable λ^* are lacking. As a result one will in general get an infinite number of candidate solutions, among which one has to search for an optimal one. In case the optimization problem has a one-dimensional state, one can use a phase plane analysis to get a survey of all candidate solutions, and then it is usually not too difficult to find out which one is optimal. In well-posed economic problems, the optimal solution often corresponds to a path converging to an equilibrium. The next example illustrates the above sketched procedure.

Example 4.14

Reconsider Example 4.4 on the linear quadratic control problem for an infinite planning horizon T with the requirement that $x(t)$ converges to zero (otherwise the cost function J becomes unbounded). That is, minimize

$$\int_0^\infty \{x^2(t) + u^2(t)\}dt$$

subject to

$$\dot{x}(t) = ax(t) + bu(t), \quad x(0) = 1, \quad (b \neq 0).$$

Assume that $a < 0$. If $u^*(.)$ is the optimal control path, $x^*(.)$ the corresponding optimal state trajectory and λ^* the solution of the differential equation

$$\dot{\lambda}^*(t) = -(2x^*(t) + a\lambda^*(t)), \tag{4.3.52}$$

where $\lambda^*(t)x^*(t)$ converges then, by Theorem 4.9, necessarily (x^*,u^*,λ^*) satisfy

$$\dot{p}^*(t) = ax^*(t) + bu^*(t), \quad x^*(0) = 1 \tag{4.3.53}$$
$$0 = 2u^*(t) + b\lambda^*(t). \tag{4.3.54}$$

In a similar way to Example 4.4 we see that x^* and λ^* should satisfy the set of differential equations

$$\begin{bmatrix} \dot{x}^*(t) \\ \dot{\lambda}^*(t) \end{bmatrix} = \begin{bmatrix} a & -\frac{1}{2}b^2 \\ -2 & -a \end{bmatrix} \begin{bmatrix} x^*(t) \\ \lambda^*(t) \end{bmatrix}, \quad x^*(0) = 1. \tag{4.3.55}$$

Next we construct a phase portrait of this system of differential equations (4.3.55). It is easily verified that $(x^e, \lambda^e) = (0,0)$ is the only equilibrium point of equations (4.3.55). Obviously, the derivative of the planar system at the equilibrium point coincides with the matrix

$$\begin{bmatrix} a & -\frac{1}{2}b^2 \\ -2 & -a \end{bmatrix}.$$

Since this matrix always has one positive eigenvalue $\sqrt{a^2 + b^2}$ and one negative eigenvalue $-\sqrt{a^2 + b^2}$, the equilibrium point is a saddle. The vertical isocline in the (x, λ)-plane is given by $\lambda = \frac{2a}{b^2}x$ and the horizontal isocline by $\lambda = -\frac{2}{a}x$. A sketch of the phase plane is given in Figure 4.5.

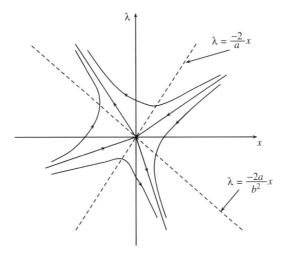

Figure 4.5   Phase plane for Example 4.14

Obviously, there is only one (x, λ) path that has the property that $x(t)$ converges to zero when $t \to \infty$. That path is the trajectory leading to the equilibrium point $(0,0)$. So, this is the only candidate optimal solution.   □

Note

If the parameter a in Example 4.14 above is positive, Theorem 4.9 can also be used to derive necessary optimality conditions by considering the equivalent problem

$$\dot{x}(t) = (a + bf)x(t) + b\tilde{u}(t)$$

$$\min_{\tilde{u}} \int_0^\infty x^2(t) + (fx(t) + \tilde{u}(t))^2 dt,$$

where f is chosen such that $a + bf < 0$.

Obviously, this approach can also be used for the general setting if matrix $A(t)$ in Theorem 4.9 does not invoke an asymptotically stable system.   □

4.4   Dynamic programming principle

Compared with Pontryagin's maximum principle, the principle of dynamic programming to solve dynamic optimal control problems is surprisingly clear and intuitive. Instead of necessary conditions it provides sufficient conditions to conclude that the dynamic optimization problem given by equations (4.2.1) and (4.2.2) will have a solution (and a way of finding the optimal control). This optimization technique was developed by Bellman and Isaacs[2] at the RAND Corporation at the end of the 1940s and early on the 1950s. Particularly in the study of optimal control problems in discrete time it has made a tremendous contribution in a wide range of applications. Its application in continuous dynamic optimal control problems has led to the invention of the famous Hamilton–Jacobi–Bellman (HJB) partial differential equation which is used for constructing an optimal feedback control law.

We shall introduce Bellman's **principle of optimality** via a simple picture as shown in Figure 4.6.

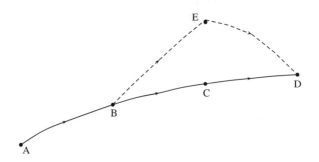

Figure 4.6   Shortest path principle

Suppose we wish to travel from the point A to the point D by the shortest possible path, and that this optimal path is given by ABCD, where the intermediate points B and C represent some intermediate stages. Now let us suppose that we start from point B and wish to travel to point D by the shortest possible path. According to Bellman's principle of optimality this optimal path is then given by BCD. Although this answer seems to be trivial we shall nevertheless prove it. Suppose there exists a different path from B to D given by the dashed line BED that is optimal. Then, since distance is additive, it implies that the path ABED is shorter in going from A to D than our optimal path ABCD. Since this is a contradiction, we conclude therefore that the path BED cannot be optimal in going from B to D.

We like to use this idea to solve our dynamic optimal control problem given by equations (4.2.1)–(4.2.2) subject to the constraint $u(.) \in \mathcal{U}$. To that end we introduce the

---

[2]Once again we stress here the point that from an historic point of view it is impossible to find out whether Bellman or Isaacs has the priority for this principle (see, for example, Breitner (2002).

intermediate optimal control problem to find a control function $u(.) \in \mathcal{U}$ that minimizes the cost function

$$J(t_0, \bar{x}, u) := \int_{t_0}^{T} g(t, x(t), u(t))dt + h(x(T)) \qquad (4.4.1)$$

where the state variable $x(t)$ satisfies the differential equation:

$$\dot{x}(t) = f(t, x(t), u(t)), \quad x(t_0) = \bar{x}. \qquad (4.4.2)$$

That is, instead of considering the original optimization problem, we consider the optimization problem starting at an intermediate point in time t_0 and assuming that the state of the system is at that time given by $\bar{x}$. Assume that the optimization problem (4.4.1) and (4.4.2) subject to the constraint $u(.) \in \mathcal{U}$ has a solution $u^*(.)$. Then, this optimal control depends on the initial state and the initial time. That is, $u^*(t) = u^*(t_0, \bar{x}, t)$. This optimal control path induces via equation (4.4.2) a corresponding optimal state trajectory $x^*(t_0, \bar{x})$. Once both these optimal control and state trajectories are known one can then compute from equation (4.4.1) the corresponding minimal cost $J^*(t_0, \bar{x})$. Following standard notation in the literature we will denote this value of the cost function by $V(t_0, \bar{x})$. So,

$$V(t_0, \bar{x}) = J^*(t_0, \bar{x}). \qquad (4.4.3)$$

Now, one can change the initial time and/or the initial state of the system. This results in different optimization problems and therefore a different cost. That is, by changing the initial time and/or initial state, the value of the cost function can be viewed at as a function of both these arguments. We will call this function $V(t, x)$ the **value function**.

We may now apply Bellman's principle of optimality to derive a partial differential equation for the value function $V(t, x)$. According to Bellman's optimality principle for $0 < \Delta t < T - t_0$ the following identity holds

$$V(t_0, \bar{x}) = \min \left\{ \int_{t_0}^{t_0 + \Delta t} g(t, x(t), u(t))dt + V(t_0 + \Delta t, x(t_0 + \Delta t)) \right\}, \qquad (4.4.4)$$

where the minimum is taken under the conditions

$$\dot{x}(t) = f(t, x(t), u(t)), \quad x(t_0) = \bar{x}, \quad u(t) \in \mathcal{U}.$$

We consider this equation for small values of Δt and, in the end, we shall let Δt tend to zero. Assume that the partial derivative of the value function $V(t, x)$ w.r.t. x exists and is, moreover, continuous. Then, since $\dot{x}$ is continuous too, we have (using the mean value theorem twice):

$$V(t_0 + \Delta t, x(t_0 + \Delta t)) = V(t_0 + \Delta t, x(t_0) + \Delta t \dot{x}(t_1))$$

$$= V(t_0 + \Delta t, x(t_0)) + \frac{\partial V}{\partial x}(t_0 + \Delta t, x(t_2))\dot{x}(t_1)\Delta t,$$

where due to the continuity assumptions $\dot{x}(t_1)$ converges to $\dot{x}(t_0) = f(t_0, \bar{x}, u_0)$ and $\frac{\partial V}{\partial x}(t_0 + \Delta t, x(t_2))$ converges to $\frac{\partial V}{\partial x}(t_0, x(t_0))$ if Δt converges to zero. Next, we insert this expression into the right-hand side of equation (4.4.4). Notice that the term $V(t_0 + \Delta t, x(t_0))$ can be taken out of the 'min' operation since it does not depend on the input $u(t)$. Taking this term to the left-hand side, we obtain

$$V(t_0, \bar{x}) - V(t_0 + \Delta t, \bar{x}) = \min \left\{ \int_{t_0}^{t_0 + \Delta t} g(t, x(t), u(t))dt + \frac{\partial V}{\partial x}(t_0 + \Delta t, x(t_2))\dot{x}(t_1)\Delta t \right\}.$$

Dividing by Δt and letting Δt approach 0, one obtains

$$-\frac{\partial V}{\partial t}(t_0, \bar{x}) = \min_{u \in \mathcal{U}} \left\{ g(t_0, \bar{x}, u_0) + \frac{\partial V}{\partial x}(t_0, \bar{x})f(t_0, \bar{x}, u_0) \right\}. \tag{4.4.5}$$

Since the time t_0 and the state $\bar{x}$ that we started with were arbitrary, we see that the value function satisfies the following conditions:

$$-\frac{\partial V}{\partial t}(t, x) = \min_{u \in \mathcal{U}} \left\{ g(t, x, u) + \frac{\partial V}{\partial x}(t, x)f(t, x, u) \right\}$$

$$V(T, x) = h(x(T)).$$

Observe that derivatives with respect to both t and x occur. For that reason this is called a **partial** differential equation. The above partial differential equation is usually called the **Hamilton–Jacobi–Bellman equation**. Once this equation is solved, the optimal control $u^*(.)$ is found for each t and x as the value that achieves the minimum in equation (4.4.5).

Theorem 4.10 (Dynamic Programming)

Consider the optimization problem given by equations (4.2.1) and (4.2.2). Assume the functions f, g and h satisfy the Basic Assumptions A. Let $u^*(t) \in \mathcal{U}$ be a control that yields a local minimum for the cost function (4.2.1), and let $x^*(t)$ be the corresponding state trajectory. Let $V(t, x) := J^*(t, x)$ be as described in equation (4.4.3) and assume that both partial derivatives of $V(t, x)$ exist, $\frac{\partial V}{\partial x}$ is continuous and, moreover, $\frac{d}{dt}V(t, x(t))$ exists. Then for $t_0 \le t \le T$

$$-\frac{\partial V}{\partial t}(t, x) = \min_{u \in \mathcal{U}} \left\{ g(t, x, u) + \frac{\partial V}{\partial x}(t, x)f(t, x, u) \right\}, \quad V(T, x) = h(x(T)) \tag{4.4.6}$$

and

$$u^*(t, x) = \arg\min_{u \in \mathcal{U}} \left\{ g(t, x, u) + \frac{\partial V}{\partial x}(t, x)f(t, x, u) \right\}. \tag{4.4.7}$$

$\square$

Proof

It has been shown above that whenever, $u^*(.)$ is an optimal control, then necessarily equation (4.4.6) holds under the stated assumptions. What is left to be shown is that if the control $u^*(.)$ is chosen as in equation (4.4.7), this control is optimal.

From equation (4.4.6) it follows that for an arbitrary $u(.) \in \mathcal{U}$,

$$-\frac{\partial V}{\partial t}(t,x) \leq g(t,x,u) + \frac{\partial V}{\partial x}(t,x)f(t,x,u), \text{ and}$$

$$-\frac{\partial V}{\partial t}(t,x^*) = g(t,x^*,u^*) + \frac{\partial V}{\partial x}(t,x^*)f(t,x^*,u^*).$$

Or, stated differently,

$$\frac{\partial V}{\partial t}(t,x) + \frac{\partial V}{\partial x}(t,x)f(t,x,u) + g(t,x,u) \geq 0, \text{ and} \qquad (4.4.8)$$

$$\frac{\partial V}{\partial t}(t,x^*) + \frac{\partial V}{\partial x}(t,x^*)f(t,x^*,u^*) + g(t,x^*,u^*) = 0. \qquad (4.4.9)$$

Since $\frac{dV(t,x(t))}{dt} = \frac{\partial V(t,x)}{\partial t} + \frac{\partial V(t,x)}{\partial x}\frac{dx(t)}{dt}$, integrating equations (4.4.8) and (4.4.9) from 0 to T yields

$$V(T,x(T)) - V(0,x(0)) + \int_0^T g(t,x,u)dt \geq 0, \text{ and} \qquad (4.4.10)$$

$$V(T,x^*(T)) - V(0,x(0)) + \int_0^T g(t,x^*,u^*)dt = 0. \qquad (4.4.11)$$

Substitution of $V(0,x(0))$ from equation (4.4.11) into equation (4.4.10) then shows that

$$\int_0^T g(t,x,u)dt + V(T,x(T)) \geq \int_0^T g(t,x^*,u^*)dt + V(T,x^*(T)),$$

from which it is clear that $u^*(.)$ is indeed optimal. □

Note

To calculate the optimal control in equation (4.4.7) one first has to solve the Hamilton–Jacobi–Bellman (HJB) equation (4.4.6). Formula equation (4.4.7) then gives the optimal control at arbitrary times t as a possibly time-dependent function of the current state. So the dynamic programming method not only solves the original optimization problem, but tells us even at arbitrary times and arbitrary states what the best action is. Obviously, this is a very nice property in case the decision maker is confronted during the control process with influences which cannot be predicted beforehand. □

To solve the optimization problem given in equations (4.2.1) and (4.2.2) by means of the method of dynamic programming one can proceed as shown in the following algorithm.

Algorithm 4.1

(i) Inspect the problem formulation. If the given problem is a maximization problem, convert it to a minimization problem by inverting the sign of the optimization criterion. Identify the state evolution function $f(t, x, u)$, the instantaneous objective function $g(t, x, u)$ and the scrap-value function $h(x(T))$.

(ii) Write down the Hamilton–Jacobi–Bellman equation. This equation contains a minimization operation.

(iii) Find, for every x and t, the value of u that achieves the minimum. This gives the optimal control u^* as a function of the time t and the state x. This function is in general expressed in terms of the value function V, which still needs to be computed.

(iv) Insert the minimizing value of u in the right-hand side of the HJB equation. The HJB equation now becomes a partial differential equation in which no longer a minimization operation appears.

(v) Solve the HJB equation.

(vi) The explicit form of the value function can now be used, together with the result of step (ii), to write down the optimal control strategy as a function of the time t and the current state x.

(vii) If it is desired to find the optimal control policy as a function of time only, solve the differential equation $\dot{x}(t) = f(t, x(t), u^*(t, x(t)))$. The optimal control policy is then $u(t) = u^*(t, x(t))$ where $x(t)$ is the solution of the differential equation. □

Usually, step (vii) in the above algorithm is skipped since, on the one hand, even when $u^*(t, x)$ is available in explicit form it may not be easy to solve the ordinary differential equation $\dot{x}(t) = f(t, x(t), u^*(t, x(t)))$ and, on the other hand, (see Note following Theorem 4.10) having the optimal policy available as a function of time and state ('feedback form') is often viewed as even better than having it available only as a function of time ('open-loop form'). The feedback strategy as obtained from the dynamic programming algorithm is optimal for any initial condition, whereas the open-loop policy is only optimal for the initial condition that is given in the problem formulation. However, to implement the feedback strategy one needs to monitor the state variable $x(t)$ continuously; so this strategy only makes sense if full information about the current state is available and can be obtained at negligible cost.

As already mentioned above, the HJB equation (4.4.6) is a partial differential equation. In fact it is in general a nonlinear partial differential equation, and so it belongs to one of the most difficult areas of scientific computing. The computational difficulties associated with partial differential equations in more than two or three variables are often prohibitive. In some examples analytical solution is possible. One of the most well-known cases is the linear quadratic case, which we will elaborate on below.

As already outlined in the beginning of this section the dynamic programming theorem just provides a set of sufficient conditions to conclude that the optimization problem has a solution. The sufficient conditions are that the dynamic programming algorithm presumes that various partial differentiability properties are satisfied by the value function. In fact in most cases where the optimal control function has discontinuities (which often occurs if, for example, the set U differs from $\mathbb{R}^m$) this will not be the case. So, in those cases the method breaks down. That is, if these conditions are not satisfied we do not know whether the problem will have a solution or not (see also section 4.5).

Note

1. Following the same lines as above it is easily verified that the recursion (4.4.6) also holds with

$$V^{inf}(t,x) := \inf_{u \in \mathcal{U}} \int_t^T g(s,x(s),u(s))ds + h(x(T)),$$

provided this infimum exists as a finite number for all $t \in [0,T]$. That is, under the same conditions as Theorem 4.10 we have that if $-\infty < V^{inf}(t,x) < \infty$, both partial derivatives of $V^{inf}(t,x)$ exist, $\frac{\partial V}{\partial x}$ is continuous and, moreover, $\frac{d}{dt}V(t,x(t))$ exists then, for $t_0 \leq t \leq T$

$$-\frac{\partial V^{inf}}{\partial t}(t,x) = \inf_{u \in U}\{g(t,x,u) + \frac{\partial V^{inf}}{\partial x}(t,x)\,f(t,x,u)\}, \quad V^{inf}(T,x) = h(x(T)).$$

2. With $T = \infty$ and $h(x(T)) = 0$ under the same conditions as in Theorem 4.10

$$V(t,x) := \min_{u \in \mathcal{U}_{re}} \int_t^\infty g(s,x(s),u(s))ds,$$

satisfies the recursion

$$-V_t(t,x) = \min_{u \in U}\{g(t,x,u) + \frac{\partial V}{\partial x}(t,x)\,f(t,x,u)\}. \tag{4.4.12}$$

where

$$u^*(t,x) = \arg\min_{u \in U}\{g(t,x,u) + \frac{\partial V}{\partial x}(t,x)\,f(t,x,u)\}. \tag{4.4.13}$$

3. In case both functions f and g do not explicitly depend on time in item 2, above, $V_t = 0$ in equation (4.4.12) (see Exercises). □

An advantage of the dynamic programming algorithm is that it immediately facilitates an interpretation of the costate variable λ^* in the maximum principle. If we consider $\lambda^*(t) := V_x(t,x)$ we see that the equations in Theorems 4.6 and 4.10 coincide if we can show that with this definition of λ^*, equation (4.3.2) is satisfied. A proof of this result is provided in the Appendix to this chapter using some additional smoothness assumptions.

Proposition 4.11 (Shadow price interpretation)

Consider the costate variable in the maximum principle. Then

$$\lambda^*(t) = \frac{\partial V(t,x^*(t))}{\partial x}. \tag{4.4.14}$$

□

The name shadow price is explained by interpreting the optimization problem as a profit maximization problem. That is, interpret the state variable, x, as representing the capital stock of a firm, the control variable, u, as representing a business decision the firm makes at each moment of time and the objective function as the profit function of the firm. Then the above Proposition 4.11 states that the use of one additional unit of capital at time t gives an additional total profit of λ^*. Therefore, if the price of capital is below this number, it makes sense for the firm to use additional capital, whereas if this number is below the price of capital it makes sense to lower its amount of capital used. So, in the optimal situation the amount of capital used is such that the price of capital equals this number λ^*.

Example 4.15

Reconsider Example 4.4 on the linear quadratic case. That is, consider the minimization of

$$\int_0^T \{x^2(t) + u^2(t)\}dt + ex^2(T) \tag{4.4.15}$$

subject to

$$\dot{x}(t) = ax(t) + bu(t), \ x(0) = 1 \ (b \neq 0). \tag{4.4.16}$$

We will follow the procedure outlined in the Algorithm following Theorem 4.10 to solve this problem.

(i) The functions in the general formulation have the following form:

$$f(t, x, u) = ax + bu$$
$$g(t, x, u) = x^2 + u^2$$
$$h(t, x, u) = ex^2.$$

(ii) The Hamilton–Jacobi–Bellman equation for the value function $V(t, x)$ is:

$$-\frac{\partial V}{\partial t}(t, x) = \min[x^2 + u^2 + \frac{\partial V}{\partial x}(t, x)(ax + bu)]$$

with the boundary condition

$$V(T, x) = ex^2.$$

(iii) Next differentiate

$$x^2 + u^2 + \frac{\partial V}{\partial x}(t, x)(ax + bu)$$

with respect to u. Finding the points where this derivative becomes zero gives

$$u^* = -\frac{1}{2}b\frac{\partial V}{\partial x}(t,x). \tag{4.4.17}$$

(iv) Substitution of equation (4.4.17) into the HJB equation gives the following partial differential equation

$$-\frac{\partial V}{\partial t}(t,x) = x^2 - \frac{1}{4}b^2\left(\frac{\partial V}{\partial x}(t,x)\right)^2 + \frac{\partial V}{\partial x}(t,x)ax, \quad V(T,x) = ex^2. \tag{4.4.18}$$

(v) To solve the above partial differential equation, try a solution of the form

$$V(t,x) = k(t)x^2 \tag{4.4.19}$$

where $k(t)$ is an as yet unknown function of time. With this guess,

$$\frac{\partial V}{\partial t}(t,x) = \dot{k}(t)x^2, \quad \text{and} \quad \frac{\partial V}{\partial x}(t,x) = 2k(t)x.$$

Substitution of this into the partial differential equation (4.4.18) gives

$$-\dot{k}(t)x^2 = x^2 - b^2k^2(t)x^2 + 2k(t)ax^2, \quad V(T,x) = ex^2. \tag{4.4.20}$$

So, a function of the form (4.4.19) satisfies equation (4.4.18) if $k(t)$ satisfies the ordinary differential equation[3]

$$\dot{k}(t) = -2ak(t) + b^2k^2(t) - 1, \quad k(T) = e. \tag{4.4.21}$$

(vi) If the Riccati equation (4.4.21) has a solution then the optimal control strategy as a function of the time and the current state is

$$u^*(t,x) = k(t)x.$$

(vii) The open-loop expression for the optimal control is obtained as follows. First substitute u^* into the differential equation (4.4.16). Solving the resulting differential equation $\dot{x}(t) = (a + bk(t))x(t)$, $x(0) = 1$ then yields the optimal state trajectory $x^*(.)$. $u^*(t) = k(t)x^*(t)$ is then the corresponding open-loop optimal control. Of

---

[3]Equations in which a square of the unknown function appears, such as equation (4.4.20), were studied by Count Jacopo Francesco Riccati in 1723, so about 25 years after the invention of the differential calculus; these quadratic differential equations are nowadays known as **Riccati equations**.

course this expression should coincide with the solution of the two-point boundary-value problem (4.2.28).

Finally notice, that $V(t, x) = k(t)x^2$ satisfies all assumptions imposed in the dynamic programming Theorem 4.10.   □

4.5   Solving optimal control problems

In the previous sections we have seen a number of theorems stating either necessary or sufficient conditions for a control path to be optimal. In practice it turns out, however, that the correct use of these theorems is sometimes tricky. Therefore we state here how one should proceed with the presented material.

Consider first the approach which uses necessary optimality conditions, that is, conditions that every optimal solution must satisfy. If there are only a few solutions that satisfy the necessary conditions, we can simply compare the value of the cost function for these candidate optimal solutions. The solution which has the lowest value is then **the candidate** sought optimal solution. Why is this still just the candidate solution? That is because the problem might not have a solution at all. For instance, if we look at the minimum value of the scalar function $f(x) = -x^2$, then the first-order necessary condition yields the only candidate solution $x = 0$. Obviously, the problem has no solution in this case. So, if one just considered the necessary condition in this example one would draw a wrong conclusion. So, the approach based on necessary optimality conditions should always be followed by an inspection of optimality of the proposed candidate solution.

Next, consider the approach that uses sufficient optimality conditions. That is, the approach that returns solutions under the assumption that a number of a priori imposed conditions are satisfied by the optimization problem. Every such solution is then optimal. The problem with this approach is that we can miss some, or maybe all, of the optimal solutions if the sufficient optimality conditions are too strong. For instance, we know that a convex function yields a global minimum at a stationary point. Now, consider the function $f(x) = \sin(x)$. This function is not convex, but it has an infinite number of global minimum locations.

4.6   Notes and references

The following books have been consulted. For section 4.1 the book by Luenberger (1969) and for section 4.2 the books of Chiang (1992) and Teo, Goh and Wong (1991). For sections 4.3 and 4.4 the lecture notes 'Dynamic Optimization' by Schumacher (2002) and 'Optimalisering van Regelsystemen' by Hautus (1980) have been extensively used.

In the literature several extensions of the basic problem setting analyzed in this chapter have been covered. For instance the inclusion of constraints on the state variables during the planning horizon and the consideration of an infinite planning horizon with adapted optimality definitions are subjects which have been thoroughly analyzed. The books by Feichtinger and Hartl (1986) and Seierstad and Sydsaeter (1987), for example, contain an extensive treatment of the maximum principle with many economic applications, whereas three classical works which give extensive technical details on optimal control problems including endpoint and/or state constraints are Fleming and Rishel (1975), Lee and Markus (1967) and Neustadt (1976). In the book by Carlson and Haurie (1991) one can find various considerations concerning infinite planning horizon optimal control problems (see also Halkin (1974) and Brock and Haurie (1976)). For linear systems there is a large amount of literature dealing with control problems in a more general setting. Two adequate textbooks are Zhou, Doyle and Glover (1996) and Trentelman Stoorvogel and Hautus (2001).

4.7 Exercises

1. Determine the Gateaux differential of the following matrix functions f at X with increment ΔX.

   (a) $f(X) = AX$;
   (b) $f(X) = X^T AX$;
   (c) $f(X) = AX^2 + X$.

2. Prove Theorem 4.3.

3. Solve the following minimization problems.

   (a) $\min\limits_{u(.)} \int_0^3 \{2x(t) + 3u^2(t)\}dt$ subject to $\dot{x}(t) = u(t),\ x(0) = -1$.

   (b) $\min\limits_{u(.)} \int_0^1 \{2x(t) + \frac{1}{2}u^2(t)\}dt - x^3(1)$ subject to $\dot{x}(t) = u(t),\ x(0) = 1$.

   (c) $\min\limits_{u(.)} \int_0^2 \{2x(t) + \frac{1}{4}u^4(t)\}dt + x(2)$ subject to $\dot{x}(t) = -u(t) + 1,\ x(0) = 0$.

4. Solve the following minimization problems.

   (a) $\min\limits_{c(.)} \int_0^2 \{n(t) + c^2(t)\}dt$ subject to $\dot{n}(t) = n(t) + 2c(t) - 4,\ n(0) = 1$.

   (b) $\min\limits_{c(.)} \int_0^T \{2n(t) + c^2(t)\}dt + 2n(T)$ subject to $\dot{n}(t) = 2n(t) + c(t),\ n(0) = 1$.

   (c) $\min\limits_{c(.)} \int_0^1 \left\{\frac{3}{2}n^2(t) + \frac{1}{2}c^2(t)\right\}dt$ subject to $\dot{n}(t) = n(t) + c(t),\ n(0) = 1$.

5. Solve the following minimization problems.

   (a) $\min\limits_{-1 \le p(.) \le 1} \int_0^1 \{-y(t)\}dt$ subject to $\dot{y}(t) = y(t) + p(t),\ y(0) = 0$.

   (b) $\min\limits_{0 \le q(.) \le 1} \int_0^1 \{-q^2(t) + q(t) - tr(t)\}dt$ subject to $\dot{r}(t) = t + 2q(t),\ r(0) = 1$.

   (c) $\min\limits_{-1 \le w(.) \le 1} \int_0^4 \{3w^2(t) + 4v(t)\}dt$ subject to $\dot{v}(t) = -v(t) + 3w(t),\ v(0) = 1$.

(d) $\min\limits_{0\le u(.)\le 1} \int_0^5 \left\{\frac{1}{2}x^2(t) - u^2(t)\right\}dt$ subject to $\dot{x}(t) = 2u(t)$, $x(0) = 3$. (Hint: first show that the costate $\lambda^*(t) \ge 0$ for $t \in [0,5]$, next show that $u^*(t) = 0$ if $\lambda^* < \frac{1}{2}$ and $u^*(t) = 1$ if $\lambda^* > \frac{1}{2}$.)

6. Assume that the problem to find the minimum of the integral

$$\int_0^T \{x^T(t)Qx(t) + u^T(t)Ru(t)\}dt + x^T(T)Q_Tx(T) \text{ subject to } \dot{x}(t) = Ax(t) + Bu(t),$$

$$x(0) = x_0,$$

has a solution. Here, Q, Q_T and R are symmetric matrices where R is additionally assumed to be positive definite. Let $S := BR^{-1}B^T$. Show that a costate function $\mu^*(.)$ exists such that the following two-point boundary-value problem has a solution:

$$\begin{bmatrix} \dot{x}^*(t) \\ \dot{\mu}^*(t) \end{bmatrix} = \begin{bmatrix} A & -S \\ -Q & -A^T \end{bmatrix} \begin{bmatrix} x^*(t) \\ \mu^*(t) \end{bmatrix}, \text{ with } \begin{bmatrix} x^*(0) \\ \mu^*(T) \end{bmatrix} = \begin{bmatrix} x_0 \\ Q_Tx^*(T) \end{bmatrix}.$$

(Hint: determine first, using the maximum principle, the necessary condition for $u^*(t)$.)

7. Let S and U be convex sets in $\mathbb{R}^n$ and $\mathbb{R}^m$, respectively. Let $f(x, u)$ be a convex function of (x, u), $x \in S$, $u \in U$. Define

$$h(x) := \min_{u \in U} f(x, u),$$

where we assume that the minimum value exists for each $x \in S$. Prove that h is convex in S. Hint: Let $x_1, x_2 \in S$, $\lambda \in [0, 1]$ such that $h(x_1) = f(x_1, u_1)$ and $h(x_2) = f(x_2, u_2)$.)

8. Consider the optimization problem

$$\min_{u(.)} \int_0^2 \{2x^2(t) + u^2(t)\}dt,$$

subject to the constraints

$$\dot{x}(t) = x(t) + 2u(t); \; x(0) = -1,$$

and $u(.) \ge 0$.

(a) Determine the necessary conditions that have to be satisfied according the maximum principle.

(b) Consider the set of differential equations

(i) $\begin{cases} \dot{x}_1(t) = x_1(t) - 2x_2(t) \\ \dot{x}_2(t) = -4x_1(t) - x_2(t). \end{cases}$

and

(ii) $\begin{cases} \dot{x}_1(t) = x_1(t) \\ \dot{x}_2(t) = -4x_1(t) - x_2(t). \end{cases}$

Make a qualitative picture of the behavior of the solutions of both sets of differential equations (i) and (ii) in the (x_1, x_2)-plane, respectively.

(c) Assume that the dynamics of a process are described for $x_2 \geq 0$ by (ii) and for $x_2 < 0$ by (i). Draw the phase-plane diagram for this process.

(d) Solve the optimization problem qualitatively.

(e) Use the approach from item (d) to determine the optimal control if $x_0 = 1$.

9. Reconsider Example 4.8. That is

$$\min_{-1 \leq u \leq 1} \int_0^2 2x^2(t) + u^2(t)dt, \text{ subject to } \dot{x}(t) = x(t) + 2u(t), \ x(0) = 1.$$

(a) Show that there are two switching curves $\lambda = 1$ and $\lambda = -1$ in the phase plane (x, λ).

(b) Show that for each of the three different control regimes $u(.) = 1$, $u(.) = -\lambda$ and $u(.) = -1$, the corresponding phase portrait has a saddle-point equilibrium. Determine the phase portrait for each of these control regimes separately.

(c) Determine the phase portrait of the state and costate variable for the optimal control problem by merging the three phase portraits in (b).

(d) Show from the phase portrait in (c) that the optimal control for the minimization problem is given by either $u^*(t) = -\lambda^*(t)$ or by $u^*(t) = -1$ during some time interval $[0, t_1]$ followed by $u^*(t) = -\lambda^*(t)$ on the remaining time interval $[t_1, 2]$.

10. Consider the problem

$$\min_{0 \leq u \leq 1} \int_0^T -\frac{1}{2}x^2(t) + x(t)dt, \text{ subject to } \dot{x}(t) = u(t), \ x(0) = 0.$$

(a) Determine the necessary conditions for optimality resulting from the maximum principle.

(b) Show that there exists one switching curve in the phase plane (x, λ).

(c) Determine for each of the two different control regimes $u(.) = 1$ and $u(.) = 0$ the corresponding phase portrait.

(d) Determine the phase portrait of the state and costate variable for the optimal control problem by merging the two phase portraits in (c).

(e) Indicate in the phase portrait of (d) the set of candidate optimal trajectories. Draw conclusions with respect to the implied optimal control trajectory.

(f) Determine analytically the optimal control if $T = 1$ and if $T = 3$.

11. Consider the optimal consumption–investment problem

$$\max_{0 \leq u \leq 2} \int_0^T e^{-rt} \ln(u(t))dt, \text{ subject to } \dot{x}(t) = 2\sqrt{x(t)} - u(t), \ x(0) = x_0.$$

Here $0 < r < 1$ is a discount factor.

(a) Give an economic interpretation of the variables and model.

(b) Rewrite the problem into the standard form and use the current-value formulation to determine the necessary conditions for optimality resulting from the maximum principle.

(c) Sketch in the (x, μ)-plane (where μ is the adjoint variable used in the current-value formulation) trajectories of solutions from the set of differential equations determined in (a).

(d) Indicate which parts of the phase portrait in (c) correspond to an optimal solution for a given x_0.

12. Consider the optimal control problem

$$\min_{u(.)} \int_0^3 \{u^2(t) - 4x^2(t)\}dt + \frac{7}{4}x^2(3), \quad \text{subject to } \dot{x}(t) = 2x(t) + u(t), \; x(0) = x_0.$$

(a) Determine the Hamilton–Jacobi–Bellman (HJB) equation associated with this problem.

(b) Show that the HJB equation has a solution of the form $V(t, x) = \left(2 - \frac{1}{t+1}\right)x^2$.

(c) Determine the optimal control for this problem.

13. Consider the optimal control problem

$$\max_{u(.)} \int_0^T e^{-rt} \sqrt{u(t)}dt + e^{-rT} \frac{\sqrt{x(T)}}{\sqrt{2r - a}}, \quad \text{subject to } \dot{x}(t) = ax(t) - u(t), \; x(0) = x_0 > 0.$$

Here $r > 0$ is a discount factor, $a < 2r$ and $u(.) \geq 0$.

(a) Determine the Hamilton–Jacobi–Bellman (HJB) equation associated with this problem.

(b) Show that the HJB equation has a solution of the form $V(t, x) = \beta e^{-rt} \sqrt{x}$, for an appropriately chosen constant β. Determine β.

(c) Determine the optimal control for this problem.

14. Consider the optimal control problem

$$\min_{u(.)} \int_0^T \frac{1}{2} e^{-rt}\{x^2(t) + u^2(t)\}dt + f_T(x), \quad \text{subject to } \dot{x}(t) = u(t), \; x(0) = x_0.$$

Here $r > 0$ is a discount factor and $f_T(x)$ is a function that will be specified later on.

(a) Determine the Hamilton–Jacobi–Bellman (HJB) equation associated with this problem.

(b) Choose f_T and a constant c such that the HJB equation has a solution of the form $V(t, x) = ce^{-rt}x^2$.

(c) Which optimal control is suggested by (b) if $T = \infty$ and $f_T(x) = 0$?

15. Consider the optimal control problem

$$\max_{u(.)} \int_0^T e^{-rt}\sqrt{u(t)}dt + e^{-rT}W(x(T)), \text{ subject to } \dot{x}(t) = ax(t) - u(t), \ x(0) = x_0 > 0.$$

Here $u(.) \geq 0$ and $W(.)$ is an as yet unspecified function.

(a) Determine the Hamilton–Jacobi–Bellman (HJB) equation associated with this problem.

(b) Show that the HJB equation has a solution of the form $V(t,x) = e^{-rt}g(x)$ only if the differential equation

$$ax\left(\frac{dg}{dx}(x)\right)^2 - rg(x)\frac{dg}{dx}(x) + \frac{1}{4} = 0$$

has a solution.

(c) Choose $a = r = 1$ and $W(x) = 1 + \frac{1}{4}x$. Determine the optimal control for this problem.

16. Consider the optimal control problem

$$\min_{u(.)} \int_0^T \{q(t)x^2(t) + r(t)u^2(t)\}dt + q_T x^2(T),$$

$$\text{subject to } \dot{x}(t) = a(t)x(t) + b(t)u(t), \ x(0) = x_0.$$

Here the functions $q(t)$, $r(t)$, $a(t)$ and $b(t)$ are piecewise continuous functions with $r(t) > 0$ and q_T is a constant. Assume that the differential equation

$$\dot{k}(t) = -2a(t)k(t) + \frac{b^2(t)}{r(t)}k^2(t) - q(t), \ k(T) = q_T,$$

has a solution on $[0, T]$.

(a) Determine the Hamilton–Jacobi–Bellman (HJB) equation associated with this problem.

(b) Show that the HJB equation has a solution of the form $V(t,x) = k(t)x^2$.

(c) Determine the optimal control for this problem.

17. Consider the optimal investment problem

$$\max_{i(.)} \int_0^\infty e^{-rt}\{U(k(t)) - ci(t)\}dt, \text{ subject to } \dot{k}(t) = \sqrt{i(t)} - \rho k(t), \ k(0) = k_0.$$

Here all parameters and functions are assumed to be positive. Moreover, $\frac{dU(k)}{dk} > 0$ and $\frac{d^2U(k)}{dk^2} < 0$.

(a) Give an economic interpretation of the above variables and model.

(b) Determine the Hamiltonian associated with this problem.

(c) Conclude from the necessary conditions that the optimal state and control trajectories, $k^*(.)$ and $i^*(.)$, respectively, satisfy the set of differential equations

$$\dot{k}(t) = \sqrt{i(t)} - \rho k(t), \quad k(0) = k_0;$$

$$\dot{i}(t) = \frac{i(t)}{c}\left\{2c(r+\rho)\sqrt{i(t)} - \frac{dU(k)}{dk}(t)\right\}.$$

(Hint: differentiate the expression implied by the necessary conditions for the optimal control $i^*(t)$ w.r.t. time and next eliminate the adjoint variable in the resulting expression.)

(d) Make a phase portrait of the state and control variable. Show that there is a unique equilibrium.

(e) Assume that in this optimization problem both the state and control trajectory have to remain bounded. Indicate in the graph of (d) how for a fixed k_0 the initial control variable should be chosen. Indicate how one can compute the optimal control on $[0, \infty)$.

18. Consider the optimal fishing policy problem

$$\max_{q(.)} \int_0^\infty e^{-rt}h(q(t))dt, \quad \text{subject to } \dot{v}(t) = v(t)(c - v(t)) - q(t), \quad v(0) = v_0.$$

Here all parameters and functions are assumed to be positive. Moreover, $\frac{dh(q)}{dq} > 0$ and $\frac{d^2h(q)}{dq^2} < 0$.

Answer the same questions as in Exercise 17, with the equations in Exercise 17(c) replaced by

$$\dot{v}(t) = v(t)(c - v(t)) - q(t), \quad v(0) = v_0;$$

$$\dot{q}(t) = \frac{\frac{dh}{dq}(q(t))}{\frac{d^2h}{dq^2}(q(t))}\{r + 2v(t) - c\}.$$

19. Consider the optimal investment problem

$$\max_{i(.)} \int_0^\infty e^{-rt}\{h(i(t)) + \frac{1}{2}k^2(t)\}dt, \quad \text{subject to } \dot{k}(t) = i(t) - \delta k(t), \quad k(0) = k_0.$$

Here all parameters and functions are assumed to be positive. Moreover, $\frac{dh(i)}{di} < 0$ and $\frac{d^2h(i)}{dh^2} > 0$.

Answer the same questions as in Exercise 17, with the equations in Exercise 17(c) replaced by

$$\dot{k}(t) = i(t) - \delta k(t), \quad k(0) = k_0;$$

$$\dot{i}(t) = \frac{1}{\frac{d^2h}{di^2}}\left\{k(t) + (r+\delta)\frac{dh}{di}(i(t))\right\}.$$

20. Consider the optimal labour recruitment policy problem

$$\max_{v(.)} \int_0^\infty e^{-rt}\{pF(L) - wL(t) - \frac{1}{2}v^2(t)\}dt, \text{ subject to } \dot{L}(t) = v(t) - qL(t), \ L(0) = L_0.$$

Here all parameters and functions are assumed to be positive. Moreover, $\frac{dF(L)}{dL} > 0$ and $\frac{d^2F(L)}{dL^2} < 0$.

Answer the same questions as in Exercise 17, with the equations in Exercise 17(c) replaced by

$$\dot{L}(t) = v(t) - qL(t), \ L(0) = L_0;$$

$$\dot{v}(t) = (r+q)v(t) - p\frac{dF}{dL}(L(t)) + w.$$

21. Consider the optimal consumption problem (Ramsey model)

$$\max_{c(.)} \int_0^\infty e^{-rt}U(c(t))dt, \text{ subject to } \dot{k}(t) = P(k(t)) - ak(t) - c(t), \ k(0) = k_0.$$

Here all parameters and functions are assumed to be positive. $P(k)$ is the production function of capital k and $U(c)$ is the utility function of consumption c. Assume that $\frac{dP(k)}{dk} > 0$, $\frac{d^2P(k)}{dk^2} < 0$, $\frac{dU(c)}{dc} > 0$ and $\frac{d^2U(c)}{dc^2} < 0$.

(a) Answer the same questions as in Exercise 17, with the equations in Exercise 17(c) replaced by

$$\dot{k}(t) = P(k(t)) - ak(t) - c(t), \ k(0) = k_0;$$

$$\dot{c}(t) = \frac{\frac{dU(c)}{dc}}{\frac{d^2U(c)}{dc^2}}\left(r + a - \frac{dP(k)}{dk}\right).$$

(b) Show that in the optimal equilibrium situation, $c^* = P(k^*) - ak^*$, where k^* is determined by $\frac{dP(k^*)}{dk} = r + a$.

(c) Consider, instead of the above dynamic optimization problem, the maximization problem

$$\max_k c(k), \text{ where } c(k) = P(k) - ak.$$

Solve this static optimization problem.

(d) Compare the answers in items (b) and (c). Give an intuitive explanation for the difference.

22. Assume that for all $t > 0$ inf $\int_t^\infty g(t,x,u)dt$ subject to $\dot{x}(t) = f(t,x,u)$, $x(0) = x_0$ exists for all x_0. Derive the HJB equation associated with this problem and state conditions under which this equation holds.

23. Show the correctness of the Note following the dynamic programming algorithm item 3. That is, consider the minimization of

$$J(t) := \int_t^\infty g(x(s), u(s))ds, \text{ subject to } \dot{x}(s) = f(x(s), u(s)), \ x(t) = x_0 \text{ and } u(.) \in \mathcal{U}_{re}.$$

Assume that this problem has a solution for all $t \geq 0$. Moreover, assume that the minimum in $J(t)$ is obtained by choosing $u(.) = u^*(t,.)$ and that at time $t = 0$ this minimum is J^*.

Show that for all $t > 0$ this minimum coincides with J^* and conclude that $V_t = 0$.

(Hint: consider $u(s) = u^*(0, s - t)$ in $J(t)$ and $u(s) = u^*(t, s + t)$ in $J(0)$.)

24. Assume that the optimization problem

$$\min_{u \in \mathcal{U}_{re}} \int_t^\infty e^{-rt} g(t, x, u) dt \text{ subject to } \dot{x}(t) = f(t, x, u), \ x(0) = x_0$$

has a solution for all x_0. Show that the HJB equation associated with this problem is

$$rV(t, x) = \min_{u \in U}\{g(t, x, u) + \frac{\partial V}{\partial x}(t, x) f(t, x, u)\}.$$

where

$$u^*(t, x) = \arg\min_{u \in U}\{g(t, x, u) + \frac{\partial V}{\partial x}(t, x) f(t, x, u)\}.$$

State conditions under which the HJB equation yields the optimal control.

4.8 Appendix

Proof of Theorem 4.6

Let $u \in \mathcal{U}$ be an arbitrary control function and introduce

$$H(t, x, u, \lambda) := f(t, x, u) + \lambda g(t, x, u).$$

Then, as in section 4.2, we consider the expression (4.2.7). Note that this formula holds for an arbitrary function $\lambda(.)$. Now choose $\lambda(.) = \lambda^*(.)$. Then

$$J(u^*) - J(u) = \int_0^T \{H(t, x^*(t), u^*(t), \lambda^*(t)) - H(t, x(t), u(t), \lambda^*(t)) + \dot{\lambda}^*(t)(x^*(t) - x(t))\}dt +$$

$$+ h(x^*(T)) - h(x(T)) + \lambda^*(T)(x(T) - x^*(T))$$

$$= \int_0^T \{H(t, x^*(t), u^*(t), \lambda^*(t)) - H(t, x^*(t), u(t), \lambda^*(t))dt +$$

$$+ \int_0^T \{H(t, x^*(t), u(t), \lambda^*(t)) - H(t, x(t), u(t), \lambda^*(t)) + \dot{\lambda}^*(t)(x^*(t) - x(t))\}dt +$$

$$+ h(x^*(T)) - h(x(T)) + \lambda^*(T)(x(T) - x^*(T)). \tag{i}$$

Next assume that at some point of continuity t_0 condition (4.3.3) is violated. So, there is a $u \in U$ such that

$$H(t_0, x^*(t_0), u, \lambda^*(t_0)) < H(t_0, x^*(t_0), u^*(t_0), \lambda^*(t_0)).$$

As a consequence of continuity, there would then exist an $\epsilon > 0$ and a $\delta > 0$ such that

$$H(t, x^*(t), u, \lambda^*(t)) + \epsilon < H(t, x^*(t), u^*(t), \lambda^*(t)) \tag{4.8.1}$$

for all $t \in [t_0 - \delta, t_0 + \delta]$. Next, for $0 < h < \delta$ define the control function $u_h(t)$ by

$$u_h(t) = \begin{cases} u^*(t) & t \in [0, t_0 - h] \cup [t_0, T] \\ u & t \in (t_0 - h, t_0). \end{cases}$$

By choosing h 'small' enough, one can easily verify, by invoking Theorem 3.13 twice, that for this control the differential equation (4.2.2) has a solution in the extended sense. So $u_h \in \mathcal{U}$. Moreover the state trajectory x_h corresponding to u_h satisfies

$$x_h(t) = x^*(t), \ t \in [0, t_0 - h],$$

and

$$x_h(t) = x^*(t) + \int_{t_0 - h}^{t} \{f(t, x_h(t), u) - f(t, x^*(t), u^*(t))\} dt, \ t \in (t_0 - h, t_0). \tag{4.8.2}$$

Since $x_h(t) \to x^*(t_0)$ if $h \to 0$, $u^*(t)$ is continuous at t_0 and f is continuous, we deduce from equation (4.8.2) that $x_h(t)$ can be approximated on $(t_0 - h, t_0)$ by

$$x_h(t) = x^*(t) + \int_{t_0 - h}^{t} \{f(t_0, x^*(t_0), u) - f(t_0, x^*(t_0), u^*(t_0))\} dt + 0(h)$$

$$= x^*(t) + (t - t_0 + h)\{f(t_0, x^*(t_0), u) - f(t_0, x^*(t_0), u^*(t_0))\} + 0(h), \ t \in (t_0 - h, t_0).$$

So, introducing $q_h = f(t_0, x^*(t_0), u) - f(t_0, x^*(t_0), u^*(t_0))$,

$$x_h(t) = x^*(t) + (t - t_0 + h)q_h + 0(h), \ t \in (t_0 - h, t_0). \tag{4.8.3}$$

On the interval $[t_0, T]$ both $x^*(t)$ and $x_h(t)$ satisfy the differential equation (4.2.2) with $u(.) = u^*(.)$. Only, the initial state at t_0 differs. Consequently, by Theorem 3.13

$$x_h(t) = x^*(t) + h\Phi(t, t_0)q_h + 0(h), \ t \in [t_0, T], \tag{4.8.4}$$

where $\Phi(t_0, t)$ is the fundamental solution of the linear differential equation

$$\dot{\Phi}(t) = f_x(t, x^*(t), u^*(t))\Phi(t), \ \Phi(t_0) = I.$$

Next, consider the expression

$$\int_0^T \{H(t, x^*(t), u_h(t), \lambda^*(t)) - H(t, x_h(t), u_h(t), \lambda^*(t)) + \dot{\lambda}^*(t)(x^*(t) - x_h(t))\} dt$$

$$+ h(x^*(T)) - h(x_h(T)) + \lambda^*(T)(x_h(T) - x^*(T)).$$

Making a first-order Taylor expansion in x for $H(t, x_h, u_h, \lambda^*)$ around x^* yields:

$$H(t, x_h(t), u_h(t), \lambda^*(t)) = H(t, x^*(t), u_h(t), \lambda^*(t)) + H_x(t, x^*(t), u_h(t), \lambda^*(t))(x_h(t) \\ - x^*(t)) + 0(x_h(t) - x^*(t)).$$

Similarly, one obtains for $h(x_h(T))$

$$h(x_h(T)) = h(x^*(T)) + h_x(x^*(T))(x_h(T) - x^*(T)) + 0(x_h(T) - x^*(T)).$$

From our analysis of $x_h(t)$ it is now easily seen that $0(x_h(t) - x^*(t)) = \theta(h)$, where $\frac{\theta(h)}{h}$ tends to zero when h tends to zero. Substitution of this and using the definition of λ^* (see equation (4.3.2)) into (ii) shows that the expression in (ii) is of order $\theta(h)$. Consequently, we have from (i) that

$$J(u^*) - J(u_h) = \int_0^T \{H(t, x^*(t), u^*(t), \lambda^*(t)) - H(t, x^*(t), u_h(t), \lambda^*(t))dt + \theta(h)$$

$$= \int_{t_0 - h}^{t_0} \{H(t, x^*(t), u^*(t), \lambda^*(t)) - H(t, x^*(t), u_h(t), \lambda^*(t))dt + \theta(h).$$

Since according to equation (4.8.1) $H(t, x^*(t), u^*(t), \lambda^*(t)) - H(t, x^*(t), u_h, \lambda^*(t)) > \epsilon$, the value of the above integral term is larger than ϵh. Therefore for all $h \in (0, \delta)$

$$\frac{J(u^*) - J(u_h)}{h} > \epsilon + \frac{\theta(h)}{h}.$$

Since $\frac{\theta(h)}{h}$ tends to zero as h tends to zero,

$$\frac{J(u^*) - J(u_h)}{h} > 0$$

for sufficiently small h. So for these values of h, $J(u^*) > J(u_h)$, which would mean that u^* is not optimal. Therefore our assumption that equation (4.3.3) fails at some point of continuity t_0 must be false, which completes the proof. $\qquad \square$

Proof of Theorem 4.9

First we show that under the conditions mentioned in the theorem, $\lambda^*(.)$ exists on $[0, \infty)$. To that end notice that the partial derivative of the Hamiltonian H w.r.t. x in equation (4.3.50) equals $g_x(t, x^*, u^*) + \lambda^* f_x(t, x^*, u^*)$. So, λ^* satisfies the linear differential equation of the form

$$\dot{\lambda} = -\lambda A(t) + b(t)$$

where both $A(t)$ and $b(t)$ are continuous and $A(t)$ is as specified above. Therefore, according to Theorem 3.5, this differential equation has a unique solution on $[0, \infty)$.

Next, consider the perturbed control function u_h as defined in the proof of the maximum principle, Theorem 4.6. We show that this control function belongs to $\mathcal{U}_{re}$.

To that end, we have to show that the induced state trajectory x_h converges. Assume that the unperturbed trajectory converges. That is,

$$\lim_{t \to \infty} x^*(t) = \bar{x}.$$

From equation (4.8.4) we have that[4]

$$x_h(t) = x^*(t) + h\Phi(t, t_0)q_h + 0(h).$$

Since due to our assumption (4.3.49) the system induced by $\dot{y} = A(t)y(t)$, $y(t_0) = y_0$, is asymptotically stable (for example Willems (1970)), it is clear that x_h indeed converges to $\bar{x}$.

Basically, the proof of the maximum principle can now be copied to derive the conclusion. This is left to the reader. There is only one technical detail left to be shown, which concerns the partial integration in equation (4.2.6). Since we assume that J^* is finite this step is justified in case $\int_0^\infty \frac{d}{dt}(\lambda^*(t)x^*(t))dt$ exists. However, by assumption $\lim_{t\to\infty} \lambda^*(t)x^*(t)$ exists. So, it is clear that the partial integration is indeed correct for J^*; but, in copying the proof we also assume that with respect to $J(u_h)$ this partial integration step is correct, which is not immediately clear. To show the correctness of this, it suffices to show that $\lim_{t\to\infty} \lambda^*(t)x_h(t)$ exists. To that end recall from section 3.2.2, Theorem 3.11, that with $\lambda^*(t_0) = \lambda_0$, λ^* satisfies the variation-of-constants formula

$$\lambda^*(t) = \lambda_0\Phi(t_0, t) + \int_0^t b(\tau)\Phi(\tau, t)d\tau.$$

Consequently,

$$\lambda^*(t)x_h(t) = \lambda^*(t)x^*(t) + h\lambda^*(t)\Phi(t, t_0)q_h + \lambda^*(t)0(h)$$

$$= \lambda^*(t)x^*(t) + h(\lambda_0\Phi(t_0, t) + \int_0^t b(\tau)\Phi(\tau, t)d\tau)\Phi(t, t_0)q_h + \lambda^*(t)0(h))$$

$$= \lambda^*(t)x^*(t) + h(\lambda_0 + \int_0^t b(\tau)\Phi(\tau, t_0)d\tau q_h) + \lambda^*(t)0(h).$$

Since the integrand is convergent, it is clear now that $\lambda^*(t)x_h(t)$ also converges. $\square$

Proof of Proposition 4.11

Let $u^*(t)$ and $x^*(t)$ be optimal. Define $\lambda(t) := V_x(t, x^*(t))$, where V is the value function satisfying the conditions of Theorem 4.10. Since $\lambda(T) = V_x(T, x^*(T))$ it follows immediately that $\lambda(T) = h_x(x^*(T))$, which shows that λ satisfies the boundary condition of equation (4.3.2). Now, under the additional assumption that $V \in C^2$ we see that

$$\dot{\lambda}(t) = \frac{d}{dt}V_x(t, x^*(t)) = V_{tx}(t, x^*(t)) + \dot{f}(t, x^*, u^*)V_{xx}(t, x^*(t)). \qquad (4.8.5)$$

---

[4]By $0(h)$ we mean that $\lim_{h\to 0} \frac{0(h)}{h} = 0$.

Now let

$$F(t,x^*) := V_t(t,x^*) + g(t,x^*,u^*) + V_x(t,x^*) f(t,x^*,u^*).$$

Then, since (x^*, u^*) are optimal, $F(t, x^*) = 0$, $t_0 \le t \le T$ (see equation (4.4.6)). Now let $\tau \in [t_0, T]$. By assumption $x^*(t)$ satisfies

$$\dot{x} = f(t,x,u^*(t)), \ x(\tau) = x^*(\tau), \ \tau \le t \le T.$$

By Theorem 3.9, for a 'close' to $x^*(\tau)$, the differential equation

$$\dot{x} = f(t,x,u^*(t)), \ x(\tau) = a, \ \tau \le t \le T$$

then has a solution. So, by equation (4.4.6) the corresponding induced trajectory satisfies

$$-\frac{\partial V}{\partial t}(\tau,a) \le g(\tau,a,u^*) + \frac{\partial V}{\partial x}(\tau,a) f(\tau,a,u^*),$$

or stated differently,

$$F(\tau,a) \ge 0, \ \forall a \in N(x^*(\tau)).$$

So, the function $a \to F(\tau, a)$ has a local minimum at $a = x^*(\tau)$. Therefore

$$F_x(\tau,x^*(\tau)) = 0.$$

Since $\tau \in [t_0, T]$ was chosen arbitrarily, it follows that

$$F_x(t,x^*(t)) = 0, \ t_0 \le t \le T.$$

Elaborating the left-hand side of this equation gives

$$V_{tx}(t,x^*) + \dot{f}(t,x^*(t),u^*(t))V_{xx}(t,x^*(t)) + g_x(t,x^*,u^*) + V_x(t,x^*(t)) f_x(t,x^*(t),u^*(t)) = 0.$$

So, using equation (4.8.5), we conclude that

$$\dot{\lambda}(t) + g_x(t,x^*,u^*) + V_x(t,x^*(t)) f_x(t,x^*(t),u^*(t)) = 0, \ t_0 \le t \le T.$$

That is, λ satisfies equation (4.3.2). □

5

Regular linear quadratic optimal control

In this chapter we treat the optimal control of linear time-invariant systems with a quadratic cost function. We consider both the finite and infinite planning horizon case.

5.1 Problem statement

Consider the linear time-invariant system:

$$\dot{x}(t) = Ax(t) + Bu(t), \ x(t_0) = x_0, \tag{5.1.1}$$

where, as usual, $x(t) \in \mathbb{R}^n$ is the state of the system, $x(t_0)$ the initial given state of the system, and $u(.) \in \mathcal{U}$ is the vector of variables that can be used to control the system. One objective might be to find a control function $u(t)$ defined on $[t_0, T]$ which drives the state to a small neighborhood of zero at time T. This is the so-called **Regulator Problem**. Conversely, if the state of the system represents a set of economic variables to which revenues are attached, which increase the larger these variables are, and u represents investment actions by which the value of these variables can be increased, the objective might be to control the value of these variables as quick as possible towards some desired level. In fact, if the system is controllable in both problem settings the objective can be accomplished in an arbitrarily short time interval. However to accomplish this, one needs a control action which is larger the shorter the time interval is within which one likes to achieve this goal. Usually, in both economic and physical systems, the use of a large control action during a short time interval is not feasible. Furthermore, linear models are often used as an approximate to the real system. Using a very large control action

LQ Dynamic Optimization and Differential Games   J. Engwerda
© 2005 John Wiley & Sons, Ltd

might actually drive the system out of the region where the given linear model is valid. So, from this perspective as well, the use of a large control action is not recommended.

Given these considerations it seems reasonable to consider the following quadratic cost criterion

$$J = \int_0^T \{x^T(t)Qx(t) + u^T(t)Ru(t)\}dt + x^T(T)Q_Tx(T), \tag{5.1.2}$$

where (without loss of generality) the matrices Q, R and Q_T are assumed to be symmetric. Moreover, we assume throughout this chapter that matrix R is positive definite. This accounts for the fact that we do not allow for any arbitrarily large control action. The matrices Q, R and Q_T can be used for instance (i) to discriminate between the two distinct goals to realize on the one hand some objective and on the other hand to attain this objective with as little as possible control action, and (ii) to discriminate between different state variables (respectively control variables) having the same value. Usually the matrix Q_T expresses the relative importance attached to the final value of the state variable. Since R is assumed to be positive definite, this problem is generally called the **regular** linear quadratic control problem.

It can be shown that if matrix R is indefinite, the control problem has no solution. Conversely, if matrix R is semi-positive definite does make sense in some applications, and has therefore also been extensively studied in literature.

The reason we study a quadratic cost function here is to a large extent motivated by the fact that the optimization problem is in this case analytically manageable. On the other hand, it is also possible to motivate this quadratic control problem from a variational point of view. To that end, assume that our basic problem is to minimize the following non-linear cost function

$$J := \int_0^T g(t, x(t), u(t))dt + h(x(T)), \tag{5.1.3}$$

where the state variable $x(t)$ satisfies the nonlinear differential equation:

$$\dot{x}(t) = f(t, x(t), u(t)), \quad x(0) = x_0. \tag{5.1.4}$$

Since the following explanation is just meant to motivate the quadratic nature of (5.1.3) we skip the technical details and assume that in the subsequent analysis all functions are sufficiently smooth and that the differential equation always has a solution. Assume that the nonlinear optimal control problem (5.1.3) and (5.1.4) has a solution. Let $u^*(.)$ be an optimal control path and $x^*(.)$ the induced optimal state trajectory. As in section 4.2, next consider a 'small' perturbation of the initial state, $\epsilon \tilde{x}_0$, followed by a perturbation $\epsilon p(.)$ from the optimal control path $u^*(.)$. Denote the corresponding state trajectory by $\tilde{x}(t, \epsilon, p)$. Then, carrying out a second-order Taylor expansion of the cost function $J(\epsilon)$ around $\epsilon = 0$, we get:

$$J(\epsilon) = J^* + \epsilon \frac{dJ(\epsilon)}{d\epsilon}(0) + \epsilon^2 \frac{d^2J(\epsilon)}{d\epsilon^2}(0) + 0(\epsilon^2).$$

Since $u^*(.)$ is optimal, by the first-order optimality conditions (see section 4.2), $\frac{dJ(\epsilon)}{d\epsilon}(0) = 0$. Furthermore, according to equation (4.2.37),

$$\frac{d^2J(\epsilon)}{d\epsilon^2}(0) = \int_0^T [q(t,p), p^T(t)]H'' \begin{bmatrix} q(t,p) \\ p(t) \end{bmatrix} dt + q^T(T,p)\frac{\partial^2 h(x^*(T))}{\partial x^2}q(T,p), \quad (5.1.5)$$

where $q(t,p) = \frac{d\tilde{x}(t,\epsilon,p)}{d\epsilon}$ at $\epsilon = 0$, and $H'' = \begin{bmatrix} H_{xx} & H_{xu} \\ H_{ux} & H_{uu} \end{bmatrix}$, is the matrix obtained from the second-order derivative of the Hamiltonian $H(t, x^*, u^*, \lambda^*) := g(t, x^*, u^*) + \lambda^*(t) f(t, x^*, u^*)$ (with λ^* the costate variable as defined in equation (4.2.13) in the Euler–Lagrange theorem). From equation (5.1.4), $\tilde{x}(t, \epsilon, p)$ satisfies the differential equation

$$\dot{\tilde{x}}(t, \epsilon, p) = f(t, \tilde{x}(t, \epsilon, p), u^*(t) + \epsilon p(t)), \quad x(0) = x_0 + \epsilon \tilde{x}_0.$$

Consequently,

$$\frac{d}{d\epsilon}\frac{d\tilde{x}}{dt}(t, \epsilon, p) = \frac{\partial f}{\partial x}(t, \tilde{x}(t, \epsilon, p), u^*(t) + \epsilon p(t))\frac{dx}{d\epsilon}(t, \epsilon, p) + \frac{\partial f}{\partial u}(t, \tilde{x}(t, \epsilon, p), u^*(t) + \epsilon p(t))p(t).$$

Assuming that the order of differentiation may be interchanged we infer that $q(t,p)$ satisfies the linear differential equation

$$\frac{dq}{dt}(t, p) = \frac{\partial f}{\partial x}(t, x^*(t), u^*(t))q(t, p) + \frac{\partial f}{\partial u}(t, x^*(t), u^*(t))p(t), \quad q(0) = \tilde{x}_0. \quad (5.1.6)$$

So we conclude that in order to minimize the effect of an initial state perturbation of the nonlinear optimally controlled system (5.1.3) and (5.1.4), application of the control obtained by solving the linear quadratic problem (5.1.5) and (5.1.6) reduces most of the additional cost incurred if no action were to take place.

Note that the linear quadratic problem (5.1.5) and (5.1.6) exhibits on the one hand cross terms in the cost function and on the other hand linear time-varying dynamics. As we have seen in section 3.6 the inclusion of cross terms does not create any difficulties, because the problem can always be converted into our standard framework (5.1.1) and (5.1.2). Taking into account time-varying dynamics for the finite-planning horizon case neither substantially changes the results. Therefore we just elaborate the time-invariant case and comment on the time-varying case.

5.2   Finite-planning horizon

In this section we consider the minimization of

$$J = \int_0^T \{x^T(t)Qx(t) + u^T(t)Ru(t)\}dt + x^T(T)Q_Tx(T), \quad (5.2.1)$$

subject to

$$\dot{x}(t) = Ax(t) + Bu(t), \ x(0) = x_0, \tag{5.2.2}$$

where R is a positive definite matrix and Q, R and Q_T are symmetric matrices.

This problem is known in literature as the **linear quadratic control problem**. We will see that the solution of this problem is closely connected to the existence of a symmetric solution to the following matrix **Riccati differential equation (RDE)**

$$\dot{K}(t) = -A^T K(t) - K(t)A + K(t)SK(t) - Q, \ K(T) = Q_T, \tag{5.2.3}$$

where $S := BR^{-1}B^T$. The fact that we are looking for a symmetric solution to this equation follows from the fact that Q_T is symmetric. For this implies that if $K(.)$ is a solution of RDE then, by taking the transposition of both sides of the equation, obviously $K^T(.)$ satisfies RDE with the same boundary value as well. Since, according to the fundamental existence–uniqueness Theorem 2.9, the solution of this RDE is unique (if it exists) it follows that $K(t) = K^T(t)$, for all $t \in [0, T]$. In this section the next theorem will be proved.

Theorem 5.1 (Linear quadratic control problem)

The linear quadratic control problem (5.2.1) and (5.2.2) has, for every initial state x_0, a solution if and only if the Riccati differential equation (5.2.3) has a symmetric solution $K(.)$ on $[0, T]$.

If the linear quadratic control problem has a solution, then it is unique and the optimal control in feedback form is

$$u^*(t) = -R^{-1}B^T K(t)x(t), \tag{5.2.4}$$

whereas in open-loop form it is

$$u^*(t) = -R^{-1}B^T K(t)\Phi(t, 0)x_0 \tag{5.2.5}$$

with Φ the solution of the transition equation

$$\dot{\Phi}(t, 0) = (A - SK(t))\Phi(t, 0); \ \Phi(0, 0) = I.$$

Moreover, $J(u^*) = x_0^T K(0)x_0$. □

The proof of this theorem will be provided in a number of subsequent steps. First it is shown that if the RDE (5.2.3) has a symmetric solution then the linear quadratic control problem has the solution stated in the theorem.

Proof of Theorem 5.1 '⇐ part'

Note that

$$\int_0^T \frac{d}{dt}\{x^T(t)K(t)x(t)\}dt = x^T(T)K(T)x(T) - x^T(0)K(0)x(0).$$

Consequently, the cost function (5.2.1) can be rewritten as

$$
J = \int_0^T \{x^T(t)Qx(t) + u^T(t)Ru(t)\}dt
$$
$$
+ \int_0^T \frac{d}{dt}\{x^T(t)K(t)x(t)\}dt + x^T(0)K(0)x(0) + x^T(T)(Q_T - K(T))x(T)
$$
$$
= \int_0^T \{x^T(t)Qx(t) + u^T(t)Ru(t) + \frac{d}{dt}\{x^T(t)K(t)x(t)\}\}dt
$$
$$
+ x^T(0)K(0)x(0) + x^T(T)(Q_T - K(T))x(T).
$$

Using equations (5.2.2) and (5.2.3), the integrand can be rewritten as follows (omitting the dependence of time)

$$
x^T Qx + u^T Ru + \frac{d}{dt}\{x^T Kx\} = x^T Qx + u^T Ru + \dot{x}^T Kx + x^T \dot{K}x + x^T K\dot{x}
$$
$$
= x^T Qx + u^T Ru + (Ax + Bu)^T Kx + x^T \dot{K}x + x^T K(Ax + Bu)
$$
$$
= x^T(Q + A^T K + KA + \dot{K})x + u^T Ru + u^T B^T Kx + x^T KBu
$$
$$
= x^T KSKx + u^T Ru + u^T B^T Kx + x^T KBu
$$
$$
= (u + R^{-1}B^T Kx)^T R(u + R^{-1}B^T Kx).
$$

As a result

$$
J = \int_0^T (u(t) + R^{-1}B^T K(t)x(t))^T R(u(t) + R^{-1}B^T K(t)x(t))dt + x^T(0)K(0)x(0).
$$

From this expression it is obvious that $J \geq x^T(0)K(0)x(0)$ for all u and that equality is achieved if $u(.)$ satisfies equation (5.2.4). $\qquad\square$

To prove the converse statement assume that $u^*(.)$ is the optimal control solution of the linear quadratic control problem; $x^*(t)$ is the induced state trajectory; and $J^*(0, x_0)$ is the associated minimum cost. One of the major problems in tackling the converse statement of Theorem 5.1 is to show that if $J^*(0, x_0)$ exists for an arbitrary initial state the minimum $J^*(t, x_0)$ exists in the optimization problem (5.2.6) below, for an arbitrary initial state x_0, and for an arbitrarily chosen $t \in [0, T]$. To prove this we will consider the following slightly different linear quadratic control optimal control problem (5.2.6). It will be shown first that if this problem has a solution it is a **quadratic form**. That is, consider the problem to find the **infimum** of the quadratic cost function

$$
J(t, x_0) = \int_t^T \{x^T(t)Qx(t) + u^T(t)Ru(t)\}dt + x^T(T)Q_T x(T), \qquad (5.2.6)
$$

subject to

$$
\dot{x}(s) = Ax(s) + Bu(s), \quad x(t) = x_0,
$$

where Q, Q_T and R satisfy the usual assumptions given in equations (5.2.1) and (5.2.2). Then under the assumption that this infimum, denoted by $J^{inf}(t, x_0)$, exists

$$J^{inf}(t, x_0(t)) = x_0^T(t)P(t)x_0(t), \tag{5.2.7}$$

where, without loss of generality, $P(t)$ is assumed to be symmetric. Its proof uses the following lemma (Molinari, 1975). The proof of this lemma is provided in the Appendix to this chapter.

Lemma 5.2

If a function V satisfies the parallelogram identity

$$V(x+y) + V(x-y) = 2\{V(x) + V(y)\}, \text{ for all } x, y \tag{5.2.8}$$

and, for every y, $W(x, y) := V(x+y) - V(x-y)$ has the property that

$$W(\lambda x, y) \text{ is continuous in } \lambda \text{ at } \lambda = 0, \tag{5.2.9}$$

then $V(x)$ is a quadratic form. $\qquad\square$

To show that $J^{inf}(t, x_0)$ satisfies the parallelogram identity (5.2.8) we use the following property.

Lemma 5.3

If $J^{inf}(t, x_0)$ exists, then $J^{inf}(t, \lambda x_0)$ also exists and, moreover,

$$J^{inf}(t, \lambda x_0) = \lambda^2 J^{inf}(t, x_0).$$

Proof

Denoting the state trajectory induced by an initial state x_0 at time t and control $u(.)$ by $x(s, x_0(t), u)$, $s \in [t, T]$, the variation-of-constants formula (3.2.7) gives that

$$x(s, x_0(t), u) = e^{A(s-t)}x_0(t) + \int_t^s e^{A(s-\tau)}Bu(\tau)d\tau. \tag{5.2.10}$$

From this it is obvious that

$$x(s, \lambda x_0(t), \lambda u) = \lambda x(s, x_0(t), u),$$

where λu is the control sequence $\lambda u(s)$, $s \in [t, T]$. Consequently, it follows straightforwardly from equation (5.2.1) that

$$J(t, \lambda x_0(t), \lambda u) = \lambda^2 J(t, x_0(t), u). \tag{5.2.11}$$

Therefore

$$\lambda^2 J^{inf}(t, x_0) = \lambda^2 \inf_u J(t, x_0, u) = \inf_u \lambda^2 J(t, x_0, u)$$
$$= \inf_u J(t, \lambda x_0, \lambda u) = \inf_{\tilde{u}} J(t, \lambda x_0, \tilde{u}) = J^{inf}(t, \lambda x_0).$$

So $\lambda^2 J^{inf}(t, x_0(t)) = J^{inf}(t, \lambda x_0(t))$. □

Note

In a similar way one can show the following property. If the infimum in the above optimization problem (5.2.6) is attained by some control function u^* if $x(t) = x_0$ (so that the infimum is actually a minimum), then λu^* is the control yielding the minimum value for this optimization problem if $x(t) = \lambda x_0$. □

Now, consider the parallelogram identity. Let $u_p^*(.)$ be a control sequence for problem (5.2.6) with initial state $x(t) = p$ for which $|J^{inf}(t, p) - J(t, p, u_p^*)| < \epsilon$, where ϵ is some positive number. Using Lemma 5.3 and Lemma 5.21 from the Appendix we have that $V := J^{inf}(t, x_0(t)) + J^{inf}(t, x_1(t))$ satisfies

$$V = \frac{1}{4}\{J^{inf}(t, 2x_0(t)) + J^{inf}(t, 2x_1(t))\}$$
$$\le \frac{1}{4}\{J(t, 2x_0(t), u^*_{(x_0+x_1)(t)} + u^*_{(x_0-x_1)(t)}) + J(t, 2x_1(t), u^*_{(x_0+x_1)(t)} - u^*_{(x_0-x_1)(t)})\}$$
$$= \frac{1}{2}\{J(t, (x_0 + x_1)(t), u^*_{x_0+x_1}) + J(t, (x_0 - x_1)(t), u^*_{x_0-x_1})\}$$
$$\le \frac{1}{2}\{J^{inf}(t, (x_0 + x_1)(t)) + J^{inf}(t, (x_0 - x_1)(t))\} + 2\epsilon$$
$$\le \frac{1}{2}\{J(t, (x_0 + x_1)(t), u^*_{x_0} + u^*_{x_1}) + J(t, (x_0 - x_1)(t), u^*_{x_0} - u^*_{x_1})\} + 2\epsilon$$
$$= J(t, x_0(t), u^*_{x_0}) + J(t, x_1(t), u^*_{x_1}) + 2\epsilon$$
$$\le J^{inf}(t, x_0(t)) + J^{inf}(t, x_1(t)) + 4\epsilon.$$

So, comparing both sides of these inequalities, we conclude that since ϵ is an arbitrary positive number, all inequalities can be replaced by equalities, establishing the fact that $J^{inf}(t, x_0(t))$ satisfies the parallelogram identity (5.2.8).

Next we show that $J^{inf}(t, x_0)$ is continuous in x_0. To that end notice that with time, t, fixed

$$J^{inf}(t, x_0 + \Delta x) - J(t, x_0, u^*_{x_0+\Delta x}) \le J^{inf}(t, x_0 + \Delta x) - J^{inf}(t, x_0)$$
$$\le J(t, x_0 + \Delta x, u^*_{x_0}) - J^{inf}(t, x_0). \tag{5.2.12}$$

With $w(s) := e^{A(s-t)}x_0 + \int_t^s e^{A(s-\tau)}Bu^*_{x_0}(\tau)d\tau$

$$J(t, x_0 + \Delta x, u^*_{x_0}) - J(t, x_0, u^*_{x_0}) = \int_t^T 2w(s)Qe^{A(s-t)}\Delta x + (e^{A(s-t)}\Delta x)^T Qe^{A(s-t)}\Delta x\,ds.$$

Since the right-hand side in equation (5.2.12) can be estimated by

$$J(t, x_0 + \Delta x, u^*_{x_0}) - J(t, x_0, u^*_{x_0}) + \epsilon,$$

where ϵ is again an arbitrary positive number, it is clear that the right-hand side in equation (5.2.12) converges to zero if $\Delta x \to 0$. Similarly it follows for the left-hand side in equation (5.2.12) that

$$J^{inf}(t, x_0 + \Delta x) - J(t, x_0, u^*_{x_0 + \Delta x}) \geq -\epsilon + J(t, x_0 + \Delta x, u^*_{x_0 + \Delta x}) - J(t, x_0, u^*_{x_0 + \Delta x}).$$

Since

$$J(t, x_0 + \Delta x, u^*_{x_0 + \Delta x}) - J(t, x_0, u^*_{x_0 + \Delta x}) = e^{A(T-t)} \Delta x$$

converges to zero if $\Delta x \to 0$ and ϵ is an arbitrary positive number the left-hand side in equation (5.2.12) also converges to zero if $\Delta x \to 0$. Consequently, $J^{inf}(t, x_0)$ is continuous and, thus, the second condition (5.2.9) in Lemma 5.2 is also satisfied and we conclude the following.

Corollary 5.4

Assume that the infimum in equation (5.2.6) exists for all x_0. Then there exists a symmetric matrix $P(t)$ such that

$$J^{inf}(t, x_0) = x_0^T P(t) x_0. \tag{5.2.13}$$

$\square$

As already mentioned above it is difficult to show directly that whenever $J^*(0, x_0)$ exists, $J^*(t, x_0)$ will also exist for $t > 0$. However, it can easily be shown that the infimum exists in equation (5.2.6) at an arbitrary point in time t if $J^*(0, x_0)$ exists.

Lemma 5.5

If $J^*(0, x_0)$ exists for all $x_0 \in \mathbb{R}^n$, then for all $t \in [0, T]$, $J^{inf}(t, x_0)$ exists for all x_0.

Proof

Let v be an arbitrary state at time t. Then, using the control $u(s) = 0$, $s \in [0, t]$, this state can be reached from the initial state $x_0 := e^{-At} v$ at time $t = 0$. By assumption, $J^*(0, x_0)$ exists. So for all $u \in \mathcal{U}$

$$J^*(0, x_0) \leq \int_0^t (e^{As} x_0)^T Q e^{As} ds + J(t, v, u). \tag{5.2.14}$$

Consequently $\inf J(t, v, u)$ exists.

$\square$

Corollary 5.6

If $J^*(0, x_0)$ exists for all x_0 then there exists a symmetric bounded matrix $P(.)$ such that $J^{inf}(t, x_0) = x_0^T P(t) x_0$, for all $t \in [0, T]$. $\qquad\square$

All components are now available to complete the proof of Theorem 5.1.

We can, for example, use the dynamic programming theorem to conclude that the converse statement of Theorem 5.1 holds. It is obvious that the partial derivative $\frac{\partial J^{inf}}{\partial x}(t, x)$ exists, and its continuity and the existence of the partial derivative of $\frac{\partial J^{inf}}{\partial t}(t, x)$ can also be shown (using some nontrivial elementary analysis; in the Exercises at the end of the chapter the reader is asked to prove these statements under the assumption that $J^*(t)$ exists for all $t \in [0, T]$). Due to the latter technicalities, and the fact that the above mentioned analysis w.r.t. the partial derivatives cannot be used anymore in case we allow the system matrices to become piecewise continuous functions of time, we proceed another way.

Proof of Theorem 5.1 '⇒ part'

Consider the Riccati differential equation

$$\dot{K}(t) = -A^T K(t) - K(t)A + K(t)SK(t) - Q, \text{ with } K(T) = Q_T. \qquad (5.2.15)$$

According to the fundamental existence–uniqueness theorem of differential equations there exists a maximum time interval $(t_1, T]$ where equation (5.2.15) has a unique solution. Assume that $t_1 > 0$. From the '⇐ part' of this theorem we conclude then that on this time interval the optimization problem (5.2.6) has a solution and actually a minimum. This minimum equals $x_0^T K(t) x_0$. On the other hand, by Corollary 5.6, this minimum equals $x_0^T P(t) x_0$. Since this holds for an arbitrary initial state x_0 we conclude that $P(t) = K(t)$ on $(t_1, T]$. However, since $P(t)$ is bounded on $[0, T]$ it follows that $K(t)$ is also bounded on $(t_1, T]$; but this implies that the Riccati differential equation (5.2.15) also has a solution on some time interval $(t_2, T]$ for some $t_2 > t_1$. That is, the interval of existence $(t_1, T]$ is not maximal. This violates our assumption. So, we conclude that $t_1 \leq 0$. Which completes the proof of the theorem. $\qquad\square$

Corollary 5.7

1. An immediate consequence of Theorem 5.1 is that, if the linear quadratic control problem (5.1.1) and (5.1.2) has a minimum for all x_0, for all $t \in [0, T]$ equation (5.2.6) actually attains a minimum by choosing $u^*(s) = K(s)x^*(s)$, with $K(.)$ and $x^*(.)$ as defined in Theorem 5.1.

2. A second consequence of Theorem 5.1 is that at an arbitrary point in time $t \in [0, T]$, for every state $x(t) \in \mathbb{R}^n$ there exists an initial state $x(0) = x_0$ such that the with this initial state x_0 corresponding optimal state trajectory has the property that $x^*(t, x_0, u^*) = x(t)$ (see Exercises). That is, at every point in time every state can be reached as the outcome of an optimal trajectory. $\qquad\square$

Example 5.1

Consider the optimization problem

$$\min J = \int_0^T -x^2(t) + u^2(t)\, dt,$$

subject to

$$\dot{x}(t) = u(t), \ x(0) = x_0.$$

According to Theorem 5.1 this problem has a solution for every initial state x_0 if and only if the following Riccati differential equation has a solution on $[0, T]$

$$\dot{k}(t) = k^2(t) + 1; \ k(T) = 0. \tag{5.2.16}$$

The solution of this differential equation (5.2.16) exists on $[0, T]$ if and only if $T < \pi$ in which case it is given by $k(t) = \tan(t - T)$. In this case the optimal control is given in feedback form by

$$u^*(t) = -k(t)x(t).$$

An interesting aspect of this example is that if the planning horizon $T = \frac{\pi}{2}$, and the initial state of the system is $x_0 = 0$, there still does exist an optimal control, i.e. $u(.) = 0$ (see Exercises). However, it can be shown (Clements and Anderson (1978), Molinari (1975) and Pars (1962)) that for any other initial state of the system an optimal control does not exist. That is,

$$\inf J = \int_0^{\frac{\pi}{2}} -x^2(t) + u^2(t)dt,$$

subject to

$$\dot{x}(t) = u(t), \ x(0) = x_0 \neq 0$$

is arbitrarily negative $(-\infty)$. Moreover, for any planning horizon $T > \frac{\pi}{2}$, the optimization does not have a solution for the initial state $x_0 = 0$ anymore.   □

Example 5.2

Consider a factory which produces an article x. At least in the short run, the number of articles that can be produced increases quadratically with the total amount of capital C invested in the firm. Assume that capital depreciates at a rate α, that new investments in the production process at time t are $I(t)$ and that the associated cost of investing $I(t)$ is also a quadratic function. This enables the following formulation of the optimization problem.

$$\max \Pi = \int_0^T pC^2(t) - rI^2(t)\, dt,$$

subject to

$$\dot{C}(t) = -\alpha C(t) + I(t),$$

where all constants p, r and α are positive.

First, we rewrite this problem into our standard framework. That is, the maximization problem is rewritten as the minimization problem

$$\min \int_0^T -pC^2(t) + rI^2(t)\, dt.$$

With $A := -\alpha$, $B := 1$, $Q := -p$, $R := r$ and $Q_T = 0$ Theorem 5.1 shows that this optimization problem has a solution if and only if the following Riccati differential equation has a solution on $[0, T]$

$$\dot{k}(t) = 2\alpha k(t) + \frac{1}{r}k^2(t) + p, \quad k(T) = 0. \tag{5.2.17}$$

This ordinary differential equation can be solved analytically using the separation of variables technique. Depending on the number of roots of the polynomial on the right-hand side of this equation (5.2.17)

$$\frac{1}{r}k^2 + 2\alpha k + p = 0 \quad \text{or, equivalently,} \quad k^2 + 2\alpha r k + pr = 0 \tag{5.2.18}$$

then four qualitatively different situations can occur. Denote the discriminant[1] of this equation (5.2.18), $\alpha^2 r^2 - pr$, by D. Then these situations are as follows.

Case 1. $D = 0$

Then the solution of the differential equation (5.2.17) is

$$k(t) = -\alpha r - \frac{1}{\frac{1}{r}t + c}, \quad \text{with} \quad c = -\frac{1}{r}\left(T + \frac{1}{\alpha}\right).$$

Case 2. $D < 0$

Then the solution of the differential equation (5.2.17) is

$$k(t) = -\alpha + \sqrt{-D}\tan\left(\frac{\sqrt{-D}}{r}t + c\right),$$

where c is the solution of $0 = -\alpha + \sqrt{-D}\tan\left(\frac{\sqrt{-D}}{r}T + c\right)$. Obviously, this solution only makes sense on the whole planning horizon $[0, T]$ provided the parameters in the

---

[1]The **discriminant** of $ax^2 + bx + c$ is the number $b^2 - 4ac$.

model are such that $\left(\frac{\sqrt{-D}}{r}t + c\right) \in \left(-\frac{\pi}{2}, \frac{\pi}{2}\right)$ for $t \in [0, T]$. Otherwise equation (5.2.17) has no solution on $[0, T]$, and consequently the optimization problem has no solution.

Case 3. $D > 0$

Let $\beta := -\alpha r - \sqrt{D}$ and $\gamma := -2\sqrt{D}$. Then the solution of the differential equation (5.2.17) is

$$k(t) = \beta + \frac{\gamma}{e^{\gamma\left(\frac{t}{r}+c\right)} - 1}, \quad \text{with} \quad c = \frac{-T}{r} + \gamma \ln \frac{\beta - \gamma}{\beta}.$$

As in the previous case, this solution only makes sense as long as the model parameters are such that $\frac{t}{r} + c \neq 0$ for $t \in [0, T]$. Otherwise the problem again has no solution.

Case 4.

The differential equation has no solution on $[0, T]$ and, therefore, the optimization problem has no solution. As we have seen in Cases 2 and 3, depending on the model parameters, such situations can occur. Obviously, this makes sense. For instance, if the price p in this model is much larger than the cost of investment, i.e. $p \gg r$, it pays to invest as much as possible in the short term and thereby gain 'infinite' profits. So, one might say that if these parameter conditions hold the model probably is not very accurate in describing reality. This situation occurs, for example, in Case 3 if and only if $T > \frac{r}{\beta} \ln \frac{\beta - \gamma}{\beta}$. $\qquad\square$

Next assume that (A, B) is controllable. Theorem 5.8, below, then shows that under the assumption that for some initial state the optimal control problem has a solution with a planning horizon T, the same optimal control with a planning horizon strictly smaller than T will have a solution for every initial state.

Theorem 5.8

Let (A, B) be controllable. Assume that there exists an initial state $x(0) = \bar{x}$ such that the linear quadratic control problem (5.2.1) and (5.2.2) has a solution. Take, without loss of generality, $t_0 = 0$. Then, for every $t_1 \in (0, T]$ the linear quadratic control problem

$$J(t_1, x_0, u) = \int_{t_1}^{T} \left\{ x^T(t)Qx(t) + u^T(t)Ru(t) \right\} dt + x^T(T)Q_T x(T), \qquad (5.2.19)$$

subject to

$$\dot{x}(t) = Ax(t) + Bu(t), \quad x(t_1) = x_0, \qquad (5.2.20)$$

has a solution for every initial state x_0.

Proof

Let $t_1 > 0$ be an arbitrarily chosen time in the interval $(0, T]$. By taking $u(.) = 0$ it is clear that $J^{inf}(t_1, x_0) < \infty$. To prove that also $J^{inf}(t_1, x_0) > -\infty$ first note that due to the controllability assumption matrix

$$W(t) := \int_0^t e^{A(t-s)} BB^T e^{A^T(t-s)} ds$$

is invertible for every $t > 0$. Consequently, the control

$$u_0(s) := B^T e^{A^T(t_1-s)} W^{-1}(t_1)(x_0 - e^{At_1}\bar{x})$$

steers the state of the system from $x(0) = \bar{x}$ at time $t = 0$ towards $x(t_1) = x_0$ at time $t = t_1$.

Since by assumption $J^*(0, \bar{x})$ exists, for every $u(.)$, $J(0, \bar{x}, u) \geq J^*(0, \bar{x})$. Consequently, denoting $w(x, u) := x^T(s)Qx(s) + u^T(s)Ru(s)$,

$$\int_0^{t_1} w(x, u_0)ds + \int_{t_1}^T w(x, u)ds + x^T(T)Q_T x(T) \geq J^*(0, \bar{x}),$$

which shows that for an arbitrary $u(.)$

$$\int_{t_1}^T w(x, u)ds + x^T(T)Q_T x(T) \geq J^*(0, \bar{x}) - \int_0^{t_1} w(x, u_0)ds.$$

Since the right-hand side of this inequality is bounded, $J^{inf}(t_1, x_0)$ exists and, moreover, satisfies the above inequality. The rest of the proof then follows along the lines of the proof of Theorem 5.1. □

Note

We should stress that Theorem 5.8 states that just for some t_1 strictly larger than zero one can conclude that the linear quadratic control problem has a solution for every initial state. Indeed, there exist examples (for example Example 5.1 with $T = \frac{\pi}{2}$) for which the linear quadratic control problem has for some initial state value a solution, whereas for other initial state values the solution for the linear quadratic control problem does not exist. □

Proposition 5.9

Assume that there exists an initial state $\bar{x} \in \mathcal{R}(A) := \text{Im}[B|AB| \cdots |A^{n-1}B]$ for which the linear quadratic control problem (5.2.1) and (5.2.2) has a solution. Then the Riccati differential equation (5.2.3) has a solution on $(t_0, T]$.

Proof

First note that $\mathcal{R}(A)$ is a linear subspace. This subspace is A-invariant since, due to the theorem of Cayley–Hamilton,

$$A\mathcal{R}(A) = A\text{Im}[B|AB| \cdots |A^{n-1}B] = \text{Im}[AB|AB| \cdots |A^n B] \subset \text{Im}[B|AB| \cdots |A^{n-1}B] = \mathcal{R}(A).$$

$$(5.2.21)$$

Next consider a basis of $\mathcal{R}(A)$ extended with a number of independent vectors such that this whole set of vectors together span $\mathbb{R}^n$. This set of vectors constitutes a basis of $\mathbb{R}^n$. With respect to this basis matrix A and B have the following structure

$$A = \begin{bmatrix} A_{11} & A_{12} \\ 0 & A_{22} \end{bmatrix}; \text{ and } B = \begin{bmatrix} B_1 \\ 0 \end{bmatrix}.$$

Now consider the Riccati differential equation (5.2.3) with respect to this basis. Denoting

$$\begin{bmatrix} K_{11} & K_{12} \\ K_{12}^T & K_{22} \end{bmatrix} := K, \; S_1 := B_1 R^{-1} B_1^T, \; \begin{bmatrix} Q_{11} & Q_{12} \\ Q_{12}^T & Q_{22} \end{bmatrix} := Q \text{ and } \begin{bmatrix} Q_{T11} & Q_{T12} \\ Q_{T12}^T & Q_{T22} \end{bmatrix} := Q_T,$$

elementary calculation shows that equation (5.2.3) has a solution $K(t)$ on $(t_0, T]$ if and only if the following three differential equations have a solution on $(t_0, T]$

$$\dot{K}_{11}(t) = -A_{11}^T K_{11}(t) - K_{11}(t)A_{11} + K_{11}(t)S_1 K_{11}(t) - Q_{11}, \; K_{11}(T) = Q_{T11}; \quad (5.2.22)$$

$$\dot{K}_{12}(t) = (K_{11}(t)S_1 - A_{11}^T)K_{12}(t) - K_{12}(t)A_{22} - Q_{12} - K_{11}(t)A_{12}, \; K_{12}(T) = Q_{T12}; \quad (5.2.23)$$

$$\dot{K}_{22}(t) = -A_{21}^T K_{22}(t) - K_{22}(t)A_{22} - Q_{22} + (K_{12}^T(t)S_1 - A_{12}^T)K_{12}(t) - K_{12}^T(t)A_{12}, \quad (5.2.24)$$

$$K_{22}(T) = Q_{T22}. \quad (5.2.25)$$

Taking a closer look at these differential equations, we observe that if equation (5.2.22) has a solution, equation (5.2.23) is just an ordinary linear differential equation which always has a solution. Similarly it is seen then that by substituting the solutions of equations (5.2.22) and (5.2.23) into equation (5.2.24) this last differential equation is also a linear differential equation, which therefore also always has a solution. So equation (5.2.3) has a solution $K(t)$ on $(t_0, T]$ if and only if the differential equation (5.2.22) has a solution on $(t_0, T]$. Now, by assumption, the linear quadratic control problem has a solution for some $x_0 \in \mathcal{R}(A)$. This implies that with respect to our basis, $x_0 = [x_{10}^T \; 0]^T$. So, denoting $[x_1(t) \; x_2(t)] := x(t)$, the linear quadratic control problem

$$J(t_0, x_0, u) = \int_{t_0}^{T} \{x^T(t)Qx(t) + u^T(t)Ru(t)\}dt + x^T(T)Q_T x(T), \quad (5.2.26)$$

subject to

$$\begin{bmatrix} \dot{x}_1(t) \\ \dot{x}_2(t) \end{bmatrix} = \begin{bmatrix} A_{11} & A_{12} \\ 0 & A_{22} \end{bmatrix} \begin{bmatrix} x_1(t) \\ x_2(t) \end{bmatrix} + \begin{bmatrix} B_1 \\ 0 \end{bmatrix} u(t), \quad x(t_0) = \begin{bmatrix} x_{10} \\ 0 \end{bmatrix}$$

has a solution for some x_{10} with (A_{11}, B_1) controllable.

It is easily verified that $x_2(t) = 0$ for all $t \in [t_0, T]$. Consequently, the linear quadratic control problem (5.2.26) is equivalent with the minimization of

$$J(t_0, x_{10}, u) = \int_{t_0}^{T} \{x_1^T(t)Q_{11}x_1(t) + u^T(t)Ru(t)\}dt + x_1^T(T)Q_{T11}x_1(T), \quad (5.2.27)$$

subject to

$$\dot{x}_1(t) = A_{11}(t) + B_1 u(t), \; x_1(t_0) = x_{10}.$$

By assumption this problem has a solution for x_{10}. Therefore, by Theorem 5.8, the Riccati equation corresponding to this linear quadratic control problem has a solution on $(t_0, T]$. That is equation (5.2.22) has a solution on $(t_0, T]$, which concludes the proof. $\qquad\square$

Corollary 5.10

Notice that always $0 \in \mathcal{R}(A)$. Consequently, if the linear quadratic control problem (5.2.1) and (5.2.2) has a solution for $x(0) = 0$, the minimization problem (5.2.19) and (5.2.20) will have a unique solution for any $t_1 > 0$ for every initial state $x(t_1) = x_0$. Or, stated differently, the Riccati differential equation (5.2.3) has a solution on $(0, T]$. $\qquad\square$

Note

1. The results of Theorem 5.1 and Corollary 5.10 also hold if the matrices A, B, Q and R are piecewise continuous functions of time (Clements and Anderson (1978)).

2. From Theorem 5.1 the following statement results. For all $T \in [0, t_1)$ the linear quadratic control problem (5.2.1) and (5.2.2) has a solution for every initial state $x(0) = x_0$ if and only if the Riccati differential equation (5.2.3) has a solution on $(0, t_1]$.

   This equivalence result will be used to formulate well-posedness assumptions in the convergence analysis of this optimal control problem if the planning horizon becomes arbitrarily long (see sections 5.5 and 7.6). $\qquad\square$

We conclude this section by considering the non-homogeneous linear quadratic control problem. That is, the minimization of

$$J = \int_0^T \{x^T(t)Qx(t) + u^T(t)Ru(t)\}dt + x^T(T)Q_Tx(T), \qquad (5.2.28)$$

subject to

$$\dot{x}(t) = Ax(t) + Bu(t) + c(t), \quad x(0) = x_0, \qquad (5.2.29)$$

under the usual assumptions that R is a positive definite matrix and Q, R and Q_T are symmetric matrices. The non-homogeneous term, $c(t)$, is assumed to be a known function which is such that the solution $x(t)$ of the differential equation is uniquely defined in the extended sense. In principle one could use the transformation outlined in section 3.6, part IV on the affine systems, to rewrite this problem as a standard linear quadratic control problem; but in that case the extended state matrix becomes time dependent (see equation (3.6.15)). If, for example, $c(.)$ is just a square integrable function this time-dependency is, in general, not piecewise continuous. So, in that case one cannot appeal to the Note following Corollary 5.10 to solve the problem. Therefore we will tackle this non-homogeneous linear quadratic control problem directly. The following result states that this problem has a solution if and only if the original linear quadratic control problem with $c(t) = 0$ has a solution. Parts of the proof are best shown using a functional analysis approach. Since this subject is somewhat outside the scope of this book, that part of the proof is deferred to the Appendix at the end of this chapter.

Theorem 5.11

Let $c(t)$ be an arbitrarily square integrable function. Consider the minimization of the linear quadratic cost function (5.2.28) subject to the state dynamics

$$\dot{x}(t) = Ax(t) + Bu(t) + c(t), \quad x(0) = x_0, \tag{5.2.30}$$

Then,

1. the linear quadratic problem (5.2.28) and (5.2.30) has a solution for all $x_0 \in \mathbb{R}^n$ if and only if the Riccati differential equation (5.2.3) has a symmetric solution $K(.)$ on $[0, T]$;

2. If the linear quadratic problem (5.2.28) and (5.2.30) has a solution for some $x_0 \in \mathcal{R}(A)$, then for all $x_0 \in \mathbb{R}^n$ the linear quadratic control problem

$$J(t_0, x_0, u, c) = \int_{t_0}^{T} \{x^T(t)Qx(t) + u^T(t)Ru(t)\}dt + x^T(T)Q_Tx(T), \tag{5.2.31}$$

subject to

$$\dot{x}(t) = Ax(t) + Bu(t) + c(t), \quad x(t_0) = x_0, \tag{5.2.32}$$

has a solution for all $t_0 \in (0, T]$.

Moreover, if the linear quadratic control problem (5.2.28) and (5.2.30) has a solution, then the optimal control in feedback form is

$$u^*(t) = -R^{-1}B^T(K(t)x(t) + m(t)),$$

where $m(t)$ is the solution of the linear differential equation

$$\dot{m}(t) = (K(t)S - A^T)m(t) - K(t)c(t), \quad m(T) = 0, \tag{5.2.33}$$

and $x(t)$ is the solution of the differential equation

$$\dot{x}(t) = (A - SK(t))x(t) - Sm(t) + c(t), \quad x(0) = x_0.$$

Proof

1. '$\Leftarrow$ **part**' Let $K(t)$ be the solution of the Riccati differential equation (5.2.3) and $m(t)$ the solution of the with this solution $K(.)$ corresponding linear differential equation (5.2.33). Next consider the value function

$$V(t) := x^T(t)K(t)x(t) + 2m^T(t)x(t) + n(t),$$

where

$$n(t) := \int_t^T \{-m^T(s)Sm(s) + 2m^T(s)c(s)\}ds.$$

Substitution of $\dot{K}$, $\dot{x}$ and $\dot{m}$ from equations (5.2.3), (5.2.29) and (5.2.33), respectively, and $\dot{n}(t) = m^T(t)Sm(t) - 2m^T(t)c(t)$ into $\dot{V}$ gives

$$\dot{V}(t) = \dot{x}^T(t)K(t)x(t) + x^T(t)\dot{K}(t)x(t) + x^T(t)K(t)\dot{x}(t) + 2\dot{m}^T(t)x(t) + 2m^T(t)\dot{x}(t) + \dot{n}(t)$$
$$= -x^T(t)Qx(T) - u^T(t)Ru(t) + [u(t) + R^{-1}B^T(K(t)x(t) + m(t))]^T$$
$$\times R[u(t) + R^{-1}B^T(K(t)x(t) + m(t))].$$

Now,

$$\int_0^T \dot{V}(s)ds = V(T) - V(0),$$

and, due to the fact that $m(T) = 0$, $V(T) = x^T(T)Q_Tx(T)$. Consequently substitution of $\dot{V}$ into this expression and rearranging terms gives

$$\int_0^T \{x^T(t)Qx(t) + u^T(t)Ru(t)\}dt + x^T(T)Q_Tx(T)$$
$$= V(0) + \int_0^T [u + R^{-1}B^T(K(t)x(t) + m(t))]^T R[u + R^{-1}B^T(K(t)x(t) + m(t))]dt.$$

Since $V(0)$ does not depend on $u(.)$ and R is positive definite, the required result follows.

1. '$\Rightarrow$ part' This part of the proof can be found in the Appendix at the end of this chapter.

2. Consider the notation from the proof of Proposition 5.9. In the same way as in that proof it follows that for any $\bar{x}_1$ there is a control sequence which steers $x_1(t_0)$ towards $\bar{x}_1$ at time t_1. Following the lines of the proof of Theorem 5.11 we then have that, for all $\bar{x}_1$, the minimization of

$$\int_{t_1}^T \{x_1^T(s)Q_{11}x_1(s) + 2x_1^T(s)Q_{12}x_2(s) + u^T(s)Ru(s)\}ds$$
$$+ \int_{t_1}^T x_2^T(s)Q_{22}x_2(s)ds + x^T(T)Q_Tx(T)$$

subject to the system

$$\dot{x}_1(t) = A_{11}x_1(t) + B_1u(t) + A_{12}(t)x_2(t) + c_1(t), \quad x_1(t_1) = \bar{x}_1$$

has a solution, where $x_2(t) = e^{A_{22}t}x_2(t_1) + \int_{t_1}^t e^{A_{22}(t_1-s)}c_2(s)ds$. Using Lemma 5.22 the stated result then follows analogously to the proof of part 1 '$\Rightarrow$'. $\qquad\square$

Example 5.3

Consider the optimization problem

$$\min_{u} \int_0^1 \{-x^2(s) + u^2(s)\}ds + \frac{1}{2}x^2(1),$$

subject to the dynamics

$$\dot{x}(t) = x(t) + u(t) + e^{-t}, \quad x(0) = 1.$$

The Riccati differential equation corresponding to this problem is

$$\dot{k}(t) = -2k(t) + k^2(t) + 1, \quad k(1) = \frac{1}{2}.$$

It is easily verified that

$$k(t) = 1 - \frac{1}{t+1}.$$

Next, consider the differential equation

$$\dot{m}(t) = (k(t) - 1)m(t) - k(t)e^{-t}, \quad m(1) = 0.$$

That is,

$$\dot{m}(t) = \frac{-1}{t+1}m(t) + \left(\frac{1}{t+1} - 1\right)e^{-t}, \quad m(1) = 0.$$

By straightforward substitution it is readily verified that the solution of this differential equation is

$$m(t) = e^{-t} - \frac{2/e}{t+1}.$$

According to Theorem 5.11, the optimization problem has a solution which is given by

$$u^*(t) = \left(\frac{1}{t+1} - 1\right)x(t) + \frac{2/e}{t+1} - e^{-t}.$$

Here $x(t)$ is given by the solution of the differential equation

$$\dot{x}(t) = \frac{1}{t+1}x(t) + \frac{2/e}{t+1}, \quad x(0) = 1. \qquad \square$$

5.3   Riccati differential equations

An important property of the Riccati differential equations we encounter in both this chapter and Chapter 7 is that their solution can be found by solving a set of linear differential equations. This is particularly important from a computational point of view because there are many efficient numerical algorithms that can accurately calculate solutions of linear differential equations. These algorithms have been implemented, for example, in the computer software MATLAB to calculate solutions of Riccati differential equations.

To show this equivalence, consider the following non-symmetric matrix Riccati differential equation

$$\dot{K}(t) = -DK(t) - K(t)A + K(t)SK(t) - Q, \tag{5.3.1}$$

where K, $Q \in \mathbb{R}^{m \times n}$, $D \in \mathbb{R}^{m \times m}$, $A \in \mathbb{R}^{n \times n}$ and $S \in \mathbb{R}^{n \times m}$. In fact the ensueing theory can be copied if one assumes the matrices D, A, S and Q to be Lebesgue integrable functions of time. For notational convenience we skip this possibility. Details on this extension can be found in Reid (1972).

The solution of this Riccati differential equation (5.3.1) is intimately connected with the next set of linear differential equations

$$\begin{bmatrix} \dot{U}(t) \\ \dot{V}(t) \end{bmatrix} = \begin{bmatrix} A & -S \\ -Q & -D \end{bmatrix} \begin{bmatrix} U(t) \\ V(t) \end{bmatrix}. \tag{5.3.2}$$

This relationship is summarized in the following theorem.

Theorem 5.12

If U, V is a solution pair of equation (5.3.2) with U nonsingular on the interval $[0, T]$, then $K(t) = VU^{-1}$ is a solution of the Riccati differential equation (5.3.1) on $[0, T]$. Conversely, if $K(t)$ is a solution of equation (5.3.1) on $[0, T]$ and $U(.)$ is a fundamental solution of

$$\dot{U}(t) = (A - SK(t))U(t)$$

then the pair $U(t)$, $V(t) := K(t)U(t)$ is a solution of equation (5.3.1) on $[0, T]$.

Proof

Assume $U(.), V(.)$ satisfies equation (5.3.2) with $U(.)$ invertible. Since $U(t)U^{-1}(t) = I$, differentiation of this identity gives $\dot{U}(t)U^{-1}(t) + U(t)\frac{d}{dt}\{U^{-1}(t)\} = 0$, or, $\frac{d}{dt}\{U^{-1}(t)\} = -U^{-1}(t)\dot{U}(t)U^{-1}(t)$. Then, $K := VU^{-1}$ satisfies

$$\begin{aligned} \dot{K}(t) &= \dot{V}(t)U^{-1}(t) - V(t)U^{-1}(t)\dot{U}(t)U^{-1}(t) \\ &= (-QU(t) - DV(t))U^{-1}(t) - V(t)U^{-1}(t)(AU(t) - SV(t))U^{-1}(t) \\ &= -Q - DK(t) - K(t)A + K(t)SK(t). \end{aligned}$$

Conversely, if $K(t)$ is a solution of equation (5.3.1) on $[0, T]$ and $U(.)$ is a fundamental solution of

$$\dot{U}(t) = (A - SK(t))U(t)$$

then the pair $U(t)$, $V(t) := K(t)U(t)$ satisfy

$$\begin{bmatrix} \dot{U}(t) \\ \dot{V}(t) \end{bmatrix} = \begin{bmatrix} (A - SK(t))U(t) \\ \dot{K}(t)U(t) + K(t)\dot{U}(t) \end{bmatrix}$$

$$= \begin{bmatrix} AU(t) - SV(t) \\ (-Q - DK(t) - K(t)A + K(t)SK(t))U(t) + K(t)(A - SK(t))U(t) \end{bmatrix}$$

$$= \begin{bmatrix} AU(t) - SV(t) \\ -QU(t) - DK(t)U(t) \end{bmatrix}$$

$$= \begin{bmatrix} A & -S \\ -Q & -D \end{bmatrix} \begin{bmatrix} U(t) \\ V(t) \end{bmatrix}. \qquad \square$$

By considering the special case $D = A^T$ in the above theorem we obtain the following result.

Corollary 5.13

The Riccati differential equation (5.2.3) has a solution on $[0, T]$ if and only if the set of linear differential equations

$$\begin{bmatrix} \dot{U}(t) \\ \dot{V}(t) \end{bmatrix} = \begin{bmatrix} A & -S \\ -Q & -A^T \end{bmatrix} \begin{bmatrix} U(t) \\ V(t) \end{bmatrix}; \quad \begin{bmatrix} U(T) \\ V(T) \end{bmatrix} = \begin{bmatrix} I \\ Q_T \end{bmatrix} \qquad (5.3.3)$$

has a solution on $[0, T]$, with $U(.)$ nonsingular.

Moreover, if equation (5.3.3) has an appropriate solution $(U(.), V(.))$, the solution of equation (5.2.3) is

$$K(t) = V(t)U^{-1}(t). \qquad \square$$

With $H := \begin{bmatrix} A & -S \\ -Q & -D \end{bmatrix}$, the solution of the above differential equation (5.3.3) is (see Lemma 3.3)

$$\begin{bmatrix} U(t) \\ V(t) \end{bmatrix} = e^{H(T-t)} \begin{bmatrix} I \\ Q_T \end{bmatrix}.$$

By determining the Jordan canonical form of matrix H one can determine then, in principle, an analytic expression for the solution (see Theorem 3.2). From this expression one can then derive an analytic solution for the Riccati differential equation (5.2.3) by calculating $V(t)U^{-1}(t)$.

Note

If $u^*(.)$ yields a minimum of the linear quadratic control problem and $x^*(.)$ is the corresponding state trajectory, according to the maximum principle (see also Exercises at the end of Chapter 4), there exists a costate function $\lambda^*(.)$ satisfying

$$\begin{bmatrix} \dot{x}^*(t) \\ \dot{\lambda}^*(t) \end{bmatrix} = \begin{bmatrix} A & -S \\ -Q & -A^T \end{bmatrix} \begin{bmatrix} x^*(t) \\ \lambda^*(t) \end{bmatrix}, \text{ with } \begin{bmatrix} x^*(0) \\ \lambda^*(T) \end{bmatrix} = \begin{bmatrix} x_0 \\ Q_T x^*(T) \end{bmatrix}.$$

Obviously, this two-point boundary-value problem can be rewritten as a one-point boundary-value problem with a constraint on the final time T

$$\begin{bmatrix} x^*(T) \\ \lambda^*(T) \end{bmatrix} = \begin{bmatrix} x^*(T) \\ Q_T x^*(T) \end{bmatrix}.$$

According to Corollary 5.7, item 2, all terminal states $x^*(T)$ can be reached by an appropriate choice of the initial state x_0. So, using this, we conclude that equation (5.3.3) has a solution. □

Example 5.4

Consider the Riccati differential equation

$$\dot{k}(t) = -2k(t) + k^2(t) - 3, \ k(8) = 4.$$

Then, with $T = 8, A = D = 1, S = 1, Q = 3$ and $Q_T = 4$ we can determine the solution of this differential equation using the same approach as Corollary 5.13. To that end, consider

$$H := \begin{bmatrix} 1 & -1 \\ -3 & -1 \end{bmatrix}.$$

Next, determine the eigenstructure of this matrix H. The eigenvalues are $\{2, -2\}$ and corresponding eigenvectors are

$$v_1 = \begin{bmatrix} 1 \\ -1 \end{bmatrix} \text{ and } v_1 = \begin{bmatrix} 1 \\ 3 \end{bmatrix},$$

respectively.

The solution of the differential equation (5.3.3) is then

$$\begin{bmatrix} u(t) \\ v(t) \end{bmatrix} = e^{H(t-8)} \begin{bmatrix} 1 \\ 4 \end{bmatrix}$$

$$= [v_1 \ v_2] \begin{bmatrix} e^{2(t-8)} & 0 \\ 0 & e^{-2(t-8)} \end{bmatrix} [v_1 \ v_2]^{-1} \begin{bmatrix} 1 \\ 4 \end{bmatrix}$$

$$= \frac{1}{4} \begin{bmatrix} -e^{2(t-8)} + 5e^{-2(t-8)} \\ e^{2(t-8)} + 15e^{-2(t-8)} \end{bmatrix}.$$

Notice that $u(t) \neq 0$ on $[0, 8]$. So, according to Corollary 5.13, the solution of the Riccati differential equation is

$$k(t) = \frac{v(t)}{u(t)} = \frac{e^{2(t-8)} + 15e^{-2(t-8)}}{-e^{2(t-8)} + 5e^{-2(t-8)}}.$$

By substitution of this solution into the Riccati differential equation one can straight-forwardly verify that this is indeed the correct solution. $\qquad\qquad\qquad\square$

5.4 Infinite-planning horizon

In this section we lift the restriction imposed in section 5.2 that the final time T in the planning horizon must be finite. We consider the problem of finding a control function $u(.) = Fx(.)$ (where F is a time-invariant matrix) for each $x_0 \in \mathbb{R}^n$ that minimizes the cost functional

$$J(x_0, u) := \int_0^\infty \{x^T(t)Qx(t) + u^T(t)Ru(t)\}dt, \qquad (5.4.1)$$

under the additional constraint that $\lim_{t\to\infty} x(t) = 0$. Here $Q = Q^T$, $R > 0$, and the state variable x is the solution of

$$\dot{x}(t) = Ax(t) + Bu(t), \quad x(0) = x_0. \qquad (5.4.2)$$

By Theorem 3.20 the imposed stabilization constraint is equivalent to the requirement that the system is stabilizable. Therefore, throughout this section, the assumption is made that the pair (A, B) is stabilizable. Furthermore, we introduce the set of linear, stabilizing, time-invariant feedback matrices, i.e.

$$\mathcal{F} := \{F \mid A + BF \text{ is stable}\}.$$

For notational convenience the notation $S := BR^{-1}B^T$ is used. A state feedback control function corresponding to a feedback matrix F and an initial state x_0 is denoted by $u^{FB}(x_0, F)$. With a small change of notation we shall write $J(x_0, F) := J(x_0, u^{FB}(x_0, F))$. Now,

$$J(x_0, F) = \int_0^\infty x^T(s)(Q + F^T RF)x(s)ds$$

$$= x_0^T \int_0^\infty (e^{(A+BF)s})^T(Q + F^T RF)e^{(A+BF)s}dsx_0.$$

Let $P := \int_0^\infty (e^{(A+BF)s})^T(Q + F^T RF)e^{(A+BF)s}ds$. Then $J(x_0, F) = x_0^T Px_0$ and, since $A + BF$ is stable,

$$0 - (Q + F^T RF) = \int_0^\infty \frac{d}{ds}\{(e^{(A+BF)s})^T(Q + F^T RF)e^{(A+BF)s}\}ds$$

$$= \int_0^\infty (A + BF)^T(e^{(A+BF)s})^T(Q + F^T RF)e^{(A+BF)s}$$

$$+ (e^{(A+BF)s})^T(Q + F^T RF)e^{(A+BF)s}(A + BF)ds$$

$$= (A + BF)^T P + P(A + BF).$$

That is, for each $F \in \mathcal{F}$, $J(x_0, F) = x_0^T \varphi(F) x_0$ with $\varphi : \mathcal{F} \to \mathbb{R}^{n \times n}$ defined by $\varphi : F \mapsto P$ where P is the unique solution of the Lyapunov equation

$$(A + BF)^T P + P(A + BF) = -(Q + F^T RF). \tag{5.4.3}$$

We will see that the next algebraic Riccati equation (ARE)

$$Q + A^T X + XA - XSX = 0. \tag{5.4.4}$$

plays an important role in the problem under consideration. Recall from section 2.7 that a solution K of this equation is called **stabilizing** if the matrix $A - SK$ is stable and, furthermore, from Theorem 2.33 that such a solution, if it exists, is unique. Theorem 5.14, below, gives the analogue of the finite-planning horizon linear quadratic control problem.

Theorem 5.14 (Infinite horizon linear quadratic control problem)

Assume that (A, B) is stabilizable and $u = Fx$, with $F \in \mathcal{F}$. The linear quadratic control problem (5.4.1) and (5.4.2) has a minimum $\hat{F} \in \mathcal{F}$ for $J(F)$ for each x_0 if and only if the algebraic Riccati equation (5.4.4) has a symmetric stabilizing solution K. If the linear quadratic control problem has a solution, then the solution is uniquely given by $\hat{F} = -R^{-1}B^T K$ and the optimal control in feedback form is

$$u^*(t) = -R^{-1}B^T Kx(t). \tag{5.4.5}$$

In open-loop form it is

$$u^*(t) = -R^{-1}B^T K\Phi(t, 0)x_0 \tag{5.4.6}$$

with Φ the solution of the transition equation

$$\dot{\Phi}(t, 0) = (A - SK)\Phi(t, 0), \Phi(0, 0) = I.$$

Moreover, $J(u^*) = x_0^T Kx_0$.   □

Proof

'$\Leftarrow$ part' This part of the proof mimics the proof of the finite-planning horizon case. Let K be the stabilizing solution of the ARE. Then

$$\int_0^\infty \frac{d}{dt}\{x^T(t)Kx(t)\}dt = 0 - x^T(0)Kx(0).$$

Consequently, making a completion of squares again within the cost function gives

$$J = \int_0^\infty \{x^T(t)Qx(t) + u^T(t)Ru(t)\}dt + \int_0^\infty \frac{d}{dt}\{x^T(t)Kx(t)\}dt + x^T(0)Kx(0)$$

$$= \int_0^\infty \{x^T(t)Qx(t) + u^T(t)Ru(t) + \frac{d}{dt}\{x^T(t)Kx(t)\}\}dt + x^T(0)Kx(0).$$

Now for an arbitrary $F \in \mathcal{F}$, with $u = Fx$, $\dot{x} = (A + BF)x$. Using this and the algebraic Riccati equation (5.4.4), the integrand can be rewritten as follows

$$x^T Q x + u^T R u + \frac{d}{dt}\{x^T K x\} = x^T Q x + x^T F^T R F x + \dot{x}^T K x + x^T K \dot{x}$$

$$= x^T Q x + x^T F^T R F x + x^T (A + BF)^T K x + x^T K(A + BF)x$$

$$= x^T (Q + A^T K + KA)x + x^T F^T R F x + x^T F^T B^T K x + x^T K B F x$$

$$= x^T K S K x + x^T F^T R F x + x^T F^T B^T K x + x^T K B F x$$

$$= x^T (F + R^{-1} B^T K)^T R(F + R^{-1} B^T K)x.$$

So

$$J = \int_0^\infty \{x^T(t)(F + R^{-1}B^T K)^T R(F + R^{-1}B^T K)x(t)\}dt + x^T(0)Kx(0).$$

From this expression it is obvious that $J \geq x^T(0)Kx(0)$ for all F and that equality is achieved if $F = -R^{-1}B^T K$.

Proof

'$\Rightarrow$ **part**' This part of the proof is based on a variational argument. First, note that the set $\mathcal{F}$ is a nonempty open set. Second, note that the smoothness of the coefficients in a Lyapunov equation is preserved by the solution of this equation (Lancaster and Rodman (1995)), which implies that J is differentiable with respect to F. Now, let $\hat{F} \in \mathcal{F}$ be a minimum of J for each x_0. Then, according to Theorem 4.1, $\delta_2 J(x_0, \hat{F}; \Delta F) = 0$ for each ΔF and for each x_0. Since $\delta_2 J(x_0, \hat{F}; \Delta F) = x_0^T \delta\varphi(\hat{F}; \Delta F)x_0$, this implies that $\delta\varphi(\hat{F}; \Delta F) = 0$ for all increments ΔF. Hence

$$\partial \varphi(\hat{F}) = 0. \tag{5.4.7}$$

Next, introduce the map $\Phi : \mathcal{F} \times \mathbb{R}^{n \times n} \rightarrow \mathbb{R}^{n \times n}$ by

$$\Phi(F, P) = (A + BF)^T P + P(A + BF) + Q + F^T R F.$$

By definition, $\Phi(F, \varphi(F)) = 0$ for all $F \in \mathcal{F}$. Taking the derivative of this equality and applying the chain rule yields

$$\partial_1 \Phi(F, \varphi(F)) + \partial_2 \Phi(F, \varphi(F))\partial\varphi(F) = 0 \text{ for all } F \in \mathcal{F}.$$

Substituting $F = \hat{F}$ in this equation, and using equation (5.4.7), shows that $\partial_1\Phi(\hat{F}, \varphi(\hat{F})) = 0$, or, equivalently,

$$\delta_1\Phi(\hat{F}, \varphi(\hat{F}); \Delta F) = 0 \text{ for all } \Delta F. \tag{5.4.8}$$

The differential of Φ with respect to its first argument with increment ΔF is (see also Example 4.2)

$$\delta_1\Phi(F, P; \Delta F) = \Delta F^T (B^T P + RF) + (PB + F^T R)\Delta F.$$

Combining this result with (5.4.8) produces

$$\Delta F^T (B^T \varphi(\hat{F}) + R\hat{F}) + (\varphi(\hat{F})B + \hat{F}^T R)\Delta F = 0 \text{ for all } \Delta F,$$

which clearly implies that $B^T \varphi(\hat{F}) + R\hat{F} = 0$, or, equivalently, $\hat{F} = -R^{-1}B^T \varphi(\hat{F})$. Now, since $\Phi(\hat{F}, \varphi(\hat{F})) = 0$, we conclude that $X := \varphi(\hat{F})$ is the stabilizing solution of the ARE (5.4.4). □

Note

Recall from section 2.7 that the unique stabilizing solution X_s of the algebraic Riccati equation (5.4.4) can be calculated by determining the graph subspace $\text{Im} \begin{bmatrix} X_1 \\ X_2 \end{bmatrix}$ of the Hamiltonian matrix $\begin{bmatrix} A & -S \\ -Q & -A^T \end{bmatrix}$ which has the property that all eigenvalues of the matrix $A - SX_2X_1^{-1}$ have a strictly negative real part. As we already noticed in section 2.7 this graph subspace is uniquely determined if it exists and $X_s = X_2X_1^{-1}$.

This approach has been elaborated in the literature in more detail. Lancaster and Rodman (1995) and Laub (1991) have shown that the existence of the stabilizing solution of the ARE can, for instance, be verified by checking whether the above Hamiltonian matrix has no purely imaginary eigenvalues, and whether a rank condition on the matrix sign of a certain matrix is satisfied. There are many algorithms for computing the matrix sign accurately and there is a comprehensive list of references in the review paper by Laub (1991) □

Example 5.5

Consider the minimization of

$$J := \int_0^\infty \{x^T(t)x(t) + 2u^2(t)\}dt,$$

subject to the system

$$\dot{x}(t) = \begin{bmatrix} 1 & 0 \\ 0 & -1 \end{bmatrix} + \begin{bmatrix} 1 \\ 0 \end{bmatrix} u(t), \ x(0) = [1, \ 2]^T.$$

This system is stabilizable. So, according to Theorem 5.14, the problem has a solution $\hat{F} \in \mathcal{F}$ if and only if the following algebraic Riccati equation has a stabilizing solution

$$\begin{bmatrix} 1 & 0 \\ 0 & 1 \end{bmatrix} + \begin{bmatrix} 1 & 0 \\ 0 & -1 \end{bmatrix} X + X \begin{bmatrix} 1 & 0 \\ 0 & -1 \end{bmatrix} - X \begin{bmatrix} \frac{1}{2} & 0 \\ 0 & 0 \end{bmatrix} X = 0.$$

Simple calculations show that

$$X = \begin{bmatrix} 2 + \sqrt{6} & 0 \\ 0 & \frac{1}{2} \end{bmatrix}$$

is the stabilizing solution of this Riccati equation. The resulting optimal control and cost are

$$u^*(t) = -\frac{1}{2}[2 + \sqrt{6}, \ 0]x(t) \text{ and } J^* = 4 + \sqrt{6},$$

respectively.   □

Notes

1. Willems (1971) considers the optimization problem (5.4.1) and (5.4.2) with $u \in \mathcal{U}_s(x_0)$ under the assumption that (A, B) is controllable. Here the class of control functions $\mathcal{U}_s(x_0)$ is defined by:

$$\mathcal{U}_s(x_0) = \left\{ u \in L_{2,loc} \mid J(x_0, u) \text{ exists in } \mathbb{R} \cup \{-\infty, \infty\}, \lim_{t \to \infty} x(t) = 0 \right\},$$

where $L_{2,loc}$ is the set of **locally square-integrable** functions, i.e.

$$L_{2,loc} = \{u[0, \infty) \mid \forall T > 0, \int_0^T u^T(s)u(s)ds < \infty\}.$$

Under these conditions, by Theorem 3.11, the solution to the differential equation (5.4.2) exists (in the extended sense) for all finite T. Combining Willems' results (1971, also Trentelman and Willems (1991)) and Theorem 5.14 we have the next result. Assume (A, B) is controllable, then the following statements are equivalent.

1. $\forall x_0 \exists \hat{u} \in \mathcal{U}_s(x_0) \forall u \in \mathcal{U}_s(x_0) \quad J(x_0, \hat{u}) \leq J(x_0, u)$;
2. $\exists \hat{F} \in \mathcal{F} \ \forall x_0 \ \forall u \in \mathcal{U}_s(x_0) \quad J\left(x_0, u^{FB}(x_0, \hat{F})\right) \leq J(x_0, u)$;
3. $\exists \hat{F} \in \mathcal{F} \forall x_0 \forall F \in \mathcal{F} \quad J\left(x_0, u^{FB}(x_0, \hat{F})\right) \leq J(x_0, u^{FB}(x_0, F))$;
4. Δ is positive definite where Δ denotes the difference between the largest and smallest real symmetric solution of the algebraic Riccati equation (5.4.3);
5. the algebraic Riccati equation (5.4.3) has a stabilizing solution.

2. Following the lines of the proof of the finite-planning horizon case, one can show the next equivalent statement of Theorem 5.14. Assume that (A, B) is stabilizable. Then, the linear quadratic control problem (5.4.1) and (5.4.2) has a minimum $u \in \mathcal{U}_{re}$ for every $x_0 \in \mathbb{R}^n$ if and only if the algebraic Riccati equation (5.4.1) and (5.4.2) has a symmetric stabilizing solution K. The reader is asked to fill out the details of the proof of this statement in one of the Exercises at the end of this chapter.   □

An important case where our control problem has a solution is when Q is positive semi-definite. This case is summarized in the next proposition. Furthermore, the proposition relates the finite-planning horizon case to the infinite-planning horizon solution if the planning horizon expands. In this proposition we allow for a larger set of control functions than we did before: all locally square-integrable functions are considered. So, unlike in Theorem 5.14, we do not suppose a priori that the state of the controlled system must converge to zero.

Proposition 5.15

Assume (A, B) is stabilizable and Q positive semi-definite. Consider the set of locally square-integrable control functions. Then both the finite and the infinite linear quadratic control problems attain a minimum within this class of control functions. Furthermore, with $Q_T = 0$, the finite-planning horizon solution converges to the infinite-planning horizon solution if the planning horizon expands. That is,

$$K_T(0) \to K, \text{ if } T \to \infty,$$

where $K_T(t) > 0$ is the solution of the Riccati differential equation (5.2.3) with $Q_T = 0$ and $K \geq 0$ is the smallest[2] semi-positive definite solution of the algebraic Riccati equation (5.4.4).

Moreover, in case (Q, A) is additionally detectable, $K \in \mathcal{F}$.

Proof

Since $Q \geq 0$ it follows straightforwardly that the Hamiltonian function $H(x, u, \lambda)$ associated with the finite-planning horizon optimal control problem is simultaneously convex in (x, u). Therefore, the finite-planning horizon problem always has a solution (see Theorem 4.8 and the ensueing discussion).

Next notice that due to the positive semi-definiteness assumption on Q again, for every $h > 0$,

$$0 \leq \int_0^T \{x^T(t)Qx(t) + u^T(t)Ru(t)\}dt \leq \int_0^{T+h} \{x^T(t)Qx(t) + u^T(t)Ru(t)\}dt.$$

So, by choosing $h = T_1$ and $h = \infty$, respectively, one obtains

$$0 \leq x_0^T K_T(0)x_0 = \min \int_0^T \{x^T(t)Qx(t) + u^T(t)Ru(t)\}dt$$

$$\leq x_0^T K_{T+T_1}(0)x_0 = \min \int_0^{T+T_1} \{x^T(t)Qx(t) + u^T(t)Ru(t)\}dt$$

$$\leq \int_0^\infty \{x^T(t)Qx(t) + u^T(t)Ru(t)\}dt =: J(x_0, u).$$

Since (A, B) is stabilizable there exists a state-feedback control $u = Fx$ which stabilizes the system. Using this control it is clear that the finite horizon minimum cost is always bounded. That is, with $H := \int_0^\infty e^{(A+BF)^T t}(Q + F^T RF)e^{(A+BF)t}dt$, $x_0^T K_T(0)x_0 \leq x_0^T Hx_0$ for all T.

---

[2]That is, every solution $X \geq 0$ of equation (5.4.4) satisfies $X \geq K$.

Since x_0 is arbitrary and K_{T_i} are symmetric, we conclude that the above inequalities must hold w.r.t. the matrices $K_{T_i}(0)$ too and, therefore, $K_{T_1}(0) \geq K_{T_2}(0)$ if $T_1 \geq T_2$ and K_T is bounded from above by H. Consequently $\lim_{T \to \infty} K_T(0)$ exists. Denote this limit by K. Then, in particular, it follows that

$$J(x_0, u) \geq x_0^T K_T(0) x_0.$$

Therefore, by letting $T \to \infty$ in this inequality,

$$J(x_0, u) \geq x_0^T K(0) x_0. \tag{5.4.9}$$

Furthermore, from equation (5.2.3), then $\lim_{T \to \infty} \dot{K}_T(0)$ also exists. It is then easily verified that $\dot{K}_T(t)$ converges to zero at $t = 0$ if $T \to \infty$. So K satisfies the algebraic Riccati equation (5.4.4).

To see that K is the minimum positive semi-definite solution of equation (5.4.4) that yields the solution for the infinite horizon problem, consider an arbitrary solution of equation (5.4.4), X. Then, with $u = -R^{-1}B^T X x$,

$$J(x_0, u, T) = x_0^T X x_0 - x^T(T) X x(T) \geq x_0^T K_T(0) x_0. \tag{5.4.10}$$

If we choose in this expression $X := K$, we infer that

$$\lim_{T \to \infty} x^T(T) K x(T) \to 0.$$

So, in particular with $u^* = -R^{-1}B^T K x$

$$J(x_0, u^*) = \lim_{T \to \infty} J(x_0, u^*, T) = \lim_{T \to \infty} \{x_0^T K x_0 - x^T(T) K x(T)\} = x_0^T K x_0.$$

This shows (see equation (5.4.9)) that u^* is optimal. On the other hand we immediately have from equation (5.4.10) that

$$x_0^T X x_0 \geq x_0^T K_T(0) x_0.$$

Taking the limit for $T \to \infty$ then shows that K is indeed the smallest positive semi-definite solution of equation (5.4.4).

Finally notice that, if $Q > 0$, it follows from the cost function that the corresponding optimal control sequence must stabilize the closed-loop system. So K is then the unique stabilizing solution of the algebraic Riccati equation (5.4.4) (see Theorem 2.33).

In principle, a similar reasoning holds if the system is detectable. Then it follows from the cost function that all states that can be observed are (strictly positive quadratically) penalized and, therefore, we conclude that those states have to converge to zero. On the other hand, since by the detectability assumption all non-observable states converge to zero anyway, the conclusion follows that the proposed control stabilizes the closed-loop system. A formal proof can be found, for example, in Zhou, Doyle and Glover (1996).   □

A simple example illustrates the fact that just requiring Q to be semi-positive definite is not enough to guarantee that the infinite-planning horizon problem will have a stabilizing optimal solution as follows. Let R be positive definite and $Q = 0$. Then for every matrix A, particularly if A is unstable, the optimal feedback is $u(.) = 0$. So, indeed an optimal solution always exists, but it does not necessarily stabilize the closed-loop system. This clearly demonstrates the role played by our assumption to restrict ourselves to the set of controls stabilizing the closed-loop system. In the Notes and references section at the end of this chapter we will discuss some theoretical results if this assumption is dropped.

Note

In Section 2.7, Proposition 2.35, it was shown that the stabilizing solution of the algebraic Riccati equation is also the maximal solution of this equation. So we conclude from Proposition 5.15 that, in general, the finite-planning horizon solution does not converge to the stabilizing solution of equation (5.4.4). That is, if one considers the limiting optimal control sequence implied by the finite-planning horizon problem, this control will (in general) **not** solve our optimization problem (5.4.1) and (5.4.2) subject to the constraint that the state converges to zero if time expands. In fact Proposition 5.15 demonstrates that the finite-planning horizon solution, with $Q_T = 0$, converges to the solution of an infinite-planning horizon problem in which there are no constraints on the final state. If (Q,A) is detectable we saw that the corresponding infinite-planning horizon solution stabilizes the system. So apparently in that case the minimal semi-positive definite and maximal solution of the algebraic Riccati equation coincide. Or, stated differently, the algebraic Riccati equation has a unique semi-positive definite solution under this additional detectability condition.

A further discussion on this convergence issue is postponed until the next section.

□

Example 5.6

1. Consider the problem to minimize

$$J := \int_0^T u^2(t)dt,$$

subject to the system dynamics

$$\dot{x}(t) = x(t) + u(t), \ x(0) = x_0.$$

For $T < \infty$, the solution is found by solving the differential equation:

$$\dot{k}(t) = -2k(t) + k^2(t); k(T) = 0.$$

Obviously $k(t) = 0$ solves this differential equation. Therefore the optimal control is: $u(.) = 0$. Next consider the case $T = \infty$. The corresponding algebraic Riccati equation is

$$0 = -2k + k^2.$$

This equation has two non-negative solutions, $k_1 = 0$ and $k_2 = 2$.

According to Proposition 5.15 the solution of the minimization problem which does not impose any further assumptions on the final state of the system is then given by $u^*(t) = -k_1x(t) = 0$.

The solution of the minimization problem with the additional constraint that the state $x(t)$ must converge to zero, if $t \to \infty$, is (see Theorem 5.14) given by $u^*(t) = -k_2x(t) = -2x(t)$.

Obviously, the solution of the finite-planning horizon problem converges to k_1.

2. Consider the next slight modification of the above example. Minimize

$$J := \int_0^T u^2(t)dt,$$

subject to the system dynamics

$$\dot{x}(t) = -x(t) + u(t), \ x(0) = x_0.$$

Notice that now, since the uncontrolled system 'matrix' A is stable, the system is now detectable.

For $T < \infty$, the solution is now obtained by solving the differential equation:

$$\dot{k}(t) = 2k(t) + k^2(t); k(T) = 0,$$

which, again, has the solution $k(t) = 0$.

The corresponding algebraic Riccati equation is:

$$0 = 2k + k^2.$$

This equation has only one non negative solution, $k_1 = 0$. So, the solution of both the problem with and without the constraint that the final state must converge to zero coincide. The optimal control is in both cases $u(.) = 0$.   □

Example 5.7

Reconsider Example 5.2. Using the same notation, consider the following discounted cost function

$$\min \int_0^\infty e^{-\rho t}\{-pC^2(t) + rI^2(t)\}dt,$$

subject to

$$\dot{C}(t) = -\alpha C(t) + I(t),$$

where $\rho > 0$ is the discount factor.

To solve this problem we first rewrite it into our standard framework. That is, introducing

$$\tilde{C}(t) = e^{-\frac{1}{2}\rho t}C(t) \text{ and } \tilde{I}(t) = e^{-\frac{1}{2}\rho t}I(t),$$

we rewrite the minimization problem as follows

$$\min \int_0^T \{-p\tilde{C}^2(t) + r\tilde{I}^2(t)\}dt,$$

subject to

$$\dot{\tilde{C}}(t) = -\left(\alpha + \frac{1}{2}\rho\right)\tilde{C}(t) + \tilde{I}(t).$$

So according to Theorem 5.14, with $A = -(\alpha + \frac{1}{2}\rho)$, $B = 1$, $Q = -p$ and $R = r$, this optimization problem has a solution for which $\tilde{C}$ converges to zero if and only if the following algebraic Riccati equation has a stabilizing solution

$$2\left(\alpha + \frac{1}{2}\rho\right)k + \frac{1}{r}k^2 + p = 0. \tag{5.4.11}$$

Obviously, this quadratic equation has a real solution k only if its discriminant is non-negative. That is,

$$D := 4\left(\alpha + \frac{1}{2}\rho\right)^2 - 4\frac{p}{r} \geq 0. \tag{5.4.12}$$

The second condition which must hold is that the solution of equation (5.4.11) stabilizes the closed-loop system. It is easily verified that the only appropriate candidate solution of equation (5.4.11) is

$$\bar{k} = \frac{r}{2}\left\{-2\left(\alpha + \frac{1}{2}\rho\right) + \sqrt{D}\right\}. \tag{5.4.13}$$

So the optimization problem has a solution if and only if both (i) $D \geq 0$ and (ii) $-(\alpha + \frac{1}{2}\rho) - \frac{1}{r}\bar{k} < 0$ hold. If a solution exists, the optimal investment policy is $I(t) = -\frac{1}{r}\bar{k}C(t)$.

Further elaboration of both conditions (i) and (ii) shows that these conditions hold if and only if $\left(\alpha + \frac{1}{2}\rho\right)^2 > \frac{p}{r}$. □

Example 5.8

Consider the following simple macroeconomic multiplier-accelerator model (for example Turnovsky (1977)).

The aggregate demand of the economy at time t, $Z(t)$, is defined by

$$Z(t) = C(t) + I(t) + G(t) \tag{5.4.14}$$

where $C(t)$ denotes consumption, $I(t)$ denotes investment and $G(t)$ denotes government expenditure.

Assume that in disequilibrium product markets adjust as follows

$$\dot{Y}(t) = \alpha(Z(t) - Y(t)), \quad \alpha > 0 \tag{5.4.15}$$

where $Y(t)$ denotes aggregate supply at time t. That is, if demand exceeds supply, producers increase supply proportional to the excess demand. If $Z(t) < Y(t)$, the adjustment is reversed.

Furthermore assume that consumption is proportional to supply,

$$C(t) = \beta Y(t), \quad 0 < \beta < 1, \tag{5.4.16}$$

and investment I satisfies

$$I(t) = \dot{K}(t), \tag{5.4.17}$$

where $K(t)$ is the capital stock. Finally assume that the desired capital stock $K^*(t)$ is proportional to income

$$K^*(t) = \gamma Y(t), \tag{5.4.18}$$

and that the adjustment of capital stock to its desired level is described by

$$\dot{K}(t) = \delta(K^*(t) - K(t)). \tag{5.4.19}$$

Substitution of Z from equation (5.4.14), I from equation (5.4.17), K from equation (5.4.19) and, finally, K^* from equation (5.4.18) into equation (5.4.15) then yields the following dynamics

$$\begin{bmatrix} \dot{Y}(t) \\ \dot{K}(t) \end{bmatrix} = \begin{bmatrix} \rho & -\alpha\delta \\ \delta\gamma & -\delta \end{bmatrix} \begin{bmatrix} Y(t) \\ K(t) \end{bmatrix} + \begin{bmatrix} \alpha \\ 0 \end{bmatrix} G(t), \tag{5.4.20}$$

where $\rho := \alpha(\beta + \delta\gamma - 1)$.

Now assume that equation (5.4.20) also describes the dynamic adjustment of the economy if there is an initial disturbance from supply and capital from their long-term equilibrium values. That is, denoting $y(t)$, $k(t)$ and $g(t)$ as the deviations of $Y(t)$, $K(t)$ and $G(t)$ from their long-term equilibrium values, we assume they satisfy

$$\begin{bmatrix} \dot{y}(t) \\ \dot{k}(t) \end{bmatrix} = \begin{bmatrix} \rho & -\alpha\delta \\ \delta\gamma & -\delta \end{bmatrix} \begin{bmatrix} y(t) \\ k(t) \end{bmatrix} + \begin{bmatrix} \alpha \\ 0 \end{bmatrix} g(t), y(0) = y_0, \quad k(0) = k_0. \tag{5.4.21}$$

Then, a government policy aimed at minimizing most of the effects of this disturbance is obtained (in terms of section 5.1 one might think of equation (5.4.21) as being the linearized system dynamics of the optimally controlled economy) by considering the next minimization problem

$$\int_0^\infty [y(t) \; k(t)] Q \begin{bmatrix} y(t) \\ k(t) \end{bmatrix} + Rg^2(t) dt, \tag{5.4.22}$$

where Q is a positive definite matrix and $R > 0$.

With $A := \begin{bmatrix} \rho & -\alpha\delta \\ \delta\gamma & -\delta \end{bmatrix}$ and $B := \begin{bmatrix} \alpha \\ 0 \end{bmatrix}$

$$[B, \ AB] = \begin{bmatrix} \alpha & \alpha\rho \\ 0 & \alpha\delta\gamma \end{bmatrix}.$$

So the system is controllable and thus, in particular, stabilizable (see Theorem 3.20). Since, moreover, Q is positive definite we know that this problem has a solution.
The solution is given by

$$u^*(t) = -\begin{bmatrix} \dfrac{\alpha}{r} & 0 \end{bmatrix} Kx(t),$$

where K is the stabilizing solution of the algebraic Riccati equation (5.4.4)

$$Q + A^T X + XA - XSX = 0.$$

According to Section 2.7 this solution can be calculated by considering the invariant subspaces of the matrix

$$\begin{bmatrix} A & -S \\ -Q & -A^T \end{bmatrix}.$$

That is, by studying the eigenstructure of the matrix

$$\begin{bmatrix} \rho & -\alpha\delta & \dfrac{-\alpha^2}{r} & 0 \\ \delta\gamma & -\delta & 0 & 0 \\ -q_{11} & -q_{12} & -\rho & -\delta\gamma \\ -q_{12} & -q_{22} & \alpha\delta & \delta \end{bmatrix}. \qquad \square$$

As in the finite-planning horizon case, we also consider the inhomogeneous infinite horizon linear quadratic control problem. Obviously, not every inhomogeneous function $c(.)$ can be allowed. We will assume that $c(.)$ belongs to the class of locally square integrable functions, $L^e_{2,loc}$ which exponentially converge to zero when the time t expands to infinity. That is, it is assumed that for every $c(.)$ there exist strictly positive constants M and α such that

$$|c(t)| \leq Me^{-\alpha t}.$$

Under this assumption we have the following analogue of Theorem 5.11.

Theorem 5.16

Let $c(.) \in L^e_{2,loc}$. Consider the minimization of the linear quadratic cost function (5.4.1) subject to the state dynamics

$$\dot{x}(t) = Ax(t) + Bu(t) + c(t), \ x(0) = x_0, \qquad (5.4.23)$$

and $u \in \mathcal{U}_s(x_0)$. Then we have the following result.

The linear quadratic problem (5.4.1) and (5.4.23) has a solution for all $x_0 \in \mathbb{R}^n$ if and only if the algebraic Riccati equation (5.4.4) has a symmetric stabilizing solution $K(.)$. Moreover, if the linear quadratic control problem has a solution, then the optimal control is

$$u^*(t) = -R^{-1}B^T(Kx^*(t) + m(t)).$$

Here $m(t)$ is given by

$$m(t) = \int_t^\infty e^{-(A-SK)^T(t-s)} Kc(s)ds, \qquad (5.4.24)$$

and $x^*(t)$ is the solution of the differential equation implied through this optimal control

$$\dot{x}^*(t) = (A - SK)x^*(t) - Sm(t) + c(t), \quad x^*(0) = x_0.$$

Proof

'$\Leftarrow$ **part**' This part mimics the proof of Theorem 5.11.

Let K be the stabilizing solution of the algebraic Riccati equation (5.4.4) and $m(t)$ as defined in equation (5.4.24). Next consider the value function

$$V(t) := x^T(t)Kx(t) + 2m^T(t)x(t) + n(t),$$

where

$$n(t) = \int_t^\infty \{-m^T(s)Sm(s) + 2m^T(s)c(s)\}ds.$$

Note that $\dot{n}(t) = m^T(t)Sm(t) - 2m^T(t)c(t)$. Substitution of $\dot{n}$, $\dot{x}$ and $\dot{m}$ (see equations (5.4.23) and (5.4.24)) into $\dot{V}$, using the fact that $A^T K + KA = -Q + KSK$ (see equation (5.4.4)) gives

$$\dot{V}(t) = \dot{x}^T(t)Kx(t) + x^T(t)K\dot{x}(t) + 2\dot{m}^T(t)x(t) + 2m^T(t)\dot{x}(t) + \dot{n}(t)$$
$$= -x^T(t)Qx(t) - u^T(t)Ru(t) + [u(t) + R^{-1}B^T(Kx(t) + m(t))]^T R[u(t)$$
$$+ R^{-1}B^T(Kx(t) + m(t))].$$

Since $m(t)$ converges exponentially to zero $\lim_{t\to\infty} n(t) = 0$ too. Since $\lim_{t\to\infty} x(t) = 0$ too,

$$\int_0^\infty \dot{V}(s)ds = -V(0).$$

Substitution of $\dot{V}$ into this expression and rearranging terms gives

$$\int_0^\infty x^T(t)Qx(t) + u^T(t)Ru(t)dt = V(0)$$
$$+ \int_0^\infty [u + R^{-1}B^T(Kx(t) + m(t))]^T R[u + R^{-1}B^T(Kx(t) + m(t))]dt.$$

Since $V(0)$ does not depend on $u(.)$ and R is positive definite, the proposed result follows.

Proof

'$\Rightarrow$ **part**' To prove this claim, we can copy the corresponding part of the proof of Theorem 5.11 if we consider the inner product $\langle f, g \rangle := \int_0^\infty f^T(s)g(s)ds$ on the space $L^e_{2,loc}$. $\quad\square$

It is easily verified from the proof that the above theorem continues to hold if the assumption $c(\cdot) \epsilon L^e_{2,loc}$ is relaxed to the assumption that $|c(t)| \leq Me^{-xt}$ where $x = \max \sigma (A - SK)$. $\quad\square$

Example 5.9

Consider the optimization problem

$$\min_u \int_0^\infty \{3x^2(t) + u^2(t)\}dt,$$

subject to the dynamics

$$\dot{x}(t) = x(t) + u(t) + e^{-t}, \quad x(0) = 3.$$

The Riccati equation associated with this problem is

$$3 + 2k - k^2 = 0.$$

The stabilizing solution of this equation is $k = 3$.

Therefore, according Theorem 5.16, the minimization problem has a solution. The optimal control is

$$u^*(t) = -3x(t) - m(t),$$

where

$$m(t) = 3 \int_t^\infty e^{2(t-s)} e^{-s} ds = e^{-t},$$

and

$$\dot{x}(t) = -2x(t), \quad x(0) = 3.$$

The resulting closed-loop trajectory of the regulated system is $3e^{-2t}$. $\quad\square$

5.5 Convergence results

In Proposition 5.15 we saw the interesting phenomenon that the Riccati differential equation solution $K(0)$ associated with the finite-planning horizon solution of

$$\dot{K}(t) = -A^T K(t) - K(t)A + K(t)SK(t) - Q, \quad K(T) = Q_T, \qquad \text{(RDE)} \quad (5.5.1)$$

converges if $Q_T = 0$ (under some additional conditions) to the smallest semi-positive definite solution of the corresponding algebraic Riccati equation

$$A^T X + XA - XSX + Q = 0. \qquad \text{(ARE)} \quad (5.5.2)$$

We will show in this section that this is, in some sense, a somewhat extraordinary case. That is, if we consider the finite-planning horizon optimization problem with Q_T different from zero then, generically, the solution $K(0)$ of the corresponding (RDE) will converge to the maximal solution of the algebraic Riccati equation (5.5.2) if the planning horizon T becomes arbitrarily large. The following scalar example illustrates this point.

Example 5.10

Consider the minimization of the scalar cost function

$$\int_0^T u^2(t)dt + q_T x^2(T), \qquad (5.5.3)$$

subject to the scalar differential equation

$$\dot{x}(t) = ax(t) + u(t), \quad x(0) = x_0 \quad \text{and} \quad a > 0. \qquad (5.5.4)$$

According to Theorem 5.1 this minimization problem (5.5.3) and (5.5.4) has a solution if and only if the following Riccati differential equation has a solution

$$\dot{k}(t) = -2ak(t) + k^2(t), \quad k(T) = q_T. \qquad (5.5.5)$$

Since the cost function is at every planning horizon T, as well as bounded from below (by zero) and bounded from above, it follows that $x_0^2 k(0)$ is bounded for every T. That is, the solution $k(t)$ does not have a finite escape time, and consequently equation (5.5.5) has a solution on $[0, T]$, for every finite T (one could also use the convexity argument of the Hamiltonian, as in the proof of Proposition 5.15, to obtain this conclusion). So, for every planning horizon T our minimization problem (5.5.3), (5.5.4) has a solution, which is given by

$$u^*(t) = -k(t)x(t).$$

Next, consider the corresponding infinite-planning horizon problem

$$\min \int_0^\infty u^2(t)dt, \qquad (5.5.6)$$

subject to equation (5.5.4). According to Proposition 5.15 the solution for this minimization problem is

$$u^*(t) = -x_s x(t),$$

where x_s is the stabilizing solution of

$$0 = -2ax + x^2. \tag{5.5.7}$$

Simple calculations show that the solutions of equation (5.5.7) are

$$x = 0 \quad \text{and} \quad x = 2a.$$

Since we assumed that $a > 0$, it follows that $x_s = 2a$.

Next, we study the convergence properties of the Riccati differential equation (5.5.5). That is, we study the behavior of $k(0)$ if the planning horizon T in this equation becomes arbitrarily large. To that purpose we first rewrite this terminal value differential equation (5.5.5) as an initial value differential equation. Introduce the new variable

$$p(t) := k(T - t).$$

Then

$$k(t) = p(T - t) \quad \text{and} \quad \frac{dk(t)}{dt} = \frac{dp(T - t)}{dt} = -\dot{p}(T - t).$$

Consequently (with $s := T - t$), the terminal value problem (5.5.5) has a solution if and only if the next initial-value problem has a solution

$$\dot{p}(s) = 2ap(s) - p^2(s), \quad p(0) = q_T. \tag{5.5.8}$$

Moreover, since the solution of the minimization problem (5.5.3) and (5.5.4) is given by $u^*(t) = p(T - t)x(t)$, the solution at time $t = 0$ is

$$u^*(0) = p(T)x(0).$$

So, to analyze the limiting behavior of the optimal control at time $t = 0$, one can equivalently analyze the dynamic behavior of the initial-value problem (5.5.8). This equation has the structure

$$\dot{p} = f(p), \quad \text{where} \quad f(p) = 2ap - p^2, \tag{5.5.9}$$

which we discussed in Chapter 3. This differential equation can be solved either explicitly (using the separation of variables technique) or qualitatively. We will do the latter.

To that end, we first look for the equilibrium points of equation (5.5.9). Obviously, they coincide with the solutions of the algebraic Riccati equation (5.5.7), i.e. $p = 0$ and $p = 2a$.

To determine the local behavior of equation (5.5.9) near the equilibrium points, we consider the derivative of $f(p)$ at the equilibrium points. Straightforward differentiation of f gives

$$f'(0) = 2a \quad \text{and} \quad f'(2a) = -2a.$$

Figure 5.1   Phase diagram for the differential equation (5.5.9)

Since $a > 0, f'(0) > 0$, and thus the equilibrium point $p = 0$ is unstable. Similarly, one concludes that $p = 2a$ is a stable equilibrium. So, we obtain the 'one-dimensional phase diagram' shown in Figure 5.1.

Notice that for any $q_T > 0$ the solution will converge towards the maximal and stabilizing solution of the algebraic Riccati equation. Only in the extraordinary case where $q_T = 0$ does convergence take place towards the minimal solution of the algebraic Riccati equation (since this is an equilibrium point of the dynamical system). In all other cases, the solution diverges.   □

To study the convergence properties for the finite-planning horizon problem in more detail, a prerequisite is that this problem has a solution for every T. Or, equivalently (see Theorem 5.1), the following assumption is valid.

Well-posedness assumption

The Riccati differential equation (5.5.1) has a solution for every $T > 0$.   □

We have seen in Proposition 5.15 that if $Q \geq 0$, this assumption is always satisfied if $Q_T = 0$. However, even in case $Q \geq 0$, the finite horizon problem does not have a solution for every Q_T. We elaborate this point for the scalar case.

Proposition 5.17

Consider the minimization of

$$\int_0^T \{qx^2(t) + ru^2(t)\}dt + q_T x^2(T),$$

subject to the scalar differential equation $\dot{x}(t) = ax(t) + bu(t)$, with $b \neq 0$.

Let $d := a^2 + qs$. Then this minimization problem has a solution, for every $T > 0$, if and only if $d \geq 0$ and $q_T > x_{min}$, where $x_{min} = \frac{a + \sqrt{d}}{s}$ is the minimal solution of the algebraic Riccati equation (5.5.2).

Proof

The proof can be given along the lines of the previous example. That is, first, rewrite the Riccati differential equation relevant to this problem

$$\dot{k}(t) = -2ak(t) + sk^2(t) - q, \ k(T) = q(T)$$

as an initial value problem (see the derivation of equation (5.5.8)) by introducing $p(t) := k(T - t)$. Then our minimization problem has a solution for every $T > 0$ if and only if the initial value problem

$$\dot{p}(t) = 2ap(t) - sp^2(t) + q, \quad p(0) = q(T) \tag{5.5.10}$$

has a solution $p(t)$ on $[0, \infty)$.

One can then pursue the same dynamic analysis as in Example 5.10. Similarly it follows that, if the discriminant, d, of the right-hand side of equation (5.5.10) is positive, then (5.5.10) has a solution on $[0, \infty)$ if and only if the above mentioned conditions hold.

If this discriminant is zero the derivative of $2ap - sp^2 + q$ becomes zero at the equilibrium point $x = \frac{a}{s}$. In this case the differential equation can be rewritten as

$$\dot{p}(t) = -s\left(p(t) - \frac{a}{s}\right)^2, \quad p(0) = q(T). \tag{5.5.11}$$

The solution of this differential equation (5.5.11) is

$$p(t) = \frac{a}{s} + \frac{1}{st + s/(sq_T - a)}.$$

So, this differential equation has no finite escape time if and only if $t + \frac{1}{sq_T - a}$ remains positive for all $t > 0$. Some elementary rewriting then shows that this condition can be reformulated as in the case $d > 0$. $\qquad\square$

The conclusion in Proposition 5.17 that, in case the scrap matrix Q_T is larger than or equal to the minimal solution of the corresponding algebraic Riccati equation, the finite-planning horizon always has a solution can be generalized to the multivariable case. To that end we consider a lemma from which this result can easily be derived. The proof of this lemma is along the lines of Knobloch and Kwakernaak (1985).

Lemma 5.18

Assume $G(t)$ is a continuous matrix and $V(t)$ is a differentiable symmetric matrix on some open interval I that satisfies

$$\dot{V}(t) = G^T(t)V(t) + V(t)G(t). \tag{5.5.12}$$

Then, if $V(t_1) \geq 0(\leq 0)$, for all $t \in I$ $V(t) \geq 0$ (≤ 0).

Proof

Consider $\dot{x}(s) = -G(s)x(s)$ and assume that $\Phi(t, \tau)$ is a corresponding fundamental matrix on $(\tau, t) \subset I$ (see section 3.1).

Introduce for a fixed τ

$$P_\tau(t) = \Phi^T(t, \tau)V(t)\Phi(t, \tau). \tag{5.5.13}$$

Then,

$$\dot{P}_\tau(t) = \dot{\Phi}^T(t,\tau)V(t)\Phi(t,\tau) + \Phi^T(t,\tau)\dot{V}(t)\Phi(t,\tau) + \Phi^T(t,\tau)(t)V(t)\dot{\Phi}(t,\tau)$$

$$= \Phi^T(t,\tau)(t)\{-G^T(t)V(t) + \dot{V}(t) - V(t)G(t)\}\Phi(t,\tau)$$

$$= 0.$$

So, $P_\tau(t) = C$, where C is a constant matrix, for all $t \geq \tau$. Since the fundamental matrix is invertible it follows in particular that, at $t = t_1$,

$$C = P_\tau(t_1) = \Phi^T(t_1,\tau)V(t_1)\Phi(t_1,\tau) \geq 0.$$

Therefore,

$$\Phi^T(t,\tau)V(t)\Phi(t,\tau) \geq 0, \quad \text{for all } t \geq \tau.$$

Using the invertibility of the fundamental matrix again, we conclude that $V(t) \geq 0$, for all $t \geq \tau$. Finally, since τ was chosen arbitrarily, it follows that $V(t) \geq 0$ for all $t \in I$. Obviously, the same arguments apply if $V(t_1) \leq 0$. $\qquad \square$

Theorem 5.19

Assume that (A,B) is stabilizable and that the algebraic Riccati equation (5.5.2) has a minimal solution X_{min}. Consider the Riccati differential equation (5.5.1) with $Q_T \geq X_{min}$. Then for all $T > 0$ this differential equation has a solution on $[0,T]$.

Proof

Notice that at least in some small interval $(T - \Delta, T]$ the solution of equation (5.5.1) exists. We will show next that, on this time interval, $K(t)$ is always larger than X_{min}. Since, obviously[3], $K(t)$ is also bounded from above independently of T, this shows that $K(t)$ cannot have a finite escape time at $t = T - \Delta$. That is, the solution also exists at $t = T - \Delta$. Without going into mathematical details, basically a repetition of this argument then shows the result (for example Perko (2001)).

So, consider $V(t) := K(t) - X_{min}$, where $K(t)$ solves equation (5.5.1) and X_{min} is the minimal solution of equation (5.5.2). Then, with $G(t) := -A + \frac{1}{2}SK_1 + \frac{1}{2}SK_2$,

$$\dot{V}(t) = V(t)G(t) + G^T(t)V(t).$$

Since $V(T) \geq 0$ by Lemma 5.18 $V(t) \geq 0$ for all $t \in (T - \Delta, T]$. $\qquad \square$

It is tempting to believe that under the conditions of Theorem 5.19 the finite-planning horizon problem will always converge. However, for the non-scalar case this guess is not correct. Oscillatary behavior may occur as the next example, presented by Callier and Willems (1981), illustrates.

---

[3]Choose $u = Fx$, where F is a stabilizing control, in the corresponding cost function.

Example 5.11

Consider the minimization of

$$\int_0^T u^T(t)u(t)dt + x^T(T)\begin{bmatrix} 1 & 0 \\ 0 & 0 \end{bmatrix}x(T)$$

subject to the differential equation

$$\dot{x}(t) = \begin{bmatrix} 1 & 1 \\ -1 & 1 \end{bmatrix}x(t) + u(t), \quad x(0) = x_0.$$

The solution of the corresponding Riccati differential equation (5.5.1) is

$$K(t) = \frac{2}{1 + e^{2(t-T)}}\begin{bmatrix} \cos^2(t-T) & -\sin(t-T)\cos(t-T) \\ -\sin(t-T)\cos(t-T) & \sin^2(t-T) \end{bmatrix}.$$

Hence, for this example, the solution of the Riccati differential equation (5.5.1) at $t = 0$ remains oscillatory if T becomes arbitrarily large.

Notice that $X_0 = \begin{bmatrix} 0 & 0 \\ 0 & 0 \end{bmatrix}$ solves the corresponding algebraic Riccati equation (5.5.2) and that $Q_T \geq X_0$. Therefore, $Q_T \geq X_{min}$ too.   □

As we saw in Chapter 2, one way to find the solution(s) of the algebraic Riccati equation is by determining the invariant subspaces of the corresponding Hamiltonian matrix

$$H = \begin{bmatrix} A & -S \\ -Q & -A^T \end{bmatrix}.$$ (5.5.14)

Corollary 5.13 showed that this matrix H also determines the solvability of the Riccati differential equation

$$\dot{K}(t) = -A^T K(t) - K(t)A + K(t)SK(t) - Q.$$ (5.5.15)

In fact, this relationship can also be obtained more directly if one uses Pontryagin's maximum principle to solve the minimization problem (5.2.1) and (5.2.2). This approach will be elaborated in Chapter 7 for the non-cooperative open-loop game. Using that setting one can rather straightforwardly analyze the convergence issue. Since that analysis can be copied for the 'one-player' case it is, therefore, omitted here. We will confine ourselves to just stating the basic convergence result here for this 'one-player' case.

In this convergence result, the so-called **dichotomic separability** of matrix H plays a crucial role. To introduce this notion consider the next ordering, $<_c$, of complex numbers. Two complex numbers $w := x_1 + y_1 i$ and $z := x_2 + y_2 i$ are ordered as: $w <_c z$ if $x_1 < x_2$; and they have the same order, $w =_c z$, if their real parts coincide. Then, if we consider the eigenvalues of matrix H and rank them (taking the multiplicities into account) in increasing order we call the **spectrum of H separable** if the order of the nth and $(n+1)$th eigenvalue differs.

Example 5.12

Assume the eigenvalues of H are as follows.

1. $\sigma(H) = \{2, 1, -2, -1\}$. Then, the order of the eigenvalues is $-2 <_c -1 <_c 1 <_c 2$. Since $-1 <_c 1$, the spectrum of H is separable.

2. $\sigma(H) = \{1, 1, -1, -1\}$. Then, the order of the eigenvalues is $-1 =_c -1 <_c 1 =_c 1$. Since $-1 <_c 1$, the spectrum of H is separable.

3. $\sigma(H) = \{1, 0, -1, 0\}$. Then, the order of the eigenvalues is $-1 <_c 0 =_c 0 <_c 1$. Then, the spectrum of H is not separable.

4. $\sigma(H) = \{1 + i, 1 - i, -1 + i, -1 - i\}$. Then, the order of the eigenvalues is $-1 + i =_c -1 - i <_c 1 + i =_c 1 - i$. Since $-1 - i <_c 1 + i$, the spectrum of H is separable.

5. $\sigma(H) = \{1, i, -i, -1\}$. Then, the order of the eigenvalues is $-1 <_c -i =_c i <_c 1$. Since $-i =_c i$, the spectrum of H is not separable. $\qquad\square$

Notice that separability of the spectrum of matrix H implies that H has an n-dimensional invariant subspace corresponding to the n smallest eigenvalues (counting multiplicities) of H. If this subspace satisfies some additional property, outlined in the next definition, matrix H is called dichotomically separable.

Definition 5.1

Matrix $H = \begin{bmatrix} A & -S \\ -Q & -A^T \end{bmatrix}$ is called **dichotomically separable** if there exist two H-invariant n-dimensional subspaces V_1 and V_2 such that $V_1 \oplus V_2 = \mathbb{R}^{2n}$,[4] and

$$V_1 \oplus \mathrm{Im} \begin{bmatrix} 0 \\ I \end{bmatrix} = \mathbb{R}^{2n}.$$

Here $0, I \in \mathbb{R}^{n \times n}$ are the zero and identity matrix, respectively. Moreover, V_1 and V_2 should be such that $\mathrm{Re}\,\lambda < \mathrm{Re}\,\mu$ for all $\lambda \in \sigma(H|V_1)$, $\mu \in \sigma(H|V_1)$. $\qquad\square$

Example 5.13

Consider the following cases.

1. $H = \mathrm{diag}\{-2, -1, 2, 1\}$. With $V_1 = \mathrm{Im} \begin{bmatrix} I \\ 0 \end{bmatrix}$ and $V_2 = \mathrm{Im} \begin{bmatrix} 0 \\ I \end{bmatrix}$, it is clear that H is dichotomically separable.

2. $H = \mathrm{diag}\{2, 1, -2, -1\}$. With $E_1 = \mathrm{Im} \begin{bmatrix} 0 \\ I \end{bmatrix}$, the stable invariant subspace, and $E_2 = \mathrm{Im} \begin{bmatrix} I \\ 0 \end{bmatrix}$, the unstable invariant subspace, it is clear that we can never find an

---

[4]A subspace V is called the direct sum of two subspaces V_1 and V_2, denoted by $V = V_1 \oplus V_2$, if $V_1 \cap V_2 = \{0\}$ and for all $v \in V$ there are $v_i \in V_i$, $i = 1, 2$, such that $v = v_1 + v_2$.

appropriate subspace V_1 and V_2 satisfying the constraint that $V_1 \oplus \mathrm{Im} \begin{bmatrix} 0 \\ I \end{bmatrix} = \mathbb{R}^{2n}$. So, H is not dichotomically separable. $\qquad\square$

The main theorem of this section on convergence now follows. Its proof is along the lines of the corresponding result presented for open-loop games in Chapter 7.

Theorem 5.20

Assume that the well-posedness assumption holds. Then, if the Hamiltonian matrix H is dichotomically separable and (with the notation of Definition 5.1)

$$\mathrm{Span} \begin{bmatrix} I \\ Q_T \end{bmatrix} \oplus V_2 = \mathbb{R}^{2n}, \tag{5.5.16}$$

the solution of the Riccati differential equation (5.5.1) at $t = 0$ converges to the stabilizing solution of the algebraic Riccati equation (5.5.2) if the planning horizon T becomes arbitrarily large. $\qquad\square$

Example 5.14

1. Consider the minimization of

$$\int_0^T u^2(t)dt + x^T(T)x(T),$$

subject to the dynamics

$$\dot{x}(t) = \begin{bmatrix} -2 & 0 \\ 0 & -1 \end{bmatrix} + \begin{bmatrix} 1 \\ 1 \end{bmatrix} u(t), \quad x(0) = x_0.$$

Then, one solution of the algebraic Riccati equation corresponding to this problem is $X = 0$. Since $Q_T = I > 0 \geq X_{min}$, by Theorem 5.19 the well-posedness assumption holds.

Furthermore it is easily verified that the eigenvalues of the with this problem corresponding Hamiltonian matrix H are $\{-2, -1, 2, 1\}$. Eigenvectors corresponding with these eigenvalues are

$$v_1 = \begin{bmatrix} 1 \\ 0 \\ 0 \\ 0 \end{bmatrix}, \quad v_2 = \begin{bmatrix} 0 \\ 1 \\ 0 \\ 0 \end{bmatrix}, \quad v_3 = \begin{bmatrix} 3 \\ 4 \\ 12 \\ 0 \end{bmatrix} \quad \text{and} \quad v_4 = \begin{bmatrix} 2 \\ 3 \\ 0 \\ 6 \end{bmatrix}, \quad \text{respectively.}$$

So, with $V_1 := \mathrm{Im}[v_1 \; v_2]$ and $V_2 := \mathrm{Im}[v_3 \; v_4]$ it is easily verified that the Hamiltonian matrix H is dichotomically separable and that $\mathrm{Span} \begin{bmatrix} I \\ I \end{bmatrix} \oplus V_2 = \mathbb{R}^{2n}$. Therefore, all conditions in Theorem 5.20 are met and we conclude that the finite-planning horizon solution converges to the stabilizing solution of the infinite-planning horizon problem.

2. Reconsider the problem

$$\min \int_0^T u^2(t)dt$$

subject to the dynamics

$$\dot{x}(t) = x(t) + u(t), \ x(0) = x_0.$$

In Example 5.6, part 1, we showed that this problem always has a solution and that the solution of the associated Riccati differential equation converges to the non-stabilizing smallest positive semi-definite solution of the corresponding algebraic Riccati equation. Here, we would like to see which condition(s) in Theorem 5.20 are violated in this example. Here,

$$H = \begin{bmatrix} 1 & -1 \\ 0 & -1 \end{bmatrix}.$$

So, the eigenvalues of H are $\{-1, 1\}$. Eigenvectors corresponding with these eigenvalues are

$$v_1 = \begin{bmatrix} 1 \\ 2 \end{bmatrix} \text{ and } v_2 = \begin{bmatrix} 1 \\ 0 \end{bmatrix},$$

respectively. With $V_1 := \text{Im}[v_1]$ and $V_2 := \text{Im}[v_2]$ it is then easily verified that H is dichotomically separable.

Next we consider condition (5.5.16). We have

$$\text{Span} \begin{bmatrix} I \\ Q_T \end{bmatrix} \oplus V_2 = \text{Span} \begin{bmatrix} I \\ 0 \end{bmatrix} \oplus \text{Span} \begin{bmatrix} I \\ 0 \end{bmatrix},$$

which clearly differs from $\mathbb{R}^2$. So, this condition is the only one that is violated in Theorem 5.20. □

5.6 Notes and references

For this chapter in particular the book by Clements and Anderson (1978) and the papers by Molinari (1975, 1977) and van den Broek Engwerda and Schumacher (2003c) have been consulted. More details on related subjects and references can be found in these references and, for example, Anderson and Moore (1989), Kwakernaak and Sivan (1972), Zhou, Doyle and Glover (1996) and Trentelman, Stoorvogel and Hautus (2001). Furthermore, a complete survey of the phase portrait of the Riccati differential equation if Q is semi-positive definite can be found in Shayman (1986).

Various extensions of the (in)finite horizon regular linear quadratic control problem have been studied in literature. Often a distinction is made between two versions of the problem, the **fixed-endpoint** version ($x(T)$ fixed) and the **free-endpoint** version ($x(T)$ free). The theory presented here for the infinite horizon case belongs to the fixed-endpoint version, whereas the finite horizon theory belongs to the free-endpoint case. For the fixed-endpoint finite horizon case one can consult Jacobson (1971, 1977). For the fixed-endpoint infinite horizon problem a rather complete solution has been provided by Willems (1971). He showed, under a controllability assumption, that this problem has an infimum if and only if the algebraic Riccati equation (5.4.4) has a solution. Moreover, he showed that this infimum is attained if and only if equation (5.4.4) has a stabilizing solution (see Note 1 following Example 5.5). Trentelman (1989) considered the free-endpoint infinite horizon problem. He showed that this problem has a solution whenever the algebraic Riccati equation (5.4.4) has a negative semi-definite solution. Moreover, under this assumption he provides a both necessary and sufficient condition for the existence of an optimal control that attains the infimum. This control is, in general, obtained using a combination of the smallest and largest solution of (5.4.4). In Soethoudt and Trentelman (1989) the fixed- and free-endpoint problems and infinite horizon problems are combined by restricting the set of feasible controls to those for which the distance between the implied state and a predefined subspace L becomes zero.

Other generalizations of this problem that have been studied in literature are, for example, the case where matrix R is just assumed to be positive semi-definite (the 'singular' problem) and the case where the dynamics of the system are not only described by a set of linear differential equations but the states are additionally subject to linear equality constraints (the 'descriptor' problem) (for example Geerts (1989) and Mehrmann (1991)). Furthermore, the linear quadratic problem has been studied for systems where the evolution of the state is described by partial differential equations (the 'infinite dimensional' problem) (for example Grabowski (1993) who showed, under a stabilizability and detectability assumption on the system, that the corresponding regular infinite-horizon linear quadratic control problem has a unique solution.)

5.7 Exercises

1. Verify which of the following minimization problems has a solution for every initial state x_0. Calculate, if possible, the optimal control and cost.

(a) $\min\limits_{u(.)} \int_0^1 \{-2x^2(t) + \frac{1}{2}u^2(t)\}dt - \frac{5}{4}x^2(1)$ subject to $\dot{x}(t) = -2x(t) + u(t)$, $x(0) = x_0$.

(b) $\min\limits_{u(.)} \int_0^1 \{2x^2(t) + u^2(t)\}dt$ subject to $\dot{x}(t) = \frac{1}{2}x(t) + u(t)$, $x(0) = x_0$.

(c) $\min\limits_{u(.)} \int_0^1 \{-2x^2(t) + \frac{1}{2}u^2(t)\}dt - 2x^2(1)$ subject to $\dot{x}(t) = -2x(t) + u(t)$, $x(0) = x_0$.

(d) $\min\limits_{u(.)} \int_0^{\frac{\pi}{4}} \{-2x^2(t) + u^2(t)\}dt + 2x^2\left(\frac{\pi}{4}\right)$ subject to $\dot{x}(t) = x(t) + u(t)$, $x(0) = x_0$.

(e) $\min\limits_{u(.)} \int_0^4 \{x^2(t) - u^2(t)\}dt + 2x^2(4)$ subject to $\dot{x}(t) = x(t) + u(t)$, $x(0) = x_0$.

(Hint: consider $u(t) = \alpha x(t)$.)

(f) $\min\limits_{u(.)} \int_0^T \{-x^2(t) + u^2(t)\}dt$ subject to $\dot{x}(t) = -x(t) + u(t)$, $x(0) = x_0$.

(g) $\min\limits_{u(.)} \int_0^T \{-x^2(t) + u^2(t)\}dt - x^2(T)$ subject to $\dot{x}(t) = -x(t) + u(t)$, $x(0) = x_0$.

2. Consider the minimization of

$$J(x_0, u) := \int_0^{\frac{\pi}{2}} \{-x^2(t) + u^2(t)\}dt \text{ subject to } \dot{x}(t) = u(t), \ x(0) = x_0.$$

(a) Show that if $x_0 \neq 0$ this problem has no solution.

(Hint: rewrite the integrand as $J(\epsilon) := \int_0^{\frac{\pi}{2}-\epsilon} \{-x^2(t) + u^2(t)\}dt + \int_{\frac{\pi}{2}-\epsilon}^{\frac{\pi}{2}} \{-x^2(t) + u^2(t)\}dt$ (where $\epsilon > 0$) and use for the time interval $\left[0, \frac{\pi}{2} - \epsilon\right]$ the implied optimal control of the optimization problem if the planning period is restricted to this time interval; and for the time interval $\left[\frac{\pi}{2} - \epsilon, \frac{\pi}{2}\right]$, $u(.) = 0$. Next, analyze $J(\epsilon)$ if $\epsilon \to 0$.)

(b) Next consider $x_0 = 0$. Show that $J(0, u) \geq 0$ for all $u(.)$. Conclude that $\min\limits_{u(.)} J(0, u)$ exists. Determine the optimal control.

(Hint: notice that $J = \int_0^{\frac{\pi}{2}} \{-x^2(t) + \dot{x}^2(t)\}dt$ where (assuming that $u(.)$ is piecewise continuous) $x(.)$ is continuous. Next make a Fourier series expansion of $x(t)$ and determine J.)

3. Consider the optimal control problem

$$\min\limits_{u(.)} \int_0^T \{-9x^2(t) + u^2(t)\}dt - \frac{7}{2}x^2(T) \text{ subject to } \dot{x}(t) = -3x(t) + u(t) + e^t, \ x(0) = 2.$$

$$(5.7.1)$$

(a) Let $T = 1$. Determine the optimal control and cost of this optimization problem.

(b) Determine all $T > 0$ for which the optimal control problem (5.7.1) has a solution for all initial states x_0.

4. Show that if $Q_i \geq 0$ and $Q_{iT} \geq 0$, $i = 1, 2$, in the finite-planning horizon regulator problem, this problem always has a unique solution. (Hint: notice that the cost is always bounded from below by zero.)

5. Tracking problem: a continuous differentiable reference trajectory, $x^*(t)$, is given which one likes to track with $x(t)$ on the time interval $[0, T]$. The dynamics of $x(t)$ are described by

$$\dot{x}(t) = Ax(t) + Bu(t), \ x(0) = x_0.$$

The problem is modeled as follows

$$\min_u \int_0^T \{(x(t) - x^*(t))^T Q(x(t) - x^*(t)) + u^T(t)Ru(t)\}dt,$$

where Q and R are positive definite matrices.

(a) Show that this tracking problem has a unique solution. Determine this solution. (Hint: rewrite the problem into the standard framework by introducing the new state variable $\tilde{x}(t) := x(t) - x^*(t)$.)

(b) Consider the problem with $T = \infty$. Formulate conditions under which this problem has a solution. Give the corresponding optimal control provided your conditions are satisfied.

6. Assume that 'with Q, Q_T symmetric and $R > 0$'

$$\int_t^T \{x^T(s)Qx(s) + u^T(s)Ru(s)\}ds + x^T(T)Q_Tx(T) \text{ subject to}$$

$$\dot{x}(s) = Ax(s) + Bu(s), \ x(0) = x_0,$$

has a minimum $J^*(t, x_0) = x_0^T P(t)x_0$ for all x_0, for all $t \in [0, T]$. Here $P(.)$ is a bounded matrix on $[0, T]$. Assume that with $x(0) = x_0$ the minimum is attained for the bounded function $u_{x_0}^*(.)$ and the corresponding optimal state trajectory is $x^*(.)$.

(a) Show that $\frac{\partial J^*(t,x)}{\partial x}$ exists for $t \in [0, T]$.

(b) Let $\Delta x := x^*(t + \Delta t) - x_0$. Show that $J^*(t, x_0) - J^*(t + \Delta, x_0)$ equals

$$\int_t^{t+\Delta t} \{x^{*T}(s)Qx^*(s) + u_{x_0}^{*T}(s)Ru_{x_0}^*(s)\}ds + 2(\Delta x)^T P(t + \Delta t)x_0 + (\Delta x)^T P(t + \Delta t)\Delta x.$$

(c) Show that $\Delta x = e^{A\Delta t}x_0 + \int_t^{t+\Delta t} e^{A(t+\Delta t-s)}Bu_{x_0}^*(s)ds$.

(d) Show that $\lim_{\Delta t \to 0} J^*(t + \Delta t, x_0) - J^*(t, x_0) = 0$.

(e) Show that $\frac{\partial J^*(t,x)}{\partial x}$ is a continuous function.

(f) Use item (b) to show that

$$\lim_{\Delta t \to 0} \frac{J^*(t + \Delta t, x_0) - J^*(t, x_0)}{\Delta t} = x_0^T Qx_0 + u_{x_0}^{*T}(t)Ru_{x_0}^* + 2\dot{x}^T(t)P(t)x_0.$$

Conclude that $\frac{\partial J^*(t,x)}{\partial t}$ exists.

(g) Use the dynamic programming theorem to show that, under the above mentioned assumptions, the Riccati differential equation (5.2.3) has a solution on $[0, T]$.

7. Show the correctness of Corollary 5.7, item 2. (Hint: notice that the optimal strategy is a linear feedback strategy.)

8. Verify, either numerically (by using, for example, MATLAB) or by hand, which of the following minimization problems has a solution for every initial state x_0. To that end first calculate the eigenstructure of matrix

$$H := \begin{bmatrix} A & -S \\ -Q & -A^T \end{bmatrix}.$$

Also calculate, if possible, the optimal control and cost.

(a) $\min_{u(.)} \int_0^1 \{x^T(t) \begin{bmatrix} 3 & 1 \\ 1 & 4 \end{bmatrix} x(t) + u^2(t)\} dt$ subject to $\dot{x}(t) = \begin{bmatrix} 1 & 0 \\ 0 & -1 \end{bmatrix} x(t) + \begin{bmatrix} 1 \\ 0 \end{bmatrix} u(t)$.

(b) $\min_{u(.)} \int_0^2 \{x^T(t) \begin{bmatrix} -4 & 1 \\ 1 & 4 \end{bmatrix} x(t) + u^2(t)\} dt + x^T(2) \begin{bmatrix} 1 & 0 \\ 0 & 1 \end{bmatrix} x(2)$ subject to

$\dot{x}(t) = \begin{bmatrix} 1 & 0 \\ 0 & -1 \end{bmatrix} x(t) + \begin{bmatrix} 1 \\ 0 \end{bmatrix} u(t)$.

(c) $\min_{u(.)} \int_0^2 \{x^T(t) \begin{bmatrix} 2 & 0 \\ 0 & 1 \end{bmatrix} x(t) + u^2(t)\} dt$ subject to $\dot{x}(t) = \begin{bmatrix} 1 & 0 \\ 1 & 1 \end{bmatrix} x(t) + \begin{bmatrix} 1 \\ 0 \end{bmatrix} u(t)$.

9. Verify which of the following minimization problems, subject to the constraint that $x(t) \to 0$ if $t \to \infty$, has a solution for every initial state x_0. Calculate, if possible, the optimal control and cost.

(a) $\min_{u(.)} \int_0^\infty \{2x^2(t) + u^2(t)\} dt$ subject to $\dot{x}(t) = \frac{1}{2}x(t) + u(t)$, $x(0) = x_0$.

(b) $\min_{u(.)} \int_0^\infty \{-x^2(t) + u^2(t)\} dt$ subject to $\dot{x}(t) = -2x(t) + u(t)$, $x(0) = x_0$.

(c) $\min_{u(.)} \int_0^\infty \{-2x^2(t) + \frac{1}{2}u^2(t)\} dt$ subject to $\dot{x}(t) = -2x(t) + u(t)$ $x(0) = x_0$.

(d) $\min_{u(.)} \int_0^\infty \{-2x^2(t) + u^2(t)\} dt$ subject to $\dot{x}(t) = x(t) + u(t)$, $x(0) = x_0$.

(e) $\min_{u(.)} \int_0^\infty \{x^2(t) - u^2(t)\} dt$ subject to $\dot{x}(t) = x(t) + u(t)$, $x(0) = x_0$.

(Hint: consider $u(t) = \alpha x(t)$.)

10. Verify, either numerically (by using, for example, MATLAB) or by hand, which of the next minimization problems, subject to the constraint that $x(t) \to 0$ if $t \to \infty$, has a solution for every initial state x_0. To that end first calculate the eigenstructure of matrix

$$H := \begin{bmatrix} A & -S \\ -Q & -A^T \end{bmatrix}.$$

Also calculate, if possible, the optimal control and cost.

(a) $\min_{u(.)} \int_0^\infty \{x^T(t) \begin{bmatrix} 3 & 1 \\ 1 & 4 \end{bmatrix} x(t) + u^2(t)\} dt$ subject to $\dot{x}(t) = \begin{bmatrix} 1 & 0 \\ 0 & -1 \end{bmatrix} x(t) + \begin{bmatrix} 1 \\ 0 \end{bmatrix} u(t)$.

(b) $\min\limits_{u(.)} \int_0^\infty \{x^T(t) \begin{bmatrix} -4 & 1 \\ 1 & 4 \end{bmatrix} x(t) + u^2(t)\}dt$ subject to $\dot{x}(t) = \begin{bmatrix} 1 & 0 \\ 0 & -1 \end{bmatrix} x(t) + \begin{bmatrix} 1 \\ 0 \end{bmatrix} u(t).$

(c) $\min\limits_{u(.)} \int_0^\infty \{x^T(t) \begin{bmatrix} 2 & 0 \\ 0 & 1 \end{bmatrix} x(t) + u^2(t)\}dt$ subject to $\dot{x}(t) = \begin{bmatrix} 1 & 0 \\ 1 & 1 \end{bmatrix} x(t) + \begin{bmatrix} 1 \\ 0 \end{bmatrix} u(t).$

11. In this exercise we show that, if (A, B) is stabilizable and the linear quadratic control problem (5.4.1), (5.4.2) has a minimum $u \in \mathcal{U}_{re}$ for each x_0, the algebraic Riccati equation (5.4.4) has a symmetric stabilizing solution K. To that end we consider the optimization problem

$$J(t, x_0, u) := \int_t^\infty \{x^T(s)Qx(s) + u^T(s)Ru(s)\}ds \text{ subject to}$$

$$\dot{x}(s) = Ax(s) + Bu(s), \; x(t) = x_0. \tag{5.7.2}$$

Furthermore, let $u^*(.)$ be the argument that minimizes $J(0, x_0, u)$ and J^* be the corresponding minimal cost.

(a) Show that $J^* = x_0^T K(0)x_0$ for some symmetric matrix $K(0)$.
(b) Show that for all $x_0 \in \mathbb{R}^n$ $\inf\limits_{u \in \mathcal{U}_{re}} J(t, x_0, u)$ exists and equals $x_0^T K(t)x_0$.
(c) Show that $K(t) = K(0)$ for all t (Hint: see Exercise 5.20.).
(d) Show that the conditions for using the dynamic programming theorem are satisfied.
(e) Show, using the Hamilton–Jacobi–Bellman equation (4.4.12), that K satisfies the algebraic Riccati equation (5.4.4).
(f) Conclude that K is a stabilizing solution of equation (5.4.4).

12. Consider for $u \in \mathcal{U}_{re}$ the optimization problem

$$J(x_0, u, T) := \int_0^T -x^2(s) + u^2(s)ds \text{ subject to } \dot{x}(s) = -x(s) + u(s), \; x(0) = x_0,$$

(Trentelman, 1989).

(a) Show that $J(x_0, u, T) = \int_0^T (x(s) - u(s))^2 ds + x^2(T) - x_0^2.$
(b) Show that $J(x_0, u, \infty) \geq -x_0^2.$
(c) Show that the algebraic Riccati equation corresponding to this problem has no stabilizing solution.
(d) Show that $\min\limits_{u \in \mathcal{U}_{re}} J(0, u, \infty) = 0.$
(e) Show that with $u_\epsilon(s) = (1 - \epsilon)x(s)$, $J(x_0, u_\epsilon, \infty) = -x_0^2 + \frac{\epsilon}{2}x_0^2.$
(f) Show that $\inf\limits_{u \in \mathcal{U}_{re}} J(x_0, u, \infty) = -x_0^2.$
(g) Show that for all $x_0 \neq 0$ $\min\limits_{u \in \mathcal{U}_{re}} J(x_0, u, \infty)$ does not exist.

5.8 Appendix

Proof of Lemma 5.2

By taking $x = y = 0$, $x = 0$ and $x = y$, respectively, in the definition (5.2.8) of V we immediately see that

$$V(0) = 0, V(-x) = V(x), \quad \text{and} \quad V(2x) = 4V(x).$$

From which it follows that $W(x, x) = V(2x) = 4V(x)$. We next show that $W(x, y)$ is a bilinear form, that is, W is linear in each of its arguments when the other is held fixed, i.e.

$$W(\lambda_1 x_1 + \lambda_2 x_2, y) = \lambda_1 W(x_1, y) + \lambda_2 W(x_2, y)$$
$$W(x, \mu_1 y_1 + \mu_2 y_2) = \mu_1 W(x, y_1) + \mu_2 W(x, y_2).$$

It is well-known (for example Greub (1967)) that if a function W is a bilinear form on $\mathbb{R}$, then there exists a matrix P such that $W(x, y) = x^T P y$. Since $V(x) = \frac{1}{4} W(x, x)$ it is obvious then that $V(x)$ is a quadratic form. Notice that $W(x, y) = W(y, x)$. Therefore it is enough to show that $W(x, y)$ is linear in its first argument for y fixed. We first show that

$$W(x_1 + x_2, y) = W(x_1, y) + W(x_2, y). \tag{5.8.1}$$

From the definition and the parallelogram identity (see equation (5.2.8))

$$V(x + y) + V(x - y) = 2\{V(x) + V(y)\}, \quad \text{for all } x, y,$$

it follows that

$$
\begin{aligned}
2\{W(x_1, y) + W(x_2, y)\} &= 2\{V(x_1 + y) - V(x_1 - y) + V(x_2 + y) - V(x_2 - y)\} \\
&= \{V(x_1 + x_2 + 2y) + V(x_1 - x_2)\} - \{V(x_1 + x_2 - 2y) \\
&\quad + V(x_1 - x_2)\} = W(x_1 + x_2, 2y)
\end{aligned}
\tag{5.8.2}
$$

Taking $x_2 = 0$ in the above equality shows that

$$2W(x_1, y) = W(x_1, 2y), \quad \text{for all } x_1.$$

So, substituting $x_1 + x_2$ in this expression for x_1 yields the equality

$$2W(x_1 + x_2, y) = W(x_1 + x_2, 2y).$$

Using this, we conclude from equation (5.8.2) that

$$2\{W(x_1, y) + W(x_2, y)\} = W(x_1 + x_2, 2y) = 2W(x_1 + x_2, y).$$

That is, equation (5.8.1) holds.

Next we show that

$$W(\lambda x, y) = \lambda W(x, y). \tag{5.8.3}$$

To that end first note that $W(0, y) = 0$ and $W(-x, y) = -W(x, y)$. Using induction on equation (5.8.1) shows that

$$W(nx, y) = nW(x, y) \text{ for all integers } n.$$

Using this property twice, one then obtains that, for arbitrary integers m, n ($m \neq 0$),

$$mW\left(\frac{n}{m}x, y\right) = W(nx, y) = nW(x, y).$$

From this it is obvious that equation (5.8.3) holds for all rational numbers λ. This linearity property then also holds for an arbitrary real number λ following from our continuity assumption on $W(\lambda x, y)$ at $\lambda = 0$. For, let $\lambda \in \mathbb{R}$. Then there exists a sequence of rational numbers r_n such that $\lim r_n = \lambda$. So, the sequence $\lambda_n := \lambda - r_n$ converges to zero when $n \to \infty$. From equation (5.8.1), and the fact that equation (5.8.3) holds for all rational numbers, we then have

$$
\begin{aligned}
W(\lambda x, y) &= W(r_n x + \lambda_n x, y) \\
&= W(r_n x, y) + W(\lambda_n x, y) \\
&= r_n W(x, y) + W(\lambda_n x, y).
\end{aligned}
$$

Therefore, invoking our continuity property, we conclude that

$$
\begin{aligned}
W(\lambda x, y) &= \lim_{n \to \infty} r_n W(x, y) + W(\lambda_n x, y) \\
&= \lambda W(x, y) + W\left(\lim_{n \to \infty} \lambda_n x, y\right) \\
&= \lambda W(x, y) + W(0, y) \\
&= \lambda W(x, y).
\end{aligned}
$$

$\square$

Lemma 5.21

The following identities hold

$$
\frac{1}{2}\{J(t, 2x_0(t), u_0 + u_1) + J(t, 2x_1(t), u_0 - u_1)\} = J(t, (x_0 + x_1)(t), u_0)
$$
$$
+ J(t, (x_0 - x_1)(t), u_1); \tag{5.8.4}
$$
$$
\frac{1}{2}\{J(t, (x_0 + x_1)(t), u_0 + u_1) + J(t, (x_0 - x_1)(t), u_0 - u_1)\} = J(t, x_0(t), u_0)
$$
$$
+ J(t, x_1(t), u_1). \tag{5.8.5}
$$

Proof

According to the variation-of-constants formula, see equation (3.1.5),

$$x(s, x_0(t), u) = e^{A(s-t)}x_0(t) + \int_t^s e^{A(s-\tau)}Bu(\tau)d\tau.$$

So, introducing $v_i(s) := e^{A(s-t)}x_i(t)$ and $w_i(s) := \int_t^s e^{A(s-\tau)}Bu_i(\tau)d\tau,\ i = 0,1,$

$$x(s, 2x_0(t), u_0 + u_1) = 2v_0(s) + w_0(s) + w_1(s);$$
$$x(s, 2x_1(t), u_0 - u_1) = 2v_1(s) + w_0(s) - w_1(s);$$
$$x(s, (x_0 + x_1)(t), u_0) = v_0(s) + v_1(s) + w_0(s);$$
$$x(s, (x_0 - x_1)(t), u_1) = v_0(s) - v_1(s) + w_1(s).$$

Substitution of this into the cost function J (see equation (5.2.1)) and comparing the terms on both sides of the equality sign in equation (5.8.4) then shows the correctness of this statement.

Similarly one obtains

$$x(s, (x_0 + x_1)(t), u_0 + u_1) = v_0(s) + v_1(s) + w_0(s) + w_1(s);$$
$$x(s, (x_0 - x_1)(t), u_0 - u_1) = v_0(s) - v_1(s) + w_0(s) - w_1(s);$$
$$x(s, x_0(t), u_0) = v_0(s) + w_0(s);$$
$$x(s, x_1(t), u_1) = v_1(s) + w_1(s).$$

Using the same arguments as above then shows the correctness of equation (5.8.5). $\square$

Proof of Theorem 5.11, item 1 '$\Rightarrow$ part'

To prove this part of the theorem we use a functional analysis approach (see, for example, Luenberger (1969) for an introduction into this theory). Define the linear mappings (operators)

$$\mathcal{P} : x \mapsto e^{At}x,$$
$$\mathcal{L} : u \mapsto \int_{t_0}^t e^{A(t-s)}Bu(s)ds \text{ and}$$
$$\mathcal{D} : c \mapsto \int_{t_0}^t e^{A(t-s)}c(s)ds.$$

Then the state trajectory x becomes

$$x = \mathcal{P}(x_0) + \mathcal{L}(u) + \mathcal{D}(c). \tag{5.8.6}$$

Consider the inner product $\langle f, g \rangle := \int_{t_0}^T f^T(s)g(s)ds + f^T(T)g(T)$, on the set $\mathbb{H}_2^n$ of all square integrable state functions from $[t_0, T]$ into $\mathbb{R}^n$ and the inner product $\langle u, v \rangle := \int_{t_0}^T u^T(s)v(s)ds$ on the set L_2^m of all square integrable control functions. Introducing

$$\bar{Q}(t) := \begin{cases} Q & \text{if } t \in [t_0, T) \quad \text{and} \\ Q_T & \text{if } t = T \end{cases}$$

the cost function J can be rewritten as

$$J(t_0, x_0, u, c) = \langle \mathcal{P}(x_0) + \mathcal{L}(u) + \mathcal{D}(c), \bar{Q}(\mathcal{P}(x_0) + \mathcal{L}(u) + \mathcal{D}(c)) \rangle_{\mathbb{H}_2^n} + \langle u, Ru \rangle_{L_2^m}$$
$$= \langle u, (\mathcal{L}^* \bar{Q} \mathcal{L} + R)(u) \rangle_{L_2^m} + 2 \langle u, \mathcal{L}^* \bar{Q}(\mathcal{P}(x_0) + \mathcal{D}(c)) \rangle_{L_2^m}$$
$$+ \langle \mathcal{P}(x_0) + \mathcal{D}(c), \bar{Q}(\mathcal{P}(x_0) + \mathcal{D}(c)) \rangle_{\mathbb{H}_2^n}.$$

Here $\mathcal{L}^*$ denotes the **adjoint operator** of $\mathcal{L}$, that is the unique linear operator that has the property that for all x, y, $\langle \mathcal{L}(x), y \rangle = \langle x, \mathcal{L}^*(y) \rangle$.

The proof of this part of the theorem basically follows from the following lemma.

Lemma 5.22

Let $\mathbb{H}$ be an arbitrary Hilbert space[5] with some inner product $\langle ., . \rangle$; $u, v \in \mathbb{H}$; $T: \mathbb{H} \mapsto \mathbb{H}$ a self-adjoint operator[6] and $\alpha \in \mathbb{R}$. Consider the functional

$$\langle u, T(u) \rangle + 2 \langle u, v \rangle + \alpha. \tag{5.8.7}$$

Then,

1. a necessary condition for equation (5.8.7) to have a minimum is that the operator T is positive semi-definite[7].

2. (5.8.7) has a unique minimum if and only if the operator T is positive definite. In this case the minimum is achieved at $u^* = -T^{-1}v$.

Proof

We follow the proof as outlined in Kun (2001).

Consider the scalar function defined by

$$f(\lambda) := \langle \lambda \tilde{u}, T(\lambda \tilde{u}) \rangle + 2 \langle \lambda \tilde{u}, v \rangle + \alpha = \lambda^2 \langle \tilde{u}, T(\tilde{u}) \rangle + 2\lambda \langle \tilde{u}, v \rangle + \alpha. \tag{5.8.8}$$

This is a real quadratic polynomial. Now, assume that there exists a $\tilde{u} \in \mathbb{H}$ with $\langle \tilde{u}, T(\tilde{u}) \rangle < 0$. Then, by choosing $u = \lambda \tilde{u}$, the value of equation (5.8.7) can be made arbitrarily small by choosing λ large. So, it is obvious that under these assumptions equation (5.8.7) does not have a minimum. So, if equation (5.8.7) has a minimum, then necessarily for all $u \in \mathbb{H}$ $\langle u, T(u) \rangle \geq 0$.

---

[5] A **Hilbert space** Ho is a vector space with an inner product $\langle ., . \rangle$ defined on it. The norm of a vector $x \in Ho$, $\|x\|$, is then defined by $\|x\| := \langle x, x \rangle^{1/2}$. Moreover, it is assumed that within this space the next property holds ('Ho is complete'): if x_n is a sequence of vectors in Ho having the property that $\|x_s - x_t\|$ converges to zero if s, t become arbitrarily large, then x_n converges to a vector $x \in Ho$. Examples of Hilbert spaces are $\mathbb{R}^n$, with the usual inner product $< x, y >= x^T y$, and L_2^n the set of all square integrable functions defined on an interval $I \subset \mathbb{R}$ with the inner product $\langle f, g \rangle = \int_I f^T(t) g(t) dt$.

[6] T is called **self-adjoint** if $T^* = T$.

[7] T is called **positive semi-definite** if $\langle u, T(u) \rangle > (\geq) 0$ for all $u \in \mathbb{H}$.

Next, suppose that there exists a $\tilde{u} \in \mathbb{H}$ with $\langle \tilde{u}, \mathcal{T}(\tilde{u}) \rangle = 0$. Then $f(\lambda)$ defined by equation (5.8.8) is either linear or constant. Hence equation (5.8.7) cannot have a unique minimum. Thus, in order to have a unique minimum, for all $u \in \mathbb{H}$ we must have that $\langle u, \mathcal{T}(u) \rangle > 0$. Or, stated differently, the operator $\mathcal{T}$ must be positive definite. However, if $\mathcal{T}$ is positive definite, it is in particular invertible. A simple completion of squares yields then that equation (5.8.7) can be rewritten as

$$\langle u + \mathcal{T}^{-1}(v), \mathcal{T}(u + \mathcal{T}^{-1}(v)) \rangle - \langle \mathcal{T}^{-1}(v), v \rangle + \alpha.$$

Since $\mathcal{T}$ is positive definite, it is clear that this function achieves a unique minimum by choosing $u = -\mathcal{T}^{-1}v$. From this the statements of the lemma are now obvious. $\qquad\square$

From item 2 of this lemma it now follows immediately that $J(t_0, x_0, u, c)$ has a unique minimum for an arbitrary choice of x_0 if and only if the operator $\mathcal{T} := \mathcal{L}^* \bar{Q} \mathcal{L} + R$ is positive definite. However, this implies that $J(t_0, x_0, u, 0)$ also has a unique minimum for an arbitrary choice of x_0. According to Theorem 5.1 this implies that the Riccati equation (5.2.3) has a solution. $\qquad\square$

6

Cooperative games

In the rest of this book we deal with the situation where there is more than one player. Each of these, N, players has a quadratic cost function (6.0.9) he/she wants to minimize

$$J_i = \int_0^T \{x^T(t)Q_i x(t) + u_i^T(t)R_i u_i(t)\}dt + x^T(T)Q_{T,i}x(T), \quad i = 1, \ldots, N. \tag{6.0.1}$$

Throughout the remaining chapters the matrices Q_i, R_i and $Q_{T,i}$ are assumed to be symmetric and R_i positive definite. Sometimes some additional positive definiteness assumptions are made with respect to the matrices Q_i and $Q_{T,i}$. In the minimization a state variable $x(t)$ occurs. This is a dynamic variable that can be influenced by all players. That is,

$$\dot{x}(t) = Ax(t) + B_1 u_1 + \cdots + B_N u_N, \quad x(0) = x_0, \tag{6.0.2}$$

where A and B_i, $i = 1, \ldots, N$, are constant matrices, and u_i is a vector of variables which can be manipulated by player i. The objectives are possibly conflicting. That is, a set of policies u_1 which is optimal for one player, may have rather negative effects on the evolution of the state variable x from another player's point of view.

In general, before one can analyze the outcome of such a decision process, a number of points have to be made more clear (see section 3.8).

1. What information is available to the players?
   (a) What do they know about the system dynamics?
   (b) What do they know about each others' cost functions?
   (c) When do the players have to announce their actions or strategies?
   (d) Can the players communicate with one another?
   (e) Are side-payments permitted?

LQ Dynamic Optimization and Differential Games   J. Engwerda
© 2005 John Wiley & Sons, Ltd

2. Can players commit themself to the proposed decisions?

3. Which strategies are used by the different players?

In this chapter we assume that players can communicate and can enter into binding agreements. Furthermore it is assumed that they cooperate in order to achieve their objectives. However, no side-payments take place. Moreover, it is assumed that every player has all the information on the state dynamics and cost functions of his opponents and all players are able to implement their decisions. Concerning the strategies used by the players we assume that there are no restrictions. That is, every $u_i(.)$ may be chosen arbitrarily from the set $\mathcal{U}$ (or, depending on the context, its equivalent) in order to have a well-posed problem.

6.1 Pareto solutions

By cooperation, in general, the cost one specific player incurs is not uniquely determined anymore. If all players decide, for example, to use their control variables to reduce the cost of player 1 as much as possible, a different minimum is attained for player 1 than in the case where all players agree collectively to help a different player in minimizing his cost. So, depending on how the players choose to 'divide' their contol efforts, a player incurs different 'minima'. So, in general, each player is confronted with a whole set of possible outcomes from which somehow one outcome (which in general does not coincide with a player's overall lowest cost) is cooperatively selected. Now, if there are two strategies γ_1 and γ_2 such that every player has a lower cost if strategy γ_1 is played, then it seems reasonable to assume that all players would prefer this strategy. We say that the solution induced by strategy γ_1 **dominates** over that case where the solution is induced by the strategy γ_2. So, dominance means that the outcome is better for all players. Proceeding along this line of thought, it seems reasonable to consider only those cooperative outcomes which have the property that if a different strategy than the one corresponding with this cooperative outcome is chosen, then at least one of the players has higher costs. Or, stated differently, to consider only solutions that are such that they cannot be improved upon by all players simultaneously. This motivates the concept of Pareto[1] efficiency.

Definition 6.1

A set of strategies $\hat{\gamma}$ is called **Pareto efficient** if the set of inequalities

$$J_i(\gamma) \leq J_i(\hat{\gamma}), \ i = 1, \ldots, N,$$

where at least one of the inequalities is strict, does not allow for any solution $\gamma \in \Gamma$. The corresponding point $(J_1(\hat{\gamma}), \ldots, J_N(\hat{\gamma})) \in \mathbb{R}^N$ is called a **Pareto solution**. The set of all Pareto solutions is called the **Pareto frontier**. $\qquad\square$

---

[1]See footnotes in Chapter 1 for some biographic details.

A Pareto solution is therefore never dominated and for that reason it is called an **undominated** solution. Typically there is always more than one Pareto solution, because dominance is a property which generally does not provide a total ordering.

It turns out that if we assume $Q_i \geq 0$, $i = 1, \ldots, N$, in our cost functions (6.0.1) there is a simple characterization for all Pareto solutions in our cooperative linear quadratic game. This will be shown using the following lemma that states how one can find Pareto solutions in general. In the subsequent analysis the following set of parameters, $\mathcal{A}$, plays a crucial role.

$$\mathcal{A} := \left\{ \alpha = (\alpha_1, \ldots, \alpha_N) | \alpha_i \geq 0 \quad \text{and} \quad \sum_{i=1}^{N} \alpha_i = 1 \right\}.$$

Lemma 6.1

Let $\alpha_i \in (0, 1)$, with $\sum_{i=1}^{N} \alpha_i = 1$. If $\hat{\gamma} \in \Gamma$ is such that

$$\hat{\gamma} \in \arg \min_{\gamma \in \Gamma} \left\{ \sum_{i=1}^{N} \alpha_i J_i(\gamma) \right\}, \tag{6.1.1}$$

then $\hat{\gamma}$ is Pareto efficient.

Proof

Let $\alpha_i \in (0, 1)$, with $\sum_{i=1}^{N} \alpha_i = 1$ and $\hat{\gamma} \in \arg \min_{\gamma \in \Gamma} \left\{ \sum_{i=1}^{N} \alpha_i J_i(\gamma) \right\}$. Assume $\hat{\gamma}$ is not Pareto efficient. Then, there exists an N-multiple of strategies $\bar{\gamma}$ such that

$$J_i(\bar{\gamma}) \leq J_i(\hat{\gamma}), \ i = 1, \ldots, N,$$

where at least one of the inequalities is strict. However, then

$$\sum_{i=1}^{N} \alpha_i J_i(\bar{\gamma}) < \sum_{i=1}^{N} \alpha_i J_i(\hat{\gamma}),$$

which contradicts the fact that $\hat{\gamma}$ is minimizing. $\qquad\square$

Note

Notice that in this lemma we neither use any convexity[2] conditions on the J_i's nor any convexity assumptions regarding the Γ_i's. $\qquad\square$

---

[2]A set S is called **convex** if for every $x, y \in S$, $\alpha_i \in \mathcal{A}$ also $\alpha_1 x + \alpha_2 y \in S$. A function $f(x)$ is **convex** if for every $x, y \in S$, $\alpha_i \in \mathcal{A}$, $f(\alpha_1 x + \alpha_2 y) \leq \alpha_1 f(x) + \alpha_2 f(y)$.

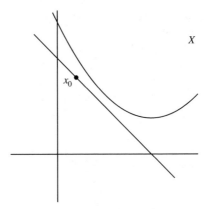

Figure 6.1   The separation Theorem 6.2

To prove a lemma stating in a sense the converse of the above lemma, we need a so-called separation theorem. The proof of this theorem can, for instance, be found in Takayama (1985). In two dimensions, geometrically this separation theorem states that for any convex set X and a point x_0 outside this set, there is a line through x_0 above which the whole set X is located. Figure 6.1 illustrates the situation.

Theorem 6.2

Let X be a nonempty convex set in $\mathbb{R}^n$. Furthermore, let $x_0 \in \mathbb{R}^n$, such that $x_0 \notin X$. Then there exists a $p \in \mathbb{R}^n, p \neq 0, |p| < \infty$, such that for all $x \in X$ $p^T x \geq p^T x_0$.   □

Note

Usually this lemma is stated under the additional assumption that X is a closed subset of $\mathbb{R}^n$. The proof in Takayama (1985), however, shows that this condition is superfluous. Note that when we demand X to be closed, the inequality is strict.   □

The next converse to Lemma 6.1 was proved by Fan, Glicksberg and Hoffman (1957). The lemma states that, under some convexity assumptions on the cost functions, all Pareto-efficient strategies can be obtained by considering the minimization problem (6.1.1). The proof is essentially taken from Takayama (1985) (see also Weeren (1995)).

Lemma 6.3

Assume that the strategy spaces Γ_i, $i = 1, \ldots, N$ are convex. Moreover, assume that the payoffs J_i are convex. Then, if $\hat{\gamma}$ is Pareto efficient, there exist $\alpha \in \mathcal{A}$, such that for all $\gamma \in \Gamma$

$$\sum_{i=1}^{N} \alpha_i J_i(\gamma) \geq \sum_{i=1}^{N} \alpha_i J_i(\hat{\gamma}).$$

Proof

Define for all $\gamma \in \Gamma$ the set $Z_\gamma \subset \mathbb{R}^N$ by

$$Z_\gamma := \{z \in \mathbb{R}^N \mid z_i > J_i(\gamma) - J_i(\hat{\gamma}), \ i = 1, \ldots, N\}, \tag{6.1.2}$$

and define Z by

$$Z := \cup_{\gamma \in \Gamma} Z_\gamma.$$

Then, because $\hat{\gamma}$ is Pareto efficient, $0 \notin Z$. Moreover, Z is convex. For, if $z \in Z_\gamma$, $\tilde{z} \in Z_{\tilde{\gamma}}$ and $\lambda \in [0, 1]$ then

$$\lambda z_i + (1 - \lambda)\tilde{z}_i > \lambda J_i(\gamma) + (1 - \lambda)J_i(\tilde{\gamma}) - J_i(\hat{\gamma})$$
$$\geq J_i(\lambda\gamma + (1 - \lambda)\tilde{\gamma}) - J_i(\hat{\gamma}),$$

and hence, $\lambda z + (1 - \lambda)\tilde{z} \in Z_{\lambda\gamma + (1-\lambda)\tilde{\gamma}} \subset Z$.

Taking $x_0 = 0$ in the separation theorem, Theorem 6.2, we infer that there exists a $p \neq 0$, such that $p^T z \geq 0$ for all $z \in Z$. From equation (6.1.2) it is clear that, for every $i = 1, \ldots, N$, we can choose z_i arbitrarily large. Since $p^T z > 0$, this implies that the ith entry of p cannot be negative for every $i = 1, \ldots, N$.

Let $z \in Z$. Then there exists a $\gamma \in \Gamma$, and $\epsilon \in \mathbb{R}^N$, with $\epsilon_i > 0$, $i = 1, \ldots, N$, such that

$$z_i = J_i(\gamma) - J_i(\hat{\gamma}) - \epsilon_i, \quad i = 1, \ldots, N.$$

Moreover, by varying $\gamma \in \Gamma$ and $\epsilon_i > 0$, $i = 1, \ldots, N$, we obtain all $z \in Z$. Hence, for all $\gamma \in \Gamma$ and for all $\epsilon_i > 0$

$$p^T z = \sum_{i=1}^{N} p_i(J_i(\gamma) - J_i(\hat{\gamma}) - \epsilon_i) \geq 0.$$

Consequently for all $\gamma \in \Gamma$

$$\sum_{i=1}^{N} p_i J_i(\gamma) \geq \sum_{i=1}^{N} p_i J_i(\hat{\gamma}).$$

In particular, with

$$\alpha_i := \frac{p_i}{\sum_{j=1}^{N} p_j},$$

we find that for all $\gamma \in \Gamma$

$$\sum_{i=1}^{N} \alpha_i J_i(\gamma) \geq \sum_{i=1}^{N} \alpha_i J_i(\hat{\gamma}). \qquad \square$$

Combining the results from Lemma 6.1 and Lemma 6.3 then leads to the following theorem.

Theorem 6.4

Let $\alpha_i > 0$, $i = 1, \ldots, N$, satisfy $\sum_{i=1}^{N} \alpha_i = 1$. If $\hat{\gamma} \in \Gamma$ is such that

$$\hat{\gamma} \in \arg\min_{\gamma \in \Gamma} \left\{ \sum_{i=1}^{N} \alpha_i J_i(\gamma) \right\},$$

then $\hat{\gamma}$ is Pareto efficient. Moreover, if Γ_i is convex and J_i is convex for all $i = 1, \ldots, N$, then for all Pareto-efficient $\hat{\gamma}$ there exist $\alpha \in \mathcal{A}$, such that

$$\hat{\gamma} \in \arg\min_{\gamma \in \Gamma} \left\{ \sum_{i=1}^{N} \alpha_i J_i(\gamma) \right\},$$ □

Note

Verkama (1994) gives a short and elegant proof of this theorem, in the case where the strategy spaces Γ_i are compact and convex subspaces of some Euclidean space $\mathbb{R}^{n_i}$, and the payoffs J_i are pseudoconcave.   □

Corollary 6.5

Assume that Γ_i is convex and J_i is strictly convex for all $i = 1, \ldots, N$. Let $\alpha \in \mathcal{A}$ and

$$\hat{\gamma}_\alpha = \arg\min_{\gamma \in \Gamma} \left\{ \sum_{i=1}^{N} \alpha_i J_i(\gamma) \right\}. \qquad (6.1.3)$$

Then if $\hat{\gamma}_{\alpha^1} \neq \hat{\gamma}_{\alpha^2}$, whenever $\alpha^1 \neq \alpha^2$, there is a bijection[3] between $\alpha \in \mathcal{A}$ and the Pareto frontier.

Proof

By assumption there is a bijection between all solutions $\hat{\gamma}_\alpha$ of the optimization problem (6.1.3) and α. We next show that if $\hat{\gamma}_{\alpha^1} \neq \hat{\gamma}_{\alpha^2}$, whenever $\alpha^1 \neq \alpha^2$, then the corresponding Pareto solutions differ. The rest of the claim then follows directly from Theorem 6.4.

We show this by contradiction. That is, assume that there is a Pareto solution J for which $J(\hat{\gamma}_{\alpha^1}) = J(\hat{\gamma}_{\alpha^2})$ for some $\alpha^1 \neq \alpha^2$. Then it follows immediately, that $\hat{\gamma}_{\alpha^2}$ solves the minimization problem

$$\min_{\gamma \in \Gamma} \left\{ \sum_{i=1}^{N} \alpha_i^1 J_i(\gamma) \right\}.$$

---

[3] A synonym for a bijection often encountered in literature is one-to-one correspondence.

However, by assumption, the above cost functional is strictly convex. So, it has a unique minimum. Therefore, $\hat{\gamma}_{\alpha^2}$ must coincide with $\hat{\gamma}_{\alpha^1}$, which violates our assumption that different optimal strategies correspond to different α^i. □

In general the Pareto frontier does not always have to be an $N-1$ dimensional surface in $\mathbb{R}^N$, as in the above corollary. This is already illustrated in the two-player case when both players have the same cost function. In that case the Pareto frontier reduces to a single point in $\mathbb{R}^2$.

Next consider, as a particular case, the linear quadratic differential game. In the Appendix to this chapter we show that the cost functions (6.0.9) are convex if $Q_i \geq 0$. An immediate corollary from Theorem 6.4 is given below.

Corollary 6.6

Consider the optimization problem (6.0.9) and (6.0.10). Under the additional assumption that $Q_i \geq 0$, $i = 1, \ldots, N$, $J_i(u)$ are convex. The set of all cooperative Pareto solutions is given by

$$(J_1(u^*(\alpha)), \ldots, J_N(u^*(\alpha))), \quad \text{where } \alpha \in \mathcal{A}, \tag{6.1.4}$$

where the corresponding Pareto-efficient strategy is obtained as

$$u^*(\alpha) = \arg\min_{u \in \mathcal{U}} \sum_{i=1}^N \alpha_i J_i, \quad \text{subject to (6.0.10)}. \tag{6.1.5}$$

□

Theorem 6.2 and Corollary 6.6 show that to find all cooperative solutions for the linear quadratic game one has to solve a regular linear quadratic optimal control problem which depends on a parameter α. From Chapter 5 we know that the existence of a solution for this problem is related to the existence of solutions of Riccati equations. In Lancaster and Rodman (1995) it is shown that if the parameters appearing in an algebraic Riccati equation are, for example, differentiable functions of some parameter α (or, more generally, depend analytically on a parameter α), and the maximal solution exists for all α in some open set V, then this maximal solution of the Riccati equation will also be a differentiable function of this parameter α on V (or, more generally, depend analytically on this parameter α). Since in the linear quadratic case the parameters depend linearly on α, this implies that in the infinite horizon case the corresponding Pareto frontier will be a smooth function of α (provided the maximal solution exists for all α). A similar statement holds for the finite-planning horizon case. In case for all $\alpha \in V$ the cooperative linear quadratic differential game has a solution or, equivalently, the corresponding Riccati differential equations have a solution, then it follows directly from Theorem 3.8 (see also Perko (2001) for a precise statement and proof) that the solution of the Riccati differential equation is a differentiable function of α, since all parameters in this Riccati differential equation are differentiable functions of α.

Example 6.1

Consider the following differential game on government debt stabilization (see van Aarle, Bovenberg and Raith (1995)). Assume that government debt accumulation, $\dot{d}(t)$, is the sum of interest payments on government debt, $rd(t)$, and primary fiscal deficits, $f(t)$, minus the seignorage (i.e. the issue of base money) $m(t)$. So,

$$\dot{d}(t) = rd(t) + f(t) - m(t), \quad d(0) = d_0.$$

Here $d(t)$, $f(t)$ and $m(t)$ are expressed as fractions of GDP and r represents the rate of interest on outstanding government debt minus the growth rate of output. The interest rate $r > 0$ is assumed to be external. Assume that fiscal and monetary policies are controlled by different institutions, the fiscal authority and the monetary authority, respectively, which have different objectives. The objective of the fiscal authority is to minimize a sum of time profiles of the primary fiscal deficit, base-money growth and government debt

$$J_1 = \int_0^\infty e^{-\delta t}\{f^2(t) + \eta m^2(t) + \lambda d^2(t)\}dt.$$

The parameters, η and λ express the relative priority attached to base-money growth and government debt by the fiscal authority. The monetary authorities are assumed to choose the growth of base money such that a sum of time profiles of base-money growth and government debt is minimized. That is

$$J_2 = \int_0^\infty e^{-\delta t}\{m^2(t) + \kappa d^2(t)\}dt.$$

Here $1/\kappa$ can be interpreted as a measure for the conservatism of the central bank with respect to the money growth. Furthermore all variables are normalized such that their targets are zero, and all parameters are positive.

Introducing $\tilde{d}(t) := e^{-\frac{1}{2}\delta t}d(t)$, $\tilde{m} := e^{-\frac{1}{2}\delta t}m(t)$ and $\tilde{f} := e^{-\frac{1}{2}\delta t}f(t)$ the above model can be rewritten as

$$\dot{\tilde{d}}(t) = \left(r - \frac{1}{2}\delta\right)\tilde{d}(t) + \tilde{f}(t) - \tilde{m}(t), \quad \tilde{d}(0) = d_0$$

where the cost functions of both players are

$$J_1 = \int_0^\infty \{\tilde{f}^2(t) + \eta\tilde{m}^2(t) + \lambda\tilde{d}^2(t)\}dt$$

and

$$J_2 = \int_0^\infty \{\tilde{m}^2(t) + \kappa\tilde{d}^2(t)\}dt.$$

If both the monetary and fiscal authority agree to cooperate in order to reach their goals, then by Corollary 6.6 the set of all Pareto solutions is obtained by considering the simultaneous minimization of

$$
J_c(\alpha) := \alpha J_1 + (1 - \alpha) J_2
$$

$$
= \int_0^\infty \{\alpha \tilde{f}^2(t) + \beta_1 \tilde{m}^2(t) + \beta_2 \tilde{d}^2(t)\} dt,
$$

where $\beta_1 = 1 + \alpha(-1 + \eta)$ and $\beta_2 = \kappa + \alpha(\lambda - \kappa)$. So, rewriting the above model into our standard framework, the cooperative game problem can be reformulated as the minimization of

$$
J_c(\alpha) = \int_0^\infty \left\{ \beta_2 \tilde{d}^2(t) + [\tilde{f} \ \tilde{m}] \begin{bmatrix} \alpha & 0 \\ 0 & \beta_1 \end{bmatrix} \begin{bmatrix} \tilde{f} \\ \tilde{m} \end{bmatrix} \right\} dt,
$$

subject to

$$
\dot{\tilde{d}}(t) = \left(r - \frac{1}{2}\delta \right) \tilde{d}(t) + [1 \ -1] \begin{bmatrix} \tilde{f} \\ \tilde{m} \end{bmatrix}, \quad \tilde{d}(0) = d_0.
$$

According to Theorem 5.14 this optimization problem has a unique solution for every α. The algebraic Riccati equation corresponding to this solution is

$$
\beta_2 + 2 \left(r - \frac{1}{2}\delta \right) x - \left(\frac{1}{\alpha} + \frac{1}{\beta_1} \right) x^2 = 0.
$$

The stabilizing solution of this Riccati equation is

$$
k = \frac{\alpha \beta_1}{\alpha + \beta_1} \left\{ r - \frac{1}{2}\delta + \sqrt{\left(r - \frac{1}{2}\delta \right)^2 + \beta_2 \left(\frac{1}{\alpha} + \frac{1}{\beta_1} \right)} \right\}.
$$

So the optimal control is

$$
\tilde{f}^*(t) = -\frac{k}{\alpha} \tilde{d}^*(t), \quad \text{and} \quad \tilde{m}^*(t) = \frac{k}{\beta_1} \tilde{d}^*(t),
$$

where $\tilde{d}^*(t)$ is the solution of the differential equation

$$
\dot{\tilde{d}}^*(t) = \left(r - \frac{1}{2}\delta - k \left(\frac{1}{\alpha} + \frac{1}{\beta_1} \right) \right) \tilde{d}^*(t) =: \beta_3 \tilde{d}^*(t), \quad \tilde{d}^*(0) = d_0.
$$

Substitution of these expressions into the cost functions gives

$$
J_1 = \int_0^\infty \{\tilde{f}^{*2}(t) + \eta \tilde{m}^{*2}(t) + \lambda \tilde{d}^{*2}(t)\} dt
$$

$$
= \left(\frac{k^2}{\alpha^2} + \eta \frac{k^2}{\beta_1^2} + \lambda \right) \int_0^\infty \tilde{d}^{*2}(t) dt
$$

$$
= \left(\frac{k^2}{\alpha^2} + \eta \frac{k^2}{\beta_1^2} + \lambda \right) \frac{-1}{2\beta_3},
$$

and

$$J_2 = \int_0^\infty \{\tilde{m}^{*^2}(t) + \kappa \tilde{d}^{*^2}(t)\}dt$$

$$= \left(\frac{k^2}{\beta_1^2} + \kappa\right) \int_0^\infty \tilde{d}^{*^2}(t)dt$$

$$= \left(\frac{k^2}{\beta_1^2} + \kappa\right) \frac{-1}{2\beta_3}.$$

Notice that the parameter β_3 is negative here. The set of all Pareto solutions is then found by varying the parameter α in J_i (simultaneously) between zero and one.

In Figure 6.2 we have plotted the set of Pareto solutions if $\eta = 0.1$, $\lambda = 0.6$, $\kappa = 0.5$, $r = 0.06$ and $\delta = 0.04$.

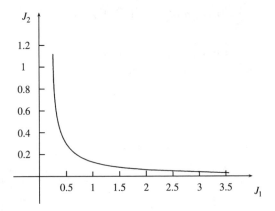

Figure 6.2 Pareto frontier Example 6.1 if $\eta = 0.1$, $\lambda = 0.6$, $\kappa = 0.5$, $r = 0.06$ and $\delta = 0.04$

To assess the optimal control for the actual system we have to transform the optimal control, we determined in terms of the transformed system, back again. This is not too difficult. For instance for the optimal fiscal policy this transformation gives

$$f^*(t) = e^{-\frac{1}{2}\delta t}\tilde{f}^*(t)$$

$$= e^{-\frac{1}{2}\delta t}\frac{-k}{\alpha}\tilde{d}^*$$

$$= e^{-\frac{1}{2}\delta t}\frac{-k}{\alpha}e^{\frac{1}{2}\delta t}d^*(t)$$

$$= -\frac{k}{\alpha}d^*(t).$$

Similarly it can be verified that $m^*(t) = \frac{k}{\beta_1}d^*(t)$ and $\dot{d}^*(t) = \left(r - k\left(\frac{1}{\alpha} + \frac{1}{\beta_1}\right)\right)$ $d^*(t)$, $d^*(0) = d_0$. That is, the dynamics of the actual optimally controlled system is obtained by using the feedback controls $f^*(t) = h_1 d(t)$ and $m^*(t) = h_2 d(t)$, respectively, where the gain parameters h_1 and h_2 coincide with the corresponding feedback gain parameters determined for the transformed system. □

Example 6.2

Consider the situation in which there are two individuals who invest in a public stock of knowledge (see also Dockner *et al.* (2000)). Let $x(t)$ be the stock of knowledge at time t and $u_i(t)$ the investment of player i in public knowledge at time t. Assume that the stock of knowledge evolves according to the accumulation equation

$$\dot{x}(t) = -\beta x(t) + u_1(t) + u_2(t), \quad x(0) = x_0, \tag{6.1.6}$$

where β is the depreciation rate. Assume that each player derives quadratic utility from the consumption of the stock of knowledge and that the cost of investment increases quadratically with the investment effort. That is, the cost function of both players is given by

$$J_i = \int_0^\infty e^{-\theta t} \{-q_i x^2(t) + r_i u_i^2(t)\} dt.$$

If both individuals decide to cooperate in order to increase the stock of knowledge, then all solutions (as a function of $\alpha \in (0, 1)$) of the minimization of the next cost functional subject to equation (6.1.6) provide Pareto solutions

$$J(\alpha) := \int_0^\infty e^{-\theta t} \{-\beta_1(\alpha)x^2(t) + \beta_2(\alpha)u_1^2(t) + \beta_3(\alpha)u_2^2(t)\} dt,$$

where $\beta_1 = \alpha q_1 + (1 - \alpha)q_2$, $\beta_2 = \alpha r_1$ and $\beta_3 = (1 - \alpha)r_2$.

Introducing the variables $\tilde{x}(t) := e^{-\frac{1}{2}\theta t} x(t)$, $\tilde{u}_i(t) := e^{-\frac{1}{2}\theta t} u_i(t)$, $i = 1, 2$ and $\tilde{u}(t) := [\tilde{u}_1(t) \ \tilde{u}_2(t)]^T$ the minimization problem can be rewritten as (see, for example, Example 6.1)

$$\min \int_0^\infty \left\{ -\beta_1(\alpha)\tilde{x}^2(t) + \tilde{u}^T(t) \begin{bmatrix} \beta_2 & 0 \\ 0 & \beta_3 \end{bmatrix} \tilde{u}(t) \right\} dt,$$

subject to

$$\dot{\tilde{x}}(t) = -a\tilde{x}(t) + [1 \ 1]\tilde{u}(t), \quad \tilde{x}(0) = x_0,$$

where $a := \beta + \frac{1}{2}\theta$.

According to Theorem 5.14 this problem has a stabilizing feedback solution if and only if, with $s := \frac{1}{\beta_2} + \frac{1}{\beta_3}$, the quadratic equation

$$\beta_1 + 2ak + sk^2 = 0 \tag{6.1.7}$$

has a solution k such that $-a - sk < 0$. The solutions of equation (6.1.7) are

$$k_{1,2} = \frac{-a \pm \sqrt{a^2 - \beta_1 s}}{s}.$$

From this it is then easily verified that equation (6.1.7) has an appropriate solution if and only if the discriminant of this equation is stricly positive, that is

$$a^2 - \beta_1 s > 0.$$

Or, stated in the original model parameters,

$$\left(\beta + \frac{1}{2}\theta\right)^2 - (\alpha q_1 + (1 - \alpha)q_2)\left(\frac{1}{\alpha r_1} + \frac{1}{(1 - \alpha)r_2}\right) > 0. \qquad (6.1.8)$$

As in Example 6.1 we therefore obtain (see Theorem 6.2) that for all $\alpha \in (0, 1)$ satisfying the above inequality (6.1.8) the next control yields a Pareto solution

$$u_1(t) = -\frac{1}{\alpha r_1} kx(t) \text{ and } u_2(t) = -\frac{1}{(1 - \alpha)r_2} kx(t),$$

where $k = \dfrac{-\left(\beta + \frac{1}{2}\theta\right) + \sqrt{\left(\beta + \frac{1}{2}\theta\right)^2 - (\alpha q_1 + (1 - \alpha)q_2)s}}{s}$ and $s := \frac{1}{\alpha r_1} + \frac{1}{(1-\alpha)r_2}$.

Unfortunately, it is unclear whether or not we have now characterized all Pareto solutions for this problem with this. □

6.2   Bargaining concepts

In the previous section we argued that it is rational to consider in a cooperative environment the set of Pareto solutions. However, as Corollary 6.6 already indicates there are, in general, a lot of Pareto solutions. This raises the question as to which one is the 'best'. By considering this question we enter the arena of what is called bargaining theory.

This theory has its origin in two papers by Nash (1950b, 1953). In these papers a bargaining problem is defined as a situation in which two (or more) individuals or organizations have to agree on the choice of one specific alternative from a set of alternatives available to them, while having conflicting interests over this set of alternatives. Nash (1953) proposes two different approaches to the bargaining problem, namely the **axiomatic** and the **strategic** approach. The axiomatic approach lists a number of desirable properties the solution must have, called the **axioms**. The strategic approach, on the other hand, sets out a particular bargaining procedure and asks what outcomes would result from rational behavior by the individual players. In this section we discuss three well-known solutions which will be motivated using the axiomatic approach. We will give a brief outline of this theory. For proofs, more background and other axiomatic bargaining solutions we refer the reader to Thomson (1994).

So, bargaining theory deals with the situation in which players can realize – through cooperation – other better outcomes than the one which becomes effective when they do not cooperate. This non-cooperative outcome is called the **threatpoint**. The question is which outcome might the players possibly agree to.

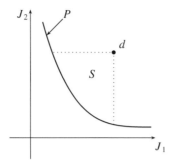

Figure 6.3 The bargaining game

In Figure 6.3 a typical bargaining game is sketched (see also Figure 6.2). The ellipse marks out the set of possible outcomes, the **feasible set** S, of the game. The point d is the threatpoint. The edge P is the set of individually rational Pareto-optimal outcomes.

We assume that, if the agents unanimously agree on a point $x = (J_1, \ldots, J_N) \in S$, they obtain x. Otherwise they obtain d. This presupposes that each player can enforce the threatpoint, when he does not agree with a proposal. The outcome x the players will finally agree on is called the solution of the bargaining problem. Since the solution also depends on the feasible set S as the threatpoint d, it will be written as $F(S, d)$. Notice that the difference for player i between the solution and the threatpoint, $J_i - d_i$, is the reduction in cost player i incurs by accepting the solution. In the sequel we will call this difference the utility gain for player i. We will use the notation $J := (J_1, \ldots, J_N)$ to denote a point in S and $x \succ y (x \prec y)$ to denote the **vector inequality**, i.e. $x_i > y_i (x_i < y_i)$, $i = 1, \ldots, N$. In axiomatic bargaining theory a number of solutions have been proposed. In Thomson (1994) a survey is given on this theory. We will present here the three most commonly used solutions: the Nash bargainig solution, the Kalai–Smorodinsky solution and the egalitarian solution.

The **Nash bargaining solution**, $N(S, d)$, selects the point of S at which the product of utility gains from d is maximal. That is,

$$N(S, d) = \arg\max_{J \in S} \prod_{i=1}^{N} (J_i - d_i), \quad \text{for } J \in S \text{ with } J \preceq d.$$

In Figure 6.4 we sketched the N solution. Geometrically, the Nash bargaining solution is the point on the edge of S (i.e. a part of the Pareto frontier) which yields the largest rectangle (N, A, B, d).

The Kalai–Smorodinsky solution, $K(S, d)$, sets utility gains from the threatpoint proportional to the player's most optimistic expectations. For each agent, the most optimistic expectation is defined as the lowest cost he can attain in the feasible set subject to the constraint that no agent incurs a cost higher than his coordinate of the threatpoint. Defining the **ideal point** as

$$I(S, d) := \max\{J_i \mid J \in S, \ J \preceq d\},$$

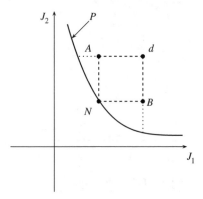

Figure 6.4 The Nash bargaining solution $N(S, d)$

the **Kalai–Smorodinsky solution** is then

$$K(S, d) := \text{ maximal point of } S \text{ on the segment connecting } d \text{ to } I(S, d).$$

In Figure 6.5 the Kalai–Smorodinsky solution is sketched for the two-player case. Geometrically, it is the intersection of the Pareto frontier P with the line which connects the threatpoint and the ideal point. The components of the ideal point are the minima each player can reach when the other player is fully altruistic under cooperation.

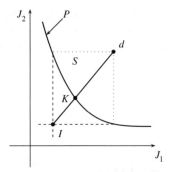

Figure 6.5 The Kalai–Smorodinsky solution $K(S, d)$

Finally, the **egalitarian solution**, $E(S, d)$, represents the idea that gains should be equally divided between the players. Thus

$$E(S, d) := \text{ maximal point in } S \text{ for which } E_i(S, d) - d_i = E_j(S, d) - d_j, \ i, j = 1, \dots, N.$$

Again, we sketched this solution for the two-player case. In Figure 6.6 we observe that geometrically this egalitarian solution is obtained as the intersection point of the 45°-line through the threatpoint d with the Pareto frontier P.

Notice that particularly in contexts where interpersonal comparisons of utility is inappropriate or impossible, the first two bargaining solutions still make sense.

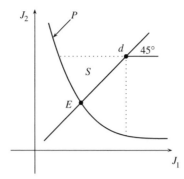

Figure 6.6   The egalitarian solution $E(S, d)$

As already mentioned above, these bargaining solutions can be motivated using an 'axiomatic approach'. In this case some people prefer to speak of an arbitration scheme instead of a bargaining game. An arbiter draws up the reasonable axioms and depending on these axioms, a solution results. Up to now, we were not very specific about the feasible set S. To understand the axiomatic bargaining theory we have to be more specific on this. We assume that our bargaining problem (S, d) belongs to the class of sets, Σ_d^N, that satisfy the properties:

1. S is convex, bounded, and closed (it contains its boundary);

2. there is at least one point of S strictly dominating d;

3. (S, d) is **d-comprehensive**, that is, if $x \in S$ and $x \preceq y \preceq d$, then $y \in S$.

Notice that condition 3 implies that the players will have an incentive to reach an agreement. In the axiomatic bargaining theory a **bargaining solution** on Σ_d^N is then a map

$$F : \sum_d^N \to \mathbb{R}^N \text{ such that } F(S, d) \in S \text{ for all } (S, d) \in \sum_d^N.$$

Nash has shown that if this function F satisfies the next four rules (axioms) on Σ_d^N, then the outcome is completely characterized by the calculation rule we sketched above to calculate the N solution. The first one states that the solution should be a Pareto solution, or stated differently, that all gains from cooperation should be exhausted.

Axiom 1 Pareto-optimality

$F(S, d) \in \{J \in S \mid \text{there is no } J' \in S \text{ with } J' \preceq J\}.$

The second axiom states that if both S and d are invariant under all exchanges of the names of the players, the solution outcome should have equal coordinates. That is, if initially players have completely symmetric roles, the outcomes of the bargaining solution should also be symmetric.

Axiom 2 Symmetry

For $x \in S$ let $\pi(x)$ be any permutation of the coordinates of x. If with $x \in S$ also $\pi(x) \in S$, then $F_i(S,d) = F_j(S,d)$ for all $i,j = 1,\ldots,N$.

The third axiom axiom states that the solution should be independent of the utility scales used to represent the players' preferences. That is

Axiom 3 Scale invariance

Let $a \succ 0$ and the transformation $\lambda : S \to S'$ be defined as $\lambda(x)_i = a_i x_i, i = 1,\ldots,N$, for all $x \in S$. Then $F(\lambda(S), \lambda(d)) = \lambda F(S,d)$.

Finally, the fourth axiom states that if the feasible set S is reduced to a new set S', which contains the solution outcome of the original game, then the solution outcome of the new game must coincide with that of the original game. This, under the assumption that the threatpoint in both games is the same.

Axiom 4 Independence of irrelevant alternatives

If $S' \subset S$ and $F(S,d) \in S'$, then $F(S',d) = F(S,d)$.

Under these assumptions on the set S and function F, Nash (1950b) has shown the following result in case $N = 2$, and this result was extended later on for more than two players (Thomson (1994)).

Theorem 6.7

The Nash bargaining solution $N(S,d)$ is the unique bargaining solution which satisfies for an arbitrary $(S,d) \in \Sigma_d^N$ the axioms of Pareto-optimality, symmetry, scale invariance and independence of irrelevant alternatives. $\qquad\square$

The fourth axiom on independence of irrelevant alternatives has been criticized by several authors. The so-called irrelevant alternatives might not be so irrelevant after all. Figure 6.7 illustrates the basic idea for the criticism.

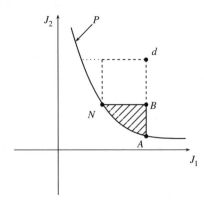

Figure 6.7 Consequences of the axiom of independence of irrelevant alternatives

When the shaded area is left out, a new bargaining game results with the same Nash solution N. However, player 2's alternatives have been reduced from the interval (A, d) to the interval (B, d). The critics argue that this should influence the solution of the bargaining problem. The Kalai–Smorodinsky solution is an answer to this criticism. Axiom four is replaced by the following axiom which states that an expansion of the feasible set 'in a direction favorable to a particular player' always benefits this player.

Axiom 5 (Restricted) monotonicity property

If for all $(S, d), (S', d) \in \Sigma_d^N$ with $S \subset S'$ and $I(S, d) = I(S', d)$ we have $F(S, d) \preceq F(S', d)$.

Kalai and Smorodinsky (1975) proved that the replacement of the fourth axiom by the restricted monotonicity property leads to the following result.

Theorem 6.8

The K-solution is the unique bargaining solution on the class of two-player bargaining problems which satisfies the axioms of Pareto-optimality, symmetry, scale invariance and restricted monotonicity. □

Finally, the egalitarian solution is also motivated from the monotonicity property point of view. The idea is that all players should benefit from **any** expansion of the feasible set S; this irrespective of whether the expansion may be biased in favor of one of them. The price paid by requiring this strong monotonicity is that the resulting solution involves interpersonal comparisons of utility (it violates scale invariance). Furthermore it satisfies only a **weak Pareto-optimality** condition, that is, it satisfies $F(S, d) \in \{J \in S | \nexists J' \in S \text{ with } J' \prec J\}$.

Strong monotonicity

If $S' \supset S$, then $F(S', d) \succeq F(S, d)$.

Kalai (1977) proved the following theorem

Theorem 6.9

The egalitarian solution $E(S, d)$ is the only bargaining solution on Σ_d^N satisfying weak Pareto-optimality, symmetry and strong monotonicity. □

The Pareto-optimality, symmetry and scale invariance axioms have also been the object of some criticism. Usually, the citicism concerning the first two axioms on Pareto-optimality and symmetry is application oriented. One likes to use these concepts in models of reality which do not capture all relevant aspects. So, using these axioms in these models may lead to disputable outcomes. The scale invariance axiom has been criticized, as we have already noted above, because it prevents basing compromises on interpersonal comparisons of utility. Such comparisons, however, are made in a wide variety of situations.

6.3 Nash bargaining solution

In the previous section we suggested, using the axiomatic approach, that the Nash bargaining solution can be viewed in many applications to be a reasonable cooperative solution. Therefore this solution is considered in this section in somewhat more detail.

As already noted in the previous section the Nash bargaining solution, $N(S, d)$, selects for a given set $S \in \Sigma_d^N$ the point at which the product of utility gains from d is maximal. That is,

$$N(S, d) = \arg\max_{J \in S} \prod_{i=1}^{N} (d_i - J_i), \text{ for } J \in S \text{ with } J \preceq d. \tag{6.3.1}$$

In Theorem 6.4 it was shown that, under the assumption that all cost functions J_i are convex, every strategy yielding a Pareto solution can be obtained by minimizing a linear combination of these cost functions. That is, if $(J_1(u^*), \ldots, J_N(u^*))$ is a Pareto solution, then

$$u^* = \arg\min_{u \in \mathcal{U}} \alpha_1 J_1(u) + \cdots + \alpha_N J_N(u), \tag{6.3.2}$$

for some $\alpha \in \mathcal{A}$. Since the N-solution is also located on the Pareto frontier, we conclude that for the N-solution there also exists an $\alpha^N \in \mathcal{A}$ for which equation (6.3.2) holds[4]. Theorem 6.10 gives a characterization of this number.

Theorem 6.10

Let X be a normed vector space. Assume that the convex functions $J_i(u)$, $i = 1, \ldots, N$, are Fréchet differentiable and u^*, defined by equation (6.3.2), is a differentiable function of α. Denote $J_i' := \left[\frac{\partial J_i(u^*)}{\partial \alpha_1}, \ldots, \frac{\partial J_i(u^*)}{\partial \alpha_N} \right]^T$. Consider the matrix of derivatives

$$J' := [J_1', \ldots, J_N'],$$

Assume that J' has at least rank $N - 1$ at the Nash bargaining solution.

Then the following relationship holds between the value of the cost functions at the Nash bargaining solution, $(J_1^N, \ldots, J_N^N)$, the threatpoint d and the weight $\alpha^N = (\alpha_1^N, \ldots, \alpha_N^N)$

$$\alpha_1^N (d_1 - J_1^N) = \alpha_2^N (d_2 - J_2^N) = \cdots = \alpha_N^N (d_N - J_N^N), \tag{6.3.3}$$

or, equivalently,

$$\alpha_j^N = \frac{\prod_{i \neq j} (d_i - J_i^N)}{\sum_{i=1}^{N} \prod_{k \neq i} (d_k - J_k^N)}, \quad j = 1, \ldots, N. \tag{6.3.4}$$

---

[4]Some confusion might arise on the notation here – a subscript N indicates the number of players whereas the superscript N indicates the fact that we are dealing with the N-solution.

Proof

Let $u^* \in X$ denote the argument that minimizes equation (6.3.2) at the Nash solution, and α^N be as above. Then the Fréchet differential of $J := \alpha_1 J_1(u) + \cdots + \alpha_N J_N(u)$ at u^* with increment h is zero, i.e. $\delta J(u^*, h) = 0$ for all $h \in X$ (see Theorem 4.1). That is

$$\alpha_1^N \delta J_1(u^*, h) + \cdots + \alpha_N^N \delta J_N(u^*, h) = 0, \quad \text{for all } h \in X. \tag{6.3.5}$$

Next consider the first-order conditions from maximizing the Nash product (6.3.1). Since this maximum occurs on the Pareto frontier and each solution on this Pareto frontier is parametrized by $\alpha \in \mathcal{A}$ (see Theorem 6.4), the N-solution can also be obtained by determining

$$\alpha^N = \arg\max_{\alpha \in \mathcal{A}} \prod_{i=1}^{N} (d_i - J_i(u^*(\alpha))).$$

Introducing the simplifying notation $s_i := d_i - J_i(u^*)$ and $J_i' := \frac{dJ_i(u^*(\alpha))}{d\alpha}$, differentiation of the above product with respect to α yields the first-order conditions

$$J_1' s_2 s_3 \ldots s_N + J_2' s_1 s_3 \ldots s_N + \cdots + J_N' s_1 s_2 \ldots s_{N-1} = 0.$$

According to Theorem 4.3, $J_i' = \partial J_i(u^*)u^{*'}$. Substitution of this relationship into the above equation gives

$$\partial J_1(u^*)u^{*'} s_2 s_3 \ldots s_N + \partial J_2(u^*)u^{*'} s_1 s_3 \ldots s_N + \cdots + \partial J_N(u^*)u^{*'} s_1 s_2 \ldots s_{N-1} = 0. \tag{6.3.6}$$

Since, by assumption, the matrix of derivatives $[J_1' \cdots J_N']$ has at least rank $N - 1$ at u^* we can find $N - 1$ linearly independent columns in this matrix. Assume, without loss of generality, that the first derivative, J_1', does not belong to this set. By equation (6.3.5)

$$\partial J_1 = -\frac{\alpha_2^N}{\alpha_1^N} \partial J_2 - \cdots - \frac{\alpha_N^N}{\alpha_1^N} \partial J_N.$$

Substitution of this expression for ∂J_1 into equation (6.3.6) gives, after some elementary manipulations, the following equalities

$$0 = \partial J_2(u^*)u^{*'} \left(s_1 - \frac{\alpha_2^N}{\alpha_1^N} s_2 \right) s_3 \ldots s_N + \cdots + \partial J_N(u^*)u^{*'} \left(s_1 - \frac{\alpha_N^N}{\alpha_1^N} s_N \right) s_2 \ldots s_{N-1}$$

$$= [J_2' \ldots J_N'] \begin{bmatrix} \left(s_1 - \frac{\alpha_2^N}{\alpha_1^N} s_2 \right) s_3 \ldots s_N \\ \vdots \\ \left(s_1 - \frac{\alpha_N^N}{\alpha_1^N} s_N \right) s_2 \ldots s_{N-1} \end{bmatrix}.$$

Since $[J_2' \cdots J_N']$ is a full column rank matrix it follows that

$$\left[\left(s_1 - \frac{\alpha_2^N}{\alpha_1^N} s_2 \right) s_3 \ldots s_N, \ldots, \left(s_1 - \frac{\alpha_N^N}{\alpha_1^N} s_N \right) s_2 \ldots s_{N-1} \right]^T = 0.$$

From this we immediately deduce equation (6.3.3). Using equation (6.3.3) it follows then that

$$\frac{\prod_{i \neq j}(d_i - J_i^N)}{\sum_{i=1}^N \prod_{k \neq i}(d_k - J_k^N)} = \frac{1}{\sum_{i=1}^N \frac{d_j - J_j^N}{d_i - J_i^N}}$$

$$= \frac{1}{\sum_{i=1}^N \frac{\alpha_i^N}{\alpha_j^N}}$$

$$= \frac{\alpha_j^N}{1}.$$

On the other hand, if equation (6.3.4) holds it follows immediately by multiplying this expression for α_j^N with $(d_j - J_j^N)$ that for all $j = 1, \ldots, N$ we obtain the same expression. That is, equation (6.3.3) holds. $\qquad\square$

Note

With $X := \mathbb{H}_2^N$, the set of all square integrable functions endowed with the norm induced by the usual inner product (see Appendix to Chapter 5, proof of Theorem 5.11), Theorem 6.10 applies for the linear quadratic differential game. In Corollary 6.6 (and the following discussion) we argued that, at least with $Q_i \geq 0$, generically the Pareto frontier will be a smooth $N - 1$ dimensional manifold. Therefore, one may expect that for this linear quadratic differential game the conditions of Theorem 6.10 are satisfied and that the relationships (6.3.3) and (6.3.4) hold. These relationships will be used in the next section to calculate the Nash bargaining solution. $\qquad\square$

Example 6.3

As an example reconsider Example 6.1. In this example it was shown that

$$J_1 = \left(\frac{k^2}{\alpha^2} + \eta \frac{k^2}{\beta_1^2} + \lambda \right) \frac{-1}{2\beta_3} \quad \text{and} \quad J_2 = \left(\frac{k^2}{\beta_1^2} + \kappa \right) \frac{-1}{2\beta_3},$$

where $\beta_1 = 1 + \alpha(-1 + \eta)$, $\beta_2 = \kappa + \alpha(\lambda - \kappa)$,

$$k = \frac{\alpha \beta_1}{\alpha + \beta_1} \left\{ r - \frac{1}{2}\delta + \sqrt{\left(r - \frac{1}{2}\delta \right)^2 + \beta_2 \left(\frac{1}{\alpha} + \frac{1}{\beta_1} \right)} \right\},$$

and

$$\beta_3 = r - \frac{1}{2}\delta - k\left(\frac{1}{\alpha} + \frac{1}{\beta_1}\right) = -\sqrt{\left(r - \frac{1}{2}\delta\right)^2 + \beta_2\left(\frac{1}{\alpha} + \frac{1}{\beta_1}\right)}.$$

From these rather involved expressions for J_1 and J_2 it is clear that they are smooth functions of α. Furthermore, it is to be expected that only under some rare combinations of the parameters will the derivatives of both J_1 and J_2 w.r.t. α simultaneously become zero. For obvious reasons we will not elaborate this point here.

The Nash bargaining solution can then be determined either by calculating the derivative of J_i, $i = 1, 2$, w.r.t. α and next solving the equation

$$J_1'(\alpha)(d_2 - J_2(\alpha)) + J_2'(\alpha)(d_1 - J_1(\alpha)) = 0,$$

or, by solving the equation

$$\alpha(d_1 - J_1(\alpha)) - (1 - \alpha)(d_2 - J_2(\alpha)) = 0. \tag{6.3.7}$$

Notice that both equations might, in principle, yield more than one solution. However, since we know that $0 < \alpha_i < 1$, $J_i < d_i$ and the product $(d_1 - J_1)(d_2 - J_2)$ is maximized at the Nash bargaining solution, from all candidate solutions resulting from these equations it can easily be verified which one yields the Nash bargaining solution we are looking for.

Choosing again the parameters $\eta = 0.1$, $\lambda = 0.6$, $\kappa = 0.5$, $r = 0.06$ and $\delta = 0.04$, solving numerically the second equation (6.3.7) yields $\alpha = 0.3426$. This corresponds with the Nash bargaining solution $(J_1, J_2) = (0.5649, 0.2519)$. $\qquad\square$

To get a better understanding of the relationship (6.3.3) Figure 6.8 shows this result for the two-player case. The figure shows that the N-solution on the Pareto frontier is geometrically obtained as the solution for which the angle of the line through $d = (d_1, d_2)$ and (J_1^N, J_2^N) on the Pareto frontier and the J_1-axis exactly equals the negative angle of the tangent of the Pareto frontier in the point (J_1^N, J_2^N) and the J_1-axis. Both angles are denoted by β in Figure 6.8.

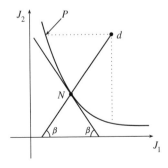

Figure 6.8 A characterization of the N-solution in the two-player case

The correctness of this statement is seen as follows. The derivative of the first line is given by $\frac{d_2-J_2^N}{d_1-J_1^N}$. So what is left to be shown is that the derivative of the tangent to the Pareto frontier at the N-solution is $-\frac{d_2-J_2^N}{d_1-J_1^N}$. We will show that the derivative of the tangent to the Pareto frontier at any point is $-\frac{\alpha_1}{\alpha_2}$. To see this recall that, for $\alpha \in \mathcal{A}$, $\alpha_1 J_1(u) + \alpha_2 J_2(u)$ is minimized w.r.t. u in order to obtain the corresponding Pareto point. So (under the assumption that J_i are differentiable and $J_2' \neq 0$) $\alpha_1 J_1'(u) + \alpha_2 J_2'(u) = 0$ at the Pareto point. This implicitly provides u^* as a function of α. In particular we see that $\frac{J_2'(u^*)}{J_1'(u^*)} = -\frac{\alpha_1}{\alpha_2}$.

Since $\alpha_2 = 1 - \alpha_1$ the Pareto frontier is parametrized by $\alpha_1 \rightarrow [J_1^*(\alpha_1), J_2^*(\alpha_1)]^T$. Consequently, we can view the Pareto frontier as a curve $J_2^* = g(J_1^*)$. The derivative of this curve is $g' = \frac{dJ_2(u^*(\alpha_1))/d\alpha_1}{dJ_1(u^*(\alpha_1))/d\alpha_1}$. From the chain rule it follows then that $g'(\alpha_1) = -\frac{\alpha_1}{\alpha_2}$[5] (see above).

Therefore, particularly at the N-solution, the slope of the tangent to the Pareto curve is $-\frac{\alpha_1}{\alpha_2}$. This equals, according to equation (6.3.3), $-\frac{d_2-J_2^N}{d_1-J_1^N}$. □

We conclude this section with an interpretation of the N-solution which typically fits in the policy coordination literature. In order to make comparisons between the possible utilities of the players one can replace the scale invariance assumption in the axiomatic approach with the assumption that interpersonal utility is comparable. Since interpersonal utility is comparable, it is possible to interpret the relationship

$$\alpha_1^N(d_1 - J_1^N) = \alpha_2^N(d_2 - J_2^N),$$

in the two-player case, in the following way. The player who gains more from playing cooperatively is more willing to accept a smaller welfare weight in the cooperative cost function than the player who gains less. Alternatively, the player who gains less may demand a higher welfare weight by threatening not to cooperate, knowing that the potential loss from no agreement is larger for the other player. This interpretation facilitates a more general interpretation of the N-solution. To that end recall that each player faces the maximization problem:

$$\max_u \ d_i - J_i(u), \quad i = 1, 2.$$

The Pareto solutions of this two-player maximization problem can be found by maximizing w.r.t. $\alpha \in \mathcal{A}$ the cost function

$$\alpha_1(d_1 - J_1(u)) + \alpha_2(d_2 - J_2(u)). \tag{6.3.8}$$

Now both players simultaneously determine α_i in the following way. They agree that the more gain a player receives the less weight they will get in the minimization problem. They formalize this agreement by giving player 1 a weight of $(d_2 - J_2)$ and player 2 a

---

[5]In Douven (1995) it is shown that in general (under some smoothness conditions) $\frac{\partial J_i^*}{\partial J_j} = -\frac{\alpha_j}{\alpha_i}$ where $(J_1^*, \ldots, J_N^*)$ is the point on the Pareto frontier obtained by the minimization of $\sum_{i=1}^N \alpha_i J_i(u)$.

weight of $(d_1 - J_1)$. If we substitute these weights in the minimization problem (6.3.8), we get:

$$\max_u \ (d_2 - J_2(u))(d_1 - J_1(u)) + (d_1 - J_1(u))(d_2 - J_2(u)).$$

which gives us back the original Nash bargaining solution.

This idea can easily be extended to the N-player case. In that case the weight α_i of player $i = 1, \cdots, N$ is determined by the product: $\prod_{j \neq i}(d_j - J_j(u))$. So, the weight a player gets in the minimization problem which determines the N-solution is characterized by the product of the gains of the other players.

6.4 Numerical solution

In this section we will briefly outline some algorithms to calculate the N-solution and the K-solution. First, consider the N-solution. A major advantage of the relationship specified in equation (6.3.4) is that numerical calculation in real problems becomes much easier. Before explaining and comparing an algorithm based on this relationship, we will first give a brief description of the traditional approach. Since each point of the Pareto frontier is uniquely determined by a point $\alpha \in \mathcal{A}$ in practice the maximization algorithm contains the following steps.

Algorithm 6.1

General algorithm to calculate the N-solution

Step 1 Start with an initial $\alpha^0 \in \mathcal{A}$. A good guess is often $\alpha^0 = \left(\frac{1}{N}, \ldots, \frac{1}{N}\right)$.

Step 2 Compute

$$u^*(\alpha^0) = \arg\min_{u \in \mathcal{U}} \left\{ \sum_{i=1}^{N} \alpha_i^0 J_i(u) \right\}.$$

Step 3 Verify whether $J_i(u^*) \leq d_i$, $i = 1, \ldots, N$. If not, use this result for making a new guess for an initial value α^0 and return to Step 2.

Step 4 Check whether this $J_i(u^*)$, $i = 1, \ldots, N$, maximizes the Nash product (6.3.1).

Step 5 If Step 4 gives an affirmative answer, terminate the algorithm, otherwise calculate a new α^0 according to a certain decision rule and go back to Step 2. □

Step 3 is included to make sure that one always starts in the bargaining set, i.e. the set where for all players $J_i(u^*)$ is preferred over d_i. If one skips this step, it is possible that one will get stuck with a nonadmissible α in Step 4. This algorithm description is typical for problems of finding maximum points of a constrained multivariable function by iterative methods. Most of these algorithms are already implemented in existing computer packages and the type of problems are generally referred to as constraint non-linear optimization problems in the numerical literature. Since, as often occurs in applications, the Pareto frontier can be very flat the solution of this kind of problem is not straight-forward, even if we have a convex surface. However, the existence of relationship (6.3.4) facilitates the following approach.

Algorithm 6.2

Algorithm to calculate the N-solution using the bargaining weights property

Step 1 Start with an initial $\alpha^0 \in \mathcal{A}$. A good guess is often $\alpha^0 = \left(\frac{1}{N}, \ldots, \frac{1}{N}\right)$.

Step 2 Compute

$$u^*(\alpha^0) = \arg\min_{u \in \mathcal{U}} \left\{ \sum_{i=1}^{N} \alpha_i^0 J_i(u) \right\}.$$

Step 3 Verify whether $J_i(u^*) \leq d_i$, $i = 1, \ldots, N$. If not, then there is an i_0 for which $J_{i_0}(u^*) > d_{i_0}$. In that case update $\alpha_{i_0}^0 := \alpha_{i_0}^0 + 0.01$, $\alpha_i^0 := \alpha_i^0 - \frac{0.01}{N-1}$, for $i \neq i_0$ and return to Step 2.

Step 4 Calculate

$$\tilde{\alpha}_j^N = \frac{\prod_{i \neq j}(d_i - J_i(u^*(\alpha^0)))}{\sum_{i=1}^{N} \prod_{k \neq i}(d_k - J_k(u^*(\alpha^0)))}, \quad j = 1, \ldots, N.$$

Step 5 If $|\tilde{\alpha}_i^N - \alpha_i^0| < 0.01$, $i = 1, \ldots, N$, then terminate the algorithm and set $\alpha^N = \tilde{\alpha}^N$. Else $\alpha_i^0 := 0.8\alpha_i^0 + 0.2\tilde{\alpha}_i^N$ and return to Step 2. □

Obviously, the numbers 0.01 and 0.8 in the above algorithm are, to a certain extent, chosen arbitrarily and depend on the goals of the user. In Step 5 we use the update $\alpha_i^0 := 0.8\alpha_i^0 + 0.2\tilde{\alpha}_i^N$ instead of the more intuitively appealing update $\alpha_i^0 := \tilde{\alpha}_i^N$. This is to prevent too large steps in the update process, which might result in a vector α^0 for which the inequalities $J_i(u^*(\alpha^0)) \leq d_i$ might no longer be satisfied.

A proof that this algorithm always converges is lacking. Simulation studies suggest, however, (see Douven (1995) and van Aarle et al. (2001)) that the algorithm works well and converges in quite a few iterations.

Algorithm 6.2 is referred to in the numerical literature as a non-linear equations problem. There are many solution methods for these kind of problems, such as the Gauss–Newton algorithm or the line-search algorithm (for example, Stoer and Bulursch (1980)). The main difference between both Algorithms 6.1 and 6.2 is that for large problems the non-linear equations problem that has to be solved in Algorithm 6.2 requires much less computer time than solving the constrained maximization problem in Algorithm 6.1. This is because in Algorithm 6.2 we do not have to verify whether the Nash product (6.3.1) really is maximized and, secondly, Step 4 of the algorithm automatically takes care of the fact that $\alpha^0 \in \mathcal{A}$.

Next consider the computation of the Kalai–Smorodinsky solution. The following global algorithm provides a method to calculate this solution.

Algorithm 6.3

Algorithm to calculate the K-solution

Step 1 Compute a control vector $v_1 \in \mathcal{U}$ such that $J_j(v_1) = d_j$, $j = 2, \cdots, N$; denote $I_1 := J_1(v_1)$. Similarly, compute a vector $v_2 \in \mathcal{U}$ such that $J_j(v_2) = d_j$, $j = 1, 3, \cdots, N$; denote $I_2 := J_2(v_2)$. etc.

Step 2 Denote the ideal point by $I := (I_1, \cdots, I_N)$. Calculate the intersection point between the Pareto curve and the line which connects I and d. $\qquad\square$

Algorithm 6.3 requires the computation of $N + 1$ non-linear constrained equations problems. In practice the computer time involved for computing each of these non-linear constrained problems is about equal to the computer time involved for computing the N-solution using Algorithm 6.2. Therefore, for large problems it takes much more time to compute the K-solution than the N-solution.

Finally, notice that the calculation of the egalitarian solution requires the solution of one non-linear constrained equations problem. The involved computer time to calculate this E-solution approximately equals that of calculating the N-solution.

6.5 Notes and references

For this chapter the following references were consulted: Douven (1995), Thomson (1994), Weeren (1995) and de Zeeuw (1984).

In literature the question of implementability of the equilibrium concepts has also been raised. That is, suppose that some equilibrium solution has been selected as embodying society's objectives, does there exist a game whose equilibrium outcome always yields the desired utility allocations? For instance, Howard (1992) showed the implementability of the Nash solution, whereas Moulin (1984) showed this for the Kalai–Smorodinsky solution.

References on bargaining theory are van Damme (1991), Fudenberg and Tirole (1991), Osborne and Rubinstein (1991) and Thomson (1994).

Another strand of literature that deals with how to assign the profits of the outcome of a cooperative game between groups of the players are the Transferable Utility (TU) games. In this literature it is assumed that joint profits are freely transferable beween the players. Well-known allocation rules are the Shapley value (1953), the nucleolus introduced by Schmeidler (1969) and the τ-value introduced by Tijs (1981). In Tijs (2004) one can find an overview on static cooperative game theory.

A survey on the use of cooperative solution concepts in the theory of differential games can be found in Haurie (2001). Some well-known references in this area are Haurie and

Tolwinski (1985) and Tolwinski, Haurie and Leitmann (1986). Typical questions addressed in this literature are how to deal with the fact that if one applies the static bargaining concepts directly to a differential game, the solution is only valid for the initial state.

6.6 Exercises

1. Consider two firms who want to maximize their profits given by

$$J_1 := 10y_1 - 4y_1^2 - y_2^2 \text{ and } J_2 := 8y_2 - 2y_2^2 - y_1^2,$$

respectively. Here, y_i is the production of firm i, $i = 1, 2$.

(a) Show that the set of Pareto solutions is $(y_1^*, y_2^*) = (\frac{5}{3} - \frac{\frac{5}{9}}{\lambda+\frac{1}{3}}, 4 - \frac{8}{4-2\lambda})$, where $\lambda \in [0, 1]$.

(b) Show that the Pareto frontier is given by $y_1^* = \frac{30}{21} + \frac{60}{21}\frac{1}{7y_2^*-16}$, where $y_2^* \in [0, 2]$.

(c) Plot the Pareto frontier.

2. Assume that the cost function of player i equals J_i, where

$$J_1 := (x_1 + x_2)^2 + (x_1 + 1)^2 \text{ and } J_2 := (x_1 - x_2)^2 + (x_2 - 1)^2,$$

respectively. Here x_i is a variable player i can control, $i = 1, 2$.

(a) Show that the Pareto frontier is parameterized by

$$(x_1^*, x_2^*) = \frac{1}{-\lambda^2 + \lambda + 1}(\lambda^2 - 3\lambda + 1, -\lambda^2 - \lambda + 1), \ \lambda \in [0, 1].$$

(b) Plot numerically the graph of the Pareto frontier.

3. Assume that the profit function of firm i equals J_i, where

$$J_1 := 84x_1 - x_1^2 - x_1x_2 - 24 \text{ and } J_2 := 92x_2 - \frac{3}{2}x_2^2 - x_1x_2,$$

respectively. Here x_i is the production of firm i, $i = 1, 2$.

(a) Show that the following solutions are Pareto-optimal solutions

$$(x_1^*, x_2^*) = \frac{1}{6\lambda(1-\lambda)-1}((1-\lambda)(252\lambda - 92), \lambda(100 - 184\lambda)), \ \lambda \in \left(\frac{1}{2} - \frac{1}{6}\sqrt{3}, \frac{1}{2} + \frac{1}{6}\sqrt{3}\right).$$

(b) Plot the set of Pareto solutions in (a) numerically.

4. Consider the minimization of

$$J_i := \int_0^\infty x^2(t) + u_i^2(t)dt, \text{ subject to } \dot{x}(t) = x(t) + u_1(t) + u_2(t), \ x(0) = x_0.$$

Here, u_i is the control variable of player i.

(a) Show that for $\lambda \in (0,1)$ the following control actions yield Pareto-optimal solutions.

$$u_1^*(t) = -\left\{1 - \lambda + \sqrt{(1-\lambda)^2 + \frac{1-\lambda}{\lambda}}\right\}x^*(t)$$

$$u_2^*(t) = -\left\{\lambda + \sqrt{\lambda^2 + \frac{\lambda}{1-\lambda}}\right\}x^*(t),$$

where $x^*(t)$ is the solution of $\dot{x}^*(t) = -acl\, x^*(t)$, $x^*(0) = x_0$, with $acl := \sqrt{1 + \frac{1}{\lambda(1-\lambda)}}$.

(b) Show that for $\lambda \in (0,1)$

$$J_1^* := \frac{1 + (1-\lambda)^2(1+acl)^2}{2acl}x_0^2 \quad \text{and} \quad J_2^* := \frac{1 + \lambda^2(1+acl)^2}{2acl}x_0^2$$

yield Pareto optimal solutions.

(c) Plot the set of Pareto solutions in (b) numerically if $x_0 = 1$, $x_0 = 2$ and $x_0 = 4$, respectively. What are the consequences of a different initial state x_0 for the graph of the Pareto solutions in general?

5. Consider the minimization of

$$J_i := \int_0^{\frac{\pi}{4}} -x^2(t) + u_i^2(t)dt, \text{ subject to } \dot{x}(t) = u_1(t) + u_2(t), \ x(0) = 1.$$

Here, u_i is the control variable of player i.

(a) Show that for $\lambda \in (0,1)$ the following control actions yield Pareto-optimal solutions.

$$u_1^*(t) = -\frac{1}{\lambda}k(t)x^*(t)$$

$$u_2^*(t) = -\frac{1}{1-\lambda}k(t)x^*(t),$$

where $k(t)$ is the solution of the differential equation

$$\dot{k}(t) = b(\lambda)k^2(t) + 1, k\left(\frac{\pi}{4}\right) = 0, \quad \text{with} \quad b(\lambda) := \frac{1}{\lambda(1-\lambda)}$$

and $x^*(t)$ is the solution of $\dot{x}^*(t) = -b(\lambda)k(t)x^*(t)$, $x^*(0) = 1$.

(b) Show (e.g. by substitution) that

$$k(t) = \sqrt{\lambda(1 - \lambda)} \tan\left(\sqrt{b(\lambda)}\left(t - \frac{\pi}{4}\right)\right) \quad \text{and} \quad x^*(t) = \cos\left(\sqrt{b(\lambda)}\left(t - \frac{\pi}{4}\right)\right).$$

(c) Show that for $\lambda \in (0, 1)$ the solutions

$$J_1^* = \left(-1 + \frac{1}{2\lambda}\right)\frac{\pi}{4} - \frac{\sqrt{1 - \lambda}}{4\sqrt{\lambda}} \sin\left(\sqrt{b(\lambda)}\frac{\pi}{2}\right)$$

$$J_2^* = \left(-1 + \frac{1}{2(1 - \lambda)}\right)\frac{\pi}{4} - \frac{\sqrt{\lambda}}{4\sqrt{1 - \lambda}} \sin\left(\sqrt{b(\lambda)}\frac{\pi}{2}\right)$$

are Pareto optimal.

(d) Plot the set of Pareto solutions in (c) numerically.

6. Consider a cost minimization problem where the Pareto frontier is described as a part of the curve

$$(x_1 - 7)^2 + 2(x_2 - 8)^2 = 18.$$

Here, x_i is the cost incurred by player i, $i = 1, 2$.

(a) Draw the graph of the above curve and determine the Pareto frontier for this cost problem.

(b) Assume that the threatpoint for this problem is $(x_1, x_2) = (7, 8)$. Determine graphically the feasible set of the bargaining game associated with this problem.

(c) Find analytically the Nash bargaining solution of the bargaining problem in (b) by solving the constrained optimization problem

$$\max(x_1 - 7)(x_2 - 8) \text{ such that } (x_1 - 7)^2 + 2(x_2 - 8)^2 = 18.$$

(d) Find analytically the Kalai–Smorodinsky solution for the bargaining problem in (b). (Hint: first determine the ideal point, and next find the intersection point of the curve with the line through the ideal point and the threatpoint.)

(e) Find analytically the egalitarian solution for the bargaining problem in (b). (Hint: first determine the $45°$ line through the threatpoint.)

7. Reconsider the cost minimization problem of Exercise 6 above. Answer the same questions as in Exercise 6(b)–(e), but now w.r.t. the threatpoint $(x_1, x_2) = (7, 7)$. Compare your answers with those of Exercise 6. Explain the differences for the Kalai–Smorodinsky solution and the egalitarian solution from the underlying bargaining axioms.

8. Assume that in Exercises 6 and 7 above the cost functions are given by the convex functions $x_i(u_1, u_2)$, where u_i is the control variable of player i, $i = 1, 2$. Determine for both exercises the weights α_i, $i = 1, 2$, in the cooperative cost function $\alpha_1 x_1(u_1, u_2) + \alpha_2 x_2(u_1, u_2)$ that yield the Nash bargaining solution. Compare the results for both exercises and give an intuitive explanation for the differences.

9. Assume that player i likes to minimize w.r.t. u_i

$$J_i = \alpha_i \int_0^T \{x^T(t)Qx(t) + u_1^T(t)R_1u_1(t) + u_2^T(t)R_2u_2(t)\}dt$$

subject to $\dot{x}(t) = Ax(t) + B_1u_1(t) + B_2(t)u_2(t), \quad x(0) = x_0,$

where $Q > 0$, $R_i > 0$ and $\alpha_i > 0$, $i = 1, 2$. Show that the Pareto frontier consists in this case of one point.

10. Consider the interaction of fiscal stabilization policies of two countries. Assume that the competitiveness between both countries is described by the differential equation

$$\dot{s}(t) = -as(t) + f_1(t) - f_2(t), \quad s(0) = s_0,$$

where $a > 0$, the variable $s(t)$ denotes the difference in prices between both countries at time t and $f_i(t)$ is the fiscal deficit set by the fiscal authority in country i, $i = 1, 2$. Each fiscal authority seeks to minimize the following intertemporal loss function that is assumed to be quadratic in the price differential and fiscal deficits,

$$J_i = \int_0^\infty \{q_i s^2(t) + r_i f_i^2(t)\}dt, \quad i = 1, 2.$$

Assume that both countries agree to cooperate in reducing the initial price differential between both countries.

(a) Indicate how the countries can determine the set of Pareto efficient strategies. Explain your answer.

(b) Reformulate the problem in (a) as a standard linear quadratic control problem and determine all Pareto efficient strategies (if possible).

(c) Consider the case that both countries are completely symmetric in their preferences. Determine the Pareto frontier in that case. What are the corresponding Pareto efficient strategies?

(d) Assume $a = 1$, $q_1 = r_1 = 1$ and $q_2 = 0.5$, $r_2 = 2$. Plot numerically the Pareto frontier.

(e) Choose a threatpoint in (d) and determine geometrically the Nash, the Kalai–Smorodinsky and egalitarian bargaining solutions, respectively. Which solution is in your opinion the most appropriate one here (if any of these)?

(f) The same question as in (d) and (e) but now for $q_2 = 2$ and $r_2 = 0.5$ and the rest of the parameters unchanged. Can you comment on the differences between the answers?

11. Consider two interacting countries where the government of each country likes to stabilize the growth of national income around its corresponding equilibrium path. Let x_i denote the deviation of national income from its equilibrium path for country i, and u_i the fiscal policy used by the government in country i to realize this goal. Moreover, assume that both countries are convinced that large discrepancies between

the deviations of growth in both countries is also not desirable (due to the migration problems this might cause, for example). This problem is formalized by government i as that they like to minimize w.r.t. u_i, $i = 1, 2$,

$$J_1 = \int_0^\infty \{x^T(t) \begin{bmatrix} 2 & -1 \\ -1 & 1 \end{bmatrix} x(t) + u_1^2(t)\} dt$$

$$\text{and } J_2 = \int_0^\infty \{x^T(t) \begin{bmatrix} 1 & -1 \\ -1 & 2 \end{bmatrix} x(t) + u_2^2(t)\} dt$$

$$\text{subject to } \dot{x}(t) = \begin{bmatrix} -3 & 1 \\ 2 & -2 \end{bmatrix} x(t) + \begin{bmatrix} 2 \\ 1 \end{bmatrix} u_1(t) + \begin{bmatrix} 1 \\ 2 \end{bmatrix} u_2(t),$$

$$x_0 = \begin{bmatrix} 1 \\ 1 \end{bmatrix}, \text{ respectively.}$$

(a) Show that for all $\lambda \in (0, 1)$ the actions (u_1^*, u_2^*) minimizing $\lambda J_1 + (1 - \lambda) J_2$ are

$$\begin{bmatrix} u_1^*(t) \\ u_2^*(t) \end{bmatrix} = - \begin{bmatrix} \dfrac{2}{\lambda} & \dfrac{1}{\lambda} \\ \dfrac{1}{1-\lambda} & \dfrac{2}{1-\lambda} \end{bmatrix} K x^*(t),$$

where K is the stabilizing solution of the algebraic Riccati equation

$$0 = - \begin{bmatrix} -3 & 1 \\ 2 & -2 \end{bmatrix}^T X - X \begin{bmatrix} -3 & 1 \\ 2 & -2 \end{bmatrix} + X \begin{bmatrix} \dfrac{4-3\lambda}{\lambda(1-\lambda)} & \dfrac{2}{\lambda(1-\lambda)} \\ \dfrac{2}{\lambda(1-\lambda)} & \dfrac{1+3\lambda}{\lambda(1-\lambda)} \end{bmatrix} X - \begin{bmatrix} 1+\lambda & -1 \\ -1 & 2-\lambda \end{bmatrix},$$

and $x^*(t)$ is the solution of the differential equation

$$\dot{x}^*(t) = \left(\begin{bmatrix} -3 & 1 \\ 2 & -2 \end{bmatrix} - \dfrac{1}{\lambda(1-\lambda)} \begin{bmatrix} 4-3\lambda & 2 \\ 2 & 1+3\lambda \end{bmatrix} K \right) x^*(t) =: A_{cl} x^*(t), \ x(0) = \begin{bmatrix} 1 \\ 1 \end{bmatrix}.$$

(b) Let

$$V_1 := \begin{bmatrix} 2 & -1 \\ -1 & 1 \end{bmatrix} + K \begin{bmatrix} \dfrac{4}{\lambda^2} & \dfrac{2}{\lambda(1-\lambda)} \\ \dfrac{2}{\lambda(1-\lambda)} & \dfrac{1}{(1-\lambda)^2} \end{bmatrix} K \text{ and}$$

$$V_2 := \begin{bmatrix} 1 & -1 \\ -1 & 2 \end{bmatrix} + K \begin{bmatrix} \dfrac{1}{\lambda^2} & \dfrac{2}{\lambda(1-\lambda)} \\ \dfrac{2}{\lambda(1-\lambda)} & \dfrac{4}{(1-\lambda)^2} \end{bmatrix} K.$$

Show that $J_i(u_1^*, u_2^*) = x_0^T M_i x_0$, where M_i is the solution of the Lyapunov equation

$$A_{cl}^T M_i + M_i A_{cl} + V_i = 0, \ i = 1, 2.$$

(c) Assume that the threatpoint of the bargaining game is the non-cooperative open-loop Nash solution (see Chapter 7) of this game: $(d_1, d_2) = (0.185, 0.365)$. Use Algorithm 6.2 to find numerically the Nash bargaining solution of this game. To calculate the involved stabilizing solution of the Riccati equation and the Lyapunov equation use, the functions ARE and LYAP of the MATLAB control toolbox.

(d) Calculate the percentage gains of cooperation, measured by $\frac{d_i - J_i^*}{d_i} * 100$, for both governments.

6.7 Appendix

Proof of Corollary 6.6

In Theorem 6.4 we did not specify the strategy spaces. In fact they can be chosen to be any convex subset of a vector space[6]. Now, our set of admissible control functions $\mathcal{U}$ is a vector space if we consider as the addition rule for two control functions $u(.)$ and $v(.)$ the function whose value at t is $(u + v)(t) := u(t) + v(t)$, and likewise for a scalar λ and a $u \in \mathcal{U}$, λu as the function whose value at t is $\lambda u(t)$. In particular the set $\mathcal{U}$ is convex.

So, all we have to show here is that the function $J_i(u)$ defined by

$$J_i(u) := \int_0^T \{x^T(t) Q_i x(t) + u_i^T(t) R_i u_i(t)\} dt + x^T(T) Q_{T,i} x(T),$$

with

$$x(t) = e^{At} x_0 + \int_0^t e^{A(t-s)} Bu(s) ds,$$

is convex. That is, $J_i(.)$ satisfies

$$J(\lambda u + (1 - \lambda)v) \leq \lambda J(u) + (1 - \lambda)J(v), \ 0 \leq \lambda \leq 1,$$

where, for notational convenience, we dropped the index i; $u = [u_1, \ldots, u_N]^T$ (and v defined similarly); and $B = [B_1, \ldots, B_N]$.

---

[6]A vector space is a nonempty set of objects, called vectors, on which are defined two operations, called addition and multiplication by scalars, subject to a number of axioms. For a proper definition see, for example, Lay (2003).

Next, introduce the notation

$$x_u(t) := e^{At}x_0 + \int_0^t e^{A(t-s)}Bu(s)ds.$$

Then,

$$J(\lambda u + (1-\lambda)v) = \int_0^T \{[x_{\lambda u+(1-\lambda)v}(t)]^T Q_i[x_{\lambda u+(1-\lambda)v}(t)] + (\lambda u_i(t) + (1-\lambda)v_i(t))^T R_i(\lambda u_i(t)$$

$$+ (1-\lambda)v_i(t))\}dt + [x_{\lambda u+(1-\lambda)v}(T)]^T Q_{T,i}[x_{\lambda u+(1-\lambda)v}(T)]$$

$$= \int_0^T \{[\lambda x_u(t) + (1-\lambda)x_v(t)]^T Q_i[\lambda x_u(t) + (1-\lambda)x_v(t)]$$

$$+ (\lambda u_i(t) + (1-\lambda)v_i(t))^T R_i(\lambda u_i(t) + (1-\lambda)v_i(t))\}dt$$

$$+ [\lambda x_u(T) + (1-\lambda)x_v(T)]^T Q_{T,i}[\lambda x_u(T) + (1-\lambda)x_v(T)].$$

Since all matrices Q_i, R_i and $Q_{T,i}$ are positive semi-definite, we can factorize them as $\tilde{Q}_i^2$, $\tilde{R}_i^2$ and $\tilde{Q}_{T,i}^2$, respectively, for some positive semi-definite matrices $\tilde{Q}_i$, $\tilde{R}_i$ and $\tilde{Q}_{T,i}$. So, introducing $\tilde{x}_w(t) := \tilde{Q}_i(e^{At}x_0 + \int_0^t e^{A(t-s)}Bw(s)ds)$, $\tilde{w}_i := \tilde{R}_iw_i(t)$ and $\tilde{x}_w(T) := \tilde{Q}_i(e^{AT}x_0 + \int_0^T e^{A(T-s)}Bw(s)ds)$, we can rewrite $J(\lambda u + (1-\lambda)v)$ as

$$\int_0^T \{[\lambda\tilde{x}_u(t) + (1-\lambda)\tilde{x}_v(t)]^T[\lambda\tilde{x}_u(t) + (1-\lambda)\tilde{x}_v(t)] + [\lambda\tilde{u}_i + (1-\lambda)\tilde{v}_i]^T$$

$$\times[\lambda\tilde{u}_i + (1-\lambda)\tilde{v}_i]\}dt + [\lambda\tilde{x}_u(T) + (1-\lambda)\tilde{x}_v(T)]^T[\lambda\tilde{x}_u(T) + (1-\lambda)\tilde{x}_v(T)].$$

Next notice that the standard Euclidean norm $\|x\|^2 := x^Tx$ satisfies the triangular inequality $\|x + y\| \le \|x\| + \|y\|$. Therefore $J(\lambda u + (1-\lambda)v)$ can be rewritten, and next estimated as

$$\int_0^T \{\|\lambda\tilde{x}_u(t) + (1-\lambda)\tilde{x}_v(t)\|^2 + \|\lambda\tilde{u}_i + (1-\lambda)\tilde{v}_i\|^2\}dt + \|\lambda\tilde{x}_u(T) + (1-\lambda)\tilde{x}_v(T)\|^2$$

$$\le \int_0^T \{(\|\lambda\tilde{x}_u(t)\| + \|(1-\lambda)\tilde{x}_v(t)\|)^2 + (\|\lambda\tilde{u}_i\| + \|(1-\lambda)\tilde{v}_i\|)^2\}dt + (\|\lambda\tilde{x}_u(T)\|$$

$$+ \|(1-\lambda)\tilde{x}_v(T)\|)^2 \le \int_0^T \{\|\lambda\tilde{x}_u(t)\|^2 + \|(1-\lambda)\tilde{x}_v(t)\|^2 + \|\lambda\tilde{u}_i\|^2 + \|(1-\lambda)\tilde{v}_i\|^2\}dt$$

$$+ \|\lambda\tilde{x}_u(T)\|^2 + \|(1-\lambda)\tilde{x}_v(T)\|^2 \le \int_0^T \{\lambda\|\tilde{x}_u(t)\|^2 + (1-\lambda)\|\tilde{x}_v(t)\|^2 + \lambda\|\tilde{u}_i\|^2$$

$$+ (1-\lambda)\|\tilde{v}_i\|^2\}dt + \lambda\|\tilde{x}_u(T)\|^2 + (1-\lambda)\|\tilde{x}_v(T)\|^2 = \int_0^T \{\lambda x_u^T(t)Q_ix_u(t)$$

$$+ (1-\lambda)x_v^T(t)Q_ix_v(t) + \lambda u_i^T(t)R_iu_i(t) + (1-\lambda)v_i^T(t)R_iv_i(t)\}dt + \lambda x_u^T(T)Q_{i,T}x_u(T)$$

$$+ (1-\lambda)x_v^T(T)Q_{i,T}x_v(T) = \lambda J(u) + (1-\lambda)J(v),$$

where we used the notation $x_w(t)$ to denote the state of the system at time t if the control $w(.)$ is used. $\qquad\square$

7

Non-cooperative open-loop information games

As already mentioned in Chapter 1, non-cooperative differential games were first introduced in Isaacs (1954), within the framework of two-person zero-sum games. Whereas
nonzero-sum differential games were introduced in the papers by Starr and Ho (1969a, b).
In the next chapters we study the special class of non-cooperative linear-quadratic differential games. The dynamics of these games are described by the linear differential
equation

$$\dot{x}(t) = Ax(t) + B_1 u_1(t) + \cdots B_N u_N(t), \quad x(0) = x_0. \tag{7.0.1}$$

Each player has a quadratic cost function:

$$J_i(u_1, \ldots, u_N) = \int_0^T \left\{ x^T(t) Q_i x(t) + \sum_{j=1}^N u_j^T(t) R_{ij} u_j(t) \right\} dt + x^T(T) Q_{iT} x(T), \quad i = 1, \ldots, N,$$

$$\tag{7.0.2}$$

in which all matrices are symmetric, and R_{ii} are positive definite.

The objective for each player is the minimization of his own cost function by choosing
appropriate controls for the underlying linear dynamical system. The non-cooperative
aspect implies that the players are assumed not to collaborate in trying to attain this
goal. Depending on the information players have on the game, denoted by $\eta_i(t), t \in [0, T]$,
and the set of strategies the players like to choose from (which depends obviously on
the information the players have on the game), denoted by Γ_i, the actions (or controls)
of the players are determined by the relations

$$u_i = \gamma_i(\eta_i), \quad \text{where } \gamma_i \in \Gamma_i.$$

Substitution of these controls into (7.0.2) and (7.0.1) shows that the cost functions J_i
depend on the information players have on the game and their strategy space. Depending
on these 'parameters' the value of the cost function obviously also depends for each
player i on the pursued actions of the other players. So, if for example we consider

LQ Dynamic Optimization and Differential Games   J. Engwerda
© 2005 John Wiley & Sons, Ltd

two players, the question arises as to which actions will be played. Now, assume that both players may propose in turn an action in a negotiation process that proceeds the actual implementation of the control. So, the players can react to each other's proposal. Then, assuming that ultimately the proposition process ends, it seems reasonable that in that final situation each player likes to play an action which he cannot improve upon anymore. That is, any unilateral deviation from the action he has in mind will lead to a worse value of his cost function J_i. We illustrate this idea in the following example.

Example 7.1

Consider two firms Ping Ping and Pong Pong producing ping-pong balls for the Chinese market. The inverse demand function is

$$p = 92 - (x_1 + x_2),$$

where p is the price of a ping-pong ball and x_i is the produced quantity (in millions) of firm i, $i = 1, 2$. The cost functions of Ping Ping and Pong Pong are

$$F_1(x_1) = 8x_1 + 24$$

and

$$F_2(x_2) = \frac{1}{2}x_2^2,$$

respectively. Consequently, the profit function π_1 of Ping Ping is

$$\pi_1(x_1, x_2) = px_1 - (8x_1 + 24)$$
$$= -x_1^2 - x_1 x_2 + 84x_1 - 24$$

and the profit function π_2 of Pong Pong is

$$\pi_2(x_1, x_2) = px_2 - \frac{1}{2}x_2^2$$
$$= -\frac{3}{2}x_2^2 - x_1 x_2 + 92x_2.$$

Therefore, for a fixed number $\bar{x}_2$ Pong Pong produces, Ping Ping's optimal production of balls is determined by maximizing $\pi_1(x_1, \bar{x}_2)$. Differentiation of π_1 shows that this maximum number is

$$x_1^* = \begin{cases} -\frac{1}{2}\bar{x}_2 + 42 & \text{if } 0 \le \bar{x}_2 \le 84, \\ 0 & \text{if } \bar{x}_2 > 84. \end{cases}$$

Similarly, one obtains for a fixed number $\bar{x}_1$ Ping Ping produces, that Pong Pong's optimal production is given by

$$x_2^* = \begin{cases} -\frac{1}{3}\bar{x}_1 + 30\frac{2}{3} & \text{if } 0 \le \bar{x}_1 \le 92, \\ 0 & \text{if } \bar{x}_1 > 92. \end{cases}$$

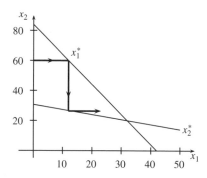

Figure 7.1   Non-cooperative Nash equilibrium

So, two curves $x_1^*(x_2)$ and $x_2^*(x_1)$ are obtained which determine the optimal 'reaction' of Ping Ping if Pong Pong produces x_2, and the optimal 'reaction' of Pong Pong if Ping Ping produces x_1, respectively. Therefore, these curves are usually called the reaction curves of both players. Both reaction curves are illustrated in Figure 7.1. Now, assume that Ping Ping has a spy who is on the board of Pong Pong, and similarly Pong Pong has a spy who is on the board of Ping Ping. Assume that Ping Ping's spy phones to say that Pong Pong wants to produce $x_2 = 60$. Ping Ping's board of directors reacts to this information by suggesting a production of $x_1 = 12$. Pong Pong's spy phones this number to his board of directors and they decide to lower their production plan to $x_2 = 26\frac{2}{3}$, etc. We illustrate this process in Figure 7.1, and see that ultimately this process converges to the intersection point $(32, 20)$ of both reaction curves.   □

Nash (1950a, 1951) argued that this is a natural concept to be used in a non-cooperative context. He defined the non-cooperative Nash equilibrium in the following way.

Definition 7.1

An admissible set of actions $(u_1^*, \ldots, u_N^*)$ is a **Nash equilibrium** for an N-player game, where each player has a cost function $J_i(u_1, \ldots, u_N)$, if for all admissible $(u_1, \ldots, u_N)$ the following inequalities hold:

$$J_i(u_1^*, \ldots, u_{i-1}^*, u_i^*, u_{i+1}^*, \ldots, u_N^*) \leq J_i(u_1^*, \ldots, u_{i-1}^*, u_i, u_{i+1}^*, \ldots, u_N^*), \quad i = 1, \ldots, N. \quad □$$

Here admissibility is meant in the sense above mentioned. That is, $u_i(.)$ belongs to some restricted set, where this set depends on the information, $\eta_i(.)$, which players have on the game and the set of strategies, Γ_i, which the players like to use to control the system.

So, the Nash equilibrium is defined such that it has the property that there is no incentive for any unilateral deviation by any one of the players. Notice that in general one cannot expect to have a unique Nash equilibrium. Moreover, it is easily verified that whenever a set of actions $(u_1^*, \ldots, u_N^*)$ is a Nash equilibrium for a game with cost functions J_i, $i = 1, \ldots, N$, these actions also constitute a Nash equilibrium for the game with cost functions $\alpha_i J_i$, $i = 1, \ldots, N$, for every choice of $\alpha_i > 0$.

In the rest of this chapter we will be dealing with the so-called **open-loop** information structure. That is, the case where every player knows at time $t \in [0, T]$ just the initial state x_0 and the model structure (usually denoted with $\eta_i(t) = x_0$). This scenario can be interpreted as being that the players simultaneously determine their actions, and then submit their actions to some authority who then enforces these plans as binding commitments.

7.1 Introduction

Particularly in economics, there is an increasing interest in the study of problems using a dynamic game theoretical setting. In the area of environmental economics and macroeconomic policy coordination this is a very natural framework in which to model problems (for example, de Zeeuw and van der Ploeg (1991), Mäler (1992), Kaitala, Pohjola and Tahvonen (1992), Dockner *et al.* (2000), Tabellini (1986), Fershtman and Kamien (1987), Petit (1989), Levine and Brociner (1994), van Aarle, Bovenberg and Raith (1995), Douven and Engwerda (1995), van Aarle, Engwerda and Plasmans (2002)). In policy coordination problems, for example, usually two basic questions arise: first, are policies coordinated and, second, which information do the participating parties have. Usually both these points are rather unclear and, therefore, strategies for different possible scenarios are calculated and compared with each other. Often, one of these scenarios is the open-loop scenario. Obviously, since according to this scenario the participating parties cannot react to each other's policies, its economic relevance is mostly rather limited. However, as a benchmark to see how much parties can gain by playing other strategies, it plays a fundamental role. Due to its analytic tractability the open-loop Nash equilibrium strategy is, in particular, very popular for problems where the underlying model can be described by a set of linear differential equations and the individual objectives can be approximated by functions which quadratically penalize deviations from some equilibrium targets. Under the assumption that the parties have a finite-planning horizon, this problem was first modeled and solved in a mathematically rigorous way by Starr and Ho (1969a). However, due to some inaccurate formulations it is, even in current literature, an often encountered misunderstanding that this problem always has a unique Nash equilibrium which can be obtained in terms of the solutions of a set of coupled matrix differential equations resembling (but more complicated than) the matrix Riccati equations which arise in optimal control theory. Eisele (1982), who extended the Hilbert space approach of this problem taken by Lukes and Russell (1971), had already noted that there were some misleading formulations in the literature. However, probably due to the rather abstract approach he took, this point was not noted in the mainstream literature. So, in other words, situations do exist where the set of coupled matrix differential equations has no solution, whereas the game does have an equilibrium. In section 7.2 we will present such an example and use the Hamiltonian approach to analyze the problem. In addition to its simplicity this approach has the advantage that it also permits an elementary study of convergence of the equilibrium actions if the planning horizon expands. As in the theory on optimal control we will show in section 7.6 that under some conditions it can be shown that these actions converge. One nice property of this converged solution is, as we will see, that it is rather easy to calculate and much easier to implement than any finite-planning horizon equilibrium solution.

Apart from this computational point of view that the equilibrium actions are much easier to implement and to analyze than those for a finite-planning horizon, there is also at least one other reason from economics to consider an infinite-planning horizon. In economic growth theory it is usually difficult to justify the assumption that a firm (or government) has a finite-planning horizon T – why should it ignore profits earned after T, or utility of generations alive beyond T.

Two remarkable points we will see in sections 7.4 and 7.6 that may happen are: first, though the problem may have a unique equilibrium strategy for an arbitrary finite-planning horizon, there may exist more than one equilibrium solution for the infinite-planning horizon case; second, the limit of this unique finite-planning horizon equilibrium solution may not be a solution for the infinite-planning horizon problem. On the other hand, we will see that it can easily be verified whether or not the limiting solution of the finite-planning horizon problem also solves the infinite-planning horizon case. Furthermore, it is shown that, if the participating parties discount their future objectives, then the finite-planning horizon equilibrium solution converges to a limit which is generically the unique solution to the infinite-plannning horizon case, if the discount factor is large enough.

Section 7.7 studies the scalar case which is of particular interest for many economic applications. Both necessary and sufficient conditions are presented under which the game will have an open-loop Nash equilibrium for all initial states. Moreover it is shown that, if these conditions are satisfied, the equilibrium is unique if and only if the system matrix A in equation (7.0.1) is non-positive. Furthermore we show that, under a mild regularity condition, the finite-planning equilibrium solution can be obtained by solving the set of Riccati differential equations and that the equilibrium solution converges to a stationary stabilizing feedback policy which also solves the infinite-planning horizon problem.

This chapter concludes by illustrating some of the developed theory in two economic examples.

7.2 Finite-planning horizon

In this section we consider N players who try to minimize their individual quadratic performance criterion (7.0.1), where the planning horizon T is finite. Each player controls a different set of inputs u_i to the single system, described by the differential equation (7.0.2). The players have an open-loop information structure, $\eta_t = x_0, t \in [0, T]$. That is, the players already have to formulate their actions at the moment the system starts to evolve and these actions cannot be changed once the system is running. Therefore, the players have to minimize their performance criterion based on the information that they only know the differential equation and its initial state. So, let

$$\mathcal{U} = \{(u_1(.), \ldots, u_N(.)) \mid (7.0.1) \text{ has a solution in the extended sense on } [0, T]\}.$$

Then,

$$\Gamma_i = \{u_i(.) \mid u_i(.) = f_i(t, x_0) \text{ and } (u_1(.), \ldots, u_N(.)) \in \mathcal{U}\}, \quad i = 1, \ldots, N.$$

Notice that this definition presupposes that players communicate their strategies in the sense that they only choose actions for which the differential equation (7.0.1) has an appropriately defined solution. This condition is trivially satisfied, for example, by assuming from the outset that the players only use bounded piecewise continuous control functions.

We are looking for the Nash equilibria of this game. That is, for the combinations of actions of all players which are secure against any attempt by one player to unilaterally alter his strategy. Or, stated differently, for those set of actions which are such that if one player deviates from his action he will only lose. In the literature on dynamic games this problem is known as the open-loop Nash nonzero-sum linear quadratic differential game and has been analyzed by several authors (for example Starr and Ho (1969a), Simaan and Cruz (1973), Abou-Kandil and Bertrand (1986), Feucht (1994), Başar and Olsder (1999), and Kremer (2002)). To avoid cumbersome notation, we will restrict the analyses to the two-player case and just state for the most important results the general N-player case. So, we concentrate on the system described by:

$$\dot{x} = Ax + B_1 u_1 + B_2 u_2, \quad x(0) = x_0, \tag{7.2.1}$$

where x is the n-dimensional state of the system, u_i is an m_i-dimensional (control) vector player i, $i = 1, 2$, can manipulate, x_0 is the initial state of the system, A, B_1, and B_2 are constant matrices of appropriate dimensions, and $\dot{x}$ denotes the time derivative of x.

The performance criteria player i, $i = 1, 2$, aims to minimize are:

$$J_1(u_1, u_2) := \int_0^T \{x^T(t)Q_1 x(t) + u_1^T(t)R_{11}u_1(t) + u_2^T(t)R_{12}u_2(t)\}dt + x^T(T)Q_{1T}x(T),$$

$$\tag{7.2.2}$$

and

$$J_2(u_1, u_2) := \int_0^T \{x^T(t)Q_2 x(t) + u_1^T(t)R_{21}u_1(t) + u_2^T(t)R_{22}u_2(t)\}dt + x^T(T)Q_{2T}x(T),$$

$$\tag{7.2.3}$$

in which all matrices are symmetric and, moreover, R_{ii} are positive definite.

Using the shorthand notation $S_i := B_i R_{ii}^{-1} B_i^T$, the following theorem can be stated (it is proved in the Appendix at the end of this chapter).

Theorem 7.1

Consider matrix

$$M := \begin{bmatrix} A & -S_1 & -S_2 \\ -Q_1 & -A^T & 0 \\ -Q_2 & 0 & -A^T \end{bmatrix}. \tag{7.2.4}$$

Assume that the two Riccati differential equations,

$$\dot{K}_i(t) = -A^T K_i(t) - K_i(t)A + K_i(t)S_i K_i(t) - Q_i, \quad K_i(T) = Q_{iT}, \quad i = 1, 2, \tag{7.2.5}$$

have a symmetric solution $K_i(.)$ on $[0, T]$. Then, the two-player linear quadratic differential game (7.2.1)–(7.2.3) has an open-loop Nash equilibrium for every initial state x_0 if and only if matrix

$$H(T) := [I \ 0 \ 0]e^{-MT} \begin{bmatrix} I \\ Q_{1T} \\ Q_{2T} \end{bmatrix} \quad (7.2.6)$$

is invertible. Moreover, if for every x_0 there exists an open-loop Nash equilibrium then the solution is unique. The unique equilibrium actions, as well as the associated state trajectory, can be calculated from the linear two-point boundary-value problem

$$\dot{y}(t) = My(t), \quad \text{with } Py(0) + Qy(T) = [x_0^T \ 0 \ 0]^T. \quad (7.2.7)$$

Here

$$P = \begin{bmatrix} I & 0 & 0 \\ 0 & 0 & 0 \\ 0 & 0 & 0 \end{bmatrix} \quad \text{and} \quad Q = \begin{bmatrix} 0 & 0 & 0 \\ -Q_{1T} & I & 0 \\ -Q_{2T} & 0 & I \end{bmatrix}.$$

Denoting $[y_0^T(t), \ y_1^T(t), \ y_2^T(t)]^T := y(t)$, with $y_0 \in \mathbb{R}^n$, and $y_i \in \mathbb{R}^{m_i}$, $i = 1, 2$, the state and equilibrium actions are

$$x(t) = y_0(t) \quad \text{and} \quad u_i(t) = -R_{ii}^{-1}B_i^T y_i(t), \quad i = 1, 2,$$

respectively. $\qquad \qquad \square$

Example 7.2

Consider the game defined by the system

$$\dot{x}(t) = 2x(t) + u_1(t) + u_2(t), \quad x(0) = 1;$$

and cost functions

$$J_1 = \int_0^3 \{x^2(t) + u_1^2(t)\}dt \quad \text{and} \quad J_2 = \int_0^3 \{4x^2(t) + u_2^2(t)\}dt + 5x^2(3).$$

Then, using our standard notation, $A = 2$, $B_i = R_i = 1$ (and thus $S_i = 1$), $i = 1, 2$, $Q_1 = 1$, $Q_2 = 4$, $Q_{1T} = 0$ and $Q_{2T} = 5$.

Since $Q_i \geq 0$ and $Q_{iT} \geq 0$, $i = 1, 2$, the two Riccati differential equations (7.2.5) have a solution on $[0, 3]$ (see Exercise 4, Chapter 5).

Furthermore, with

$$M = \begin{bmatrix} 2 & -1 & -1 \\ -1 & -2 & 0 \\ -4 & 0 & -2 \end{bmatrix},$$

the exponential of $-M$ can be obtained by determining the eigenvalues and corresponding eigenvectors of this matrix. Elementary calculations show that, with

$$S := \begin{bmatrix} -1 & 0 & 5 \\ -1 & -1 & -1 \\ -4 & 1 & -4 \end{bmatrix} \quad \text{and, consequently,} \quad S^{-1} = \frac{1}{30} \begin{bmatrix} -5 & -5 & -5 \\ 0 & -24 & 6 \\ 5 & -1 & -1 \end{bmatrix},$$

matrix $-M$ can be factorized as

$$-M = S \begin{bmatrix} 3 & 0 & 0 \\ 0 & 2 & 0 \\ 0 & 0 & -3 \end{bmatrix} S^{-1}.$$

Consequently,

$$H(3) = [1 \quad 0 \quad 0] S \begin{bmatrix} e^9 & 0 & 0 \\ 0 & e^6 & 0 \\ 0 & 0 & e^{-9} \end{bmatrix} S^{-1} \begin{bmatrix} 1 \\ 0 \\ 5 \end{bmatrix}$$

$$= e^9 \neq 0.$$

So, $H(3)$ is invertible. Therefore, by Theorem 7.1, the game has a unique open-loop Nash equilibrium.

The equilibrium actions are obtained by solving the two-point boundary-value problem (see Theorem 7.1)

$$\dot{y}(t) = My(t), \quad \text{with} \quad Py(0) + Qe^{3M}y(0) = [1\ 0\ 0]^T.$$

From the boundary condition, $Py(0) + Qe^{3M}y(0) = [1\,0\,0]^T$, it follows that

$$y(0) = [P + Qe^{3M}]^{-1}[1 \quad 0 \quad 0]^T$$

$$= \left[\begin{bmatrix} 1 & 0 & 0 \\ 0 & 0 & 0 \\ 0 & 0 & 0 \end{bmatrix} + \begin{bmatrix} 0 & 0 & 0 \\ 0 & 1 & 0 \\ -5 & 0 & 1 \end{bmatrix} S \begin{bmatrix} e^{-9} & 0 & 0 \\ 0 & e^{-6} & 0 \\ 0 & 0 & e^9 \end{bmatrix} S^{-1} \right]^{-1} \begin{bmatrix} 1 \\ 0 \\ 0 \end{bmatrix}$$

$$= \begin{bmatrix} 1 \\ 1 - 1/e^3 \\ 4 + 1/e^3 \end{bmatrix}.$$

So,

$$y(t) = e^{Mt}y(0) = S \begin{bmatrix} e^{-3t} & 0 & 0 \\ 0 & e^{-2t} & 0 \\ 0 & 0 & e^{3t} \end{bmatrix} S^{-1} \begin{bmatrix} 1 \\ 1 - 1/e^3 \\ 4 + 1/e^3 \end{bmatrix}$$

$$= \begin{bmatrix} e^{-3t} \\ e^{-3t} - \frac{1}{e^3}e^{-2t} \\ 4e^{-3t} + \frac{1}{e^3}e^{-2t} \end{bmatrix}.$$

Hence, the equilibrium actions are

$$u_1^*(t) = -e^{-3t} + \frac{1}{e^3}e^{-2t} \quad \text{and} \quad u_2^*(t) = -4e^{-3t} - \frac{1}{e^3}e^{-2t}.$$

The resulting closed-loop system is $x(t) = e^{-3t}$. $\qquad\qquad\qquad\qquad$ ☐

Note

1. Notice that assumption (7.2.5) is equivalent to the statement that for both players a linear quadratic control problem associated with this game problem should be solvable on $[0, T]$. That is, the optimal control problem that arises if the action of his opponent(s) are known must be solvable for each player.

2. Generically one may expect that, if an open-loop Nash equilibrium exists, the set of Riccati differential equations (7.2.5) will have a solution on the closed interval $[0, T]$. For, suppose that $(u_1^*(.), u_2^*(.))$ is a Nash equilibrium. Then, in particular

$$J_1(u_1^*, u_2^*) \leq J_1(u_1, u_2^*).$$

Or, stated differently,

$$\tilde{J}_1(u_1^*) \leq \tilde{J}_1(u_1),$$

where

$$\tilde{J}_1 = \int_0^T \{\tilde{x}^T(t)Q_1\tilde{x}(t) + u_1^T(t)R_{11}u_1(t)\}dt + \tilde{x}^T(T)Q_{1T}\tilde{x}(T),$$

and

$$\dot{\tilde{x}} = A\tilde{x} + B_1 u_1 + B_2 u_2^*, \quad \tilde{x}(0) = x_0.$$

According to Theorem 5.8 this implies that the Riccati differential equation (7.2.5) has a solution on the half-open interval $(0, T]$ (with $i = 1$). In a similar way it also follows that (7.2.5) has a solution on $(0, T]$ for $i = 2$. So in most cases one expects that, if there is an open-loop Nash equilibrium, the assumption that the set of Riccati differential equations will have a solution on the closed interval $[0, T]$ will hold.

Finally notice that the assumption that these solutions of (7.2.5) exist on $[0, T]$, together with the assumption that the boundary-value problem (7.2.7) has a solution, implies that there exists an open-loop Nash equilibrium. Therefore, the existence of an open-loop Nash equilibrium is almost equivalent to the existence of a solution of the set of Riccati differential equations (7.2.5) together with the existence of a solution of the two-point boundary-value problem (7.2.7). $\qquad\qquad\qquad\qquad$ ☐

Next consider the set of coupled asymmetric Riccati-type differential equations:

$$\dot{P}_1 = -A^T P_1 - P_1 A - Q_1 + P_1 S_1 P_1 + P_1 S_2 P_2; \quad P_1(T) = Q_{1T} \tag{7.2.8}$$

$$\dot{P}_2 = -A^T P_2 - P_2 A - Q_2 + P_2 S_2 P_2 + P_2 S_1 P_1; \quad P_2(T) = Q_{2T} \tag{7.2.9}$$

Note

1. In the sequel the notation $P_i(t, T)$ is sometimes used to stress the fact that we are dealing with solutions of equations (7.2.8) and (7.2.9) which have a boundary value at $t = T$.

2. Notice that the solutions $P_i(t)$ are, in general, not symmetric since both equations (7.2.8) and (7.2.9) contain just a term $P_i S_j P_j$ and no corresponding term $P_j S_j P_i$.

3. As for the standard Riccati equations which we encountered in the linear quadratic optimal control problem, the solutions of equations (7.2.8) and (7.2.9) are time-invariant. That is, if $P_i(t, T)$ exists on $[T - \delta, T]$, for some $\delta > 0$, then also $P_i(t + \delta, T + \delta)$ exists and equals $P_i(t, T)$ for $t \in [T - \delta, T]$. This follows by a direct inspection of the differential equations. □

Let $P_i(t)$ satisfy this set of Riccati differential equations (7.2.8) and (7.2.9) and assume that player i uses the action

$$u_i(t) = -R_{ii}^{-1} B_i^T P_i(t) \Phi(t, 0) x_0$$

where $\Phi(t, 0)$ is the solution of the transition equation

$$\dot{\Phi}(t, 0) = (A - S_1 P_1(t) - S_2 P_2(t)) \Phi(t, 0), \quad \Phi(0, 0) = I.$$

Define $\psi_i(t) := P_i(t) \Phi(t, 0) x_0$. Then

$$\dot{\psi}_i(t) = \dot{P}_i(t) \Phi(t, 0) x_0 + P_i(t) \dot{\Phi}(t, 0) x_0.$$

Substitution of $\dot{P}_i$ from equations (7.2.8) and (7.2.9) and $\dot{\Phi}(t, 0)$ gives

$$\dot{\psi}_i = (-A^T P_i - Q_i) \Phi(t, 0) x_0 = -A^T \psi_i - Q_i \Phi(t, 0) x_0.$$

It is now easily verified that $x(t) := \Phi(t, 0) x_0$, $\psi_1(t)$ and $\psi_2(t)$ satisfy the two-point boundary-value problem (7.2.7), for an arbitrary choice of x_0. So from Theorem 7.1 we conclude the following.

Theorem 7.2

Assume:

1. that the set of coupled Riccati differential equations

$$\dot{P}_1 = -A^T P_1 - P_1 A - Q_1 + P_1 S_1 P_1 + P_1 S_2 P_2; \quad P_1(T) = Q_{1T},$$

$$\dot{P}_2 = -A^T P_2 - P_2 A - Q_2 + P_2 S_2 P_2 + P_2 S_1 P_1; \quad P_2(T) = Q_{2T}$$

has a solution P_i, $i = 1, 2$, on $[0, T]$; and

2. that the two Riccati differential equations,

$$\dot{K}_i(t) = -A^T K_i(t) - K_i(t)A + K_i(t)S_i K_i(t) - Q_i(t), K_i(T) = Q_{iT}, \quad i = 1, 2,$$

have a symmetric solution $K_i(.)$ on $[0, T]$.

Then the linear quadratic differential game (7.2.1)–(7.2.3) has unique open-loop Nash equilibrium for every initial state. Moreover, the set of equilibrium actions is given by:

$$u_i^*(t) = -R_{ii}^{-1} B_i^T P_i(t)\Phi(t, 0)x_0, \quad i = 1, 2.$$

Here $\Phi(t, 0)$ satisfies the transition equation

$$\dot{\Phi}(t, 0) = (A - S_1 P_1 - S_2 P_2)\Phi(t, 0); \quad \Phi(t, t) = I. \qquad \square$$

With

$$P := \begin{bmatrix} P_1 \\ P_2 \end{bmatrix}; \quad D := \begin{bmatrix} A^T & 0 \\ 0 & A^T \end{bmatrix}; \quad S := [S_1 \ S_2] \quad \text{and} \quad Q := \begin{bmatrix} Q_1 \\ Q_2 \end{bmatrix},$$

the set of coupled Riccati equations (7.2.8) and (7.2.9) can be rewritten as the non-symmetric matrix Riccati differential equation

$$\dot{P} = -DP - PA + PSP - Q; \quad P^T(T) = [Q_{1T}, \ Q_{2T}].$$

From Chapter 5.3 we know that the solution of such a Riccati differential equation can be obtained by solving a set of linear differential equations. In particular, if this linear system of differential equations (7.2.10) can be analytically solved we also obtain an analytic solution for (P_1, P_2). Due to this relationship it is possible to compute solutions of (7.2.8) and (7.2.9) in an efficient reliable way using standard computer software packages like MATLAB. From Theorem 5.12 we get the following corollary.

Corollary 7.3

The set of coupled Riccati differential equations (7.2.8) and (7.2.9) has a solution on $[0, T]$ if and only if the set of linear differential equations

$$\begin{bmatrix} \dot{U}(t) \\ \dot{V}_1(t) \\ \dot{V}_2(t) \end{bmatrix} = M \begin{bmatrix} U(t) \\ V_1(t) \\ V_2(t) \end{bmatrix}; \quad \begin{bmatrix} U(T) \\ V_1(T) \\ V_2(T) \end{bmatrix} = \begin{bmatrix} I \\ Q_{1T} \\ Q_{2T} \end{bmatrix} \qquad (7.2.10)$$

has a solution on $[0, T]$, with $U(.)$ nonsingular. Moreover, if equations (7.2.10) have an appropriate solution $(U(.), V_1(.), V_2(.))$, the solution of (7.2.8) and (7.2.9) is obtained as $P_i(t) := V_i(t)U^{-1}(t), \ i = 1, 2.$ \qquad $\square$

Example 7.3

Reconsider Example 7.2. By Theorem 7.2 the game has an equilibrium if, additional to the regulator Riccati differential equations (7.2.5), the set of coupled Riccati differential equations (7.2.8) and (7.2.9) has a solution on $[0,3]$.

According to Corollary 7.3 the solution of this set of coupled Riccati differential equations can be obtained by solving the linear differential equation

$$\dot{y}(t) = \begin{bmatrix} 2 & -1 & -1 \\ -1 & -2 & 0 \\ -4 & 0 & -2 \end{bmatrix} y(t), \quad y(3) = \begin{bmatrix} 1 \\ 0 \\ 5 \end{bmatrix}.$$

From Example 7.2 it is easily seen that the solution of this differential equation is

$$y(t) = \begin{bmatrix} -1 & 0 & 5 \\ -1 & -1 & -1 \\ -4 & 1 & -4 \end{bmatrix} \begin{bmatrix} e^{-3(t-3)} & 0 & 0 \\ 0 & e^{-2(t-3)} & 0 \\ 0 & 0 & e^{3(t-3)} \end{bmatrix} \frac{1}{30} \begin{bmatrix} -5 & -5 & -5 \\ 0 & -24 & 6 \\ 5 & -1 & -1 \end{bmatrix} \begin{bmatrix} 1 \\ 0 \\ 5 \end{bmatrix}$$

$$= \begin{bmatrix} e^{-3(t-3)} \\ e^{-3(t-3)} - e^{-2(t-3)} \\ 4e^{-3(t-3)} + e^{-2(t-3)} \end{bmatrix}.$$

Obviously, $e^{-3(t-3)} \neq 0$ on $[0,3]$. So, according to Corollary 7.3, the solution of the set of coupled differential equations (7.2.8) and (7.2.9) is

$$P_1(t) = (e^{-3(t-3)} - e^{-2(t-3)})/e^{-3(t-3)} = 1 - e^{t-3},$$
$$P_2(t) = (4e^{-3(t-3)} + e^{-2(t-3)})/e^{-3(t-3)} = 4 + e^{t-3}.$$

Next, consider the transition equation

$$\dot{\Phi}(t) = (2 - P_1(t) - P_2(t))\Phi(t), \quad \Phi(0) = 1.$$

Substitution of $P_i(t)$, $i=1,2$, into this transition equation shows that $\dot{\Phi}(t) = -3\Phi(t)$, $\Phi(0) = 1$. So, $\Phi(t) = e^{-3t}$. Consequently, according to Theorem 7.2, the equilibrium actions are

$$u_1^*(t) = -P_1(t)e^{-3t} = -e^{-3t} + \frac{1}{e^3}e^{-2t}$$

and

$$u_2^*(t) = -P_2(t)e^{-3t} = -4e^{-3t} - \frac{1}{e^3}e^{-2t}.$$

These actions, of course, coincide with those we obtained in Example 7.2. □

The next example shows that situations exist where the set of Riccati differential equations (7.2.8) and (7.2.9) does not have a solution, whilst there does exist an open-loop Nash equilibrium for the game.

Example 7.4

Let

$$A = \begin{bmatrix} -1 & 0 \\ 0 & -5/22 \end{bmatrix}, \quad B_1 = \begin{bmatrix} 1 & 0 \\ 0 & 1 \end{bmatrix}, \quad B_2 = \begin{bmatrix} 1 \\ 0 \end{bmatrix}, \quad Q_1 = \begin{bmatrix} 1 & 0 \\ 0 & 0 \end{bmatrix},$$

$$Q_2 = \begin{bmatrix} 1 & 1 \\ 1 & 2 \end{bmatrix}, \quad R_{11} = \begin{bmatrix} 1 & 1 \\ 1 & 2 \end{bmatrix}^{-1} \quad \text{and} \quad R_{22} = 1.$$

First notice that, since Q_i are positive semi-definite, the two Riccati equations (7.2.5) have a solution whenever Q_{iT} are positive semi-definite (see Chapter 5, Exercise 4 again).

Now, choose $T = 0.1$. Then, numerical calculation shows that

$$H(0.1) = \begin{bmatrix} 1.1155 & 0.0051 \\ 0.0051 & 1.0230 \end{bmatrix} + \begin{bmatrix} 0.1007 & 0.1047 \\ 0.0964 & 0.2002 \end{bmatrix} Q_{1T} + \begin{bmatrix} 0.1005 & 0 \\ 0.0002 & 0 \end{bmatrix} Q_{2T}$$

$$:= V \begin{bmatrix} I \\ Q_{1T} \\ Q_{2T} \end{bmatrix}.$$

Now, choose

$$Q_{1T} = \begin{bmatrix} 1 & h_1 \\ h_1 & h_1^2 + 1 \end{bmatrix},$$

with

$$h_1 = \frac{-V(1,2) - V(2,3) - V(2,5) * 10}{V(2,4)},$$

and

$$Q_{2T} = \begin{bmatrix} 10 & h_2 \\ h_2 & \frac{h_2^2+1}{10} \end{bmatrix},$$

with

$$h_2 = \frac{-V(2,2) - V(2,3) * Q_{1T}(1,2) - V(2,4) * Q_{1T}(2,2)}{V(2,5)}.$$

Then, clearly, both Q_{1T} and Q_{2T} are positive definite whereas the last row of $H(0.1)$ contains, by construction, only zeros. That is,

$$H(0.1) = \begin{bmatrix} 2.1673 & -752.6945 \\ 0 & 0 \end{bmatrix}$$

is not invertible.

So, according to Theorem 7.1, the game does not have a unique open-loop Nash equilibrium for every initial state. Using the converse statement of Theorem 7.2, this implies that the corresponding set of Riccati differential equations has no solution.

Next consider $H(0.11)$. Numerical calculation shows that with the system parameters as chosen above, $H(0.11)$ is invertible. So, by Theorem 7.1 again, the game has an open-loop Nash equilibrium at $T = 0.11$. Since in this example $P_i(t, 0.1)$ does not exist for all $t \in [0, 0.1]$, we conclude (see Note following equation (7.2.9)) that $P_i(t, 0.11)$ does not exist for all $t \in [0.01, 0.11]$. So, the game does have a solution, whereas the set of Riccati differential equations (7.2.8) and (7.2.9) has no solution.   □

Notice that the above theorems are in fact local results. That is, they just make   statements concerning the existence of an equilibrium strategy for a fixed endpoint T in time.

Next, we consider for a fixed time interval $[0, t_1]$ the question under which conditions for every $T \in [0, t_1]$ there will exist an open-loop Nash equilibrium on the interval $[0, T]$. In the Appendix to this chapter it is shown that, under the assumption that both corresponding individual regular linear quadratic control problems have a solution, the open-loop game has a solution if and only if the coupled set of Riccati equations (7.2.8) and (7.2.9) has a solution.

Theorem 7.4

Assume that the two Riccati differential equations (7.2.5) have a solution on $[0, t_1]$. Then, for all $T \in [0, t_1]$ the game defined on the interval $[0, T]$ has an open-loop Nash equilibrium for all x_0 if and only if the set of Riccati differential equations (7.2.8) and (7.2.9) has a solution on the interval $[0, t_1]$.   □

From Note 2 following Example 7.2, we infer that in the above theorem the assumption that both individual linear quadratic control problems should have a solution can be dropped if we lower our ambitions and only require the existence of a solution of the game during an open time interval. Combining the results of Theorem 7.1, Theorem 7.4 and Note 2, yields the following proposition.

Proposition 7.5

The following statements are equivalent.

1. For all $T \in [0, t_1)$ there exists for all x_0 a unique open-loop Nash equilibrium for the two-player linear quadratic differential game (7.2.1) and (7.2.2).

2. The following two conditions hold on $[0, t_1)$.

    (a) $H(t)$ is invertible for all $t \in [0, t_1)$.
    (b) The two Riccati differential equations (7.2.5) have a solution $K_i(0, T)$ for all $T \in [0, t_1)$.

3. The next two conditions hold on $[0, t_1)$.

(a) The set of coupled Riccati differential equations (7.2.8) and (7.2.9) has a solution $(P_1(0,T), P_2(0,T))$ for all $T \in [0,t_1)$.

(b) The two Riccati differential equations (7.2.5) have a solution $K_i(0,T)$ for all $T \in [0,t_1)$.

Moreover, if either one of the above conditions is satisfied the equilibrium is unique.

$\square$

Example 7.5

Consider the finite-planning horizon version of the dynamic duopoly game with sticky prices (see Example 3.24):

$$\dot{p}(t) = c - s_1 p(t) - s_2(v_1 + v_2), \quad p(0) = p_0,$$

where $p(t)$ denotes the price of the product and v_i is the output rate of company i, together with the profit functions

$$J_i(v_1, v_2) = \int_0^T \left\{ p(t)v_i(t) - c_i v_i(t) - \frac{1}{2}v_i^2(t) \right\} dt.$$

Here all parameters c, s_1, s_2, p_0 and c_i are positive. Introducing as state variable $x(t) := [p(t)\ 1]^T$ and $u_i(t) := v_i(t) + 2[-\frac{1}{2}\frac{1}{2}c_i]x(t)$, the above maximization problems can be rewritten as

$$\min_{u_i} J_i(u_1, u_2) = \frac{1}{2} \int_0^T \left\{ x^T(t) \begin{bmatrix} -1 & c_i \\ c_i & -c_i^2 \end{bmatrix} x(t) + u_i^2(t) \right\} dt$$

subject to

$$\dot{x}(t) = \begin{bmatrix} -(s_1 + 2s_2) & c + s_2(c_1 + c_2) \\ 0 & 0 \end{bmatrix} + \begin{bmatrix} -s_2 \\ 0 \end{bmatrix} u_1 + \begin{bmatrix} -s_2 \\ 0 \end{bmatrix} u_2.$$

So, using Proposition 7.5, it is then easily verified that for all $T \in [0,t_1)$ the game has for all p_0 a unique open-loop Nash equilibrium if with

$$A := \begin{bmatrix} -(s_1 + 2s_2) & c + s_2(c_1 + c_2) \\ 0 & 0 \end{bmatrix}; \quad S_i := \begin{bmatrix} 2s_2^2 & 0 \\ 0 & 0 \end{bmatrix} \quad \text{and} \quad Q_i := \frac{1}{2}\begin{bmatrix} -1 & c_i \\ c_i & -c_i^2 \end{bmatrix}$$

the following sets of differential equations have a solution $P_i(.)$ and $K_i(.)$ on $(0,t_1]$, $i = 1,2$, respectively.

1. The set of coupled Riccati differential equations

$$\dot{P}_1 = -A^T P_1 - P_1 A - Q_1 + P_1 S_1 P_1 + P_1 S_2 P_2; \quad P_1(T) = 0,$$
$$\dot{P}_2 = -A^T P_2 - P_2 A - Q_2 + P_2 S_2 P_2 + P_2 S_1 P_1; \quad P_2(T) = 0.$$

2. The two Riccati differential equations

$$\dot{K}_i(t) = -A^T K_i(t) - K_i(t)A + K_i(t)S_i K_i(t) - Q_i(t), K_i(T) = 0, \quad i = 1, 2.$$

Moreover, the equilibrium actions are then:

$$v_i = [s_2 \ 0]P_i x(t) - 2\left[-\frac{1}{2} \ \frac{1}{2}c_i\right]x(t).$$

Let $K_i(t) =: \begin{bmatrix} k_{i1} & k_{i2} \\ k_{i2} & k_{i3} \end{bmatrix}$. Elementary calculation shows that K_i is obtained by solving, for $i = 1, 2$, the three differential equations:

$$\dot{k}_{i1}(t) = 2ak_{i1}(t) + hk_{i1}^2(t) + \frac{1}{2}, \quad k_{i1}(T) = 0,$$

$$\dot{k}_{i2}(t) = (a + hk_{i1}(t))k_{i2}(t) - bk_{i1}(t) - \frac{1}{2}c_i, \quad k_{i2}(T) = 0,$$

$$\dot{k}_{i3}(t) = -2bk_{i2}(t) + hk_{i2}^2(t) + \frac{1}{2}c_i^2, \quad k_{i3}(T) = 0,$$

where we introduced the notations $a := s_1 + 2s_2$, $b := c + s_2(c_1 + c_2)$ and $h := 2s_2^2$. Notice that once we determined $k_{i1}(.)$ from the first differential equation, the other two differential equations are linear in k_{i2} and k_{i3}, respectively. Since $a > 0$ one can easily verify that the first differential equation for k_{i1} always has a solution. In fact one can explicitly determine this continuous solution (see Example 8.3). Consequently, these differential equations always have a unique solution on $(0,T]$, $i = 1, 2$. Therefore, the game has an equilibrium for all p_0 if $P_i(.)$, $i = 1, 2$, exists on $(0,T]$.

If $c_1 = c_2$ the game is completely symmetric. By considering the differential equation for $P_1 - P_2$, it is easily shown that for this special case P_1 and P_2 coincide. Furthermore, it follows directly from the fact that both S_i and Q_i are symmetric that P_i will be symmetric too. Using these facts in set 1 (above), straightforward calculations then show that the solution for P_i is obtained by solving the same set of equations as for k_{ij}. Only the parameter h has to be replaced by $h := 4s_2^2$ in these equations. Similarly it then follows rather straightforwardly that the game always has a unique equilibrium in this case. In Figure 7.2 we sketch the equilibrium state and control trajectories if one chooses $s_1 = s_2 = c_1 = c_2 = 1$, $c = 5$, $T = 10$ and $p(0) = 2$. These are obtained by solving

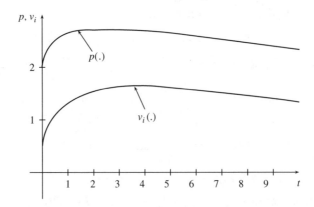

Figure 7.2 Some optimal state and control trajectories for Example 7.5

numerically the two-point boundary-value problem (7.2.7). We see that the trajectories initially converge rapidly towards some steady-state value ($p(t)$ approximately 2.7 and $v_i(t)$ approximately 1.6). Since the considered planning horizon is finite, we observe that at the end of the planning horizon all trajectories show a steady decline again. □

We conclude this section by considering the so-called two-person zero-sum differential game. In this game there is a single objective function, which one player likes to minimize and the other player likes to maximize. So in a two-person zero-sum game the gains of one player incur a loss to the other player. Examples which fit into this framework are, for example, the rope-pulling game, the division of a finite resource between two players and the pursuit–evasion game. By applying the theorems and proposition above to this special case, the analogues are obtained for the open-loop zero-sum linear quadratic game. As an example we will elaborate the analogue of Proposition 7.5.

Proposition 7.6 (Open-loop zero-sum differential game)

Consider the differential game described by

$$\dot{x}(t) = Ax(t) + B_1 u_1(t) + B_2 u_2(t), \quad x(0) = x_0,$$

with, for player one, the quadratic cost function:

$$J_1(u_1, u_2) = \int_0^T \{x^T(t)Qx(t) + u_1^T(t)R_1 u_1(t) - u_2^T(t)R_2 u_2(t)\}dt + x^T(T)Q_T x(T),$$

and, for player two, the opposite objective function

$$J_2(u_1, u_2) = -J_1(u_1, u_2);$$

where the matrices Q, Q_T and R_i, $i = 1, 2$, are symmetric. Moreover, assume that R_i, $i = 1, 2$, are positive definite. Then, for all $T \in [0, t_1)$, this zero-sum linear quadratic differential game has for every initial state an open-loop Nash equilibrium if and only if the following two conditions hold on $[0, t_1)$.

1. The Riccati differential equation

$$\dot{P}(t) = -A^T P - PA - Q + P(S_1 - S_2)P, \quad P(T) = Q_T \tag{7.2.11}$$

   has a symmetric solution $P(0, T)$ for all $T \in [0, t_1)$.

2. The two Riccati differential equations

$$\dot{K}_1(t) = -A^T K_1(t) - K_1(t)A + K_1(t)S_1 K_1(t) - Q, \quad K_1(T) = Q_T \tag{7.2.12}$$
$$\dot{K}_2(t) = -A^T K_2(t) - K_2(t)A + K_2(t)S_2 K_2(t) + Q, \quad K_2(T) = -Q_T, \tag{7.2.13}$$

   have a solution $K_i(0, T)$ for all $T \in [0, t_1)$.

Moreover, if the above conditions are satisfied the equilibrium is unique. In that case the equilibrium actions are

$$u_1^*(t) = -R_1^{-1}B_1^T P(t)\Phi(t,0)x_0 \quad \text{and} \quad u_2^*(t) = R_2^{-1}B_2^T P(t)\Phi(t,0)x_0.$$

Here $\Phi(t,0)$ satisfies the transition equation

$$\dot{\Phi}(t,0) = (A - (S_1 - S_2)P(t))\Phi(t,0); \quad \Phi(t,t) = I.$$

Proof

According to Proposition 7.5 this game has for every $T \in [0,t_1)$ an open-loop Nash equilibrium for every initial state if and only if the following two coupled Riccati differential equations have a solution $P_i(0,T)$, $i = 1,2$, for all $T \in [0,t_1)$

$$\dot{P}_1(t) = -A^T P_1(t) - P_1(t)A - Q + P_1(t)S_1P_1(t) + P_1(t)S_2P_2(t), \; P_1(T) = Q_T \quad (7.2.14)$$
$$\dot{P}_2(t) = -A^T P_2(t) - P_2(t)A + Q + P_2(t)S_2P_2(t) + P_2(t)S_1P_1(t), \; P_2(T) = -Q_T \quad (7.2.15)$$

and the two Riccati differential equations (7.2.12) and (7.2.13) have a solution $K_i(0,T)$ for all $T \in [0,t_1)$. Adding (7.2.14) to (7.2.15) gives the following differential equation in $(P_1 + P_2)(t)$:

$$\frac{d(P_1 + P_2)(t)}{dt} = -A^T(P_1 + P_2)(t) - (P_1 + P_2)(t)A + (P_1 + P_2)(t)(S_1P_1(t) + S_2P_2(t)),$$
$$(P_1 + P_2)(T) = 0.$$

Obviously $(P_1 + P_2)(.) = 0$ satisfies this differential equation. Since the solution to this differential equation is unique, we conclude that $P_1(t) = -P_2(t)$. Substitution of this into equation (7.2.14) then shows that equations (7.2.14) and (7.2.15) have a solution $P_i(0,T)$ for all $T \in [0,t_1)$ if and only if equation (7.2.11) has a solution $P(0,T)$ for all $T \in [0,t_1)$. The symmetry of $P(0,T)$ follows directly from the symmetry of Q_T (see section 5.2). The corresponding equilibrium strategies then follow directly from Theorem 7.2. $\qquad\square$

7.3 Open-loop Nash algebraic Riccati equations

In this section we consider the set of algebraic equations corresponding to the Riccati differential equations (7.2.8) and (7.2.9). Using the shorthand notation $S_i := B_i R_i^{-1}B_i$ again, this set of coupled algebraic Riccati equations (ARE) is given by

$$0 = Q_1 + A^T P_1 + P_1 A - P_1 S_1 P_1 - P_1 S_2 P_2; \quad (7.3.1)$$
$$0 = Q_2 + A^T P_2 + P_2 A - P_2 S_2 P_2 - P_2 S_1 P_1. \quad (7.3.2)$$

As we will see in the next section this set of coupled equations plays an important role in solving the infinite horizon open-loop Nash differential game. Some elementary rewriting shows that this set of coupled Riccati equations can be rewritten as a 'standard' algebraic Riccati equation. With

$$Q := \begin{bmatrix} Q_1 \\ Q_2 \end{bmatrix}; \quad A_2 := \begin{bmatrix} A & 0 \\ 0 & A \end{bmatrix}; \quad A_1 := A; \quad S := [S_1 \quad S_2] \quad \text{and} \quad X := \begin{bmatrix} P_1 \\ P_2 \end{bmatrix}$$

the above set of algebraic Riccati equations (7.3.1) and (7.3.2) can be rewritten as

$$Q + A_2^T X + X A_1 - X S X = 0. \tag{7.3.3}$$

Obviously, this algebraic Riccati equation differs from the Riccati equations we have encountered so far in that matrix A_2 differs from A_1, the dimensions of both these matrices A_i, $i = 1, 2$, differ and matrix X is not square anymore. Remember that we called a solution of the symmetric Riccati equation (5.4.4) stabilizing if the matrix $A - SX$ is stable. Taking a purely algebraic point of view one can then discern two analogues of this stabilizing solution concept here. First, one can call a solution X of the nonsymmetric algebraic Riccati equation (7.3.3) stabilizing if all eigenvalues of matrix $A - SX$ have a negative real part. However, by transposition of this equation (7.3.3), for the same reason one can call a solution X stabilizing if the matrix $A_2 - S^T X^T$ is stable. Obviously, from our application point of view, this approach is less appealing. In fact one can associate with these transposed matrices a so-called dual system which sometimes works well in the analysis of system properties. Although this relationship will not be exploited in the analysis pursued in this chapter it turns out that both notions of stabilizability play a role later on. In particular, the latter stabilizability property plays a role in characterizing uniqueness of equilibria. Definition 7.2(a) introduces the concept of stabilizability as one would expect. Definition 7.2(b)(ii) states that the spectrum of the controlled dual system should be in the closed right-half of the complex plane.

Definition 7.2

A solution (P_1, P_2) of the set of algebraic Riccati equations (7.3.1) and (7.3.2) is called

(a) stabilizing, if $\sigma(A - S_1 P_1 - S_2 P_2) \subset \mathbb{C}^-$;

(b) strongly stabilizing if

  (i) it is a stabilizing solution, and
  (ii)
$$\sigma\left(\begin{bmatrix} -A^T + P_1 S_1 & P_1 S_2 \\ P_2 S_1 & -A^T + P_2 S_2 \end{bmatrix} \right) \subset \mathbb{C}_0^+. \tag{7.3.4}$$

$\square$

According to section 2.7 the set of solutions of the symmetric algebraic Riccati equation (5.4.4) is tightly connected to the set of invariant subspaces of an associated Hamiltonian matrix H. In fact the analysis performed there can be copied to a large extent for the asymmetric case. To that end, we first rewrite equation (7.3.3) as

$$[I \ X] \begin{bmatrix} Q & A_2^T \\ A_1 & -S \end{bmatrix} \begin{bmatrix} I \\ X \end{bmatrix} = 0.$$

Then, following the analysis of section 2.7, equation (7.3.3) has a solution X if and only if there exists a matrix $\Lambda \in \mathbb{R}^{n \times n}$ such that

$$\begin{bmatrix} A_1 & -S \\ -Q & -A_2^T \end{bmatrix} \begin{bmatrix} I \\ X \end{bmatrix} = \begin{bmatrix} I \\ X \end{bmatrix} \Lambda.$$

Along the lines of the proofs of Theorem 2.29 and Theorem 2.30, with matrix

$$M := \begin{bmatrix} A & -S_1 & -S_2 \\ -Q_1 & -A^T & 0 \\ -Q_2 & 0 & -A^T \end{bmatrix}, \tag{7.3.5}$$

the following two analogues of these two theorems result.

Theorem 7.7

Let $V \subset \mathbb{R}^{3n}$ be an n-dimensional invariant subspace of M, and let $X_i \in \mathbb{R}^{n \times n}$, $i = 0, 1, 2$, be three real matrices such that

$$V = \text{Im} \begin{bmatrix} X_0 \\ X_1 \\ X_2 \end{bmatrix}.$$

If X_0 is invertible, then $P_i := X_i X_0^{-1}$, $i = 1, 2$, is a solution to the set of coupled Riccati equations (7.3.1) and (7.3.2) and $\sigma(A - S_1 P_1 - S_2 P_2) = \sigma(M \mid_V)$. Furthermore, the solution (P_1, P_2) is independent of the specific choice of basis of V. $\square$

Theorem 7.8

Let $P_i \in \mathbb{R}^{n \times n}$, $i = 1, 2$, be a solution to the set of coupled Riccati equations (7.4.1) and (7.4.2). Then there exist matrices $X_i \in \mathbb{R}^{n \times n}$, $i = 0, 1, 2$, with X_0 invertible, such that $P_i = X_i X_0^{-1}$. Furthermore, the columns of $\begin{bmatrix} X_0 \\ X_1 \\ X_2 \end{bmatrix}$ form a basis of an n-dimensional invariant subspace of M. $\square$

So, a similar relationship holds as in the symmetric case (cf. section 2.7). That is, the set of solutions of the algebraic Riccati equations (7.3.1) and (7.3.2) is tightly connected to the set of invariant subspaces of matrix M (which can be calculated from its generalized eigenspaces).

Introducing a separate notation for the set of M-invariant subspaces

$$\mathcal{M}^{inv} := \{ \mathcal{T} \mid M\mathcal{T} \subset \mathcal{T} \}$$

it follows from the above Theorems 7.7 and 7.8 that the following set of graph subspaces plays a crucial role

$$\mathcal{P}^{pos} := \left\{ \mathcal{P} \in \mathcal{M}^{inv} \mid \mathcal{P} \oplus \text{Im} \begin{bmatrix} 0 & 0 \\ I & 0 \\ 0 & I \end{bmatrix} = \mathbb{R}^{3n} \right\}.$$

Notice that elements in the set $\mathcal{P}^{pos}$ can be calculated using the set of matrices

$$K^{pos} := \left\{ K \in R^{3n \times n} \mid \text{Im } K \oplus \text{Im} \begin{bmatrix} 0 & 0 \\ I & 0 \\ 0 & I \end{bmatrix} = R^{3n} \right\}.$$

Every element of $\mathcal{P}^{pos}$ defines exactly one solution of (ARE). By Corollary 2.25 the set of M-invariant subspaces contains only a finite number of elements if and only if the geometric multiplicities of all eigenvalues of M is one. Therefore if all eigenvalues of M have a geometric multiplicity one, equations (7.3.1) and (7.3.2) will have at most a finite number of solutions.

Another conclusion which immediately follows from Theorems 7.7 and 7.8 is Corollary 7.9.

Corollary 7.9

Equations (7.3.1) and (7.3.2) have a set of stabilizing solutions (P_1, P_2) if and only if there exists an M-invariant subspace $\mathcal{P}$ in $\mathcal{P}^{pos}$ such that $\text{Re } \lambda < 0$ for all $\lambda \in \sigma(M \mid_{\mathcal{P}})$.

$\square$

To illustrate some of the above mentioned properties, reconsider Example 7.4.

Example 7.6

It can be shown analytically that both $\frac{5}{22}$ and $\frac{1}{2}$ are eigenvalues of M with algebraic multiplicity 2 and 1, respectively. Numerical calculations show that the other eigenvalues of M are $1.8810, 0.1883$ and -1.7966. Rearranging the eigenvalues as $\{\frac{5}{22}, 1.8810, 0.1883, \frac{1}{2}, -1.7966\}$, M has the following corresponding generalized eigenspaces:

$\mathcal{T}_{11} = \text{Span}\{T_{11}\}$ where $T_{11} = [0, \ 0, \ 0, \ 0, \ 0, \ 1]^T$,

$\mathcal{T}_{12} = \text{Span}\{T_{11} \ T_{12}\}$ where $T_{12} = [-0.2024, \ 0.6012, \ -0.2620, \ -0.0057, \ 0.5161, \ 0]^T$,

$\mathcal{T}_2 = \text{Span}\{T_2\}$ where $T_2 = [-0.3726, \ -0.2006, \ 0.4229, \ 0, \ 0.6505, \ 0.4679]^T$,

$\mathcal{T}_3 = \text{Span}\{T_3\}$ where $T_3 = [0.0079, \ -0.0234, \ 0.0097, \ 0, \ -0.0191, \ -0.9995]^T$,

$\mathcal{T}_4 = \text{Span}\{T_4\}$ where $T_4 = [0.0580, \ -0.1596, \ 0.1160, \ 0, \ -0.2031, \ 0.9573]^T$, and

$\mathcal{T}_5 = \text{Span}\{T_5\}$ where $T_5 = [-0.7274, \ -0.1657, \ -0.2601, \ 0, \ -0.3194, \ -0.5232]^T$.

Notice that $\mathcal{T}_{12} \notin \mathcal{P}^{pos}$ since it violates the rank condition. For the same reason it is clear that no invariant subspace of $\mathcal{P}^{pos}$ can contain $\mathcal{T}_{11}$. Therefore, there exist at most $\binom{4}{2} = 6$ different two-dimensional graph subspaces. Obviously, none of the potential graph subspaces implied by these solutions will stabilize the closed-loop system matrix. It can be numerically easily verified that all these six solutions satisfy the rank condition and, thus, are indeed graph subspaces. So, the set of algebraic Riccati equations (7.3.1) and (7.3.2) has six different solutions, none of which is stabilizing.

As an example consider $\begin{bmatrix} X \\ Y \\ Z \end{bmatrix} := [T_2 \ T_3]$. This yields the solution

$$P_1 = YX^{-1} = \begin{bmatrix} 0.4229 & 0.0097 \\ 0 & 0 \end{bmatrix} \begin{bmatrix} -0.3726 & 0.0079 \\ -0.2006 & -0.0234 \end{bmatrix}^{-1}$$

and

$$P_2 = ZX^{-1} = \begin{bmatrix} 0.6505 & -0.0191 \\ 0.4679 & -0.9995 \end{bmatrix} \begin{bmatrix} -0.3726 & 0.0079 \\ -0.2006 & -0.0234 \end{bmatrix}^{-1}.$$

The eigenvalues of the closed-loop system matrix $A - S_1 P_1 - S_2 P_2$ are $\{1.8810, 0.1883\}$. It is easily verified that the rank of the first two rows of every other candidate solution is also two. Consequently the set of algebraic Riccati equations (7.3.1) and (7.3.2) has six different solutions, none of which is stabilizing. □

By replacing matrix A by $-A$ in the above example the sign of the eigenvalues of matrix M reverses. It is then numerically easily verified that matrix M has six different graph subspaces which all provide stabilizing solutions of the algebraic Riccati equation. So, in general, the set of algebraic Riccati equations (7.3.1) and (7.3.2) does not have a unique stabilizing solution. The following theorem shows, however, that it does have a unique strongly stabilizing solution (provided there exists such a solution).

Theorem 7.10

1. The set of algebraic Riccati equations (7.3.1) and (7.3.2) has a strongly stabilizing solution (P_1, P_2) if and only if matrix M has an n-dimensional stable graph subspace and M has $2n$ eigenvalues (counting algebraic multiplicities) in $\mathbb{C}_0^+$.

2. If the set of algebraic Riccati equations (7.3.1) and (7.3.2) has a strongly stabilizing solution, then it is unique.

Proof

1. Assume that equations (7.3.1) and (7.3.2) have a strongly stabilizing solution (P_1, P_2). Then (see Theorem 2.33 or Kremer (2002)), with

$$T := \begin{bmatrix} I & 0 & 0 \\ -P_1 & I & 0 \\ -P_2 & 0 & I \end{bmatrix} \quad \text{and consequently} \quad T^{-1} = \begin{bmatrix} I & 0 & 0 \\ P_1 & I & 0 \\ P_2 & 0 & I \end{bmatrix},$$

we have that

$$TMT^{-1} = \begin{bmatrix} A - S_1 P_1 - S_2 P_2 & S_1 & S_2 \\ 0 & P_1 S_1 - A^T & P_1 S_2 \\ 0 & P_2 S_1 & P_2 S_2 - A^T \end{bmatrix}.$$

Since (P_1, P_2) is a strongly stabilizing solution by Definition 7.2 matrix M has exactly n stable eigenvalues and $2n$ eigenvalues (counted with algebraic multiplicities) in $\mathbb{C}_0^+$. Furthermore, obviously, the stable subspace is a graph subspace. The converse statement is obtained similarly using the result of Theorem 7.7.

2. Using the result from item 1 above, Corollary 7.9 shows that there exists exactly one stabilizing solution. So, our solution (P_1, P_2) must be unique, which concludes the proof. $\qquad\qquad\qquad\square$

7.4 Infinite-planning horizon

In this section we assume that the performance criterion player $i = 1, 2$, likes to minimize is:

$$\lim_{T \to \infty} J_i(x_0, u_1, u_2, T) \qquad (7.4.1)$$

where

$$J_i = \int_0^T \{x^T(t)Q_i x(t) + u_i^T(t)R_i u_i(t)\}dt,$$

subject to the familiar dynamic state equation

$$\dot{x}(t) = Ax(t) + B_1 u_1(t) + B_2 u_2(t), \quad x(0) = x_0. \qquad (7.4.2)$$

Here matrix R_i is again positive definite and Q_i symmetric, $i = 1, 2$.

We assume that the matrix pairs (A, B_i), $i = 1, 2$, are stabilizable. So, in principle, each player is capable of stabilizing the system on their own.

The scrap value has been dropped here for the same reason as in the linear quadratic control problem. The inclusion of player j's control efforts into player i's cost function has been dropped because, as in the finite-planning horizon case, this term drops out in the analysis. So, for convenience, this term is skipped here from the outset.

The information both players have at the beginning of the game is similar to the finite-planning horizon case. Each player only knows the initial state of the system. The admissible control actions are now functions of time, where time runs from zero to infinity. Since we only like to consider those outcomes of the game that yield a finite cost to both players and the players are assumed to have a common interest in stabilizing the system, we restrict ourselves to functions belonging to the set (see section 5.3)

$$\mathcal{U}_s(x_0) = \left\{ u \in L_{2,loc} \mid J_i(x_0, u) \text{ exists in } \mathbb{R} \cup \{-\infty, \infty\}, \lim_{t \infty} x(t) = 0 \right\},$$

where $L_{2,loc}$ is the set of **locally square-integrable** functions, i.e.

$$L_{2,loc} = \left\{ u[0, \infty) \mid \forall T > 0, \int_0^T u^T(s)u(s)ds < \infty \right\}.$$

Notice that $\mathcal{U}_s(x_0)$ depends on the inital state of the system. For simplicity of notation we omit, however, this dependency. Moreover, a proviso similar to the one we made for the finite-planning horizon case applies here. That is, the restriction to this set of control functions requires some form of communication between the players.

In the Appendix at the end of this chapter the following theorem is proved.

Theorem 7.11

Consider matrix

$$M = \begin{bmatrix} A & -S_1 & -S_2 \\ -Q_1 & -A^T & 0 \\ -Q_2 & 0 & -A^T \end{bmatrix}. \tag{7.4.3}$$

If the linear quadratic differential game (7.4.1) and (7.4.2) has an open-loop Nash equilibrium for every initial state, then the following statements are true.

1. M has at least n stable eigenvalues (counted with algebraic multiplicities). In particular, there exists a p-dimensional stable M-invariant subspace S, with $p \geq n$, such that

$$\mathrm{Im} \begin{bmatrix} I \\ V_1 \\ V_2 \end{bmatrix} \subset S,$$

for some $V_i \in \mathbb{R}^{n \times n}$.

2. The two algebraic Riccati equations,

$$Q_i + A^T K_i + K_i A - K_i S_i K_i = 0, \tag{7.4.4}$$

have a symmetric solution $K_i(.)$ such that $A - S_i K_i$ is stable, $i = 1, 2$.

Conversely, if the two algebraic Riccati equations (7.4.4) have a stabilizing solution and $v^T(t) =: [x^T(t), \psi_1^T(t), \psi_2^T(t)]$ is an asymptotically stable solution of

$$\dot{v}(t) = Mv(t), \quad x(0) = x_0,$$

then

$$u_i^* := -R_i^{-1} B_i^T \psi_i(t), \quad i = 1, 2,$$

provides an open-loop Nash equilibrium for the linear quadratic differential game (7.4.1) and (7.4.2). $\square$

Note

From this theorem one can draw a number of conlusions concerning the existence of open-loop Nash equilibria. A general conclusion is that this number depends critically on the eigenstructure of matrix M. We will distinguish some cases. To that end, let s denote the number (counting algebraic multiplicities) of stable eigenvalues of M.

1. If $s < n$, for some initial state there still may exist an open-loop Nash equilibrium. Consider, for example, the case that $s = 1$. Then, for every $x_0 \in$ Span $[I, \ 0, \ 0]v$, where v is an eigenvector corresponding with the stable eigenvalue, the game has a Nash equilibrium.

2. If $s \geq 2$, the situation might arise that for some initial states there exist an infinite number of equilibria. A situation in which there are an infinite number of Nash equilibrium actions occurs if, for example, v_1 and v_2 are two independent eigenvectors in the stable subspace of M for which $[I, \ 0, \ 0]v_1 = \mu[I, \ 0, \ 0]v_2$, for some scalar μ. In such a situation,

$$x_0 = \lambda[I, \ 0, \ 0]v_1 + (1 - \lambda)\mu[I, \ 0, \ 0]v_2,$$

for an arbitrary scalar $\lambda \in \mathbb{R}$. The resulting equilibrium control actions, however, differ for each λ (apart from some exceptional cases).

3. If matrix M has a stable graph subspace, S, of dimension $s > n$, for every initial state x_0 there exists, generically, an infinite number of open-loop Nash equilibria. For, let $\{b_1, \ldots, b_s\}$ be a basis for S. Denote $d_i := [I, \ 0, \ 0]b_i$, and assume (without loss of generality) that Span$[d_1, \ldots, d_n] = \mathbb{R}^n$. Then, $d_{n+1} = \lambda_1 d_1 + \cdots + \lambda_n d_n$ for some λ_i, $i = 1, \ldots, n$. Elementary calculations (see also the proof of Theorem 7.16) show then that every x_0 can be written as both $x_{0,1} = \alpha_1 d_1 + \cdots + \alpha_n d_n$ and as $x_{0,2} = \beta_1 d_1 + \cdots + \beta_{n+1} d_{n+1}$, where $\beta_{n+1} \neq 0$. So, $x_0 = \lambda x_{0,1} + (1 - \lambda)x_{0,2}$ for an arbitrary choice of the scalar λ. Since the vectors $\{b_1, \ldots, b_{n+1}\}$ are linearly independent, it follows that for every λ, the vector $b(\lambda) := \lambda(\alpha_1 b_1 + \cdots + \alpha_n b_n) + (1 - \lambda)(\beta_1 b_1 + \cdots + \beta_{n+1} b_{n+1})$ differs. So, generically, the corresponding equilibrium control actions induced by this vector $b(\lambda)$ (via $\dot{v}(t) = Mv(t)$, $v(0) = b(\lambda)$) will differ. □

It is tempting to believe that, under the conditions posed in the above Theorem 7.11, matrix M will always have an n-dimensional graph subspace S. In the Appendix at the end of this chapter it is shown that, if the equilibrium control actions allow for a **feedback synthesis**[1], this claim is correct.

Theorem 7.12

Assume the linear quadratic differential game (7.4.1) and (7.4.2) has an open-loop Nash equilibrium for every initial state and the equilibrium control actions allow for a feedback synthesis. Then the following statements are true.

1. M has at least n stable eigenvalues (counted with algebraic multiplicities). In particular, for each such Nash equilibrium there exists a uniquely determined n-dimensional stable M-invariant subspace

$$\text{Im} \begin{bmatrix} I \\ V_1 \\ V_2 \end{bmatrix}$$

for some $V_i \in \mathbb{R}^{n \times n}$.

---

[1] That is, the closed-loop dynamics of the game can be described by: $\dot{x}(t) = Fx(t)$; $x(0) = x_0$ for some constant matrix F.

2. The two algebraic Riccati equations,

$$Q_i + A^T K_i + K_i A - K_i S_i K_i = 0,$$

have a symmetric solution $K_i(.)$ such that $A - S_i K_i$ is stable, $i = 1, 2$. □

By Theorem 7.7, Theorem 7.12 item 1 implies that the set of coupled algebraic Riccati equations (7.3.1) and (7.3.2) has a stabilizing solution. The next theorem shows that the converse statement of this result always holds.

Theorem 7.13

Assume:

1. that the set of coupled algebraic Riccati equations

$$0 = A^T P_1 + P_1 A + Q_1 - P_1 S_1 P_1 - P_1 S_2 P_2; \qquad (7.4.5)$$
$$0 = A^T P_2 + P_2 A + Q_2 - P_2 S_2 P_2 - P_2 S_1 P_1 \qquad (7.4.6)$$

has a set of solutions P_i, $i = 1, 2$, such that $A - S_1 P_1 - S_2 P_2$ is stable; and

2. that the two algebraic Riccati equations,

$$0 = A^T K_i + K_i A - K_i S_i K_i + Q_i, \qquad (7.4.7)$$

have a symmetric solution $K_i(.)$ such that $A - S_i K_i$ is stable, $i = 1, 2$.

Then the linear quadratic differential game (7.4.1) and (7.4.2) has an open-loop Nash equilibrium for every initial state. Moreover, one set of equilibrium actions is given by:

$$u_i^*(t) = -R_i^{-1} B_i^T P_i \Phi(t, 0) x_0, \quad i = 1, 2. \qquad (7.4.8)$$

Here $\Phi(t, 0)$ satisfies the transition equation

$$\dot{\Phi}(t, 0) = (A - S_1 P_1 - S_2 P_2)\Phi(t, 0); \quad \Phi(t, t) = I. \qquad □$$

Corollary 7.14

An immediate consequence of Corollary 7.9 is that if M has a stable invariant graph subspace and the two algebraic Riccati equations (7.4.7) have a stabilizing solution, then the game has an open-loop Nash equilibrium that permits a feedback synthesis for every initial state. □

Combining the results of both previous theorems yields the following.

Corollary 7.15

The infinite-planning horizon two-player linear quadratic differential game (7.4.1) and (7.4.2) has, for every initial state, an open-loop Nash set of equilibrium actions (u_1^*, u_2^*) which permit a feedback synthesis if and only if:

1. there exist P_1 and P_2 which are solutions of the set of coupled algebraic Riccati equations (7.4.5) and (7.4.6) satisfying the additional constraint that the eigenvalues of $A_{cl} := A - S_1 P_1 - S_2 P_2$ are all situated in the left-half complex plane, and

2. the two algebraic Riccati equations (7.4.7) have a symmetric solution $K_i(.)$ such that $A - S_i K_i$ is stable, $i = 1, 2$.

If (P_1, P_2) is a set of stabilizing solutions of the coupled algebraic Riccati equations (7.4.5) and (7.4.6), the actions

$$u_i^*(t) = -R_i^{-1} B_i^T P_i \Phi(t, 0) x_0, \quad i = 1, 2,$$

where $\Phi(t, 0)$ satisfies the transition equation $\dot{\Phi}(t, 0) = A_{cl} \Phi(t, 0)$; $\Phi(0, 0) = I$, yield an open-loop Nash equilibrium.

The costs, by using these actions, for the players are

$$x_0^T M_i x_0, \quad i = 1, 2, \tag{7.4.9}$$

where M_i is the unique solution of the Lyapunov equation

$$A_{cl}^T M_i + M_i A_{cl} + Q_i + P_i^T S_i P_i = 0. \tag{7.4.10}$$

Proof

Everything has been proved in Theorems 7.12 and 7.13 except for the statement on the cost incurred by the players by playing the equilibrium actions u_i^*, $i = 1, 2$.

To prove equation (7.4.9) notice that with $A_{cl} := A - S_1 P_1 - S_2 P_2$, $x(t) = e^{A_{cl} t} x_0$. Furthermore, since all eigenvalues of matrix A_{cl} are located in $\mathbb{C}^-$, by Corollary 2.32 equation (7.4.10) has a unique solution M_i. Consequently, using equation (7.4.10),

$$\begin{aligned}
J_i(u_1^*, u_2^*) &= \int_0^\infty x(t)(Q_i + P_i^T S_i P_i) x(t) dt \\
&= -\int_0^\infty x(t)(A_{cl}^T M_i + M_i A_{cl}) x(t) dt \\
&= -\int_0^\infty \frac{d[x^T(t) M_i x(t)]}{dt} dt \\
&= x^T(0) M_i x(0) - \lim_{t \to \infty} x^T(t) M_i x(t) \\
&= x^T(0) M_i x(0). \qquad \square
\end{aligned}$$

Once again we stress here the point that if for every initial state there exists more than one open-loop Nash equilibrium that permits a feedback synthesis then the stable subspace has a dimension larger than n. Consequently (see Note 3 following Theorem 7.11) for every initial state there will exist, generically, an infinite number of open-loop Nash equilibria. This, even though there exist only a finite number of open-loop Nash equilibria permitting a feedback synthesis. This point was first noted by Kremer (2002) if matrix A is stable.

The above reflections raise the question whether it is possible to find conditions under which the game has a unique equilibrium for every initial state – Theorem 7.16 gives such conditions. Moreover, it shows that in that case the unique equilibrium actions can be synthesized as a state feedback. The proof of this theorem is provided in the Appendix at the end of this chapter.

Theorem 7.16

The linear quadratic differential game (7.4.1) and (7.4.2) has a unique open-loop Nash equilibrium for every initial state if and only if

1. the set of coupled algebraic Riccati equations (7.4.5) and (7.4.6) has a strongly stabilizing solution, and

2. the two algebraic Riccati equations (7.4.7) have a stabilizing solution.

Moreover, the unique equilibrium actions are given by (7.4.8).   □

Example 7.7

1. Consider the system

$$\dot{x}(t) = -2x(t) + u_1(t) + u_2(t), \quad x(0) = x_0;$$

and cost functions

$$J_1 = \int_0^\infty \{x^2(t) + u_1^2(t)\}dt \quad \text{and} \quad J_2 = \int_0^\infty \{4x^2(t) + u_2^2(t)\}dt.$$

Then,

$$M = \begin{bmatrix} -2 & -1 & -1 \\ -1 & 2 & 0 \\ -4 & 0 & 2 \end{bmatrix}.$$

The eigenvalues of M are $\{-3, 2, 3\}$. An eigenvector corresponding to the eigenvalue -3 is $[5, \ 1, \ 4]^T$.

So, according to Theorem 7.10 item 1, the set of algebraic Riccati equations (7.4.5) and (7.4.6) corresponding to this game has a strongly stabilizing solution. Furthermore, since $q_i > 0$, $i = 1, 2$, the two algebraic Riccati equations (7.4.7) have a stabilizing solution. Consequently, this game has a unique open-loop Nash equilibrium for every initial state x_0.

2. Reconsider the game in item 1, but with the system dynamics replaced by

$$\dot{x}(t) = 2x(t) + u_1(t) + u_2(t), \quad x(0) = x_0.$$

According to Example 7.2 the matrix M corresponding to this game has the eigenvalues $\{-3, -2, 3\}$. Since M has two stable eigenvalues, it follows from Theorem 7.10 item 1 that the set of algebraic Riccati equations (7.4.5) and (7.4.6) corresponding to this game does not have a strongly stabilizing solution. So (see Theorem 7.16) the game does not have for every initial state a unique open-loop Nash equilibrium.

On the other hand, since $[1, \ 1, \ 4]^T$ is an eigenvector corresponding to $\lambda = -3$, it follows from Corollaries 7.9 and 7.15 that the game does have an open-loop Nash equilibrium for every initial state that permits a feedback synthesis. It will be shown later on in Theorem 7.31 that for every initial state there are actually an infinite number of open-loop Nash equilibria. □

In applications it is often assumed that both players discount their future welfare loss. That is, the performance criterion player $i = 1, 2$ likes to minimize is:

$$\lim_{T \to \infty} J_i(u_1, u_2) := \lim_{T \to \infty} \int_0^T e^{-rt} \{x^T(t)Q_i x(t) + u_i^T(t)R_i u_i(t)\} dt,$$

where $r \geq 0$ is the discount factor. We will next, briefly, outline the consequences of such a modeling in terms of our standard framework.

Introducing $\tilde{x}(t) := e^{-\frac{1}{2}rt}x(t)$ and $\tilde{u}_i(t) := e^{-\frac{1}{2}rt}u_i(t)$ (see section 3.6), the above minimization problem can be rewritten as:

$$\min_{\tilde{u}_i} \lim_{T \to \infty} \int_0^T \{\tilde{x}^T(t)Q_i \tilde{x}(t) + \tilde{u}_i^T(t)R_i \tilde{u}_i(t)\} dt, \tag{7.4.11}$$

subject to

$$\dot{\tilde{x}} = \left(A - \frac{1}{2}rI\right)\tilde{x} + B_1 \tilde{u}_1 + B_2 \tilde{u}_2, \quad \tilde{x}(0) = x_0. \tag{7.4.12}$$

The matrix M corresponding to this problem is

$$\tilde{M} := \begin{bmatrix} (A - \frac{1}{2}rI) & -S_1 & -S_2 \\ -Q_1 & -(A^T - \frac{1}{2}rI) & 0 \\ -Q_2 & 0 & -(A^T - \frac{1}{2}rI) \end{bmatrix}$$

and the Hamiltonian matrices associated with the algebraic Riccati equations are

$$\tilde{H}_i = \begin{bmatrix} -(A - \frac{1}{2}rI) & S_i \\ Q_i & (A - \frac{1}{2}rI)^T \end{bmatrix}, \quad i = 1, 2.$$

Using, for example, Geršgorin's theorem (Lancaster and Tismenetsky (1985)) it is clear that if r is chosen large enough, matrix $\tilde{M}$ will have $2n$ stable and n unstable eigenvalues and, similarly, $\tilde{H}_i$ n stable and n unstable eigenvalues. So according to Theorem 7.16, provided the n-dimensional stable eigenspace of $\tilde{M}$ and $\tilde{H}$, respectively, are graph subspaces, there exists a unique open-loop Nash equilibrium.

As in the finite-planning horizon case we conclude this section by considering the zero-sum game. By applying the above results to this special case, analogues are obtained for the open-loop zero-sum linear quadratic game. As an example we will elaborate in the next proposition the analogues of Corollary 7.15 and Theorem 7.16.

Proposition 7.17   (Open-loop zero-sum differential game)

Consider the differential game described by

$$\dot{x}(t) = Ax(t) + B_1 u_1(t) + B_2 u_2(t), \quad x(0) = x_0, \tag{7.4.13}$$

with, for player one, the quadratic cost functional:

$$J_1(u_1, u_2) = \int_0^\infty \{x^T(t)Qx(t) + u_1^T(t)R_1 u_1(t) - u_2^T(t)R_2 u_2(t)\}dt, \tag{7.4.14}$$

and, for player two, the opposite objective function

$$J_2(u_1, u_2) = -J_1(u_1, u_2);$$

where the matrices Q and R_i, $i = 1, 2$, are symmetric. Moreover, assume that R_i, $i = 1, 2$, are positive definite.

Then this infinite-planning horizon zero-sum linear quadratic differential game has, for every initial state, the following.

1. An open-loop Nash equilibrium which permits a feedback synthesis if and only if the following two conditions hold

   (a) the coupled algebraic Riccati equations

   $$A^T P_1 + P_1 A + Q - P_1 S_1 P_1 - P_1 S_2 P_2 = 0, \tag{7.4.15}$$
   $$A^T P_2 + P_2 A - Q - P_2 S_2 P_2 - P_2 S_1 P_1 = 0, \tag{7.4.16}$$

   have a set of solutions P_i, $i = 1, 2$, such that $A - S_1 P_1 - S_2 P_2$ is stable; and
   (b) the two algebraic Riccati equations

   $$A^T K_1 + K_1 A - K_1 S_1 K_1 + Q = 0, \tag{7.4.17}$$
   $$A^T K_2 + K_2 A - K_2 S_2 K_2 - Q = 0, \tag{7.4.18}$$

   have a symmetric solution K_i such that $A - S_i K_i$ is stable, $i = 1, 2$;

moreover, the corresponding equilibrium actions are

$$u_1^*(t) = -R_1^{-1}B_1^T P_1 x(t) \quad \text{and} \quad u_2^*(t) = -R_2^{-1}B_2^T P_2 x(t),$$

where $x(t)$ satisfies the differential equation

$$\dot{x}(t) = (A - S_1 P_1 - S_2 P_2)x(t); \quad x(0) = x_0.$$

2. A unique open-loop Nash equilibrium if and only if the set of coupled algebraic Riccati equations (7.4.15) and (7.4.16) has a strongly stabilizing solution and the two algebraic Riccati equations (7.4.17) and (7.4.18) have a symmetric stabilizing solution. The corresponding equilibrium actions are as described in item 1. □

This proposition is useful in the area of robust control design. To illustrate this point, Corollary 7.18 presents a first result which can be used to design a controller $u(.)$ that encorporates a priori knowledge about disturbances $w(.)$ that have a finite energy acting on the system. In literature (Başar and Bernhard (1995)), this problem is known as the **soft-constrained open-loop differential game**. This problem formulation originates from the H_∞ disturbance attenuation control problem (see Example 3.25). A more detailed discussion of this problem setting is postponed until Chapter 9, where the multi-player feedback information case is analyzed.

Corollary 7.18

Consider the problem to find

$$\inf_{u \in \mathcal{U}_s} \sup_{w \in L_2^q(0,\infty)} \int_0^\infty \{x^T(t)Qx(t) + u^T(t)Ru(t) - w^T(t)Vw(t)\}dt$$

subject to

$$\dot{x}(t) = Ax(t) + Bu(t) + Ew(t), \quad x(0) = x_0$$

where $R > 0$ and $V > 0$. Let $S := BR^{-1}B^T$ and $M := EV^{-1}E^T$.
This problem has a solution for every initial state x_0 if

1. the coupled algebraic Riccati equations

$$A^T P_1 + P_1 A + Q - P_1 S P_1 - P_1 M P_2 = 0,$$
$$A^T P_2 + P_2 A - Q - P_2 M P_2 - P_2 S P_1 = 0,$$

have a strongly stabilizing solution; and

2. the two algebraic Riccati equations

$$A^T K_1 + K_1 A - K_1 S K_1 + Q = 0,$$
$$A^T K_2 + K_2 A - K_2 M K_2 - Q = 0,$$

have a symmetric solution K_i such that $A - SK_1$ and $A - MK_2$ are stable.

Moreover, a worst-case control for the player is

$$u^*(t) = -R^{-1}B^T P_1 x(t)$$

whereas the corresponding worst-case disturbance is

$$w^*(t) = -V^{-1}E^T P_2 x(t).$$

Here $x(t)$ satisfies the differential equation

$$\dot{x}(t) = (A - SP_1 - MP_2)x(t); \quad x(0) = x_0.$$

Proof

From Proposition 7.17 item 2 it follows that with

$$J(u, w) := \int_0^\infty \{x^T(t)Qx(t) + u^T(t)Ru(t) - w^T(t)Vw(t)\}dt,$$

the following inequalities hold:

$$J(u^*, w) \le J(u^*, w^*) \le J(u, w^*), \forall u, w \in U_s.$$

Since $w^* \in L_2^q(0, \infty)$ and for all $u \in U_s$, $w \in L_2^q(0, \infty)$ the game has a proper solution, it follows that the above inequalities also hold for all $u \in U_s$, $w \in L_2^q(0, \infty)$ (see also Theorem 3.11). That is, the game has a saddle-point solution. Therefore, by Theorem 3.26,

$$\inf_{u \in U_s} \sup_{w \in L_2^q(0,\infty)} J(u, w) = \sup_{w \in L_2^q(0,\infty)} \inf_{u \in U_s} J(u, w) = J(u^*, w^*).$$

From this the conclusion is obvious.   □

Next, consider a special case of this game – the case where matrix A is stable. Then the conditions under which the zero-sum game has an equilibrium can be further simplified. In fact it can be shown that the game has a unique equilibrium under this additional condition. This can be shown using the next lemma.

Lemma 7.19

Consider the zero-sum differential game. Then

$$\sigma(M) = \sigma(-A) \cup \sigma\left(\begin{bmatrix} A & -S_1 + S_2 \\ -Q & -A^T \end{bmatrix}\right).$$

Proof

The result follows straightforwardly from the fact that

$$M - \lambda I = S \begin{bmatrix} A - \lambda I & -S_1 + S_2 & -S_2 \\ -Q & -A^T - \lambda I & 0 \\ 0 & 0 & -A^T - \lambda I \end{bmatrix} S^{-1},$$

where $S = \begin{bmatrix} I & 0 & 0 \\ 0 & I & 0 \\ 0 & -I & I \end{bmatrix}$. $\qquad\qquad\square$

Proposition 7.20

Consider the open-loop zero-sum differential game as described in Proposition 7.17. Assume, additionally, that matrix A is stable. Then, with $S_i := B_i R_i^{-1} B_i^T$, this game has for every initial state a unique open-loop Nash equilibrium if and only if the next two conditions hold.

1. The algebraic Riccati equation

$$A^T P + PA + Q - P(S_1 - S_2)P = 0 \qquad (7.4.19)$$

has a (symmetric) solution P such that $A - (S_1 - S_2)P$ is stable.

2. The two algebraic Riccati equations (7.4.17) and (7.4.18) have a symmetric solution K_i such that $A - S_i K_i$ is stable, $i = 1, 2$.

Moreover, the corresponding unique equilibrium actions are

$$u_1^*(t) = -R_1^{-1} B_1^T P x(t) \quad \text{and} \quad u_2^*(t) = R_2^{-1} B_2^T P x(t).$$

Here $x(t)$ satisfies the differential equation

$$\dot{x}(t) = (A - (S_1 - S_2)P)x(t); \quad x(0) = x_0.$$

The cost for player 1 is $J_1 := x_0^T P x_0$ and for player 2, $-J_1$.

Proof

According to Theorem 7.17 this game has a unique open-loop Nash equilibrium for every initial state if and only if the following two coupled algebraic Riccati equations have a set of strongly stabilizing solutions P_i, $i = 1, 2$,

$$A^T P_1 + P_1 A + Q - P_1 S_1 P_1 - P_1 S_2 P_2 = 0, \qquad (7.4.20)$$
$$A^T P_2 + P_2 A - Q - P_2 S_2 P_2 - P_2 S_1 P_1 = 0, \qquad (7.4.21)$$

and the two algebraic Riccati equations (7.4.17) and (7.4.18) have a stabilizing solution K_i. Adding equations (7.4.20) to (7.4.21) yields the following differential equation in $(P_1 + P_2)$:

$$A^T(P_1 + P_2) + (P_1 + P_2)(A - S_1P_1 - S_2P_2) = 0.$$

This is a Lyapunov equation of the form $A^TX + XB^T = 0$, with $X := P_1 + P_2$ and $B = A - S_1P_1 - S_2P_2$. Now, independent of the specification of (P_1, P_2), B is always stable. So, whatever P_i, $i = 1, 2$ are, this Lyapunov equation has a unique solution $X(P_1, P_2)$. Obviously, $P_1 + P_2 = 0$ satisfies this Lyapunov equation. So, necessarily, we conclude that $P_1 = -P_2$. Substitution of this into equation (7.4.20) shows then that equations (7.4.20) and (7.4.21) have a stabilizing solution P_i if and only if equation (7.4.19) has a stabilizing solution P.

The corresponding equilibrium strategies then follow directly from Theorem 7.17. The symmetry and uniqueness properties of P follow immediately from the fact that P is a stabilizing solution of an ordinary Riccati equation (see Theorem 2.33). Furthermore, using equation (7.4.19), the cost for player 1 can be rewritten as

$$
\begin{aligned}
J_1 &= \int_0^\infty x^T(t)\{Q + PS_1P - PS_2P\}x(t)dt \\
&= \int_0^\infty x^T(t)\{2PS_1P - 2PS_2P - A^TP - PA\}x(t)dt \\
&= -\int_0^\infty x^T(t)\{(A - (S_1 - S_2)P)^TP + P(A - (S_1 - S_2)P)\}x(t)dt \\
&= -\int_0^\infty \frac{d[x^T(t)Px(t)]}{dt}dt \\
&= x_0^TPx_0.
\end{aligned}
$$

To show that the set of coupled Riccati equations (7.4.20) and (7.4.21) has a strongly stabilizing solution we show that matrix M has an n-dimensional stable graph subspace. To that end notice that since all eigenvalues of $-A$ are unstable it follows from Lemma 7.19 that, to satisfy the requirement that M must have n stable eigenvalues, the Hamiltonian matrix

$$
\begin{bmatrix} A & -S_2 + S_1 \\ -Q & -A^T \end{bmatrix}
$$

must have n stable eigenvalues. However, since the spectrum of a Hamiltonian matrix is symmetric w.r.t. the imaginary axis (see Note following Theorem 2.30), we infer that this Hamiltonian matrix must have n unstable eigenvalues as well. So, we conclude that matrix M has n stable eigenvalues and $2n$ unstable eigenvalues. As argued above, the set of coupled Riccati equations (7.4.20) and (7.4.21) has a stabilizing solution. Consequently, by Theorems 7.10 and 7.16, there is a unique open-loop Nash equilibrium. $\qquad\square$

Note

If matrix A is not stable, it may happen that there exist an infinite number of open-loop Nash equilibria in Proposition 7.17. (see for example Exercise 14 at the end of this chapter). □

In case it is additionally assumed in the above zero-sum game that Q is positive semi-definite, equation (7.4.19) has a stabilizing solution (see Proposition 5.15). Moreover in that case, by considering $K := -K_2$, the fact that equation (7.4.18) should have a stabilizing solution can be rephrased as that the following Riccati equation should have a solution K such that $A + S_2 K$ is stable:

$$A^T K + KA + KS_2 K + Q = 0.$$

Below, in Proposition 7.21 and Corollary 7.22, we summarize the consequences of this semi-definiteness assumption for Proposition 7.17 and Corollary 7.18, respectively. More results (and, in particular, converse statements) on the soft-constrained differential game considered in Corollary 7.22 can be found in Başar and Bernhard (1995).

Proposition 7.21

Consider the open-loop zero-sum differential game as described in Proposition 7.17. Assume that matrix A is stable and $Q \geq 0$. Then this game has, for every initial state, a unique open-loop Nash equilibrium if and only if the algebraic Riccati equations

$$A^T P + PA + Q - P(S_1 - S_2)P = 0$$

and

$$A^T K + KA + KS_2 K + Q = 0.$$

have a solution $\bar{P}$ and $\bar{K}$, respectively, such that $A - (S_1 - S_2)\bar{P}$ and $A + S_2\bar{K}$ are stable. Moreover, the corresponding unique equilibrium actions are

$$u_1^*(t) = -R_1^{-1}B_1^T \bar{P}x(t) \quad \text{and} \quad u_2^*(t) = R_2^{-1}B_2^T \bar{P}x(t).$$

Here $x(t)$ satisfies the differential equation

$$\dot{x}(t) = (A - (S_1 - S_2)\bar{P})x(t); \quad x(0) = x_0.$$

The costs involved with these actions for player 1 are $J_1 := x_0^T \bar{P}x_0$ and for player 2, $-J_1$. □

Corollary 7.22

Consider the problem to find

$$\bar{J} := \inf_{u \in \mathcal{U}_s} \sup_{w \in L_2^q(0,\infty)} \int_0^\infty \{x^T(t)Qx(t) + u^T(t)Ru(t) - w^T(t)Vw(t)\}dt$$

subject to

$$\dot{x}(t) = Ax(t) + Bu(t) + Ew(t), \quad x(0) = x_0,$$

where $R > 0$, $Q \geq 0$, $V > 0$ and A is stable. Let $S := BR^{-1}B^T$ and $M := EV^{-1}E^T$. This problem has a solution for every initial state x_0 if the algebraic Riccati equations

$$A^T P + PA + Q - P(S - M)P = 0$$

and

$$A^T K + KA + KMK + Q = 0.$$

have a solution $\bar{P}$ and $\bar{K}$, respectively, such that $A - (S - M)\bar{P}$ and $A + M\bar{K}$ are stable. Furthermore, a worst-case control for the player is

$$u^*(t) = -R^{-1}B^T \bar{P}x(t)$$

whereas the corresponding worst-case disturbance is

$$w^*(t) = V^{-1}E^T \bar{P}x(t).$$

Here $x(t)$ satisfies the differential equation

$$\dot{x}(t) = (A - (S - M)\bar{P})x(t); \quad x(0) = x_0.$$

Moreover, $\bar{J} = x_0^T \bar{P}x_0.$ □

From the above corollary we infer in particular that if $x_0 = 0$, the best open-loop worst-case controller is $u = 0$, whereas the worst-case signal in that case is $w = 0$. This is independent of the choice of V, under the supposition that the Riccati equations have an appropriate solution. So if a stable system is in equilibrium (i.e. $x_0 = 0$) in this open-loop framework the best reaction to potential unknown disturbances is not to react. Example 7.8, below, elaborates for a scalar game the best worst-case reaction for general initial conditions.

Example 7.8

Consider the following model which we introduced in Example 3.25.

$$\dot{x}(t) = -\alpha\delta x(t) + \delta g(t) + w(t), \quad x(0) = x_0. \tag{7.4.22}$$

Here x measures the deviation of supply from its long-term equilibrium value, g is the deviation of public sector demand from its long-term equilibrium value and w represents some unknown, though 'finite energy' (i.e. $w(.) \in L_2$), disturbance which affects the system. $w(.)$ might for instance model in this context the effect of non-linear terms that enter the system but which are not modeled explicitly here due to their expected small impact.

The objective is to design a public sector policy such that both the public sector demand and supply will remain close to their long-term equilibrium values in spite of the unpredictable disturbances $w(t)$. This is formalized as to find for a fixed $\lambda > 0$ the solution to the soft-constrained differential game

$$\inf_{g\in\mathcal{U}_s}\sup_{w\in L_2} L_\lambda(g,w) \tag{7.4.23}$$

subject to equation (7.4.22) where

$$L_\lambda(g,w) := \int_0^\infty \{x^2(t) + \phi g^2(t)\}dt - \lambda\int_0^\infty w^2(t)dt.$$

The interpretation of the parameter $\lambda > 0$ in this context is that it indicates the expectation of the government as to how severe the disturbance will be. If λ is very large, the government expects that there will be almost no disturbance acting on the system, whereas if λ becomes small the government expects a more severe impact of disturbances. This might also be interpreted as that if λ is chosen large, the government takes a more risky attitude towards the potential disturbance that might affect the system: it neglects any potential disturbance in setting its control policy in that case.

According to Corollary 7.22, for a fixed λ, equation (7.4.23) has a solution if both

$$-2\alpha\delta p + 1 - \left(\frac{\delta^2}{\phi} - \frac{1}{\lambda}\right)p^2 = 0$$

has a solution $\bar{p}$ such that $-\alpha\delta - \left(\frac{\delta^2}{\phi} - \frac{1}{\lambda}\right)\bar{p} < 0$, and

$$-2\alpha\delta k + 1 + \frac{1}{\lambda}k^2 = 0$$

has a solution $\bar{k}$ such that $-\alpha\delta + \frac{1}{\lambda}\bar{k} < 0$. Obviously this is the case if and only if $\alpha^2\delta^2 + \left(\frac{\delta^2}{\phi} - \frac{1}{\lambda}\right) > 0$ and $\alpha^2\delta^2 - \frac{1}{\lambda} > 0$. That is, if

$$\lambda > \frac{1}{\alpha^2\delta^2}. \tag{7.4.24}$$

It can be shown (Başar and Bernhard (1995)) that for all $\lambda \leq \frac{1}{\alpha^2\delta^2}$, expression (7.4.23) has no solution. From this it follows that there exists a lowerbound $\check{\lambda} := \frac{1}{\alpha^2\delta^2}$ for the risk-attitude parameter λ for which a robust control policy exists. However, notice that at this lowerbound $\check{\lambda}$ there does not exist an admissible control policy. A robust control policy corresponding with a risk attitude $\lambda > \frac{1}{\alpha^2\delta^2}$ is

$$g^*(t) = -\frac{\delta}{\phi}\bar{p}_\lambda e^{(-\alpha\delta - (\frac{\delta^2}{\phi} - \frac{1}{\lambda})\bar{p}_\lambda)t}x_0, \tag{7.4.25}$$

where

$$\bar{p}_\lambda = \frac{\alpha\delta - \sqrt{\alpha^2\delta^2 + \frac{\delta^2}{\phi} - \frac{1}{\lambda}}}{-\left(\frac{\delta^2}{\phi} - \frac{1}{\lambda}\right)}. \tag{7.4.26}$$

Using this control, for every admissible realization of the disturbance $\bar{w}(.)$ the performance of the government can be estimated by:

$$\int_0^\infty \{\bar{x}^2(t) + \phi g^{*^2}(t) - \lambda \bar{w}^2(t)\} dt \leq \bar{p}_\lambda x_0^2.$$

Here $\bar{x}(.)$ denotes the trajectory implied by the disturbance $\bar{w}(.)$ and control $g^*(.)$. Notice that the more stable the uncontrolled system is, the better this estimate will be (see also equation (7.4.27), below). By a direct inspection of Theorem 5.14 one verifies that the solution of the 'noise-free' regulator problem corresponding to this problem (i.e., the problem with $w = 0$) coincides with the solution one obtains by considering $\lambda \to \infty$ in equations (7.4.25) and (7.4.26).

To assess the difference between the 'noise-free' and the soft-constrained controller we calculate the derivative of $g^*(t)$ in equation (7.4.25) with respect to λ. According to equation (7.4.26)

$$\bar{p}_\lambda = \frac{\alpha^2\delta^2 - \left(\alpha^2\delta^2 + \frac{\delta^2}{\phi} - \frac{1}{\lambda}\right)}{-\left(\frac{\delta^2}{\phi} - \frac{1}{\lambda}\right)\left(\alpha\delta + \sqrt{\alpha^2\delta^2 + \frac{\delta^2}{\phi} - \frac{1}{\lambda}}\right)} = \frac{1}{\alpha\delta + \sqrt{\alpha^2\delta^2 + \frac{\delta^2}{\phi} - \frac{1}{\lambda}}} > 0. \qquad (7.4.27)$$

Therefore

$$\frac{d\bar{p}_\lambda}{d\lambda} = \frac{-1}{\left(\alpha\delta + \sqrt{\alpha^2\delta^2 + \frac{\delta^2}{\phi} - \frac{1}{\lambda}}\right)^2} \frac{\frac{1}{2}}{\sqrt{\alpha^2\delta^2 + \frac{\delta^2}{\phi} - \frac{1}{\lambda}}} \frac{1}{\lambda^2} < 0.$$

Consequently, assuming $x_0 > 0$,

$$\frac{dg^*(t)}{d\lambda} = -\frac{\delta}{\phi} e^{\left(-\alpha\delta - \left(\frac{\delta^2}{\phi} - \frac{1}{\lambda}\right)\bar{p}_\lambda\right)t} x_0 \left(\frac{d\bar{p}_\lambda}{d\lambda} - \frac{t}{\lambda^2}\bar{p}_\lambda^2 + \frac{t}{\lambda}\bar{p}_\lambda \frac{d\bar{p}_\lambda}{d\lambda}\right) > 0. \qquad (7.4.28)$$

From equation (7.4.28) it follows that in the 'noise-free' case the government uses at any point in time a less active public sector policy to regulate the supply back to its equilibrium level. Under the worst-case scenario the closed-loop system is

$$\dot{x}(t) = -\sqrt{\alpha^2\delta^2 + \frac{\delta^2}{\phi} - \frac{1}{\lambda}} x(t), \quad x(0) = x_0.$$

From this it follows that, the larger λ is, the quicker the supply converges to its equilibrium value. So, in conclusion, in this open-loop setting we see that the more the government is convinced that there will be disturbance effects on its out of equilibrium system the more active control policy it will use. The resulting worst-case closed-loop system will be less stable. □

Note

The fact that at the lower-bound risk-attitude parameter in the above example a control policy does not exist is not a coincidence. It can be shown (Başar and Bernhard (1995)) that in soft-constrained robust control problems at this infimum an admissible control policy never exists. So, *the* smallest risk-attitude with an admissible corresponding control policy does not exist. □

7.5 Computational aspects and illustrative examples

This section presents a numerical algorithm to verify whether the infinite-planning horizon open-loop game has a Nash equilibrium that permits a feedback synthesis. Moreover, if such an equilibrium exists, this algorithm immediately yields the appropriate control actions. The usual assumptions apply: matrix R_i is positive definite and Q_i symmetric, $i = 1, 2$, and both the matrix pairs (A, B_i), $i = 1, 2$, are stabilizable. Provided these assumptions hold the algorithm reads as follows.

Algorithm 7.1

Step 1 Calculate the eigenstructure of $H_i := \begin{bmatrix} A & -S_i \\ -Q_i & -A^T \end{bmatrix}$.

If H_i, $i = 1, 2$, has an n-dimensional stable graph subspace, then proceed. Otherwise go to Step 5.

Step 2 Calculate matrix $M := \begin{bmatrix} A & -S_1 & -S_2 \\ -Q_1 & -A^T & 0 \\ -Q_2 & 0 & -A^T \end{bmatrix}$. Next calculate the spectrum of M.

If the number of negative eigenvalues (counted with algebraic multiplicities) is less than n, go to Step 5.

Step 3 Calculate all M-invariant subspaces $\mathcal{P} \in \mathcal{P}^{pos}$ for which Re $\lambda < 0$ for all $\lambda \in \sigma(M|_{\mathcal{P}})$. If this set is empty, go to Step 5.

Step 4 Let $\mathcal{P}$ be an arbitrary element of the set determined in Step 3.

Calculate $3n \times n$ matrices X, Y and Z such that Im $\begin{bmatrix} X \\ Y \\ Z \end{bmatrix} = \mathcal{P}$.

Denote $P_1 := YX^{-1}$ and $P_2 := ZX^{-1}$. Then

$$u_i^*(t) := -R_i^{-1} B_i^T P_i e^{A_{cl}t} x_0$$

is an open-loop Nash equilibrium strategy that permits a feedback synthesis. Here $A_{cl} := A - S_1 P_1 - S_2 P_2$. The spectrum of the corresponding closed-loop matrix A_{cl} equals $\sigma(M|_{\mathcal{P}})$. The involved cost for player i is $x_0^T M_i x_0$, where M_i is the unique solution of the Lyapunov equation:

$$A_{cl}^T M_i + M_i A_{cl} + Q_i + P_i^T S_i P_i = 0.$$

If the set determined in Step 3 contains more elements one can repeat this step to calculate different equilibria. If the set determined in Step 3 contains exact one element verify whether M has $2n$ eigenvalues in $\mathbb{C}_0^+$. If this is the case the game has a unique equilibrium for every initial state.

Step 5 End of algorithm. □

The algorithm may yield infinitely many different solutions P_i. However, notice that there are at the most $\begin{bmatrix} 2n \\ n \end{bmatrix}$ different structures for the eigenvalues of the closed-loop system.

Step 1 in the algorithm verifies whether the two algebraic Riccati equations (7.4.7) have a stabilizing solution. Step 2 and 3 verify whether matrix M has an n-dimensional stable graph subspace. Finally, Step 4 then determines all n-dimensional stable graph subspaces of matrix M. According to Corollary 7.15 each of these graph subspaces implies an open-loop Nash equilibrium that allows for a feedback synthesis. Moreover, according to Theorem 7.16, there is for every initial state a unique equilibrium if and only if M has a unique stable graph subspace and matrix M has $2n$ eigenvalues (counting algebraic multiplicities) in $\mathbb{C}_0^+$.

The next examples illustrate the algorithm. We begin with a simple scalar example.

Example 7.9

Let $A = 3$; $B_i = Q_i = 2$ and $R_i = 1$, $i = 1, 2$. Since $Q_i > 0$, Step 1 in the algorithm can be skipped (see Proposition 5.15), so we proceed with Step 2. That is, determine matrix M. Obviously

$$M = \begin{bmatrix} 3 & -4 & -4 \\ -2 & -3 & 0 \\ -2 & 0 & -3 \end{bmatrix}.$$

The second step is to calculate the eigenvalues of M. The eigenvalues of M are: -5, -3 and 5. The corresponding eigenvectors are

$$\begin{bmatrix} 1 \\ 1 \\ 1 \end{bmatrix}, \begin{bmatrix} 0 \\ -1 \\ 1 \end{bmatrix} \text{ and } \begin{bmatrix} -4 \\ 1 \\ 1 \end{bmatrix}, \text{ respectively.}$$

From this we observe that, although M has two stable eigenvalues, $\mathcal{P}^{pos}$ contains only one element: the eigenspace $\text{Im} \begin{bmatrix} 1 \\ 1 \\ 1 \end{bmatrix}$ corresponding to the eigenvalue -5.

According to Step 4 of the algorithm, $P_1 = P_2 = 1$ provide a stabilizing solution of the set of coupled algebraic Riccati equations. These solutions yield the open-loop Nash equilibrium actions $(u_1^*(t), u_2^*(t)) = (-2x(t), -2x(t))$, and the closed-loop system $\dot{x}(t) = -5x(t)$, $x(0) = x_0$. The corresponding equilibrium cost for both players is then straightforwardly obtained from equation (7.4.10) as $J_i = \frac{3}{5}x_0^2$, $i = 1, 2$.

Notice that M has two stable eigenvalues. So, according to Theorem 7.16 there are initial states for which the game has more than one equilibrium. We will elaborate this point in section 7.7. □

The next example illustrates that the infinite-planning horizon game may have more than one equilibrium that permits a feedback synthesis, even if matrix A is stable.

Example 7.10

Let $A = \begin{bmatrix} -0.1 & 0 \\ 0 & -2 \end{bmatrix}$, $B_1 = \begin{bmatrix} 1 & 0 \\ 0 & 1 \end{bmatrix}$, $B_2 = \begin{bmatrix} 1 \\ 0 \end{bmatrix}$, $Q_1 = \begin{bmatrix} 1 & 0 \\ 0 & 0.1 \end{bmatrix}$, $Q_2 = \begin{bmatrix} 1 & 1 \\ 1 & 2 \end{bmatrix}$,

$R_1 = \begin{bmatrix} 2 & -1 \\ -1 & 1 \end{bmatrix}$ and $R_2 = 1$. Since matrix A is stable each pair (A, B_i), $i = 1, 2$, is stabilizable. Furthermore, both Q_i and R_i are positive definite. So, all the assumptions we made on the system and the performance criteria apply. Again, Step 1 in Algorithm 7.1 is trivially satisfied (see Proposition 5.15).

Our first step is therefore to fix matrix M. This yields:

$$M = - \begin{bmatrix} 0.1 & 0 & 1 & 1 & 1 & 0 \\ 0 & 2 & 1 & 2 & 0 & 0 \\ 1 & 0 & -0.1 & 0 & 0 & 0 \\ 0 & 0.1 & 0 & -2 & 0 & 0 \\ 1 & 1 & 0 & 0 & -0.1 & 0 \\ 1 & 2 & 0 & 0 & 0 & -2 \end{bmatrix}.$$

Next we determine the spectrum of M. Numerical calculations show that $M = TJT^{-1}$, where J is a diagonal matrix with entries $\{2; -2.2073; -1.0584; 2.0637; -0.1648; 1.4668\}$ and

$$T = \begin{bmatrix} 0 & 0.2724 & -0.6261 & -0.0303 & -0.1714 & 0.3326 \\ 0 & 0.7391 & 0.5368 & -0.0167 & 0.3358 & 0.0633 \\ 0 & 0.1181 & -0.5405 & 0.0154 & -0.6473 & -0.2433 \\ 0 & 0.0176 & 0.0176 & 0.0262 & 0.0155 & 0.0119 \\ 0 & 0.4384 & -0.0771 & 0.0239 & 0.6207 & -0.2897 \\ 1 & 0.4161 & 0.1463 & 0.9987 & 0.2311 & 0.8614 \end{bmatrix}.$$

So M has six different eigenvalues, three of them are negative. Therefore, there are at most $\begin{bmatrix} 3 \\ 2 \end{bmatrix} = 3$ different equilibrium strategies that permit a feedback synthesis.

We proceed with Step 3 of the algorithm. Introduce the following notation $T =: [T_1 \ T_2 \ T_3 \ T_4 \ T_5 \ T_6]$. First consider $\mathcal{P}_1 := \text{Im}[T_2 \ T_3]$. The first 2×2 block of this matrix equals $\begin{bmatrix} 0.2724 & -0.6261 \\ 0.7391 & 0.5368 \end{bmatrix}$. This matrix is invertible. So, $\mathcal{P}_1$ is an element of $\mathcal{P}^{pos}$ and $\sigma(M|_{\mathcal{P}_1}) = \{-2.2073, -1.0584\}$. In a similar way it can be verified that also $\mathcal{P}_2 := \text{Im}[T_2 \ T_5]$ and $\mathcal{P}_3 := \text{Im}[T_3 \ T_5]$ are appropriate graph subspaces. So, Step 3 yields three M-invariant subspaces satisfying all conditions.

In Step 4 the equilibrium strategies permitting a feedback synthesis are calculated. According to Step 3 there are three such equilibrium strategies. We will calculate the equilibrium strategy that permits a feedback synthesis resulting from $\mathcal{P}_3$. To that end we factorize $\mathcal{P}_3$ as follows

$$\mathcal{P}_3 = \begin{bmatrix} -0.6261 & -0.1714 \\ 0.5368 & 0.3358 \\ -0.5405 & -0.6473 \\ 0.0176 & 0.0155 \\ -0.0771 & 0.6207 \\ 0.1463 & 0.2311 \end{bmatrix} =: \begin{bmatrix} X \\ Y \\ Z \end{bmatrix},$$

where X, Y and Z are 2×2 matrices.

Then,

$$P_1 := YX^{-1} = \begin{bmatrix} -0.5405 & -0.6473 \\ 0.0176 & 0.0155 \end{bmatrix} \begin{bmatrix} -0.6261 & -0.1714 \\ 0.5368 & 0.3358 \end{bmatrix}^{-1}$$

and

$$P_2 := ZX^{-1} = \begin{bmatrix} -0.0771 & 0.6207 \\ 0.1463 & 0.2311 \end{bmatrix} \begin{bmatrix} -0.6261 & -0.1714 \\ 0.5368 & 0.3358 \end{bmatrix}^{-1}.$$

The corresponding open-loop Nash strategy is then $u_i^*(t) := -R_i^{-1} B_i^T P_i e^{A_{cl}t} x_0$. Here the closed-loop matrix $A_{cl} := A - S_1 P_1 - S_2 P_2$ is (approximately) $\begin{bmatrix} -1.7536 & -0.8110 \\ 1.3629 & 0.5310 \end{bmatrix}$. It is easily verified that the spectrum of this matrix A_{cl} indeed equals $\{-1.0584, -0.1648\}$ (numerical rounding up). The cost for both players in this equilibrium is obtained by solving the corresponding Lyapunov equation (7.4.10). This gives a cost $J_1 = x_0^T \begin{bmatrix} 20.6 & 25.4 \\ 25.4 & 32.5 \end{bmatrix} x_0$ and $J_2 = x_0^T \begin{bmatrix} 31.6 & 36.9 \\ 36.9 & 43.6 \end{bmatrix} x_0$. $\qquad\square$

Example 7.11 illustrates a game which has an infinite number of open-loop Nash equilibria that permit a feedback synthesis. To construct such a game, recall from Corollary 2.25 that the matrices A, B_i, Q_i and R_i, $i = 1, 2$, should be chosen such that matrix M has an eigenvalue with negative real part whose geometric multiplicity is larger than one.

Example 7.11

Let $A = \begin{bmatrix} \frac{1}{2} & 0 \\ 0 & \frac{1}{4} \end{bmatrix}$, $B_1 = B_2 = \begin{bmatrix} 1 & 0 \\ 0 & 1 \end{bmatrix}$, $Q_1 = \frac{1}{2}\begin{bmatrix} 1 & 1 \\ 1 & 3 \end{bmatrix}$, $Q_2 = \frac{1}{2}\begin{bmatrix} 2 & -\frac{7}{9} \\ -\frac{7}{9} & 1 \end{bmatrix}$, $R_1^{-1} = \frac{1}{2}\begin{bmatrix} 1 & -\frac{7}{90} \\ -\frac{7}{90} & 1 \end{bmatrix}$ and $R_2^{-1} = \frac{1}{2}\begin{bmatrix} 1 & -\frac{1}{10} \\ -\frac{1}{10} & \frac{3}{4} \end{bmatrix}$.

Notice that (A, B_i), $i = 1, 2$, is stabilizable. Furthermore, both Q_i and R_i are positive definite. So, the assumptions we made on the system and the performance criteria are fulfilled. Moreover, it follows from Proposition 5.15 that the algebraic Riccati equations (7.4.7) have stabilizing solutions.

Next, we determine the eigenstructure of matrix M. Numerical calculations show that $M = TJT^{-1}$, where J is a diagonal matrix with entries $\{-1; -1; -0.4983; -0.2525; 1.004 + 0.0227i; 1.004 - 0.0227i\}$ and, with $v_1 := -0.1096 - 0.4478i$; $v_2 := -0.4572 + 0.1112i$; $v_3 := 0.1905 + 0.1093i$; $v_4 := 0.5929 + 0.0350i$; $v_5 := -0.0405 + 0.3279i$; $v_6 := 0.1454 - 0.1864i$,

$$T = \begin{bmatrix} -0.3862 & -0.3860 & 0.0002 & 0.0045 & v_1 & \bar{v}_1 \\ 0.0044 & 0.0461 & -0.0026 & -0.0005 & v_2 & \bar{v}_2 \\ -0.3818 & -0.3399 & 0.7063 & -0.0080 & v_3 & \bar{v}_3 \\ -0.2486 & -0.1650 & -0.0156 & 0.5979 & v_4 & \bar{v}_4 \\ -0.7759 & -0.8079 & -0.7077 & -0.0189 & v_5 & \bar{v}_5 \\ 0.2032 & 0.2309 & -0.0056 & -0.8013 & v_6 & \bar{v}_6 \end{bmatrix}.$$

Matrix M has five different eigenvalues. Three of them are negative and the geometric multiplicity of the eigenvalue -1 is two. Therefore the first and third column of T, for example, constitute an invariant subspace of M corresponding to the eigenvalues $\{-1, -0.4983\}$. This gives rise to the next solution set of the coupled algebraic Riccati equations (7.4.5) and (7.4.6):

$$P_1 = \begin{bmatrix} -0.3818 & 0.7063 \\ -0.2486 & -0.0156 \end{bmatrix} \begin{bmatrix} -0.3862 & 0.0002 \\ 0.0044 & -0.0026 \end{bmatrix}^{-1} = \begin{bmatrix} -2.0840 & -267.4941 \\ 0.7120 & 5.9446 \end{bmatrix}$$

and

$$P_2 = \begin{bmatrix} -0.7759 & -0.7077 \\ 0.2032 & -0.0056 \end{bmatrix} \begin{bmatrix} -0.3862 & 0.0002 \\ 0.0044 & -0.0026 \end{bmatrix}^{-1} = \begin{bmatrix} 5.0900 & -268.2408 \\ -0.5021 & 2.0934 \end{bmatrix}.$$

The corresponding closed-loop system matrix is: $A - S_1 P_1 - S_2 P_2 = \begin{bmatrix} -1.0004 & -0.0375 \\ 0.0058 & -0.4978 \end{bmatrix}.$

The corresponding costs are (approximately) $J_1 = x_0^T \begin{bmatrix} 3 & 332 \\ 332 & 36053 \end{bmatrix} x_0$ and $J_2 = x_0^T \begin{bmatrix} 5 & -320 \\ -320 & 36218 \end{bmatrix} x_0$, respectively.

On the other hand, one can also consider the second and the third column of T. They also form an element of $\mathcal{P}^{pos}$, corresponding to the same eigenvalues $\{-1, -0.4983\}$. Numerical calculations show again that the corresponding solution set of the coupled algebraic Riccati equation is:

$$P_1 = \begin{bmatrix} -0.3399 & 0.7063 \\ -0.1650 & -0.0156 \end{bmatrix} \begin{bmatrix} -0.3860 & 0.0002 \\ 0.0461 & -0.0026 \end{bmatrix}^{-1} = \begin{bmatrix} -31.3603 & -269.6804 \\ 1.1421 & 5.9767 \end{bmatrix}$$

and

$$P_2 = \begin{bmatrix} -0.8079 & -0.7077 \\ 0.2309 & -0.0056 \end{bmatrix} \begin{bmatrix} -0.3860 & 0.0002 \\ 0.0461 & -0.0026 \end{bmatrix}^{-1} = \begin{bmatrix} 34.4236 & 270.4314 \\ -0.3466 & 2.1050 \end{bmatrix}.$$

The corresponding closed-loop system matrix is now $A - S_1 P_1 - S_2 P_2 = \begin{bmatrix} -1.0045 & -0.0378 \\ 0.0605 & -0.4937 \end{bmatrix}$, whereas the corresponding costs are $J_1 = x_0^T \begin{bmatrix} 506 & 4305 \\ 305 & 36645 \end{bmatrix} x_0$ and $J_2 = x_0^T \begin{bmatrix} 571 & 4572 \\ 4572 & 36628 \end{bmatrix} x_0$, respectively.

However, these are not the only possibilities to construct an invariant subspace of $\mathcal{P}^{pos}$ which corresponds to the eigenvalues $\{-1, -0.4983\}$. As a matter of fact, any linear combination of the first two columns of matrix T and the third column of T does the job. For instance if we consider the sum of the first column and the second column of matrix

T, together with the third column of matrix T, we obtain the next solution of the coupled algebraic Riccati equation:

$$P_1 = \begin{bmatrix} -0.7217 & 1.4126 \\ -0.4136 & -0.0156 \end{bmatrix} \begin{bmatrix} -0.7722 & 0.0004 \\ 0.0505 & -0.0052 \end{bmatrix}^{-1} = \begin{bmatrix} -16.6593 & -268.5826 \\ 0.9261 & 5.9606 \end{bmatrix}$$

and

$$P_2 = \begin{bmatrix} -1.5838 & -1.4154 \\ 0.4341 & -0.0112 \end{bmatrix} \begin{bmatrix} -0.7722 & 0.0004 \\ 0.0505 & -0.0052 \end{bmatrix}^{-1} = \begin{bmatrix} 19.6938 & 269.3314 \\ -0.4247 & 2.0992 \end{bmatrix},$$

with the corresponding closed-loop system matrix $A - S_1P_1 - S_2P_2 = \begin{bmatrix} -1.0025 & -0.0377 \\ 0.0330 & -0.4958 \end{bmatrix}$.

The corresponding costs are in this case $J_1 = x_0^T \begin{bmatrix} 146 & 2301 \\ 2301 & 36344 \end{bmatrix} x_0$ and $J_2 = x_0^T$ $\begin{bmatrix} 182 & 2568 \\ 2568 & 36327 \end{bmatrix} x_0$, respectively.

In this way one can construct an infinite number of different sets of solutions for the set of coupled algebraic Riccati equations (7.4.5) and (7.4.6). Therefore, the game has an infinite number of equilibria that permit a feedback synthesis. All these equilibria have in common that the eigenvalues of the closed-loop system matrix are $\{-1, -0.4983\}$.   □.

Finally, we demonstrate how to cope with the case that matrix M has complex eigenvalues with a negative real part. If M has a complex eigenvalue $\lambda = a + bi$ and $z = x + iy$ is a corresponding eigenvector, recall from Theorem 2.17, that $S = \text{Im}[x\, y]$ is a two-dimensional invariant subspace of M.

Example 7.12

Consider the same system matrices as in Example 7.11, except matrix A which we replace by $A = \begin{bmatrix} -\frac{1}{2} & 0 \\ 0 & -\frac{1}{4} \end{bmatrix}$.

Then, as in the previous example, condition 2 of Corollary 7.15 is satisfied. To find the solutions of the coupled algebraic Riccati equations (7.4.5) and (7.4.6) we first fix again the eigenstructure of matrix M. Numerical calculations show that $M = TJT^{-1}$, where J is a diagonal matrix with entries $\{-1.004 + 0.0227i; -1.004 - 0.0227i; 0.2525; 0.4983; 1; 1\}$ and, with $v_1 := -0.1096 + 0.4478i;$ $v_2 := -0.4572 - 0.1112i;$ $v_3 := -0.1905 + 0.1093i;$ $v_4 := -0.5929 + 0.0350i;$ $v_5 := 0.0405 + 0.3279i;$ $v_6 := -0.1454 - 0.1864i,$

$$T = \begin{bmatrix} v_1 & \bar{v}_1 & -0.0045 & 0.0002 & 0.3867 & -0.3698 \\ v_2 & \bar{v}_2 & 0.0005 & -0.0026 & -0.0270 & -0.0876 \\ v_3 & \bar{v}_3 & -0.0080 & -0.7063 & -0.3597 & 0.4574 \\ v_4 & \bar{v}_4 & 0.5979 & 0.0156 & -0.2038 & 0.4217 \\ v_5 & \bar{v}_5 & -0.0189 & 0.7077 & -0.7944 & 0.6714 \\ v_6 & \bar{v}_6 & -0.8013 & 0.0056 & 0.2185 & -0.1333 \end{bmatrix}.$$

So M has two complex eigenvalues with a negative real part. Let x be the real part of the eigenvector corresponding with the eigenvalue $-1.004 + 0.0227i$ and y the imaginary part of this eigenvector

$$x^T := [-0.1096, \ -0.4572, \ -0.1905, \ -0.5929, \ 0.0405, \ -0.1454] \text{ and}$$

$$y^T := [0.4478, \ -0.1112, \ 0.1093, \ 0.0350, \ 0.3279, \ -0.1864].$$

The invariant subspace corresponding to these eigenvalues $\{-1.004 + 0.0227i; -1.004 - 0.0227i\}$ is then $S = \text{Im}[x \ y]$. By Theorem 7.16, the unique equilibrium actions are $u_i^*(t) = -R_i^{-1}B_i^T P_i x(t)$, where

$$P_1 = \begin{bmatrix} -0.1905 & 0.1093 \\ -0.5929 & 0.0350 \end{bmatrix} \begin{bmatrix} -0.1096 & 0.4478 \\ -0.4572 & -0.1112 \end{bmatrix}^{-1} = \begin{bmatrix} 0.3280 & 0.3381 \\ 0.3775 & 1.2064 \end{bmatrix}$$

and

$$P_2 = \begin{bmatrix} 0.0405 & 0.3279 \\ -0.1454 & -0.1864 \end{bmatrix} \begin{bmatrix} -0.1096 & 0.4478 \\ -0.4572 & -0.1112 \end{bmatrix}^{-1} = \begin{bmatrix} 0.6703 & -0.2493 \\ -0.3183 & 0.3942 \end{bmatrix}.$$

The with these actions corresponding closed-loop system matrix is then $A_{cl} := \begin{bmatrix} -1.004 & 0.0222 \\ -0.0231 & -1.003 \end{bmatrix}$. The eigenvalues of this matrix are $\{-1.004 + 0.0227i; -1.004 - 0.0227i\}$.

The corresponding equilibrium costs are

$$J_1 = x_0^T \begin{bmatrix} 0.298 & 0.3703 \\ 0.3703 & 1.1314 \end{bmatrix} x_0 \quad \text{and} \quad J_2 = x_0^T \begin{bmatrix} 0.6455 & -0.2636 \\ -0.2636 & 0.2929 \end{bmatrix} x_0. \qquad \square$$

Although the equilibria of the previous Examples 7.11 and 7.12 are of course difficult to compare, the consequences for the number of equilibria of having an initially stable instead of an unstable uncontrolled system seems to be enormous. This, although the absolute distance between the entries of the unstable and stable matrix is not too large.

7.6 Convergence results

In this section we study the limiting behavior of the open-loop Nash equilibrium. That is, we compare the behavior of the open-loop Nash equilibrium actions, when the planning horizon T tends to infinity, with the equilibrium actions for the infinite-planning horizon open-loop case. More precisely, we assume that the performance criterion which player i, $i = 1, 2$, likes to minimize is:

$$J_i(x_0, u_1, u_2, T) \tag{7.6.1}$$

where

$$J_i = \int_0^T \{x^T(t)Q_i x(t) + u_i^T(t)R_i u_i(t)\}dt + x^T(T)Q_{iT}x(T).$$

This, subject to the dynamic state equation

$$\dot{x}(t) = Ax(t) + B_1 u_1(t) + B_2 u_2(t), \quad x(0) = x_0; \qquad (7.6.2)$$

$u_i(.) \in \mathcal{U}_s$; matrix R_i is positive definite and Q_i symmetric, $i = 1, 2$; and the matrix pairs (A, B_i), $i = 1, 2$, are stabilizable. Denoting a set of Nash equilibrium actions for the finite-planning horizon game by $u_i^*(T)$, $i = 1, 2$, and those for the infinite-planning horizon game by u_i^*, $i = 1, 2$, we will study the relationship between

$$\lim_{T \to \infty} u_i^*(T) \text{ and } u_i^*, \quad i = 1, 2.$$

According to section 7.4, in general the infinite-planning horizon game does not have a unique equilibrium for every initial state. There are situations where the game does not have an equilibrium for every initial state and situations where the game has an infinite number of equilibria for every initial state (see section 7.5). In particular, when the infinite-planning horizon game has more than one equilibrium and the finite-horizon equilibrium action $u_i^*(T)$ converge to an equilibrium action $\hat{u}_i^*$ of the infinite-planning horizon game, $i = 1, 2$, this might be a reason to select this equilibrium outcome $\hat{u}_i^*$ above all other infinite-planning horizon equilibria.

To study convergence properties for the finite-planning horizon problem, it seems reasonable to require that the game (7.6.1) and (7.6.2) has a properly defined solution for every finite-planning horizon T. Therefore in this section we will make the following well-posedness assumptions (see Proposition 7.5).

Well-posedness assumptions

1. The set of coupled Riccati differential equations

$$\dot{P}_1 = -A^T P_1 - P_1 A - Q_1 + P_1 S_1 P_1 + P_1 S_2 P_2; \quad P_1(T) = Q_{1T}, \qquad (7.6.3)$$

$$\dot{P}_2 = -A^T P_2 - P_2 A - Q_2 + P_2 S_2 P_2 + P_2 S_1 P_1; \quad P_2(T) = Q_{2T} \qquad (7.6.4)$$

has a solution $P_i(0)$, $i = 1, 2$, for all $T > 0$, and

2. the two Riccati differential equations,

$$\dot{K}_i(t) = -A^T K_i(t) - K_i(t)A + K_i(t)S_i K_i(t) - Q_i(t), \quad K_i(T) = Q_{iT}, \quad i = 1, 2, \qquad (7.6.5)$$

have a symmetric solution $K_i(.)$ for all $T > 0$. $\qquad \square$

Of course, these assumptions are difficult to verify in practice. Therefore, it would be nice to have general conditions under which these Riccati differential equations have a solution on $(0, \infty)$. The next example illustrates this point. It gives an example of a game where for an infinite-planning horizon the game has a unique solution whereas for some finite-planning horizon the game does not have a unique solution for every initial state.

Example 7.13

Reconsider Example 7.4, with matrix A replaced by

$$A := \begin{bmatrix} -1 & 0 \\ 0 & -1 \end{bmatrix}.$$

With, for example, $T = 0.1$ again, one can construct positive definite matrices Q_{1T} and Q_{2T} in a similar way such that the corresponding game does not have a unique solution for every initial state. That is, consider $V := [I\,0\,0]e^{-0.1M}$ and

$$Q_{1T} = \begin{bmatrix} 1 & h_1 \\ h_1 & h_1^2 + 1 \end{bmatrix} \quad \text{and} \quad Q_{2T} = \begin{bmatrix} 10 & h_2 \\ h_2 & \frac{h_2^2+1}{10} \end{bmatrix},$$

with h_1 and h_2 as defined in Example 7.4. Then the finite-planning horizon game does not have a Nash equilibrium for every planning horizon.

On the other hand, numerical calculations show that $M = TJT^{-1}$, where

$$J = \operatorname{diag}\left\{1, \sqrt{2 + \sqrt{2}}, \sqrt{2 - \sqrt{2}}, -\sqrt{2 - \sqrt{2}}, -\sqrt{2 + \sqrt{2}}, 1\right\},$$

and

$$T = \begin{bmatrix} 0 & 0.6929 & 0.2870 & -0.2870 & -0.6929 & 0 \\ 0 & 0.2870 & -0.6929 & 0.6929 & -0.2870 & 0 \\ 2 & -0.8173 & 1.2232 & -0.1626 & -0.2433 & 0 \\ 1 & 0 & 0 & 0 & 0 & 0 \\ 1 & -1.1559 & -1.7299 & 0.2299 & -0.3441 & 0 \\ 0 & -1.4945 & -4.6831 & 0.6224 & -0.4449 & 1 \end{bmatrix}.$$

Since matrix $\begin{bmatrix} -0.2870 & -0.6929 \\ 0.6929 & -0.2870 \end{bmatrix}$ is invertible, (A, B_i) are stabilizable, (Q_i, A) are detectable and $Q_i \geq 0$, according to Algorithm 7.1 (see also Proposition 5.15) the infinite-planning horizon game has a unique open-loop Nash equilibrium for every initial state. □

To derive general convergence results, we will assume that the eigenstructure of matrix M satisfies a dichotomic separability condition, which we first define. The definition is analogous to the definition of this notion which we introduced in section 5.4 for the Hamiltonian matrix H, and for convenience we repeat it below.

Definition 7.3

Matrix M is called **dichotomically separable** if there exist an n-dimensional subspace V_1 and a $2n$-dimensional subspace V_2 such that $MV_i \subset V_i$, $i = 1, 2$, $V_1 \oplus V_2 = \mathbb{R}^{3n}$ and

$$V_1 \oplus \operatorname{Im} \begin{bmatrix} 0 & 0 \\ I & 0 \\ 0 & I \end{bmatrix} = \mathbb{R}^{3n}.$$

Moreover, V_i should be such that $\operatorname{Re} \lambda < \operatorname{Re} \mu$ for all $\lambda \in \sigma(M \mid V_1)$, $\mu \in \sigma(M \mid V_2)$.

Similarly (see Definition 5.1), matrix

$$H_i := \begin{bmatrix} A & -S_i \\ -Q_i & -A^T \end{bmatrix} \tag{7.6.6}$$

is called dichotomically separable if, for $i = 1, 2$, there exist n-dimensional H_i-invariant subspaces W_{i1}, W_{i2} such that $W_{i1} \oplus W_{i2} = \mathbb{R}^{2n}$ and

$$W_{i1} \oplus \text{Im} \begin{bmatrix} 0 \\ I \end{bmatrix} = \mathbb{R}^{2n}.$$

Moreover, W_{i1}, W_{i2} should be such that Re $\lambda <$ Re μ for all $\lambda \in \sigma(M \mid W_{i1})$, $\mu \in \sigma(M \mid W_{i2})$. □

Example 7.14

1. Consider $M = \text{diag}\{-2, -1, 0, 1 + i, 1 - i, 2\}$. With $V_1 = \text{Im} \begin{bmatrix} I \\ 0 \\ 0 \end{bmatrix}$ and $V_2 = \text{Im} \begin{bmatrix} 0 & 0 \\ I & 0 \\ 0 & I \end{bmatrix}$, it is clear that M is dichotomically separable.

2. Consider $M = \text{diag}\{-3, 1, -1, 2, -2, 4\}$. With

$$V_1 = \text{Im} \begin{bmatrix} 1 & 0 \\ 0 & 0 \\ 0 & 0 \\ 0 & 0 \\ 0 & 1 \\ 0 & 0 \end{bmatrix} \quad \text{and} \quad V_2 = \text{Im} \begin{bmatrix} 0 & 0 & 0 & 0 \\ 1 & 0 & 0 & 0 \\ 0 & 1 & 0 & 0 \\ 0 & 0 & 1 & 0 \\ 0 & 0 & 0 & 0 \\ 0 & 0 & 0 & 1 \end{bmatrix},$$

$V_1 \oplus \text{Im} \begin{bmatrix} 0 & 0 \\ I & 0 \\ 0 & I \end{bmatrix} \neq \mathbb{R}^{3n}$. So, M is not dichotomically separable. □

Recall from the Note following Theorem 2.30 that the spectrum of a Hamiltonian matrix H is symmetric with respect to the imaginary axis. As a consequence we have the following Lemma.

Lemma 7.23

If the Hamiltonian matrix H is dichotomically separable, H has no eigenvalues on the imaginary axis. □

In the following analysis the solutions of the Riccati differential equations (7.6.3) and (7.6.4) and (7.6.5) at time $t = 0$ play an important role. Therefore we introduce a separate notation for them. In the sequel we will denote them by $P_i(0, T)$ and $K_i(0, T)$, $i = 1, 2$, respectively. Observe that $P_i(0, T)$ can be viewed as the solution $k(t)$ of an autonomous vector differential equation $\dot{k} = f(k)$, with $k(0) = k_0$ for some fixed k_0, and where f is a smooth function. Elementary analysis shows then that $P_i(0, T)$ converges to a limit $\bar{P}$ only if this limit $\bar{P}$ satisfies $f(\bar{P}) = 0$ (and similarly for K_i). Therefore, we immediately deduce from Theorems 7.11, 7.12, 2.29 and 2.30, respectively, the following necessary condition for convergence.

Lemma 7.24

$P_i(0, T)$ and $K_i(0, T)$ can only converge to a limit $\bar{P}_i(0)$ and $\bar{K}_i(0)$, respectively, if the corresponding algebraic Riccati equations have a solution. $\square$

Notice that dichotomic separability of M implies that $\mathcal{P}^{pos}$ is nonempty. On the other hand, the next example illustrates that there are situations where $\mathcal{P}^{pos}$ is nonempty, whereas matrix M is not dichotomically separable.

Example 7.15

Consider $A = \begin{bmatrix} 1 & 1 \\ 0 & 2 \end{bmatrix}$; $B_i = I$, $i = 1, 2$; $Q_1 = \begin{bmatrix} 1 & 0 \\ 0 & 2 \end{bmatrix}$; $Q_2 = I$; $R_1^{-1} = \begin{bmatrix} 2 & 1 \\ 1 & 1 \end{bmatrix}$ and $R_2^{-1} = \begin{bmatrix} 2 & 0 \\ 0 & 1 \end{bmatrix}$.

Then, the eigenvalues of matrix M are

$$\{-2.9102; -1.9531 + 0.2344i; -1.9531 - 0.2344i; -1; 1.9548; 2.8617\}.$$

So, M is not dichotomically separable. However, using Algorithm 7.1, it can be shown that the game has two equilibria which permit a feedback synthesis. $\square$

Theorem 7.25, below, shows that under the assumption that matrix M is dichotomic separable the finite-planning horizon solution converges. Moreover, it converges to the solution induced by the solution of the coupled algebraic Riccati equations which 'stabilize' the closed-loop system most. We put the word stabilize here in quotes because it is a little bit misleading. It may happen that this solution (P_1, P_2) is such that the closed-loop system matrix $A - S_1 P_1 - S_2 P_2$ is not stable. The terminology 'for which the real part of the eigenvalues of the closed-loop system matrix become as small as possible' is more accurate.

This result is in line with the convergence results for the regulator problem. For, if the Hamiltonian matrix H is dichotomically separable, by Lemma 7.23, H has precisely n eigenvalues with a real part strictly smaller than zero. So, for the regulator problem the corresponding algebraic Riccati equation always has exactly one stabilizing solution. Dichotomic separability of M, however, does not automatically imply that the n-'smallest' eigenvalues have a negative real part. Therefore, if the n-'smallest' eigenvalues contain an eigenvalue with a positive real part the corresponding set of solutions for the algebraic Riccati equations will not induce a stabilizing closed-loop system. Corollary 7.25 and Example 7.16 discuss some consequences of this observation.

Theorem 7.25

Assume that the well-posedness assumptions (7.6.3)–(7.6.5) hold.

If M and H_i, $i = 1, 2$, are dichotomically separable; and (with the notation of Definition 7.3)

$$\text{Span} \begin{bmatrix} I \\ Q_{1T} \\ Q_{2T} \end{bmatrix} \oplus V_2 = \mathbb{R}^{3n},$$

and, for $i = 1, 2$,

$$\text{Span} \begin{bmatrix} I \\ Q_{iT} \end{bmatrix} \oplus W_{i2} = \mathbb{R}^{2n},$$

then

$$P_i(0, T) \rightarrow P_i := X_i X_0^{-1}, \quad i = 1, 2.$$

Here X_0, X_1, X_2 are defined by $V_1 := \text{Span}[X_0^T, X_1^T, X_2^T]^T$. Consequently, the finite-planning open-loop Nash equilibrium actions satisfy

$$\lim_{T \to \infty} u_i^*(T) \rightarrow -R_i^{-1} B_i^T P_i x(t), \quad i = 1, 2,$$

where $\dot{x}(t) = (A - S_1 P_1 - S_2 P_2) x(t)$, $x(0) = x_0$. Furthermore, $K_i(0, T) \rightarrow K_i$, where K_i is the stabilizing solution of the algebraic Riccati equation

$$0 = A^T K_i + K_i A - K_i S_i K_i + Q_i. \tag{7.6.7}$$

$\square$

The proof of this theorem is provided in the Appendix at the end of this chapter. Combining the results from Theorems 7.25 and 7.12 then yields the following corollary.

Corollary 7.26

Assume the following conditions are satisfied.

1. All conditions mentioned in Theorem 7.25;

2. $\text{Re } \lambda < 0, \forall \lambda \in \sigma(M \mid V_1)$.

Then, if the planning horizon T in the differential game (7.6.1) and (7.6.2) tends to infinity, the unique open-loop Nash equilibrium solution converges to a Nash solution of the corresponding infinite-planning horizon game. This converged solution u_i^* leads to the next feedback synthesis

$$u_i^*(t) = -R_i^{-1} B_i^T P_i x(t), \quad i = 1, 2.$$

Here, P_i, $i = 1, 2$, are given by

$$P_1 = Y_0 X_0^{-1}, \quad \text{and} \quad P_2 = Z_0 X_0^{-1}$$

if a basis of V_1 is represented as follows

$$\text{Im} \begin{bmatrix} X_0 \\ Y_0 \\ Z_0 \end{bmatrix} = V_1.$$

□

The next example illustrates the phenomenon that situations exist in which the finite-planning horizon game always has an equilibrium and, even stronger, this strategy converges if the planning horizon expands, whereas the corresponding infinite-planning horizon game does not have an equilibrium for all initial states.

Example 7.16

Let $A = \begin{bmatrix} -1 & 0 \\ 0 & -5/22 \end{bmatrix}$, $B_1 = \begin{bmatrix} 1 & 0 \\ 0 & 1 \end{bmatrix}$, $B_2 = \begin{bmatrix} 1 \\ 0 \end{bmatrix}$, $Q_1 = \begin{bmatrix} 1 & 0 \\ 0 & 0.01 \end{bmatrix}$, $Q_2 = \begin{bmatrix} 1 & 1 \\ 1 & 2 \end{bmatrix}$,

$R_1 = \begin{bmatrix} 2 & -1 \\ -1 & 1 \end{bmatrix}$ and $R_2 = 1$. Notice that (A, B_i), $i = 1, 2$, is stabilizable and both Q_i and R_i are positive definite. So our standard assumptions on the system and the performance criteria are fulfilled.

Next we calculate M and its spectrum. Numerical calculations show that $M = TJT^{-1}$, where

$$J = \begin{bmatrix} \frac{5}{22} & 0 & 0 & 0 & 0 & 0 \\ 0 & -1.7978 & 0 & 0 & 0 & 0 \\ 0 & 0 & 1.8823 & 0 & 0 & 0 \\ 0 & 0 & 0 & 0.0319 & 0 & 0 \\ 0 & 0 & 0 & 0 & 0.4418 + 0.1084i & 0 \\ 0 & 0 & 0 & 0 & 0 & 0.4418 - 0.1084i \end{bmatrix},$$

and

$$T = \begin{bmatrix} 0 & 0.7271 & 0.3726 & 0.0410 & -0.0439 - 0.0222i & -0.0439 + 0.0222i \\ 0 & 0.1665 & 0.2013 & -0.1170 & 0.1228 + 0.0699i & 0.1228 - 0.0699i \\ 0 & 0.2599 & -0.4222 & 0.0423 & -0.0833 - 0.0236i & -0.0833 + 0.0236i \\ 0 & 0.0008 & -0.0012 & -0.0060 & -0.0033 - 0.0049i & -0.0033 + 0.0049i \\ 0 & 0.3194 & -0.6504 & -0.0786 & 0.1522 + 0.0558i & 0.1522 - 0.0558i \\ 1 & 0.5235 & -0.4684 & -0.9882 & -0.5289 - 0.8149i & -0.5289 + 0.8149i \end{bmatrix}.$$

Using the notation $T =: [T_1, T_2, T_3, T_4, T_5, T_6]$ it is seen that M is dichotomically separable if we choose $V_1 := [T_2, T_4]$ and $V_2 := [T_1, T_3, \text{Re}(T_5), \text{Im}(T_5)]$. Notice that $\sigma(M|_{V_1}) = \{-1.7978, 0.0319\}$ and $\sigma(M|_{V_2}) = \{\frac{5}{22}, 1.8823, 0.4418 \pm 0.1084i\}$.

Now choose $Q_{1T} := Q_{2T} := \begin{bmatrix} 0 & 0 \\ 0 & 0 \end{bmatrix}$. Numerical calculation shows that with these choices for the final cost, the determinant of $H(t)$ always differs from zero. That is, $H(t)$

is invertible for every positive t. So, the finite-planning horizon problem has a unique equilibrium for every T. This also implies that the well-posedness assumptions (7.6.3)–(7.6.5) are satisfied. Since, moreover, M is dichotomically separable and Span $\begin{bmatrix} I \\ Q_{1T} \\ Q_{2T} \end{bmatrix} \oplus V_2 = \mathbb{R}^{3n}$, it is clear from Theorem 7.25 that the equilibrium solution converges. This converged solution can be calculated from $[X_0^T, Y_0^T, Z_0^T]^T := [T_2, T_4]$. The converged solutions of the Riccati equations are

$$K_1 := Y_0 X_0^{-1} = \begin{bmatrix} 0.2599 & 0.0423 \\ 0.0008 & -0.0060 \end{bmatrix} \begin{bmatrix} 0.7271 & 0.0410 \\ 0.1665 & -0.1170 \end{bmatrix}^{-1}$$

and

$$K_2 := Z_0 X_0^{-1} = \begin{bmatrix} 0.3194 & -0.0786 \\ 0.5235 & -0.9882 \end{bmatrix} \begin{bmatrix} 0.7271 & 0.0410 \\ 0.1665 & -0.1170 \end{bmatrix}^{-1}.$$

The converged open-loop strategies are then

$$u_i^*(t) = -R_i^{-1} B_i^T P_i x(t), \quad i = 1, 2.$$

The corresponding eigenvalues of the closed-loop system are $\{1.7978, -0.0319\}$. So, the converged equilibrium solution is not a 'stabilizing solution'.

Since M has only one stable eigenvalue, according to Theorem 7.11, there are initial states for which the infinite-planning horizon game does not have an equilibrium.

So we conclude that although the finite-planning horizon game always has a solution and, even stronger, the corresponding equilibrium solution converges if the planning horizon expands, this converged strategy is not an equilibrium solution for every initial state of the infinite-planning horizon game. □

7.7 Scalar case

This section examines the scalar case. To stress this point the system parameters are put in lower case e.g., a instead of A. To avoid considering various special cases it is assumed throughout that $b_i \neq 0$, $i = 1, 2$.

First, we consider the finite-planning horizon game. The 'invertibility' of

$$h(T) = [1 \quad 0 \quad 0] e^{-MT} \begin{bmatrix} 1 \\ q_{1T} \\ q_{2T} \end{bmatrix}$$

is crucial here (see Theorem 7.1). To verify whether this number always differs from zero we calculate, first, the exponential of matrix $-M$.

Lemma 7.27

Consider matrix M in equation (7.2.6). Let[2] $\mu := \sqrt{a^2 + s_1 q_1 + s_2 q_2}$. Then, the characteristic polynomial of $-M$ is $(a - \lambda)(\mu - \lambda)(-\mu - \lambda)$. Moreover, an eigenvector of M corresponding to $\lambda = a$ is $[0, -s_2, s_1]^T$. If $\mu \neq 0$ and not both $q_1 = q_2 = 0$, the exponential of matrix $-M$, e^{-Ms}, is given by

$$
V \begin{bmatrix} e^{\mu s} & 0 & 0 \\ 0 & e^{as} & 0 \\ 0 & 0 & e^{-\mu s} \end{bmatrix} V^{-1}. \tag{7.7.1}
$$

Here

$$
V = \begin{bmatrix} a - \mu & 0 & a + \mu \\ -q_1 & -s_2 & -q_1 \\ -q_2 & s_1 & -q_2 \end{bmatrix}
$$

and its inverse

$$
V^{-1} = \frac{1}{\det V} \begin{bmatrix} -(s_1 q_1 + s_2 q_2) & -s_1(a + \mu) & -s_2(a + \mu) \\ 0 & -2q_2 \mu & 2q_1 \mu \\ (s_1 q_1 + s_2 q_2) & s_1(a - \mu) & s_2(a - \mu) \end{bmatrix},
$$

with the determinant of V, $\det V = 2\mu(s_1 q_1 + s_2 q_2)$.

Proof

Straightforward multiplication shows that M can be factorized as $M = V \text{diag}\{\mu, -a, -\mu\} V^{-1}$. So (see Section 3.1), the exponential of matrix $-M$, e^{-Ms}, is as stated above. ☐

Now, assume that $q_i > 0$ and $q_{iT} \geq 0$, $i = 1, 2$. Then the Riccati equations (7.2.5) and (7.4.7) associated with the (in)finite-planning horizon regulator problem have a (stabilizing) solution (see Proposition 5.15). Consequently, the finite-planning horizon game has a unique equilibrium solution for every initial state if and only if $h(T)$ differs from zero. We will show that this condition holds for any $T > 0$. Using the expressions in Lemma 7.27 for V and V^{-1} it follows that

$$
h(T) = \frac{1}{\det V} [(s_1 q_1 + s_2 q_2)\{(\mu - a)e^{\mu T} + (a + \mu)e^{-\mu T}\}
$$
$$
+ (\mu^2 - a^2)(e^{\mu T} - e^{-\mu T})(s_1 q_{1T} + s_2 q_{2T})].
$$

Since $s_1 q_1 + s_2 q_2 > 0$, $\mu > a$ and $e^{\mu T} > e^{-\mu T}$, a simple inspection of the terms in $h(T)$ above shows that $h(T)$ is positive for every $T \geq 0$. This implies in particular that $h(T)$

---

[2]Notice that μ can be a pure imaginary complex number.

differs from zero for every $T \in [0, t_1]$, whatever $t_1 > 0$ is. So, using Proposition 7.5, one arives at the following conclusion.

Theorem 7.28

Assume $q_i > 0$ and $q_{iT} \geq 0$, $i = 1, 2$. Then, the scalar finite-planning horizon game has a unique open-loop Nash equilibrium solution for any planning horizon T. Moreover, the equilibrium actions are (in feedback form)

$$u_1^*(t) = -\frac{1}{r_1} b_1 p_1(t) x(t)$$

$$u_2^*(t) = -\frac{1}{r_2} b_2 p_2(t) x(t)$$

where $p_1(t)$ and $p_2(t)$ are the solutions of the coupled asymmetric Riccati-type differential equations

$$\dot{p}_1 = -2ap_1 - q_1 + p_1^2 s_1 + p_1 s_2 p_2; \quad p_1(T) = q_{1T}$$

$$\dot{p}_2 = -2ap_2 - q_2 + p_2^2 s_2 + p_2 s_1 p_1; \quad p_2(T) = q_{2T}.$$

Here $s_i = \frac{b_i^2}{r_i}, i = 1, 2.$ ◻

Next, we consider the infinite-planning horizon game without the assumption that the parameters q_i should be positive. Theorems 7.29 and 7.32, below, give both necessary and sufficient conditions for the existence of an open-loop Nash equilibrium for this game.

Theorem 7.29

Assume that $s_1 q_1 + s_2 q_2 \neq 0$. Then, the infinite-planning horizon game has for every initial state a Nash equilibrium if and only if the following conditions hold.

1. $a^2 + s_1 q_1 + s_2 q_2 > 0$,

2. $a^2 + s_i q_i > 0$, $i = 1, 2$.

Moreover, if there exists an equilibrium for every initial state, then there is exactly one equilibrium that allows a feedback synthesis. The with this equilibrium corresponding set of actions is:

$$u_i^*(t) = -\frac{b_i}{r_i} p_i x(t), \quad i = 1, 2,$$

where $p_1 = \frac{(a+\mu)q_1}{s_1 q_1 + s_2 q_2}$, $p_2 = \frac{(a+\mu)q_2}{s_1 q_1 + s_2 q_2}$, $\mu = \sqrt{a^2 + s_1 q_1 + s_2 q_2}$ and $\dot{x}(t) = (a - s_1 p_1 - s_2 p_2) x(t)$, with $x(0) = x_0$. The corresponding cost for player i is: $J_i = x_0^2 \frac{q_i + s_i p_i^2}{\mu}$, $i = 1, 2$.

Proof

First assume that conditions 1 and 2 hold. Let p_i be as mentioned above. From condition 2 it follows that the algebraic Riccati equations

$$2ak_i - s_ik_i^2 + q_i = 0, \quad i = 1, 2, \tag{7.7.2}$$

have a stabilizing solution $k_i = \frac{a+\sqrt{a^2+s_iq_i}}{s_i}$, $i = 1, 2$. Moreover, it is straightforwardly verified that, if $a^2 + s_1q_1 + s_2q_2 > 0$, p_1 and p_2 satisfy the set of coupled algebraic Riccati equations

$$2ap_1 + q_1 - s_1p_1^2 - s_2p_1p_2 = 0 \quad \text{and} \quad 2ap_2 + q_2 - s_2p_2^2 - s_1p_1p_2 = 0,$$

whereas $a - s_1p_1 - s_2p_2 = -(a^2 + s_1q_1 + s_2q_2) < 0$. So, according to Theorem 7.13, $u_i^*(.)$ provides a Nash equilibrium.

Next consider the converse statement. By Theorem 7.11, part 2, the algebraic Riccati equations (7.7.2) have a stabilizing solution k_i, $i = 1, 2$. From this it follows immediately that condition 2 must hold. On the other hand if $a^2 + s_1q_1 + s_2q_2 \leq 0$, according to Lemma 7.27, matrix M will not have a stable M-invariant subspace S such that

$$\text{Im} \begin{bmatrix} 1 \\ v_1 \\ v_2 \end{bmatrix} \subset S.$$

So according to Theorem 7.11, part 1, in this case the game does not have an open-loop Nash equilibrium for all initial states.

Finally, from Algorithm 7.1 it follows straightforwardly that (p_1, p_2) yields the only Nash equilibrium that permits a feedback synthesis of the game. □

By a direct inspection of the conditions in Theorem 7.29, one sees that a property similar to the finite-planning horizon game holds for the infinite-planning horizon game.

Corollary 7.30

Assume that $q_i > 0$, $i = 1, 2$. Then, the infinite-horizon game has an open-loop Nash equilibrium for every initial state. Furthermore, there is exactly one of them that permits a feedback synthesis. □

Using Theorem 7.16 we can indicate the number of equilibria precisely.

Theorem 7.31

Assume that

1. $s_1q_1 + s_2q_2 \neq 0$;
2. $a^2 + s_1q_1 + s_2q_2 > 0$;
3. $a^2 + s_iq_i > 0$, $i = 1, 2$.

Then, for every initial state, the infinite-planning horizon game has

(a) a unique equilibrium if $a \leq 0$.

(b) an infinite number of equilibria if $a > 0$.

Proof

If $a \leq 0$ matrix M has a one-dimensional stable graph subspace (see Lemma 7.27) and two eigenvalues in $\mathbb{C}_0^+$. The claim then follows directly from Theorem 7.16.

If $a > 0$, according to Theorem 7.11, part 2, for all $\lambda \in \mathbb{R}$ every set of control actions

$$u_{i,\lambda}^*(t) = -\frac{b_i}{r_i} \psi_i(t),$$

yields an open-loop Nash equilibrium. Here $\psi_i(t)$, $i = 1, 2$, are obtained as the solutions of the differential equation

$$\dot{z}(t) = Mz(t), \quad \text{where } z_0 := \begin{bmatrix} x_0 \\ \psi_1(0) \\ \psi_2(0) \end{bmatrix} = \frac{x_0}{a - \mu} \begin{bmatrix} a - \mu \\ -q_1 \\ -q_2 \end{bmatrix} + \lambda \begin{bmatrix} 0 \\ -s_2 \\ s_1 \end{bmatrix}, \quad \lambda \in \mathbb{R}.$$

That is,

$$\psi_1(t) = \frac{-q_1 x_0}{a - \mu} e^{-\mu t} - \lambda s_2 e^{-at},$$

$$\psi_2(t) = \frac{-q_2 x_0}{a - \mu} e^{-\mu t} + \lambda s_1 e^{-at}. \qquad \square$$

Note

It is not difficult to show, using the characterization of the equilibria in the above proof, that in case $a > 0$ all equilibrium actions yield the same closed-loop system $\dot{x} = -\mu x(t)$, $x(0) = x_0$. However, the cost associated with every equilibrium differs for each player. $\square$

Example 7.17

Reconsider Example 7.9 where $a = 3$, $q_i = 2$, $i = 1, 2$ and $s_i = 4$. According to Theorem 7.31 this game has an infinite number of equilibrium actions. Furthermore (see Theorem 7.29), $p_1 = p_2 = 1$ induce the unique equilibrium actions of the infinite-planning horizon game that permits a feedback synthesis. (p_1, p_2) is not a strongly stabilizing solution of the set of coupled algebraic Riccati equations (7.4.5) and (7.4.6). This can be seen by considering, for example, the eigenvalues of

$$\begin{bmatrix} p_1 s_1 - a & p_1 s_2 \\ p_2 s_1 & p_2 s_2 - a \end{bmatrix} = \begin{bmatrix} 1 & 4 \\ 4 & 1 \end{bmatrix},$$

which are -3 and 5.

For every $\alpha \in \mathbb{R}$ the equilibrium actions

$$u_1^*(t) = -2(e^{-5t}x_0 - \alpha e^{-3t})$$
$$u_2^*(t) = -2(e^{-5t}x_0 + \alpha e^{-3t})$$

yield an open-loop Nash equilibrium. This can be verified by following the lines of the proof of Theorem 7.31 or by a direct verification that $J_1(u_1, u_2^*)$ is minimal for u_1^* (and similarly w.r.t. J_2), using the results of Chapter 5.

The corresponding cost for player 1 is $\frac{3}{5}x_0^2 - \alpha x_0 + \frac{2}{3}\alpha^2$ and for player 2 $\frac{3}{5}x_0^2 + \alpha x_0 + \frac{2}{3}\alpha^2$. In particular we see that the minimal cost is attained for different values of α for both players ($\alpha = \frac{3}{4}x_0$ and $\alpha = -\frac{3}{4}x_0$, respectively). So whenever $x_0 \neq 0$ there is not a unique Pareto efficient equilibrium strategy within this set of Nash equilibria. In this case the set of all Pareto efficient strategies are obtained by choosing $-\frac{3}{4}x_0 \leq \alpha \leq \frac{3}{4}x_0$ (assuming, without loss of generality, that $x_0 > 0$). □

In Theorem 7.29 we excluded the case $s_1q_1 + s_2q_2 = 0$. This was because the analysis of this case requires a different approach. The results for this case are summarized below.

Theorem 7.32

Assume $s_1q_1 + s_2q_2 = 0$. Then, the infinite-planning horizon game has, for every initial state, the following properties.

1. A unique set of Nash equilibrium actions

$$u_i^* := \frac{b_i}{r_i}\frac{q_i}{2a}x(t), \quad i = 1, 2, \quad \text{if } a < 0.$$

The corresponding closed-loop system and cost are

$$\dot{x}(t) = ax(t) \quad \text{and} \quad J_i = -\frac{q_i s_i}{4a^2}\left(\frac{2a}{s_i} + \frac{q_i}{2a}\right)x_0^2, \quad i = 1, 2.$$

2. Infinitely many Nash equilibrium actions (which all permit a feedback synthesis) parameterized by

$$(u_1^*, u_2^*) := \left(-\frac{1}{b_1}(a + y)x(t), -\frac{1}{b_2}(a - y)x(t)\right),$$

where $y \in \mathbb{R}$ is arbitrarily, if $a > 0$ and $q_1 = q_2 = 0$.

The corresponding closed-loop system and cost are

$$\dot{x}(t) = -ax(t), \quad J_1 = -\frac{1}{2s_1a}(q_1s_1 + (a + y)^2)x_0^2, \quad \text{and}$$

$$J_2 = -\frac{1}{2s_2a}(q_2s_2 + (a - y)^2)x_0^2.$$

3. No equilibrium in all other cases.

Proof

Item 1 can be shown along the lines of the proofs of Theorem 7.29 and 7.31.

To prove item 2 notice first that, under the assumption that $q_1 = q_2 = 0$, equation (7.4.7) reduces to $k_i(2a - s_ik_i) = 0$. Since $a > 0$ it is clear that $k_i = \frac{2a}{s_i}$, $i = 1, 2$, yields a stabilizing solution to this equation.

Next consider equations (7.4.5) and (7.4.6). These equations reduce in this case to

$$p_1(2a - s_1p_1 - s_2p_2) = 0$$
$$p_2(2a - s_1p_1 - s_2p_2) = 0.$$

Obviously, with $p_1 = p_2 = 0$, the closed-loop system $a - s_1p_1 - s_2p_2 = a > 0$. On the other hand, it is clear that for every choice of (p_1, p_2) such that $s_1p_1 + s_2p_2 = 2a$, the closed-loop system becomes stable. By Theorem 7.13 there then exist an infinite number of open-loop Nash equilibria that permit a feedback synthesis. Introducing $y := a - s_2p_2$ then yields the stated conclusion.

Next, consider item 3. First we deal with the case $a > 0$ and either q_1 or q_2 differs from zero. Then $\mu = a$. From Lemma 7.27 it then follows immediately that the first entry of all eigenvectors corresponding with the eigenvalue $-a$ of M is zero. According to Theorem 7.11, item 1, the game therefore does not have an equilibrium for every initial state in this case.

Finally, consider the only case that has not been covered yet: the case $a = 0$. Equation (7.4.7) now reduces to:

$$s_ik_i^2 = q_i, \quad i = 1, 2. \tag{7.7.3}$$

Since $s_1q_1 + s_2q_2 = 0$, it follows that equation (7.7.3) only has a solution if $q_1 = q_2 = 0$; but then $k_i = 0$, $i = 1, 2$, yielding a closed-loop system which is not stable. So, according to Theorem 7.11, item 2, neither does the game have a solution for all initial states in this game. Which completes the proof. $\square$

For completeness, we also present the generalization of Theorems 7.29 and 7.31 for the N-player situation. The proof can be given along the lines of those theorems. To that purpose notice that the characteristic polynomial of matrix M is $(-a - \lambda)^{N-1}(\lambda^2 - (a^2 + s_1q_1 + \cdots + s_Nq_N))$. Moreover, it is easily verified that every eigenvector corresponding with the eigenvalue a of M has as its first entry a zero. From this the results summarized below then follow analogously.

Theorem 7.33

Assume that $\sigma := s_1q_1 + \cdots + s_Nq_N \neq 0$. Then, the infinite-planning horizon game has for every initial state a Nash equilibrium if and only if the following conditions hold.

1. $a^2 + \sigma > 0$,

2. $a^2 + s_iq_i > 0$, $i = 1, \ldots, N$.

If an equilibrium exists for every initial state, then there is exactly one equilibrium which permits a feedback synthesis. The corresponding set of equilibrium actions is:

$$u_i^*(t) = -\frac{b_i}{r_i} p_i x(t), \quad i = 1, 2,$$

where $p_i = \frac{(a+\mu)q_i}{\sigma}$, $\mu = \sqrt{a^2 + \sigma}$ and $\dot{x}(t) = (a - s_1 p_1 - \cdots - s_N p_N)x(t)$, with $x(0) = x_0$. The corresponding cost for player i for this equilibrium is: $J_i = x_0^2 \frac{q_i + s_i p_i^2}{\mu}$, $i = 1, \ldots, N$.

Moreover, if there is an equilibrium for every initial state, this equilibrium is unique if $a \leq 0$. If $a > 0$ infinitely many equilibria exist. $\qquad\square$

As a final topic of this section, the next theorem shows that in the scalar case the equilibrium actions always converge to the actions implied by the unique equilibrium solution of the infinite-planning horizon game that permits a feedback synthesis. This is on the assumption that the state weight parameters in the cost functions are positive.

Theorem 7.34

Assume that $q_i > 0$ and $q_{iT} \geq 0$, $i = 1, 2$.

Then, the open-loop Nash equilibrium actions from Theorem 7.28 converge to the actions:

$$u_i^*(t) = -\frac{b_i}{r_i} p_i x(t), \quad i = 1, 2,$$

where $p_1 = \frac{(a+\mu)q_1}{s_1 q_1 + s_2 q_2}$, $p_2 = \frac{(a+\mu)q_2}{s_1 q_1 + s_2 q_2}$ and $\dot{x}(t) = (a - s_1 p_1 - s_2 p_2)x(t)$, with $x(0) = x_0$. These actions are obtained as the unique equilibrium solution to the infinite-planning horizon game that permits a feedback synthesis.

Proof

Since $s_1 q_1 + s_2 q_2 > 0$, it is clear from expression (7.7.1) that M is dichotomically separable. Furthermore the well-posedness assumption is satisfied by Theorem 7.28. Notice that $-\mu < -|a| < 0$. So according to Corollary 7.26 the open-loop Nash actions converge whenever q_{iT}, $i = 1, 2$, are such that

$$\det \begin{bmatrix} 1 & 0 & a + \mu \\ q_{1T} & -s_2 & -q_1 \\ q_{2T} & s_1 & -q_2 \end{bmatrix} = s_1 q_1 + s_2 q_2 + s_1(a + \mu)q_{1T} + s_2(a + \mu)q_{2T} \neq 0.$$

However, due to our assumptions, this condition is trivially satisfied. The rest of the claim then follows straightforwardly from Corollary 7.26 and Theorem 7.29. $\qquad\square$

7.8 Economics examples

We conclude this chapter with two worked economics examples. The first example is taken from van Aarle, Bovenberg and Raith (1995) and models the strategic interaction between monetary authorities who control monetization and fiscal authorities who control primary fiscal deficits. The second example studies duopolistic competition in a homogeneous product over time. This is on the assumption that its current desirability is an exponentially weighted function of accumulated past consumption, which implies that the current price of the product adapts sluggishly to the price dictated to accomodate the current level of consumption (see Fershtman and Kamien (1987)). For both examples we will calculate the open-loop Nash equilibrium strategies.

7.8.1 A simple government debt stabilization game

Van Aarle, Bovenberg and Raith (1995) analyze the following differential game on government debt stabilization. They assume that government debt accumulation $(\dot{d})$ is the sum of interest payments on government debt $(rd(t))$ and primary fiscal deficits $(f(t))$ minus the seignorage (or the issue of base money) $(m(t))$:

$$\dot{d}(t) = rd(t) + f(t) - m(t), \quad d(0) = d_0. \tag{7.8.1}$$

Here $d(t), f(t)$ and $m(t)$ are expressed as fractions of GDP and r represents the rate of interest on outstanding government debt minus the growth rate of output; r is assumed to be given. They assume that fiscal and monetary policies are controlled by different institutions, the fiscal authority and the monetary authority, respectively, which have different objectives. The objective of the fiscal authority is to minimize a sum of time profiles of the primary fiscal deficit, base-money growth and government debt:

$$L^F = \int_0^\infty \{(f(t) - \bar{f})^2 + \eta(m(t) - \bar{m})^2 + \lambda(d(t) - \bar{d})^2\}e^{-\delta t}dt. \tag{7.8.2}$$

The monetary authority, on the other hand, sets the growth of base money so as to minimize the loss function:

$$L^M = \int_0^\infty \{(m(t) - \bar{m})^2 + \kappa(d(t) - \bar{d})^2\}e^{-\delta t}dt. \tag{7.8.3}$$

Here $\frac{1}{\kappa}$ can be interpreted as a measure for the conservatism of the central bank with respect to the money growth. Furthermore, all variables denoted with a bar are assumed to be fixed targets which are given a priori.

Introducing $x_1(t) := (d(t) - \bar{d})e^{-\frac{1}{2}\delta t}, x_2(t) := (r\bar{d} + \bar{f} - \bar{m})e^{-\frac{1}{2}\delta t}, u_1(t) := (f(t) - \bar{f})e^{-\frac{1}{2}\delta t}$ and $u_2(t) := (m(t) - \bar{m})e^{-\frac{1}{2}\delta t}$ the above game can be rewritten in our notation with:

$$A = \begin{bmatrix} r - \frac{1}{2}\delta & 1 \\ 0 & -\frac{1}{2}\delta \end{bmatrix}, \quad B_1 = \begin{bmatrix} 1 \\ 0 \end{bmatrix}, \quad B_2 = \begin{bmatrix} -1 \\ 0 \end{bmatrix}, \quad Q_1 = \begin{bmatrix} \lambda & 0 \\ 0 & 0 \end{bmatrix}, \quad Q_2 = \begin{bmatrix} \kappa & 0 \\ 0 & 0 \end{bmatrix},$$

$R_1 = 1$ and $R_2 = 1$.

It is not difficult to see that the eigenvalues of M (see Algorithm 7.1) are: $\{\frac{1}{2}\delta, \frac{1}{2}\delta,$
$-\frac{1}{2}\delta, \frac{1}{2}\delta - r, -l, l\}$, where $l := \sqrt{\kappa + \lambda + (r - \frac{1}{2}\delta)^2}$. The corresponding eigenspaces are:

$\mathcal{T}_1 = \mathrm{Span}\{T_1\}$ where $T_1 = [0\,0\,0\,0\,0\,1]^T$,

$\mathcal{T}_2 = \mathrm{Span}\{T_2\}$ where $T_2 = [0\,0\,0\,1\,0\,0]^T$,

$\mathcal{T}_3 = \mathrm{Span}\{T_3\}$ where $T_3 = [-(r-\delta)\delta,\ \delta(\lambda+\kappa+r(r-\delta)),\ \lambda\delta,\ \lambda,\ \kappa\delta,\ \kappa]^T$,

$\mathcal{T}_4 = \mathrm{Span}\{T_4\}$ where $T_4 = [0,\ 0,\ -r,\ -1,\ r,\ 1]^T$,

$\mathcal{T}_5 = \mathrm{Span}\{T_5\}$ where $T_5 = \left[\frac{1}{2}\delta - r + l,\ 0,\ \lambda,\ \frac{\lambda}{\frac{1}{2}\delta + l},\ \kappa,\ \frac{\kappa}{\frac{1}{2}\delta + l}\right]^T$, and

$\mathcal{T}_6 = \mathrm{Span}\{T_6\}$ where $T_6 = \left[\left(\frac{1}{2}\delta - r - l\right)\left(\frac{1}{2}\delta - l\right),\ 0,\ \lambda\left(\frac{1}{2}\delta - l\right),\ \lambda,\ \kappa\left(\frac{1}{2}\delta - l\right),\ \kappa\right]^T$.

Notice that (A, B_i), $i = 1, 2$, is stabilizable and (Q_i, A), $i = 1, 2$, is detectable. Consequently, the algebraic Riccati equation

$$0 = A^T K_i + K_i A - K_i S_i K_i + Q_i,$$

has a stabilizing solution K_i, $i = 1, 2$ (see Proposition 5.15).

So, the equilibrium actions which permit a feedback synthesis are, according to Corollary 7.15 and Algorithm 7.1, obtained by considering the eigenspaces of matrix M corresponding to the eigenvalues $-\frac{1}{2}\delta$ and $-l$. This gives

$$u_i^*(t) = -B_i^T P_i x(t),$$

with:

$$P_1 := \lambda \begin{bmatrix} \delta & 1 \\ 1 & \frac{1}{\frac{1}{2}\delta + l} \end{bmatrix} \begin{bmatrix} -(r-\delta)\delta & \frac{1}{2}\delta - r + l \\ \delta(\lambda + \kappa + r(r-\delta)) & 0 \end{bmatrix}^{-1} \quad \text{and} \quad P_2 := \frac{\kappa}{\lambda} P_1.$$

In particular this implies that the equilibrium strategies satisfy the relationship $u_2^*(t) = -\frac{\kappa}{\lambda} u_1^*(t)$. Or, stated differently,

$$m(t) - \bar{m}(t) = -\frac{\kappa}{\lambda}(f(t) - \bar{f}(t)). \tag{7.8.4}$$

Substitution of the equilibrium strategies into the system equation gives the closed-loop system

$$\dot{x}(t) = \begin{bmatrix} -l & p \\ 0 & -\frac{1}{2}\delta \end{bmatrix} x(t), \tag{7.8.5}$$

where $p = \frac{(\delta - r)(l - \frac{1}{2}\delta)}{\lambda + \kappa - r(\delta - r)}$. Notice that we implicitly assumed here that $\lambda + \kappa + r(r - \delta)$ differs from zero – a technical assumption which is not crucial. Furthermore, the above equilibrium is unique if the policymakers are impatient (i.e. $\delta > 2r$). If $\delta < 2r$ all strategies $\tilde{u}_i(t) := u_i^*(t) + \lambda r e^{(\frac{1}{2}\delta - r)t}$, where λ is an arbitrary scalar, yield open-loop Nash equilibrium actions. The resulting closed-loop system is the same for every choice of λ. However, the associated cost differs. As in Example 7.17 for every value of d_0 there is an interval of feasible strategies in this case.

Though it is not our intention to analyze these outcomes in detail, we give some preliminary conclusions here. First, observe from equation (7.8.4) that the difference in actual stabilization efforts between the monetary and fiscal authorities is completely determined by the ratio $\frac{\kappa}{\lambda}$ of the weight both authorities attach to debt stabilization relative to other factors in their cost function, respectively. That is, if the fiscal authority also rates the stabilization of debt as important as well as its primary objective of fiscal deficit stabilization, and the monetary authority is primarily interested in price stability (i.e. κ is small), the consequence will be that the monetary authority will be much less active than the fiscal authority.

Furthermore, equation (7.8.5) shows that the speed of adjustment, l, of the government debt towards its long-term equilibrium value is determined (apart from some extraneous factors) by the absolute priority both authorities attach to debt stabilization relative to other factors in their cost function, respectively. Finally, the parameter p in equation (7.8.5) indicates how the authorities react to a distortion between the fundamental equilibrium values, measured by $r\bar{d} + \bar{f} - \bar{m}$, in their efforts to stabilize the government debt towards its long-term equilibrium $\bar{d}$. From equation (7.8.5) it follows that

$$\dot{x}_1(t) = -lx_1(t) + p(r\bar{d} + \bar{f} - \bar{m})e^{-\frac{1}{2}\delta t}; \quad x_1(0) = d_0 - \bar{d}.$$

The solution of this differential equation is

$$x_1(t) = \alpha e^{-lt} + \frac{p(r\bar{d} + \bar{f} - \bar{m})}{l - \frac{1}{2}\delta} e^{-\frac{1}{2}\delta t},$$

where $\alpha = d_0 - \bar{d} - \frac{p(r\bar{d}+\bar{f}-\bar{m})}{l-\frac{1}{2}\delta}$. Since $x_1(t) = (d(t) - \bar{d})e^{-\frac{1}{2}\delta t}$ it follows that

$$d(t) = \alpha e^{(\frac{1}{2}\delta - l)t} + \bar{d} + \frac{p(r\bar{d} + \bar{f} - \bar{m})}{l - \frac{1}{2}\delta}.$$

This shows that, assuming that there is a tension between the desired financing, $\bar{f} + r\bar{d}$, and the desired monetary accomodation, $\bar{m}$, the consequence for the debt accumulation $d(t)$ is that it converges to a constant level

$$\bar{d} + \frac{p(r\bar{d} + \bar{f} - \bar{m})}{l - \frac{1}{2}\delta}. \tag{7.8.6}$$

7.8.2 A game on dynamic duopolistic competition

In this subsection we analyze the dynamic duopoly game with sticky prices which we introduced in Example 3.24. Its finite-planning horizon version was already elaborated in

Example 7.6. The problem we address here is to find the open-loop Nash equilibria of the game defined by the revenue functions

$$J_i(u_1, u_2) = \int_0^\infty e^{-rt}\left\{ p(t)u_i(t) - cu_i(t) - \frac{1}{2}u_i^2(t) \right\} dt, \qquad (7.8.7)$$

subject to the dynamic constraint

$$\dot{p}(t) = s\{a - (u_1(t) + u_2(t)) - p(t)\}, \quad p(0) = p_0. \qquad (7.8.8)$$

Recall that in this model, $r > 0$ denotes the discount rate of future profits and $s \in (0, \infty)$ is the adjustment speed parameter of the market price, $p(t)$, towards the price dictated by the demand function. That is, for larger values of s the market price adjusts along the demand function more quickly. The cost functions of the companies are assumed to be $C(u_i) := cu_i + u_i^2$, where $c \in (0, a)$ is a fixed parameter. Furthermore, the inverse demand function is assumed to be given by $\tilde{p} = a - (u_1 + u_2)$.

To determine the open-loop equilibrium actions for this game (7.8.7) and (7.8.8) we first reformulate it into our standard framework. To that end, consider the new variables

$$x_1(t) := e^{-\frac{1}{2}rt}p(t), \quad x_2(t) := e^{-\frac{1}{2}rt} \quad \text{and} \quad \tilde{u}_i(t) := e^{-\frac{1}{2}rt}u_i(t), \quad i = 1, 2. \qquad (7.8.9)$$

With these variables and $x^T(t) := [x_1(t)\ x_2(t)]$, the model (7.8.7) and (7.8.8) can be rewritten as the problem to find the solution to

$$-\min_{u_i(\cdot)} \int_0^\infty [x_1(t)\ x_2(t)\ \tilde{u}_i(t)] \begin{bmatrix} 0 & 0 & -\frac{1}{2} \\ 0 & 0 & \frac{1}{2}c \\ -\frac{1}{2} & \frac{1}{2}c & \frac{1}{2} \end{bmatrix} \begin{bmatrix} x_1(t) \\ x_2(t) \\ \tilde{u}_i(t) \end{bmatrix} dt, \qquad (7.8.10)$$

subject to the dynamics

$$\dot{x}(t) = \begin{bmatrix} -s - \frac{1}{2}r & as \\ 0 & -\frac{1}{2}r \end{bmatrix} x(t) + \begin{bmatrix} -s \\ 0 \end{bmatrix} \tilde{u}_1(t) + \begin{bmatrix} -s \\ 0 \end{bmatrix} \tilde{u}_2(t). \qquad (7.8.11)$$

Following the analysis of Section 3.6, case III, the standard formulation is then obtained by considering the new control variables

$$v_i(t) := \tilde{u}_i(t) + 2\left[-\frac{1}{2}\ \frac{1}{2}c \right]x(t). \qquad (7.8.12)$$

Using these variables, the above problem (7.8.10) and (7.8.11) is equivalent to the problem to obtain

$$-\min_{u_i(\cdot)} \int_0^\infty x^T(t)\left(\begin{bmatrix} 0 & 0 \\ 0 & 0 \end{bmatrix} - 2\begin{bmatrix} -\frac{1}{2} \\ \frac{1}{2}c \end{bmatrix}\left[-\frac{1}{2}\ \frac{1}{2}c \right] \right)x(t) + \frac{1}{2}v_i^2(t)dt, \qquad (7.8.13)$$

subject to the dynamics

$$\dot{x}(t) = \left(\begin{bmatrix} -s - \frac{1}{2}r & as \\ 0 & -\frac{1}{2}r \end{bmatrix} - 4\begin{bmatrix} -s \\ 0 \end{bmatrix}\left[-\frac{1}{2}\ \frac{1}{2}c \right] \right)x(t) + \begin{bmatrix} -s \\ 0 \end{bmatrix} v_1(t) + \begin{bmatrix} -s \\ 0 \end{bmatrix} v_2(t).$$

$$(7.8.14)$$

So, the problem reduces to derive the open-loop Nash equilibria for the game defined by

$$J_i := \int_0^\infty \{x^T(t)Q_i x(t) + v_i^T(t)R_i v_i(t)\} dt,$$

and the dynamics

$$\dot{x}(t) = Ax(t) + B_1 v_1(t) + B_2 v_2(t), \quad x^T(0) = [p_0 \quad 1],$$

where

$$A = \begin{bmatrix} -\frac{1}{2}r - 3s & (a+2c)s \\ 0 & -\frac{1}{2}r \end{bmatrix}; \quad B_i = \begin{bmatrix} -s \\ 0 \end{bmatrix}; \quad Q_i = \begin{bmatrix} -\frac{1}{2} & \frac{1}{2}c \\ \frac{1}{2}c & -\frac{1}{2}c^2 \end{bmatrix} \quad \text{and} \quad R_i = \frac{1}{2}.$$

Following Algorithm 7.1, we again first compute the eigenvalues of matrix M. These are $\{-\frac{1}{2}r, -\lambda_1, \frac{1}{2}r, \frac{1}{2}r, \frac{1}{2}r + 3s, \lambda_1\}$, where

$$\lambda_1 = \sqrt{\left(\frac{1}{2}r + 3s\right)^2 - 2s^2}.$$

So matrix M has two stable and four unstable eigenvalues. Therefore, the only candidate open-loop Nash equilibrium is obtained by considering the eigenspaces of M corresponding to the eigenvalues $-\frac{1}{2}r$ and $-\lambda_1$. The with these eigenvalues corresponding eigenspaces are

$T_1 = \text{Span}\{T_1\}$ where

$$T_1 = \left[\frac{2r(2cs - (a+2c)(r+3s))}{a-c}, -\frac{r(6r+14s)}{a-c}, r, as - cs - cr, r, as - cs - cr\right]^T, \quad \text{and}$$

$T_2 = \text{Span}\{T_2\}$ where

$$T_2 = \left[-2\left(\frac{1}{2}r + 3s + \lambda_1\right)\left(\frac{1}{2}r + \lambda_1\right), 0, \frac{1}{2}r + \lambda_1, v, \frac{1}{2}r + \lambda_1, v\right]^T,$$

respectively, with $v := as - c\left(\frac{1}{2}r + s + \lambda_1\right)$. In particular we see that the eigenspace corresponding to $\{-\frac{1}{2}r, -\lambda_1\}$ is a graph subspace.

Next consider the Hamiltonian matrix $\begin{bmatrix} A & -S_i \\ -Q_i & -A^T \end{bmatrix}$. This matrix has two negative real eigenvalues $-\frac{1}{2}r$ and $-\left(\left(\frac{1}{2}r + 3s_i\right)^2 - s_i^2\right)$, respectively. Furthermore is the eigenspace corresponding to these eigenvalues a graph subspace. So the algebraic Riccati equation

$$0 = A^T K_i + K_i A - K_i S_i K_i + Q_i,$$

has a stabilizing solution K_i, $i = 1, 2$ (see the Note following Example 5.5).

Since (A, B_i), $i = 1, 2$, are stabilizable by Theorem 7.16 the unique open-loop equilibrium actions are

$$v_i = -2[-s \quad 0]P_i x(t),$$ (7.8.15)

where

$$P_i = \begin{bmatrix} r & \frac{1}{2}r + \lambda_1 \\ as - cs - cr & as - c(\frac{1}{2}r + s + \lambda_1) \end{bmatrix}$$

$$\times \begin{bmatrix} \frac{2r(2cs-(a+2c)(r+3s))}{a-c} & -2(\frac{1}{2}r + 3s + \lambda_1)(\frac{1}{2}r + \lambda_1) \\ -\frac{r(6r+14s)}{a-c} & 0 \end{bmatrix}^{-1}$$

$$=: \begin{bmatrix} f_o & g_o \\ h_o & l_o \end{bmatrix}, \quad i = 1, 2.$$

Elementary calculations show that

$$f_o = -\frac{1}{2(\frac{1}{2}r + 3s + \lambda_1)} \quad \text{and} \quad g_o = \frac{c - a}{6r + 14s} - \frac{2cs - (a + 2c)(r + 3s)}{(\frac{1}{2}r + 3s + \lambda_1)(6r + 14s)}.$$ (7.8.16)

Using these equilibrium actions, the resulting closed-loop system is

$$\dot{x}(t) = \begin{bmatrix} -\lambda_1 & (a + 2c)s - 4s^2 g_o \\ 0 & -\frac{1}{2}r \end{bmatrix} x(t).$$

Next, we reformulate the results in our original model parameters. To that end we first notice that from the above differential equation in x one obtains the following differential equation for the price $p(t)$:

$$\dot{p}(t) = \left(\frac{1}{2}r - \lambda_1\right)p(t) + (a + 2c)s - 4s^2 g_o.$$

It is easily verified that its solution is

$$p(t) = \alpha e^{(\frac{1}{2}r - \lambda_1)t} + \frac{s(a - 4sg_o + 2c)}{-\frac{1}{2}r + \lambda_1},$$

where $\alpha = p_0 - \frac{s(a - 4sg_o + 2c)}{-\frac{1}{2}r + \lambda_1}$.

So, the price converges to the constant level

$$\frac{s(a - 4sg_o + 2c)}{-\frac{1}{2}r + \lambda_1}$$ (7.8.17)

with a 'convergence speed' of $-\frac{1}{2}r + \lambda_1$.

Furthermore, using expressions (7.8.12) and (7.8.9), respectively, it follows by straightforward substitution into equation (7.8.16) that the equilibrium actions, stated in the original model parameters, are:

$$u_i^*(t) = (2sf_o + 1)p(t) + 2sg_o - c, \quad i = 1, 2. \tag{7.8.18}$$

Finally, we like to compare the limiting outcome of the game if $s \to \infty$ (suggesting an instantaneous adaptation of the prices) with the Nash equilibrium of the corresponding static model. This corresponding static model is

$$\min_{u_i} pu_i - cu_i - \frac{1}{2}u_i^2,$$

with $p = a - u_1 - u_2$. By straightforward differentiation one obtains the reaction functions

$$u_1 = \frac{1}{3}(a - c - u_2) \quad \text{and} \quad u_2 = \frac{1}{3}(a - c - u_1).$$

From this it then follows straightforwardly that the Nash equilibrium price for this model is $p_s := \frac{1}{2}a + \frac{1}{2}c$.

By considering the limit for $s \to \infty$ in expression (7.8.17) one shows that the limiting price in the open-loop game converges to $p_o := \theta a + (1 - \theta)c$, with $\theta = \frac{3}{7}$. So taking account of dynamics results in an equilibrium price which takes the producer's cost more into account.

7.9 Notes and references

For this chapter in particular the works by Feucht (1994), Kremer (2002) and Engwerda (1998a, b) have been consulted. In particular the work of Feucht contains some additional existence results concerning the finite-planning horizon game for just one specific initial state. Moreover, his work contains some sufficient conditions on the system parameters under which one can conclude that the finite-planning horizon game will have a Nash solution. These conditions are such that the set of coupled Riccati differential equations reduce to one non-symmetric $n \times n$ Riccati differential equation (where n is the dimension of the system), irrespective of the number of involved players. Consequently the solution of such a game can be solved rather efficiently (see Chapter 5.3). Engwerda (1998b) used these conditions to present some sufficient conditions under which both the finite and infinite horizon game always have a solution. Moreover he showed that if matrix A is additionally stable, the equilibrium strategy of the finite-planning horizon game then converges under some mild restriction to the unique equilibrium strategy of the infinite-planning horizon game.

Concerning the numerical stability of Algorithm 7.1, we notice that various suggestions have been made in literature to calculate solutions of Riccati equations in a numerical reliable way (for example, Laub (1979, 1991), Paige and Van Loane (1981), van Dooren (1981) and Mehrmann (1991)). Also see Abou-Kandil et al. (2003), for a more general survey on various types of Riccati equations. These methods can also be used to improve the numerical stability of Algorithm 7.1. Particularly if one is considering the implementation of large-scale models one should consult this literature.

The results obtained in this chapter can be generalized straightforwardly to the N-player case. All results concerning matrix M should then be substituted by

$$M = \begin{bmatrix} A & -S \\ -Q & -A_2^T \end{bmatrix},$$

where $S := [S_1, \ldots, S_N]$; $Q := [Q_1, \ldots, Q_N]^T$ and $A_2 = \text{diag}\{A\}$.

A different mathematical approach to tackle the present case open-loop dynamic games which has been successfully exploited in literature is the Hilbert-space method. Lukes and Russel (1971), Simaan and Cruz (1973), Eisele (1982) and, more recently, Kun (2001) and Kremer (2002) took this approach. A disadvantage of this approach is that there are some technical conditions under which this theory can be used. On the other hand, particularly on an infinite horizon, it sometimes leads more directly to results and therefore helps in having more intuition about the problem. For instance, the result under which the game has a unique equilibrium for all initial states was first proved by Kremer (2002) using this approach under the additional assumption that matrix A is stable.

Readers interested in more results and details on the open-loop soft-constrained differential game are referred to Başar and Bernhard (1995). Furthermore, one can find in Kun (2001) and Jank and Kun (2002) some first results on the corresponding multi-player game for a finite-planning horizon.

7.10 Exercises

1. Consider the scalar differential game

$$\min_{u_i} J_i := \int_0^T \{q_i x^2(t) + r_i u_i^2(t)\} dt + q_{iT} x^2(T), \quad i = 1, 2,$$

subject to

$$\dot{x}(t) = ax(t) + b_1 u_1(t) + b_2 u_2(t), \quad x(0) = x_0.$$

Determine which of the next differential games has an open-loop Nash equilibrium for every initial state x_0. In case an equilibrium exists, compute the equilibrium actions and involved cost.

(a) $T = 2$, $a = 0$, $b_1 = 1$, $b_2 = 1$, $q_1 = 3$, $q_2 = 1$, $r_1 = 4$, $r_2 = 4$, $q_{1T} = 0$ and $q_{2T} = 0$.

(b) $T = 1$, $a = 1$, $b_1 = 2$, $b_2 = 1$, $q_1 = 1$, $q_2 = 4$, $r_1 = 1$, $r_2 = 1$, $q_{1T} = 0$ and $q_{2T} = 0$.

(c) $T = 1$, $a = 0$, $b_1 = 1$, $b_2 = 1$, $q_1 = -1$, $q_2 = 2$, $r_1 = 1$, $r_2 = 1$, $q_{1T} = 0$ and $q_{2T} = 1$.

2. Reconsider the differential games from Exercise 1.

(a) Show that for all $T \in [0,2)$ the game considered in Exercise 1(a) has a unique open-loop Nash equilibrium. Show that the solutions of the coupled Riccati differential equations (7.2.8) and (7.2.9) are

$$p_1(t) = 3\frac{e^{-(t-2)} - e^{t-2}}{e^{-(t-2)} + e^{t-2}} \quad \text{and} \quad p_2(t) = \frac{e^{-(t-2)} - e^{t-2}}{e^{-(t-2)} + e^{t-2}}.$$

Moreover, show that the open-loop equilibrium strategies are

$$u_1^*(t) = -\frac{3}{4}\left(\frac{e^2}{e^2 + e^{-2}}e^{-t} - \frac{e^{-2}}{e^2 + e^{-2}}e^t\right)$$

$$\text{and } u_2^*(t) = -\frac{1}{4}\left(\frac{e^2}{e^2 + e^{-2}}e^{-t} - \frac{e^{-2}}{e^2 + e^{-2}}e^t\right).$$

Compare your answer with that of Exercise 1(a).

(b) Show that for all $T \in [0,1)$ the game considered in Exercise 1(b) has a unique open-loop Nash equilibrium. Show that the solution of the coupled Riccati differential equations (7.2.8) and (7.2.9) are

$$p_1(t) = \frac{1}{2}\frac{e^{-3(t-1)} - e^{3(t-1)}}{e^{-3(t-1)} + 2e^{3(t-1)}} \quad \text{and} \quad p_2(t) = 2\frac{e^{-3(t-1)} - e^{3(t-1)}}{e^{-3(t-1)} + 2e^{3(t-1)}}.$$

Moreover, show that the open-loop equilibrium strategies are

$$u_1^*(t) = u_2^*(t) = -\frac{2e^3}{e^3 + 2e^{-3}}e^{-3t} + \frac{2e^{-3}}{e^3 + 2e^{-3}}e^{3t}.$$

Compare your answer with that of Exercise 1(b).

(c) Show that for all $T \in [0,1)$ the game considered in Exercise 1(c) has a unique open-loop Nash equilibrium. Show that the solution of the coupled Riccati differential equations (7.2.8) and (7.2.9) are

$$p_1(t) = \frac{-e^{-(t-1)} + 1}{e^{-(t-1)}} \quad \text{and} \quad p_2(t) = \frac{2e^{-(t-1)} - 1}{e^{-(t-1)}}.$$

Moreover, show that the open-loop equilibrium strategies are

$$u_1^*(t) = -\frac{1}{e} + e^{-t} \quad \text{and} \quad u_2^*(t) = \frac{1}{e} - 2e^{-t}.$$

Compare your answer with that of Exercise 1(c).

3. Consider the zero-sum game with state dynamics

$$\dot{x}(t) = x(t) + u_1(t) + u_2(t), \quad x(0) = 27,$$

and cost function for player one:

$$J = \int_0^T \left\{ x^2(t) + u_1^2(t) - \frac{1}{2} u_2^2(t) \right\} dt - \frac{1}{4} x^2(T).$$

(a) Show that for all $T \in [0, 1)$ the game has for every initial state a unique open-loop Nash equilibrium if and only if the following three differential equations have a solution for all $T \in [0, 1)$:

$$\dot{p}(t) = -2p(t) - 1 - p^2(t), \quad p(T) = -\frac{1}{4};$$

$$\dot{k}_1(t) = -2k_1(t) + k_1^2(t) - 1, \quad k_1(T) = -\frac{1}{4};$$

$$\dot{k}_2(t) = -2k_2(t) + 2k_2^2(t) + 1, \quad k_2(T) = -\frac{1}{4}.$$

(b) Show that for $T = 1$ the solutions of the differential equations in (a) are

$$p(t) = -1 + \frac{1}{t + \frac{1}{3}}, \quad k_1(t) = 1 - \sqrt{2} + \frac{2\sqrt{2}}{1 - c_1 e^{2\sqrt{2}(t-1)}}$$

$$\text{and } k_2(t) = \frac{1}{2} + \frac{1}{2}\tan(t + c_2),$$

for appropriately chosen c_1 and c_2, respectively.

(c) Show that the solution of the differential equation

$$\dot{\Phi}(t) = (1 - 3p(t))\Phi(t), \quad \Phi(0) = 27,$$

is $\Phi(t) = \left(t + \frac{1}{3}\right)^{-3} e^{4t}$.

(d) Determine the constants c_1 and c_2 in Part (b). Show that the game has a unique open-loop Nash equilibrium for $T = 1$. Determine the corresponding equilibrium actions.

(e) Consider the scrap value $-cx^2(T)$ instead of $-\frac{1}{4}x^2(T)$ in the cost function J. Can you find a constant c such that the game does not have an open-loop Nash equilibrium for an arbitrary initial state for all $T \in [0, 1)$?

4. Consider two identical countries who have an agreement to keep the net transfer of pollution to each other's country at a zero level. Assume that the dynamics of the net transfer of pollution from country one to country two, $x(t)$, is described by the dynamics

$$\dot{x}(t) = ax(t) + u_1(t) + u_2(t), \quad x(0) = x_0.$$

Assume, moreover, that both countries agreed that if one country has a positive net transfer of pollution it has to pay a fine to the other country that is quadratically proportional to this amount $x(t)$. Let $x_0 > 0$. The minimization problem considered by country one is then formalized as

$$\min_{u_1} J(T) := \int_0^T \{x^2(t) + u_1^2(t) - u_2^2(t)\}dt,$$

whereas country two likes to maximize $J(T)$ with respect to u_2.

(a) Give an interpretation of the control variables $u_i(t)$, $i = 1, 2$.

(b) Show that the game has an open-loop Nash equilibrium if the following differential equations have a solution on $[0, T]$.

$$\dot{p}(t) = -2ap(t) - 1, \quad p(T) = 0;$$
$$\dot{k}_1(t) = -2ak_1(t) + k_1^2(t) - 1, \quad k_1(T) = 0;$$
$$\dot{k}_2(t) = -2ak_2(t) + k_2^2(t) + 1, \quad k_2(T) = 0.$$

(c) Determine the equilibrium actions provided the conditions under item (b) are satisfied. Can you give an intuitive explanation for the result obtained?

(d) Consider the differential equations in item (b). Show that $k_1(t)$ exists for all $t \in [0, T]$. Next show that

$$k_2(t) = \begin{cases} a + \sqrt{1-a^2}\tan(\sqrt{1-a^2}(t-T)+c_1), & \text{if } 1-a^2 > 0; \\ a - \frac{1}{t-T+c_2}, & \text{if } 1-a^2 = 0; \\ a - c + \frac{2c}{1-c_3 e^{2c(t-T)}}, & \text{if } 1-a^2 < 0, \ c := \sqrt{a^2-1}. \end{cases}$$

(e) Determine the constants c_i, $i = 1, 2, 3$ in item (d). Show that k_2 does not exist for all $t \in [0, T]$ if either (i) $-\sqrt{1-a^2}T + \arctan\frac{-a}{1-a^2} < -\frac{\pi}{2}$, (ii) $a = 1$ and $T < 1$ or (iii) $\frac{a+c}{a-c} > e^{2cT}$, respectively.

(f) Combine the results of parts (d) and (e) and draw your conclusions with respect to the existence of an open-loop Nash equilibrium for this game. Can you give an intuitive explanation for this result?

5. Consider the linear quadratic differential game

$$\dot{x} = Ax + B_1 u_1 + B_2 u_2, \quad x(0) = x_0,$$

with

$$J_i(u_1, u_2) := \frac{1}{2}\int_0^T [x^T(t), \ u_1^T(t), \ u_2^T(t)]M_i \begin{bmatrix} x(t) \\ u_1(t) \\ u_2(t) \end{bmatrix} dt + x^T(T)Q_{iT}x(T), \quad (7.10.1)$$

where $M_i = \begin{bmatrix} Q_i & V_i & W_i \\ V_i^T & R_{1i} & N_i \\ W_i^T & N_i^T & R_{2i} \end{bmatrix}$ and $R_{ii} > 0, \quad i = 1, 2.$

Assume that

$$G := \begin{bmatrix} [0 & I & 0]M_1 \\ [0 & 0 & I]M_2 \end{bmatrix} \begin{bmatrix} 0 & 0 \\ I & 0 \\ 0 & I \end{bmatrix} = \begin{bmatrix} R_{11} & N_1 \\ N_2^T & R_{22} \end{bmatrix}$$

is invertible. Let

$$A_2 := \mathrm{diag}\{A, A\}; \; B := [B_1, \; B_2]; \; \tilde{B}^T := \mathrm{diag}\{\tilde{B}_1^T, \tilde{B}_2^T\}; \; \tilde{B}_1^T := \begin{bmatrix} B_1^T \\ 0 \end{bmatrix}; \; \tilde{B}_2^T := \begin{bmatrix} 0 \\ B_2^T \end{bmatrix};$$

$$Z_i := [I \; 0 \; 0]M_i \begin{bmatrix} 0 & 0 \\ I & 0 \\ 0 & I \end{bmatrix} = [V_i, W_i], \; i = 1, 2; \; Z := \begin{bmatrix} [0 & I & 0]M_1 \\ [0 & 0 & I]M_2 \end{bmatrix} \begin{bmatrix} I \\ 0 \\ 0 \end{bmatrix} \begin{bmatrix} V_1^T \\ W_2^T \end{bmatrix};$$

$$\tilde{A} := A - BG^{-1}Z; \; \tilde{S}_i := BG^{-1}\tilde{B}_i^T; \; \tilde{Q}_i := Q_i - Z_i G^{-1}Z; \; \tilde{A}_2^T := A_2^T - \begin{bmatrix} Z_1 \\ Z_2 \end{bmatrix} G^{-1}\tilde{B}^T \text{ and}$$

$$\tilde{M} := \begin{bmatrix} \tilde{A} & -\tilde{S} \\ -\tilde{Q} & -\tilde{A}_2^T \end{bmatrix}, \text{ where } \tilde{S} := [\tilde{S}_1, \; \tilde{S}_2], \; \tilde{Q} := \begin{bmatrix} \tilde{Q}_1 \\ \tilde{Q}_2 \end{bmatrix}.$$

(a) Show that

$$\tilde{M} = \begin{bmatrix} A & 0 & 0 \\ -Q_1 & -A^T & 0 \\ -Q_2 & 0 & -A^T \end{bmatrix} + \begin{bmatrix} -B \\ Z_1 \\ Z_2 \end{bmatrix} G^{-1} [Z, \; \tilde{B}_1^T, \; \tilde{B}_2^T].$$

(b) Assume that the two Riccati differential equations

$$\dot{K}_i(t) = -A^T K_i(t) - K_i(t)A + (K_i(t)B_i + V_i)R_{ii}^{-1}(B_i^T K_i(t) + V_i^T) - Q_i; \quad K_i(T) = Q_{iT},$$

(7.10.2)

have a symmetric solution $K_i(.)$ on $[0, T]$, $i = 1, 2$. Show that this linear quadratic differential game has an open-loop Nash equilibrium for every initial state if and only if matrix

$$\tilde{H}(T) = [I \; 0 \; 0]e^{-\tilde{M}T} \begin{bmatrix} I \\ Q_{1T} \\ Q_{2T} \end{bmatrix}$$

is invertible.

(c) Assume that:

  (i) the set of (coupled) Riccati differential equations

$$\dot{P}(t) = -\tilde{A}_2^T P(t) - P(t)\tilde{A} + P(t)BG^{-1}\tilde{B}^T P(t) - \tilde{Q}; \; P^T(T) = [Q_{1T}^T, \; Q_{2T}^T]$$

has a solution P on $[0, T]$, and

(ii) the two Riccati differential equations (7.10.2) have a symmetric solution $K_i(.)$ on $[0, T]$.

Show that the differential game has a unique open-loop Nash equilibrium for every initial state. Moreover, show that the equilibrium actions are

$$\begin{bmatrix} u_1^*(t) \\ u_2^*(t) \end{bmatrix} = -G^{-1}(Z + \tilde{B}^T P(t))\tilde{\Phi}(t,0)x_0,$$

where $\tilde{\Phi}(t,0)$ is the solution of the transition equation

$$\dot{\tilde{\Phi}}(t,0) = (A - BG^{-1}(Z + \tilde{B}^T P(t)))\tilde{\Phi}(t,0); \quad \tilde{\Phi}(0,0) = I.$$

(d) Show that for all $T \in [0, t_1)$ there exists a unique open-loop Nash equilibrium for the game if and only if the Riccati differential equations (i) and (ii) in part (c) have an appropriate solution for all $T \in [0, t_1)$.

6. Consider the set of coupled algebraic Riccati equations (7.3.1) and (7.3.2). Determine numerically all solutions of this set of equations for the following parameter choices.

(a) $A = 1$, $Q_1 = 1$, $Q_2 = 2$, $S_1 \neq 3$, $S_2 = 4$.
(b) $A = 1$, $Q_1 = -1$, $Q_2 = 2$, $S_1 = 3$, $S_2 = 4$.
(c) $A = 1$, $Q_1 = 1$, $Q_2 = -2$, $S_1 = 3$, $S_2 = 4$.
(d) $A = 2$, $Q_1 = 1$, $Q_2 = -2$, $S_1 = 4$, $S_2 = 4$.
(e) $A = 1$, $Q_1 = 0$, $Q_2 = 0$, $S_1 = 3$, $S_2 = 4$.

7. Consider

$$Q_1 = \begin{bmatrix} 2 & 0 \\ 0 & 1 \end{bmatrix}, \quad Q_2 = \begin{bmatrix} 1 & 0 \\ 0 & 2 \end{bmatrix}, \quad S_1 = \begin{bmatrix} 1 & 0 \\ 0 & 0 \end{bmatrix} \quad \text{and} \quad S_2 = \begin{bmatrix} 0 & 0 \\ 0 & 1 \end{bmatrix}.$$

Answer the same question as in Exercise 6 if

(a) $A = \begin{bmatrix} 1 & 0 \\ 0 & -2 \end{bmatrix}$, (b) $A = \begin{bmatrix} 1 & 1 \\ 0 & -2 \end{bmatrix}$, (c) $A = \begin{bmatrix} 1 & 1 \\ 3 & -2 \end{bmatrix}$ and (d) $A = \begin{bmatrix} 1 & 1 \\ -3 & -2 \end{bmatrix}$.

8. Reconsider Exercise 7. Show that the set of coupled algebraic Riccati equations (7.3.1) and (7.3.2) in parts (a)–(d) has one, three, three and one stabilizing solution(s), respectively.

9. Assume that $S_1 > 0$, $S_2 = \alpha S_1$ and $Q_1 + \alpha Q_2 > 0$, for some scalar $\alpha \in \mathbb{R}$.

(a) Show that matrix M and $M_2 := \begin{bmatrix} A & -S_1 & 0 \\ -Q_1 - \alpha Q_2 & -A^T & 0 \\ -Q_2 & 0 & -A^T \end{bmatrix}$ have the same eigenvalues.

$\left(\text{Hint: show that } M_2 = SMS^{-1}, \text{ where } S == \begin{bmatrix} I & 0 & 0 \\ 0 & I & \alpha I \\ 0 & 0 & I \end{bmatrix} \right).$

(b) Show that matrix $H := \begin{bmatrix} A & -S_1 \\ -Q_1 - \alpha Q_2 & -A^T \end{bmatrix}$ has a stable invariant graph subspace $\begin{bmatrix} I \\ X \end{bmatrix}$. (Hint: see Proposition 5.15.)

(c) Assume that the matrices A and Λ are stable. Show that the matrix equation $Y\Lambda + A^T Y = -Q_2$ has a solution.

(d) Show that if matrix A is stable the set of coupled algebraic Riccati equations (7.3.1) and (7.3.2) has exactly one stabilizing solution.

10. Consider

$$A = \begin{bmatrix} 0 & -1 \\ 1 & 0 \end{bmatrix}, \ Q_1 = \begin{bmatrix} 0 & 0 \\ 0 & 0 \end{bmatrix}, \ Q_2 = \begin{bmatrix} 1 & 0 \\ 0 & 1 \end{bmatrix}, \ S_1 = \begin{bmatrix} 1 & 0 \\ 0 & 1 \end{bmatrix} \text{ and } S_2 = \begin{bmatrix} 0 & 0 \\ 0 & 0 \end{bmatrix}.$$

Show analytically that the set of coupled algebraic Riccati equations (7.3.1) and (7.3.2) has no solution.

11. Consider the games corresponding to the parameters in Exercise 6. Assume that the initial state of the system is x_0.

(a) Show that only the games (a) and (e) in Exercise 6 have a set of open-loop equilibrium actions which permit a feedback synthesis for every initial state x_0.

(b) Show that the involved costs for game 6(a) are $J_1 = 0.2156x_0^2$ and $J_2 = 0.6691x_0^2$, respectively.

(c) Show that the involved costs for game 6(e) are $J_1 = \frac{6\lambda^2}{(3\lambda + 4\mu)^2}$ and $J_2 = \frac{8\mu^2}{(3\lambda + 4\mu)^2}$, λ, $\mu \in \mathbb{R}$, respectively, with $3\lambda + 4\mu \neq 0$.

(d) Show that the set of all open-loop Nash equilibria in part (c) satisfies $J_2 = \frac{1}{4}(3J_1 - 2\sqrt{6}\sqrt{J_1} + 6)$, where $J_1 \geq 0$.

(e) Determine all Pareto efficient open-loop Nash equilibria for the set of open-loop Nash equilibria in part (c).

12. Consider the games corresponding to the parameters in Exercise 7. Assume $B_1 = [1 \ 0]^T$ and $B_2 = [0 \ 1]^T$.

(a) Verify for which games the matrix pairs (A, B_i), $i = 1, 2$, are stabilizable.

(b) Determine for which games the corresponding algebraic Riccati equations (7.4.7) have a stabilizing solution.

(c) Show that the open-loop Nash equilibria $(J_1^*, J_2^*) := (x_0^T M_1 x_0, x_0^T M_2 x_0)$ that permit a feedback synthesis for every initial state are:

for game (b) $M_1 = \begin{bmatrix} 2.6187 & 0.6029 \\ 0.6029 & 0.3634 \end{bmatrix}$, $M_2 = \begin{bmatrix} 0.3226 & 0.0340 \\ 0.0340 & 0.4559 \end{bmatrix}$; $M_1 = \begin{bmatrix} 4.7805 & 1.7768 \\ 1.7768 & 1 \end{bmatrix}$,

$M_2 = \begin{bmatrix} 3.8019 & 1.9420 \\ 1.9420 & 1.5 \end{bmatrix}$; and $M_1 = \begin{bmatrix} 2.9708 & 1.1345 \\ 1.1345 & 1 \end{bmatrix}$, $M_2 = \begin{bmatrix} 0.3814 & 0.3364 \\ 0.3364 & 1.5 \end{bmatrix}$;

for game (c) $M_1 = \begin{bmatrix} 3.4454 & 0.7933 \\ 0.7933 & 0.3569 \end{bmatrix}$, $M_2 = \begin{bmatrix} 0.6561 & 0.3321 \\ 0.3321 & 0.4780 \end{bmatrix}$; $M_1 = \begin{bmatrix} 2.6299 & 0.2076 \\ 0.2076 & 0.1400 \end{bmatrix}$,

$M_2 = \begin{bmatrix} 34.8639 & 8.1160 \\ 8.1160 & 2.2464 \end{bmatrix}$; and $M_1 = \begin{bmatrix} 2.3982 & 0.6049 \\ 0.6049 & 0.4988 \end{bmatrix}$, $M_2 = \begin{bmatrix} 4.7085 & -0.1528 \\ -0.1528 & 0.4869 \end{bmatrix}$;

for game (d) $M_1 = \begin{bmatrix} 2.0336 & 0.3003 \\ 0.3003 & 0.3027 \end{bmatrix}$, $M_2 = \begin{bmatrix} 0.9539 & -0.1493 \\ -0.1493 & 0.4076 \end{bmatrix}$,

respectively.

(d) Show that for game (b) and (c) there is no equilibrium that Pareto dominates another equilibrium for every initial state.

(e) Consider game (b) with initial state $x_0 = [1\ 0]^T$. Is there an equilibrium which Pareto dominates another equilibrium?

(f) Which of the games (a) and (d) has a unique open-loop Nash equilibrium for every initial state?

13. Reconsider Exercise 4 with $T = \infty$ and $a < 0$.

(a) Show that this game has a unique open-loop Nash equilibrium for every initial state if and only if $a < -1$.

(b) Show that if $a < -1$ the equilibrium actions are $u_1^*(t) = \frac{1}{2a}e^{at}x_0$ and $u_2^*(t) = -u_1^*(t)$. Moreover, $J_1^* = \frac{-1}{2a}x_0^2$ and $J_2^* = -J_1^*$.

14. Consider the zero-sum game with state dynamics

$$\dot{x}(t) = 2x(t) + u_1(t) + u_2(t), \quad x(0) = 1,$$

and cost function for player one:

$$J = \int_0^\infty \left\{ x^2(t) + u_1^2(t) - \frac{1}{2}u_2^2(t) \right\} dt.$$

Show by construction that this game has an infinite number of open-loop Nash equilibria.

15. Consider the linear quadratic differential game

$$\dot{x} = Ax + B_1u_1 + B_2u_2, \quad x(0) = x_0,$$

with

$$J_i(u_1, u_2) := \frac{1}{2} \int_0^\infty [x^T(t),\ u_1^T(t),\ u_2^T(t)]M_i \begin{bmatrix} x(t) \\ u_1(t) \\ u_2(t) \end{bmatrix} dt, \qquad (7.10.3)$$

where $M_i = \begin{bmatrix} Q_i & V_i & W_i \\ V_i^T & R_{1i} & N_i \\ W_i^T & N_i^T & R_{2i} \end{bmatrix}$ and $R_{ii} > 0$, $i = 1, 2$.

Consider the notation we used in Exercise 7.5. Assume that matrix G is invertible.

(a) Show that if this game has an open-loop Nash equilibrium for every initial state, then the same conclusions as in Theorems 7.11 and 7.12, part 1, hold with matrix M replaced by matrix $\tilde{M}$.

(b) Consider the algebraic Riccati equations

$$A^T K_i + K_i A - (K_i B_i + V_i) R_{ii}^{-1} (B_i^T K_i + V_i^T) + Q_i = 0, \quad i = 1, 2. \tag{7.10.4}$$

Show that if the game has an open-loop Nash equilibrium for every initial state, the two algebraic Riccati equations (7.10.4) have a symmetric stabilizing solution K_i.

(c) Next, consider the set of (coupled) algebraic Riccati equations

$$0 = -\tilde{A}_2^T P - P\tilde{A} + PBG^{-1}\tilde{B}^T P - \tilde{Q}. \tag{7.10.5}$$

Assume that (7.10.5) has a stabilizing solution P and the set of algebraic Riccati equations (7.10.4) have a stabilizing solution K_i, $i = 1, 2$.

Show that the game has an open-loop Nash equilibrium for every initial state and that

$$\begin{bmatrix} u_1^*(t) \\ u_2^*(t) \end{bmatrix} = -G^{-1}(H + \tilde{B}^T P)\tilde{\Phi}(t, 0)x_0, \tag{7.10.6}$$

where $\tilde{\Phi}(t, 0)$ is the solution of the transition equation

$$\dot{\tilde{\Phi}}(t, 0) = (A - BG^{-1}(H + \tilde{B}^T P))\tilde{\Phi}(t, 0); \quad \tilde{\Phi}(0, 0) = I,$$

provides a set of equilibrium actions.

(d) Show that if (i) matrix $\tilde{M}$ has n stable eigenvalues and $2n$ unstable eigenvalues, (ii) the stable subspace is a graph subspace and (iii) the algebraic Riccati equations (7.10.4) and (7.10.5) have a stabilizing solution, then for every initial state the game has a unique open-loop Nash equilibrium. Moreover, the equilibrium actions are given by equation (7.10.6).

16. (a) Show that the equilibrium actions for the corresponding finite-planning horizon game in Exercise 6(a) converge. Determine the converged actions.

(b) Reconsider Exercise 4, with $a < -1$. Are the conditions of Theorem 7.34 met for this game? Show that the open-loop equilibrium actions for this game converge to the equilibrium actions of the corresponding infinite-horizon planning game which we considered in Exercise 13.

17. Reconsider the games associated with the parameters in Exercise 7.

(a) Determine which of the matrices M associated with these games are dichotomically separable.

(b) Consider the Hamiltonian matrices $H_i := \begin{bmatrix} A & -S_i \\ -Q_i & -A^T \end{bmatrix}$, $i = 1, 2$ associated with these games. Which of these matrices are not dichotomically separable?

(c) Consider game (b). Use a symbolic toolbox to show that $\det([I\ 0\ 0]e^{-Mt}[I\ 0\ 0]^T) > 0$ for all $t > 0$. Here M is defined as usual.

(d) Show that the equilibrium actions of the finite game converge. Determine the converged equilibrium actions.

18. Consider the non-cooperative game

$$\dot{x}(t) = \begin{bmatrix} -0.2 & 0 \\ 0 & -2 \end{bmatrix} x(t) + \begin{bmatrix} 1 & 0 \\ 0 & 1 \end{bmatrix} u_1(t) + \begin{bmatrix} 1 \\ 0 \end{bmatrix} u_2(t), \quad x_0 = \begin{bmatrix} 1 \\ 2 \end{bmatrix},$$

with

$$J_1 = \int_0^\infty \left\{ x^T(t) \begin{bmatrix} 1 & 0 \\ 0 & 0.1 \end{bmatrix} x(t) + u_1(t) \begin{bmatrix} 2 & -1 \\ -1 & 1 \end{bmatrix} u_1(t) \right\} dt$$

and

$$J_2 = \int_0^\infty \left\{ x^T(t) \begin{bmatrix} 1 & 1 \\ 1 & 2 \end{bmatrix} x(t) + u_2^2(t) \right\} dt$$

(a) Determine numerically all open-loop Nash equilibria for this game that permit a feedback synthesis.

(b) Determine for all Nash equilibria in (a) the corresponding cost, using the MATLAB LYAP command. Compare the cost for all equilibria. Is there an equilibrium which is dominated by another equilibrium?

(c) What can you say about the convergence of the outcome of the corresponding finite-planning horizon equilibrium if the planning horizon expands using Corollary 7.26?

19. Consider the interaction of the fiscal stabilization policies of two countries. Assume that the competitiveness between both countries is described by the following differential equation

$$\dot{s}(t) = -as(t) + f_1(t) - f_2(t), \quad s(0) = s_0,$$

where $a > 0$, the variable $s(t)$ denotes the difference in prices between both countries at time t and $f_i(t)$ is the fiscal deficit set by the fiscal authority in country i, $i = 1, 2$. Each fiscal authority seeks to minimize the following intertemporal loss function which is assumed to be quadratic in the price differential and fiscal deficits,

$$J_i = \int_0^T e^{-\delta t} \{q_i s^2(t) + r_i f_i^2(t)\} dt, \quad i = 1, 2.$$

Here δ is a discount factor. Assume that both countries do not cooperate over their policies aimed at reducing the initial price differential between both countries.

(a) Present both necessary and sufficient conditions under which this game has an open-loop Nash equilibrium for all initial states, for an arbitrarily chosen planning horizon somewhere in the interval $[0, 5]$.

(b) Determine the open-loop Nash equilibrium in (a) (you are not explicitly asked to solve the differential equations involved!).

(c) Consider the case that both countries are completely symmetric in their preferences. Determine the open-loop Nash equilibrium strategies in that case. (Hint: first show that $P_1 = P_2$ in that case by considering the difference between both Riccati differential equations.)

(d) Assume $a = 1$, $q_1 = r_1 = 1$, $q_2 = 0.5$, $r_2 = 2$, $\delta = 0.05$ and $s_0 = 1$. Plot (numerically) the corresponding open-loop Nash equilibrium actions and closed-loop system.

(e) Choose $T = \infty$. Determine (numerically), for the parameters chosen in (d), the corresponding open-loop Nash equilibrium actions and closed-loop system on the interval $[0, 10]$.

(f) Compare the answers in (d) and (e). Can you comment on the differences between the answers?

20. Consider the following model on advertising in a market consisting of two suppliers who sell their product at a fixed price p_i, $i = 1, 2$. Assume that the demand for the product, $d(t)$, is given by

$$\dot{d}(t) = -ad(t) + u_1(t) + u_2(t), \quad d(0) = d_0.$$

Here, $u_i(t)$ is the effect of an advertisement by company i on the demand. $a > 0$ is assumed to be the 'forgetting' factor of consumers about the product. Assume that the cost of advertising increases quadratically with u_i. Company 1 is assumed to be the 'owner' of the market. Only by using an advertising campaign which overrides the advertising efforts of Company 1, can Company 2 gain access to this market. The demand for the product from Company 1 is $d(t) + u_1(t) - u_2(t)$, whereas Company 2 is only able to sell the instantaneous excess demand $u_2(t) - u_1(t)$ at the price p_2. We consider the profit functions Π_i, $i = 1, 2$, with

$$\Pi_1 = \int_0^\infty e^{-2rt}\{p_1(y(t) + u_1(t) - u_2(t)) - u_1^2(t)\}dt,$$

$$\Pi_2 = \int_0^\infty e^{-2rt}\{p_2(u_2(t) - u_1(t)) - u_2^2(t)\}dt,$$

where $r > 0$ is a discounting factor.

(a) Show that with $x(t) := e^{-rt}[d(t) \ 1]^T$ and $v_i(t) := e^{-rt}(u_i(t) - p_i(t))$ the model can be rewritten into the standard framework

$$\min_{v_i} \int_0^\infty \left\{ x^T(t) \begin{bmatrix} 0 & -p_i/2 \\ -p_i/2 & -p_i^2/4 \end{bmatrix} x(t) + v_i^2(t) \right\} dt,$$

subject to

$$\dot{x}(t) = \begin{bmatrix} -a-r & (p_1+p_2)/2 \\ 0 & -r \end{bmatrix} + \begin{bmatrix} 1 \\ 0 \end{bmatrix} v_1 + \begin{bmatrix} 1 \\ 0 \end{bmatrix} v_2, \quad x(0) = \begin{bmatrix} d_0 \\ 1 \end{bmatrix}.$$

(b) Show that this game always has a unique open-loop Nash equilibrium and that the stabilizing solutions for the associated coupled algebraic Riccati equations (7.3.1) and (7.3.2) are

$$P_i = \begin{bmatrix} 0 & \frac{-p_i}{2(a+2r)} \\ \frac{-p_i}{2(a+2r)} & y_i \end{bmatrix}, \quad i = 1, 2,$$

where $y_i = \frac{-p_i}{r(a+2r)^2}(\frac{p_1+p_2}{2}(1+4r+2a) + \frac{p_i}{2}(a+2r)^2), \ i = 1, 2.$

(c) Show that the unique equilibrium strategies are $u_i^* = \frac{p_i}{2(2r+a)} + p_i, \ i = 1, 2.$

(d) Show that the equilibrium demand function satisfies $d^*(t) = e^{-at}d_0 + \frac{p_1+p_2}{2a}\frac{2r+a+1}{2r+a}.$

(e) Determine conditions for the parameters a, r and p_i under which Company 2 will actually decide to enter this market.

21. Consider the transboundary acid-rain problem for two countries outlined in Example 3.23. That is,

$$\min_{e_i} \int_0^\infty e^{-rt}\{\delta_i d_i^2(t) + \gamma_i(e_i(t) - \bar{e}_i)^2\} dt,$$

subject to

$$\begin{bmatrix} d_1(t) \\ d_2(t) \end{bmatrix} = \begin{bmatrix} b_{11} \\ b_{21} \end{bmatrix} e_1(t) + \begin{bmatrix} b_{12} \\ b_{22} \end{bmatrix} e_2(t) - \begin{bmatrix} c_1 \\ c_2 \end{bmatrix}; \quad \begin{bmatrix} d_1(0) \\ d_2(0) \end{bmatrix} = \begin{bmatrix} d_{10} \\ d_{20} \end{bmatrix}.$$

Assume that all parameters and constants are nonnegative and $r > 0$.

(a) Let $x(t) := e^{-\frac{1}{2}t}[d_1(t), \ d_2(t), \ 1]^T$ and $u_i(t) := e^{-rt}(e_i(t) - \bar{e}_i)$. Show that the game can be rewritten into standard form as

$$\min_{u_i} \int_0^\infty \{x^T(t)Q_i x(t) + \gamma_i u_i^2(t)\} dt,$$

subject to

$$\dot{x}(t) = Ax(t) + B_1 u_1(t) + B_2(t); \quad x(0) = x_0.$$

Determine the matrices Q_i, A and B_i.

(b) Show that the eigenvalues of the matrix M associated with this game in Algorithm 7.1 are $\{-\frac{1}{2}r, -\lambda_1, -\lambda_2, \frac{1}{2}r, \lambda_1, \lambda_2\}$, where the algebraic multiplicity of $\frac{1}{2}r$ is 4 and λ_i are some complex numbers.

(c) Show that λ_i in part (b) are positive real numbers if and only if

$$r^4 + 4r^2(\delta_1\beta_1 + \delta_2\beta_4) + 16\delta_1\delta_2\beta_1\beta_4 > 4\delta_1\delta_2\beta_2\beta_3,$$

where $\beta_1 := \frac{b_{11}^2}{\gamma_1}$, $\beta_2 := \frac{b_{12}b_{22}}{\gamma_2}$, $\beta_3 := \frac{b_{11}b_{21}}{\gamma_1}$ and $\beta_4 := \frac{b_{22}^2}{\gamma_2}$.

(d) Use a symbolic toolbox to verify that under the conditions of part (c),

$$\lambda_1 = -\frac{1}{2}\sqrt{2s + t} \text{ and } \lambda_2 = -\frac{1}{2}\sqrt{-2s + t},$$

where $s := \sqrt{(\delta_1\beta_1 - \delta_2\beta_4)^2 + 4\delta_1\delta_2\beta_2\beta_3}$ and $t := r^2 + 2\delta_1\beta_1 + 2\delta_2\beta_4$.

(e) Let $\mu_1 := \sqrt{2s + t}$, $\mu_2 := \sqrt{-2s + t}$, $\alpha_1 := -c_1 + b_{11}\bar{e}_1 + b_{12}\bar{e}_2$, $\alpha_2 := -c_2 + b_{21}\bar{e}_1 + b_{22}\bar{e}_2$.

Furthermore define

$$v_{1i} := \frac{(r + \mu_i)^2}{4\alpha_1\delta_1}, \quad v_{2i} := \frac{(\mu_i^2 - 4\beta_1\delta_1 - r^2)(2\delta_2\beta_4 + r^2 + r\mu_i) + 8\delta_1\delta_2\beta_2\beta_3}{8\alpha_1\beta_2\delta_1\delta_2}, \quad v_{3i} := \frac{r + \mu_i}{2\alpha_1},$$

$$v_{4i} := \frac{(r + \mu_i)(\mu_i^2 - 4\beta_1\delta_1 - r^2)}{8\alpha_1\beta_2\delta_1}, \quad v_5 := \frac{\alpha_2(\beta_4\delta_2 - \beta_1\delta_1 + s)}{2\alpha_1\beta_2\delta_1}, \quad v_6 := \frac{\alpha_2(\beta_4\delta_2 - \beta_1\delta_1 - s)}{2\alpha_1\beta_2\delta_1},$$

$w_1 := \delta_2(\alpha_2\beta_2 - \alpha_1\beta_4)$ and $w_2 := r(\alpha_1\beta_3 - \alpha_2\beta_1)$.

Show that an eigenvector corresponding to $-\lambda_1$, $-\lambda_2$ and $-\frac{r}{2}$ is

$$[v_{11}, \ v_{21}, \ 0, \ v_{31}, \ 0, \ 1, \ 0, \ v_{41}, \ v_5]^T,$$

$$[v_{12}, \ v_{22}, \ 0, \ v_{32}, \ 0, \ 1, \ 0, \ v_{42}, \ v_6]^T \text{ and}$$

$$\left[\frac{rw_1}{\delta_1 w_2}, \ 1, \ -\frac{\delta_2(\beta_1\beta_4 - \beta_2\beta_3)}{w_2}, \ \frac{w_1}{w_2}, \ 0, \ \frac{\alpha_1 w_1}{rw_2}, \ 0, \ \frac{\delta_2}{r}, \ \frac{\alpha_2\delta_2}{r^2}\right]^T,$$

respectively.

(f) Formulate both necessary and sufficient conditions under which the game has a unique open-loop Nash equilibrium for every initial state.

(g) Let $X := \begin{bmatrix} v_{11} & v_{12} & \frac{rw_1}{\delta_1 w_2} \\ v_{21} & v_{22} & 1 \\ 0 & 0 & -\frac{\delta_2(\beta_1\beta_4 - \beta_2\beta_3)}{w_2} \end{bmatrix}.$

Show that under the conditions of item (f) the equilibrium actions are

$$u_1^*(t) = -\frac{b_{11}}{\gamma_1} \left[v_{31}, \ v_{32}, \ \frac{w_1}{w_2} \right] X^{-1} x(t) \quad \text{and} \quad u_2^*(t) = -\frac{b_{22}}{\gamma_2} \left[v_{41}, \ v_{42}, \ \frac{\delta_2}{r} \right] X^{-1} x(t).$$

7.11 Appendix

Proof of Theorem 7.1

'$\Rightarrow$ **part**' Suppose that $(u_1^*(.), u_2^*(.))$ is a Nash equilibrium. Then, according the maximum principle, the Hamiltonian

$$H_i = \left(x^T Q_i x + u_1^T R_{i1} u_1 + u_2^T R_{i2} u_2 \right) + \psi_i^T (Ax + B_1 u_1 + B_2 u_2),$$

is minimized by player i with respect to u_i. This gives the necessary conditions

$$u_1^*(t) = -R_{11}^{-1} B_1^T \psi_1(t),$$
$$u_2^*(t) = -R_{22}^{-1} B_2^T \psi_2(t),$$

where the n-dimensional vectors $\psi_1(t)$ and $\psi_2(t)$ satisfy

$$\dot{\psi}_1(t) = -Q_1 x(t) - A^T \psi_1(t), \quad \text{with } \psi_1(T) = Q_{1T} x(T) \tag{7.11.1}$$
$$\dot{\psi}_2(t) = -Q_2 x(t) - A^T \psi_2(t), \quad \text{with } \psi_2(T) = Q_{2T} x(T) \tag{7.11.2}$$

and

$$\dot{x}(t) = Ax(t) - S_1 \psi_1(t) - S_2 \psi_2(t); \quad x(0) = x_0.$$

In other words, if the problem has an open-loop Nash equilibrium then the differential equation

$$\frac{d}{dt} \begin{bmatrix} x(t) \\ \psi_1(t) \\ \psi_2(t) \end{bmatrix} = M \begin{bmatrix} x(t) \\ \psi_1(t) \\ \psi_2(t) \end{bmatrix} \tag{7.11.3}$$

with boundary conditions $x(0) = x_0$, $\psi_1(T) - Q_{1T} x(T) = 0$ and $\psi_2(T) - Q_{2T} x(T) = 0$, has a solution. Let $y(t) := [x^T(t), \ \psi_1^T(t), \ \psi_2^T(t)]^T$. Then the above reasoning shows that, if there is a Nash equilibrium, then for every x_0 the next linear two-point boundary-value problem has a solution.

$$\dot{y}(t) = My(t), \quad \text{with } Py(0) + Qy(T) = [x_0^T \ 0 \ 0]^T. \tag{7.11.4}$$

Here

$$P = \begin{bmatrix} I & 0 & 0 \\ 0 & 0 & 0 \\ 0 & 0 & 0 \end{bmatrix} \quad \text{and} \quad Q = \begin{bmatrix} 0 & 0 & 0 \\ -Q_{1T} & I & 0 \\ -Q_{2T} & 0 & I \end{bmatrix}.$$

Some elementary rewriting shows that the above two-point boundary-value problem (7.11.4) has a solution for every initial state x_0 if and only if

$$(P + Q e^{MT}) y(0) = [x_0^T \quad 0 \quad 0]^T,$$

or, equivalently,

$$(P e^{MT} + Q) e^{MT} y(0) = [x_0^T \quad 0 \quad 0]^T, \tag{7.11.5}$$

is solvable for every x_0.

Denoting $z := e^{MT} y(0)$ and $[W_1 \; W_2 \; W_3] := [I \; 0 \; 0] e^{-MT}$, the question whether (7.11.5) is solvable for every x_0 is equivalent to the question whether

$$\begin{bmatrix} W_1 & W_2 & W_3 \\ -Q_{1T} & I & 0 \\ -Q_{2T} & 0 & I \end{bmatrix} z = \begin{bmatrix} x_0 \\ 0 \\ 0 \end{bmatrix} \tag{7.11.6}$$

is solvable for every x_0. Or, equivalently, whether

$$\begin{bmatrix} I & -W_2 & -W_3 \\ 0 & I & 0 \\ 0 & 0 & I \end{bmatrix} \begin{bmatrix} W_1 & W_2 & W_3 \\ -Q_{1T} & I & 0 \\ -Q_{2T} & 0 & I \end{bmatrix} z = \begin{bmatrix} I & -W_2 & -W_3 \\ 0 & I & 0 \\ 0 & 0 & I \end{bmatrix} \begin{bmatrix} x_0 \\ 0 \\ 0 \end{bmatrix}$$

has a solution for every x_0. Elementary spelling out of both sides of this equation shows that a solution exists for every x_0 if and only if the equation $H(T)z_1 = x_0$ has a solution for every x_0, where H is given by equation (7.2.6). Obviously this is the case if and only if $H(T)$ is invertible. By direct substitution, for example, one verifies that

$$z = \begin{bmatrix} I \\ Q_{1T} \\ Q_{2T} \end{bmatrix} H^{-1}(T) x_0$$

satisfies the equation (7.11.6). Since $z = e^{MT} y(0)$ it then follows that the unique solution $y(0)$ of equation (7.11.5) is

$$y_0 = e^{-MT} z = e^{-MT} \begin{bmatrix} I \\ Q_{1T} \\ Q_{2T} \end{bmatrix} H^{-1}(T) x_0. \tag{7.11.7}$$

However, this implies that the solution of the two-point value problem (7.11.4), y, is also uniquely determined. That is, if for all x_0 there exists an open-loop Nash equilibrium, then there is for each x_0 exactly one equilibrium solution. This solution can be determined by solving the two-point boundary-value problem (7.11.4).

'$\Leftarrow$ **part**' By assumption the Riccati differential equations (7.2.5) have a solution on $[0, T]$. Since $H(T)$ is invertible, it is clear from the '$\Rightarrow$ part' of the proof that the two-point boundary-value problem (7.11.4) has a unique solution for every x_0. Denote the solution $y(t)$ of this two-point boundary-value problem (7.11.4) by $[x^T(t), \ \psi_1(t), \ \psi_2(t)]^T$, where the dimension of x is n and of ψ_i is m_i. Now consider

$$m_i(t) := \psi_i(t) - K_i(t)x(t).$$

Then, $m_i(T) = 0$. Furthermore, differentiation of $m_i(t)$ gives

$$
\begin{aligned}
\dot{m}_i(t) &= \dot{\psi}_i(t) - \dot{K}_i(t)x(t) - K_i(t)\dot{x}(t) \\
&= -Q_ix(t) - A^T\psi_i(t) - [-A^TK_i(t) - K_i(t)A + K_i(t)S_iK_i(t) - Q_i]x(t) \\
&\quad - K_i(t)[Ax(t) - S_1\psi_1(t) - S_2\psi_2(t)] \\
&= -A^T[m_i(t) + K_i(t)x(t)] + A^TK_i(t)x(t) - K_i(t)S_iK_i(t)x(t) \\
&\quad + K_i(t)S_1[m_1(t) + K_1(t)x(t)] + K_i(t)S_2[m_2(t) + K_2(t)x(t)] \\
&= -A^Tm_i(t) + K_i(t)[S_1m_1(t) + S_2m_2(t)] + K_i(t)[-S_iK_i(t) + S_1K_1(t) + S_2K_2(t)]x(t).
\end{aligned}
$$

Next, consider

$$u_i^* = -R_{ii}^{-1}B_i^T(K_ix + m_i), \quad i = 1, 2.$$

By Theorem 5.11 the minimization problem

$$\min_{u_1} J_1(u_1, u_2^*) = \int_0^T \{x^T(t)Q_1x(t) + u_1^T(t)R_{11}u_1(t) + u_2^{*T}(t)R_{12}u_2^*(t)\}dt + x^T(T)Q_{1T}x(T),$$

where

$$\dot{x} = Ax + B_1u_1 + B_2u_2^*, \quad x(0) = x_0,$$

has a unique solution. This solution is

$$\tilde{u}_1(t) = -R_{11}^{-1}B_1^T(K_1(t)\tilde{x}(t) + \tilde{m}_1(t)),$$

where $\tilde{m}_1(t)$ is the solution of the linear differential equation

$$\dot{\tilde{m}}_1(t) = (K_1(t)S_1 - A^T)\tilde{m}_1(t) - K_1(t)B_2u_2^*(t), \quad \tilde{m}_1(T) = 0, \tag{7.11.8}$$

and $\tilde{x}(t)$ is the solution of the differential equation implied through this optimal control

$$\dot{\tilde{x}}(t) = (A - S_1K_1)\tilde{x}(t) - S_1\tilde{m}_1(t) + B_2u_2^*(t), \quad \tilde{x}(0) = x_0. \tag{7.11.9}$$

Substitution of $u_2^*(t)$ into equations (7.11.8) and (7.11.9) shows that the variables $\tilde{m}_1$ and $\tilde{x}$ satisfy

$$\dot{\tilde{m}}_1(t) = (K_1(t)S_1 - A^T)\tilde{m}_1(t) + K_1(t)S_2K_2(t)x(t) + K_1(t)S_2m_2(t), \quad \tilde{m}_1(T) = 0,$$

and

$$\dot{\tilde{x}}(t) = (A - S_1 K_1(t) - S_2 K_2(t))\tilde{x}(t) + S_2 K_2(t)(\tilde{x}(t) - x(t)) - S_1 \tilde{m}_1(t) - S_2 m_2(t), \ \tilde{x}(0) = x_0.$$

It is easily verified that a solution of this set of differential equations is given by $\tilde{x}(t) = x(t)$ and $\tilde{m}_1(t) = m_1(t)$. Since the solution to this set of differential equations is unique this implies that $u_1(t) = u_1^*(t)$. Or, stated differently,

$$J_1(u_1^*, u_2^*) \le J_1(u_1, u_2^*), \quad \text{for all } u_1 \in \Gamma_1.$$

Similarly it can then be proved that the inequality

$$J_2(u_1^*, u_2^*) \le J_1(u_1^*, u_2), \quad \text{for all } u_2 \in \Gamma_2,$$

also holds. Which shows that (u_1^*, u_2^*) is a Nash equilibrium. □

Proof of Theorem 7.4

One part of this conjecture is rather immediate. Assume that we know that the set of Riccati differential equations (7.2.8) and (7.2.9) has a solution on $[0, t_1]$. Then (see Note 3 following equation (7.2.9)) a solution $P_i(t, T)$ also exists to this set of equations on the interval $[0, T]$, for every point $T \in [0, t_1]$. So, according to Theorem 7.2 there will exist an open-loop Nash equilibrium for the game defined on the interval $[0, T]$, for every choice of $T \in [0, t_1]$.

Next, consider the other part of this conjecture. That is, assume that the open-loop game has a Nash equilibrium on the interval $[0, T]$ for all $T \in [0, t_1]$. Then, in particular it follows from Theorem 7.1 and (7.11.7) that

$$y_0 = e^{MT} \begin{bmatrix} I \\ Q_{1T} \\ Q_{2T} \end{bmatrix} H^{-1}(T) x_0.$$

Since $y(t) = e^{Mt} y_0$, it follows that the entries of $y(t)$ can be rewritten as

$$x(t) = [I \ 0 \ 0] e^{-M(T-t)} \begin{bmatrix} I \\ Q_{1T} \\ Q_{2T} \end{bmatrix} H^{-1}(T) x_0, \tag{7.11.10}$$

$$\psi_1(t) = [0 \ I \ 0] e^{-M(T-t)} \begin{bmatrix} I \\ Q_{1T} \\ Q_{2T} \end{bmatrix} H^{-1}(T) x_0, \tag{7.11.11}$$

$$\psi_2(t) = [0 \ 0 \ I] e^{-M(T-t)} \begin{bmatrix} I \\ Q_{1T} \\ Q_{2T} \end{bmatrix} H^{-1}(T) x_0. \tag{7.11.12}$$

Since $H(t) = [I \ 0 \ 0] e^{-Mt} \begin{bmatrix} I \\ Q_{1T} \\ Q_{2T} \end{bmatrix}$, we see that (7.11.10) can be rewritten as $x(t) = H(T - t) H^{-1}(T) x_0$.

Since by assumption $H(T)$ is invertible for all $T \in [0, t_1]$ (see Theorem 7.1) in particular matrix $H(T - t)$ is invertible. Therefore, it follows that $H^{-1}(T)x_0 = H^{-1}(T - t)x(t)$. Substitution of this expression into the equations for ψ_i, $i = 1, 2$, in (7.11.11) and (7.11.12) gives:

$$\psi_1(t) = G_1(T - t)H^{-1}(T - t)x(t) \tag{7.11.13}$$

$$\psi_2(t) = G_2(T - t)H^{-1}(T - t)x(t) \tag{7.11.14}$$

for some continuously differentiable matrix functions G_i, $i = 1, 2$, and $H^{-1}(.)$. Now, denote

$$P_i(t) := G_i(T - t)H^{-1}(T - t), \quad i = 1, 2. \tag{7.11.15}$$

Then, from equations (7.11.13) and (7.11.14) it follows that $\dot{\psi}_i = \dot{P}_i x + P_i \dot{x}$, $i = 1, 2$. According to equations (7.11.1) and (7.11.2) $\psi_1(t)$ and $\psi_2(t)$ satisfy

$$\dot{\psi}_1(t) = -Q_1 x(t) - A^T \psi_1(t), \quad \text{with } \psi_1(T) = Q_{1T}x(T),$$

$$\dot{\psi}_2(t) = -Q_2 x(t) - A^T \psi_2(t), \quad \text{with } \psi_2(T) = Q_{2T}x(T)$$

and

$$\dot{x}(t) = Ax(t) - S_1\psi_1(t) - S_2\psi_2(t); \quad x(0) = x_0.$$

Substitution of $\dot{\psi}_i$ and $\psi_i, i = 1, 2$, into these formulae yields

$$(\dot{P}_1 + A^T P_1 + P_1 A + Q_1 - P_1 S_1 P_1 - P_1 S_2 P_2)e^{-Mt}x_0 = 0$$
$$\text{with } (P_1(T) - Q_{1T})e^{-MT}x_0 = 0, \text{ and}$$
$$(\dot{P}_2 + A^T P_2 + P_2 A + Q_2 - P_2 S_2 P_2 - P_2 S_1 P_1)e^{-Mt}x_0 = 0$$
$$\text{with } (P_2(T) - Q_{2T})e^{-MT}x_0 = 0,$$

for arbitrarily chosen x_0.

From this it follows that $P_i(t), i = 1, 2$, satisfy the set of Riccati differential equations (7.2.8) and (7.2.9). $\qquad\square$

Proof of Theorem 7.11

In the proof of this theorem we need some preliminary results. We state these in some separate lemmas.

Lemma 7.35

Assume that $A \in \mathbb{R}^{n \times n}$ and $\sigma(A) \subset \mathbb{C}^+$. Then,

$$\lim_{t \to \infty} Me^{At}N = 0 \quad \text{if and only if} \quad MA^iN = 0, \ i = 0, \ldots, n - 1. \qquad\square$$

Using this lemma we can then prove the following result.

Lemma 7.36

Let $x_0 \in \mathbb{R}^p$, $y_0 \in \mathbb{R}^{n-p}$ and $Y \in \mathbb{R}^{(n-p) \times p}$. Consider the differential equation

$$\frac{d}{dt} \begin{bmatrix} x(t) \\ y(t) \end{bmatrix} = \begin{bmatrix} A_{11} & A_{12} \\ A_{21} & A_{22} \end{bmatrix} \begin{bmatrix} x(t) \\ y(t) \end{bmatrix}, \quad \begin{bmatrix} x(0) \\ y(0) \end{bmatrix} = \begin{bmatrix} x_0 \\ y_0 \end{bmatrix}.$$

If $\lim_{t \to \infty} x(t) = 0$, for all $\begin{bmatrix} x_0 \\ y_0 \end{bmatrix} \in \mathrm{Span} \begin{bmatrix} I \\ Y \end{bmatrix}$, then;

1. $\dim E^s \geq p$, and

2. there exists a matrix $\bar{Y} \in \mathbb{R}^{(n-p) \times p}$ such that $\mathrm{Span} \begin{bmatrix} I \\ \bar{Y} \end{bmatrix} \subset E^s$.

Proof

1. Using the Jordan canonical form, matrix A can be factorized as

$$A = S \begin{bmatrix} \Lambda_s & 0 \\ 0 & \Lambda_u \end{bmatrix} S^{-1},$$

where $\Lambda_s \in \mathbb{R}^{q \times q}$ contains all stable eigenvalues of A and $\Lambda_u \in \mathbb{R}^{(n-q) \times (n-q)}$ contains the remaining eigenvalues. Now, assume that $q < p$. By assumption

$$\lim_{t \to \infty} \begin{bmatrix} I_q & 0 & 0 \\ 0 & I_{p-q} & 0 \end{bmatrix} S \begin{bmatrix} e^{\Lambda_s t} & 0 \\ 0 & e^{\Lambda_u t} \end{bmatrix} S^{-1} \begin{bmatrix} I_q & 0 \\ 0 & I_{p-q} \\ Y_0 & Y_1 \end{bmatrix} = 0,$$

where I_k denotes the $k \times k$ identity matrix and $Y := [Y_0 \ Y_1]$. Denoting the entries of S by S_{ij}, $i, j = 1, 2, 3$, and the entries of $S^{-1} \begin{bmatrix} I_p & 0 \\ 0 & I_{q-p} \\ Y_0 & Y_1 \end{bmatrix}$ by $\begin{bmatrix} P_1 & P_2 \\ Q_1 & Q_2 \\ R_1 & R_2 \end{bmatrix}$ we can rewrite the above equation as

$$\lim_{t \to \infty} \begin{bmatrix} S_{11} \\ S_{21} \end{bmatrix} e^{\Lambda_s t} [P_1 \ P_2] + \begin{bmatrix} S_{12} & S_{13} \\ S_{22} & S_{23} \end{bmatrix} e^{\Lambda_u t} \begin{bmatrix} Q_1 & Q_2 \\ R_1 & R_2 \end{bmatrix} = 0.$$

Since $\lim_{t \to \infty} e^{\Lambda_s t} = 0$, we conclude that

$$\lim_{t \to \infty} \begin{bmatrix} S_{12} & S_{13} \\ S_{22} & S_{23} \end{bmatrix} e^{\Lambda_u t} \begin{bmatrix} Q_1 & Q_2 \\ R_1 & R_2 \end{bmatrix} = 0.$$

So, from Lemma 7.35, it follows in particular that

$$\begin{bmatrix} S_{12} & S_{13} \\ S_{22} & S_{23} \end{bmatrix} \begin{bmatrix} Q_1 & Q_2 \\ R_1 & R_2 \end{bmatrix} = 0. \tag{7.11.16}$$

Next observe that, on the one hand,

$$H := \begin{bmatrix} S_{11} & S_{12} & S_{13} \\ S_{21} & S_{22} & S_{23} \end{bmatrix} \begin{bmatrix} P_1 & P_2 \\ Q_1 & Q_2 \\ R_1 & R_2 \end{bmatrix} = \begin{bmatrix} I_q & 0 & 0 \\ 0 & I_{p-q} & 0 \end{bmatrix} \begin{bmatrix} I_q & 0 \\ 0 & I_{p-q} \\ Y_0 & Y_1 \end{bmatrix} = I_p.$$

On the other hand we have that, using equation (7.11.16),

$$H = \begin{bmatrix} S_{11} \\ S_{21} \end{bmatrix} [P_1 \quad P_2].$$

So, combining both results, gives

$$H = I_p = \begin{bmatrix} S_{11} \\ S_{21} \end{bmatrix} [P_1 \quad P_2].$$

However, the matrix on the right-hand side is obviously not a full rank matrix, so we have a contradiction. Therefore, our assumption that $q < p$ must have been wrong, which completes the first part of the proof.

2. From the first part of the lemma we conclude that $q \geq p$. Using the same notation, assume that $\begin{bmatrix} S_{11} & S_{12} \\ S_{21} & S_{22} \\ S_{31} & S_{32} \end{bmatrix}$ and $\begin{bmatrix} S_{13} \\ S_{23} \\ S_{33} \end{bmatrix}$ constitute a basis for E^s and E^u, respectively.

Denote $S^{-1} \begin{bmatrix} I_p \\ Y \end{bmatrix} =: \begin{bmatrix} P \\ Q \\ R \end{bmatrix}$. Since

$$\lim_{t \to \infty} [I_p \quad 0 \quad 0] S \begin{bmatrix} e^{\Lambda_s t} & 0 \\ 0 & e^{\Lambda_u t} \end{bmatrix} S^{-1} \begin{bmatrix} I_p \\ Y \end{bmatrix} = 0,$$

we conclude that

$$\lim_{t \to \infty} [S_{11} \quad S_{12}] e^{\Lambda_s t} \begin{bmatrix} P \\ Q \end{bmatrix} + S_{13} e^{\Lambda_u t} R = 0.$$

So, from Lemma 7.35 again, we infer that $S_{13}R = 0$; but, since

$$H := [S_{11} \ S_{12} \ S_{13}] \begin{bmatrix} P \\ Q \\ R \end{bmatrix} = I_p,$$

it follows that

$$H = [S_{11} \ S_{12}] \begin{bmatrix} P \\ Q \end{bmatrix} = I_p.$$

So, necessarily matrix $[S_{11} \ S_{12}]$ must be a full row rank matrix. From this observation the conclusion then follows straightforwardly. $\square$

Next we proceed with the proof of Theorem 7.11.

'$\Rightarrow$ **part**' Suppose that u_1^*, u_2^* are a Nash solution. That is,

$$J_1(u_1, u_2^*) \geq J_1(u_1^*, u_2^*) \quad \text{and} \quad J_2(u_1^*, u_2) \geq J_2(u_1^*, u_2^*).$$

From the first inequality we see that for every $x_0 \in \mathbb{R}^n$ the (nonhomogeneous) linear quadratic control problem to minimize

$$J_1 = \int_0^\infty \{x^T(t)Q_1 x(t) + u_1^T(t)R_1 u_1(t)\} dt,$$

subject to the (nonhomogeneous) state equation

$$\dot{x}(t) = Ax(t) + B_1 u_1(t) + B_2 u_2^*(t), \quad x(0) = x_0,$$

has a solution. This implies (see Theorem 5.16) that the algebraic Riccati equation (7.4.7) has a stabilizing solution (with $i = 1$). In a similar way it also follows that the second algebraic Riccati equation must have a stabilizing solution. Which completes the proof of part 2.

To prove part 1 we consider Theorem 5.16 in some more detail. According to Theorem 5.32 the minimization problem

$$\min_{u_1} J_1(x_0, u_1, u_2^*) = \int_0^\infty \{x_1^T(t)Q_1 x_1(t) + u_1^T(t)R_1 u_1(t)\} dt,$$

where

$$\dot{x}_1 = Ax_1 + B_1 u_1 + B_2 u_2^*, \quad x_1(0) = x_0,$$

has a unique solution. Its solution is

$$\tilde{u}_1(t) = -R_1^{-1} B_1^T (K_1 x_1(t) + m_1(t)). \tag{7.11.17}$$

Here $m_1(t)$ is given by

$$m_1(t) = \int_t^\infty e^{-(A - S_1 K_1)^T (t-s)} K_1 B_2 u_2^*(s) ds, \tag{7.11.18}$$

and K_1 is the stabilizing solution of the algebraic Riccati equation

$$Q_1 + A^T X + XA - XS_1 X = 0. \tag{7.11.19}$$

Notice that, since the optimal control $\tilde{u}_1$ is uniquely determined, and by definition the equilibrium control u_1^* solves the optimization problem, $u_1^*(t) = \tilde{u}_1(t)$. Consequently,

$$\frac{d(x(t) - x_1(t))}{dt} = Ax(t) + B_1 u_1^*(t) + B_2 u_2^*(t) - ((A - S_1 K_1)x_1(t) - S_1 m_1(t) + B_2 u_2^*(t))$$

$$= Ax(t) - S_1(K_1 x_1(t) + m_1(t)) - Ax_1(t) + S_1 K_1 x_1(t) + S_1 m_1(t)$$

$$= A(x(t) - x_1(t)).$$

Since $x(0) - x_1(0) = x_0 - x_0 = 0$ it follows that $x_1(t) = x(t)$.

In a similar way we obtain from the minimization of J_2, with u_1^* now entering into the system as an external signal, that

$$u_2^*(t) = -R_2^{-1}B_2^T(K_2x(t) + m_2(t)),$$

where

$$m_2(t) = \int_t^\infty e^{-(A-S_2K_2)^T(t-s)}K_2B_1u_1^*(s)ds \qquad (7.11.20)$$

and K_2 is the stabilizing solution of the algebraic Riccati equation

$$Q_2 + A^TX + XA - XS_2X = 0. \qquad (7.11.21)$$

By straightforward differentiation of equations (7.11.18) and (7.11.20), respectively, we obtain

$$\dot{m}_1(t) = -(A - S_1K_1)^T m_1(t) - K_1B_2u_2^*(t) \qquad (7.11.22)$$

and

$$\dot{m}_2(t) = -(A - S_2K_2)^T m_2(t) - K_2B_1u_1^*(t). \qquad (7.11.23)$$

Next, introduce

$$\psi_i(t) := K_ix(t) + m_i(t), \quad i = 1, 2. \qquad (7.11.24)$$

Using equations (7.11.22) and (7.11.19) we get

$$\begin{aligned}
\dot{\psi}_1(t) &= K_1\dot{x}(t) + \dot{m}_1(t) \\
&= K_1(A - S_1K_1)x(t) - K_1S_1m_1(t) + K_1B_2u_2^*(t) - K_1B_2u_2^*(t) - (A - S_1K_1)^T m_1(t) \\
&= (-Q_1 - A^TK_1)x(t) - K_1S_1m_1(t) - (A - S_1K_1)^T m_1(t) \\
&= -Q_1x(t) - A^T(K_1x(t) + m_1(t)) \\
&= -Q_1x(t) - A^T\psi_1(t). \qquad (7.11.25)
\end{aligned}$$

In a similar way it follows that $\dot{\psi}_2(t) = -Q_2x(t) - A^T\psi_2(t)$. Consequently, with $v^T(t) := [x^T(t), \psi_1^T(t), \psi_2^T(t), v(t)]$ satisfies

$$\dot{v}(t) = \begin{bmatrix} A & -S_1 & -S_2 \\ -Q_1 & -A^T & 0 \\ -Q_2 & 0 & -A^T \end{bmatrix} v(t), \quad \text{with } v_1(0) = x_0.$$

Since by assumption, for arbitrary x_0, $v_1(t)$ converges to zero it is clear from Lemma 7.36 by choosing consecutively $x_0 = e_i$, $i = 1, \ldots, n$, that matrix M must have at least n stable eigenvalues (counting algebraic multiplicities). Moreover, the other statement follows from the second part of this lemma. Which completes this part of the proof.

'⇐ **part**' Let u_2^* be as claimed in the theorem, that is

$$u_2^*(t) = -R_2^{-1}B_2^T\psi_2.$$

We next show that necessarily u_1^* then solves the optimization problem

$$\min_{u_1} \int_0^\infty \{\tilde{x}^T(t)Q_1\tilde{x}(t) + u_1^T R_1 u_1(t)\}dt,$$

subject to

$$\dot{\tilde{x}}(t) = A\tilde{x}(t) + B_1 u_1(t) + B_2 u^*(t), \quad \tilde{x}(0) = x_0.$$

Since, by assumption, the algebraic Riccati equation

$$Q_1 + A^T K_1 + K_1 A - K_1 S_1 K_1 = 0 \tag{7.11.26}$$

has a stabilizing solution, according to Theorem 5.16, the above minimization problem has a solution. This solution is given by

$$\tilde{u}_1^*(t) = -R^{-1}B_1^T(K_1\tilde{x} + m_1),$$

where

$$m_1 = \int_t^\infty e^{-(A-S_1K_1)^T(t-s)}K_1 B_2 u_2^*(s)ds.$$

Next, introduce

$$\tilde{\psi}_1(t) := K_1\tilde{x}(t) + m_1(t).$$

Then, in a similar way to equation (7.11.25) we obtain

$$\dot{\tilde{\psi}}_1 = -Q_1\tilde{x} - A^T\tilde{\psi}_1.$$

Consequently, $x_d(t) := x(t) - \tilde{x}(t)$ and $\psi_d(t) := \psi_1(t) - \tilde{\psi}_1(t)$ satisfy

$$\begin{bmatrix} \dot{x}_d(t) \\ \dot{\psi}_d(t) \end{bmatrix} = \begin{bmatrix} A & -S_1 \\ -Q_1 & -A^T \end{bmatrix} \begin{bmatrix} x_d(t) \\ \psi_d(t) \end{bmatrix}, \quad \begin{bmatrix} x_d(0) \\ \psi_d(0) \end{bmatrix} = \begin{bmatrix} 0 \\ p \end{bmatrix},$$

for some $p \in \mathbb{R}^n$.

Notice that matrix $\begin{bmatrix} A & -S_1 \\ -Q_1 & -A^T \end{bmatrix}$ is the Hamiltonian matrix associated with the algebraic Riccati equation (7.11.26). Recall that the spectrum of this matrix is symmetric with respect to the imaginary axis. Since, by assumption, the Riccati equation (7.11.26) has a stabilizing solution, we know that its stable invariant subspace is given by

Span$[I\ K_1]^T$. Therefore, with E^u representing a basis for the unstable subspace, we can write

$$\begin{bmatrix} 0 \\ p \end{bmatrix} = \begin{bmatrix} I \\ K_1 \end{bmatrix} v_1 + E^u v_2,$$

for some vectors v_i, $i = 1, 2$. However, it is easily verified that due to our asymptotic stability assumption both $x_d(t)$ and $\psi_d(t)$ converge to zero if $t \to \infty$. So, v_2 must be zero. From this it now follows directly that $p = 0$. Since the solution of the differential equation is uniquely determined, and $[x_d(t)\ \psi_d(t)] = [0\ 0]$ solve it, we conclude that $\tilde{x}(t) = x(t)$ and $\tilde{\psi}_1(t) = \psi_1(t)$. Or, stated differently, u_1^* solves the minimization problem.

In a similar way it is shown that for u_1 given by u_1^*, player two's optimal control is given by u_2^*, which proves the claim. □

Proof of Theorem 7.12

Part 2 of this theorem was already proved in Theorem 7.11.

To prove part 1 we use the notation of Theorem 7.11. In particular we recall that the optimal strategies $u_i^*(t)$, $i = 1, 2$, satisfy (see equations (7.11.17) and (7.11.24))

$$u_i^*(t) = -R_i^{-1} B_i^T \psi_{i, x_0}(t), \tag{7.11.27}$$

where $\psi_{i, x_0}(t) = K_i x(t) + m_i(t)$, $i = 1, 2$. So, $x_{\bar{u}}(s)$ satisfies

$$\dot{x}_{\bar{u}}(t) = A x_{\bar{u}}(t) - S_1 \psi_{1, x_0}(t) - S_2 \psi_{2, x_0}(t); \quad x(0) = x_0. \tag{7.11.28}$$

Notice that by assumption, for arbitrary x_0, both $x(t)$ and $\psi_{i, x_0}(t)$ converge to zero. Next introduce the matrices $P_i = [\psi_{i, e_1}(0), \ldots, \psi_{i, e_n}(0)]$, $i = 1, 2$, and $X = [x_{\bar{u}, e_1}, \ldots, x_{\bar{u}, e_n}]$ where e_i denotes the ith unit vector in $\mathbb{R}^n$ and $x_{\bar{u}, e_i}$ denotes the optimal trajectory corresponding to the initial state $x(0) = e_i$. From equation (7.11.28), equations (7.11.18) and (7.11.20) and the assumption that $u_i(t) = F_i x(t)$, for some constant matrix F_i, it now follows immediately that X satisfies the equation

$$\dot{X}(t) = A X(t) - S_1 Z_1(t) - S_2 Z_2(t); \quad X(0) = I, \tag{7.11.29}$$

with $Z_i = K_i X(t) + \int_t^\infty e^{-(A - S_i K_i)^T (t-s)} K_i B_j F_j X(s) ds$, $i \neq j = 1, 2$. Now, due to our assumption on the considered control functions, the solution $X(s)$ of this equation is an exponential function. So, $X(s + t) = X(s)X(t)$, for any s, t. Using this, we have that

$$Z_1(t) = K_1 X(t) + \int_t^\infty e^{-(A - S_1 K_1)^T (t-s)} K_1 B_2 F_2 X(s) ds$$

$$= K_1 X(t) + \int_0^\infty e^{(A - S_1 K_1)^T (v)} K_1 B_2 F_2 X(v + t) dv$$

$$= \left(K_1 + \int_0^\infty e^{(A - S_1 K_1)^T (v)} K_1 B_2 F_2 X(v) dv \right) X(t)$$

$$=: P_1 X(t).$$

This implies, on the one hand, that

$$\dot{Z}_i(t) = P_i \dot{X}(t), \tag{7.11.30}$$

and, on the other hand, that the differential equation (7.11.29) can be rewritten as

$$\dot{X}(t) = AX(t) - S_1 P_1 X(t) - S_2 P_2 X(t); \quad X(0) = I.$$

So, obviously $X(t) = e^{(A - S_1 P_1 - S_2 P_2)t}$ solves this equation (7.11.29). Furthermore since, due to our assumptions, $X(t)$ converges to zero it follows that matrix $A - S_1 P_1 - S_2 P_2$ is stable. Moreover,

$$M \begin{bmatrix} I \\ P_1 \\ P_2 \end{bmatrix} X(t) = \begin{bmatrix} A - S_1 P_1 - S_2 P_2 \\ -Q_1 + A^T P_1 \\ -Q_2 + A^T P_2 \end{bmatrix} X(t)$$

$$= \begin{bmatrix} \dot{X}(t) \\ \dot{Z}_1(t) \\ \dot{Z}_2(t) \end{bmatrix}$$

$$= \begin{bmatrix} I \\ P_1 \\ P_2 \end{bmatrix} (A - S_1 P_1 - S_2 P_2) X(t).$$

Taking $t = 0$ in the above equality then shows the part of the claim that still had to be proved. □

Proof of Theorem 7.13

Consider u_2^* as defined in equation (7.4.8). We will next show that

$$\min_{u_1 \in \mathcal{U}_s} \lim_{T \to \infty} J_1(x_0, u_1, u_2^*, T)$$

is obtained by choosing

$$u_1(t) = u_1^*(t) = -R_1^{-1} B_1^T P_1 \Phi(t, 0) x_0.$$

Since a similar reasoning shows that $\lim_{T \to \infty} J_2(x_0, u_1^*, u_2, T) \geq \lim_{T \to \infty} J_2(x_0, u_1^*, u_2^*, T)$, for all $u_1 \in \mathcal{U}_s$, we have by definition that (u_1^*, u_2^*) is an open-loop Nash equilibrium.

By Theorem 5.16 the minimization problem

$$\min_{u_1} J_1(x_0, u_1, u_2^*) = \int_0^\infty \{\tilde{x}^T(t) Q_1 \tilde{x}(t) + u_1^T(t) R_1 u_1(t)\} dt,$$

where

$$\dot{\tilde{x}} = A\tilde{x} + B_1 u_1 + B_2 u_2^*, \quad \tilde{x}(0) = x_0,$$

has a unique solution. Its solution is

$$\tilde{u}_1(t) = -R_1^{-1}B_1^T(K_1\tilde{x}(t) + \tilde{m}_1(t)).$$

Here $\tilde{m}_1(t)$ is given by

$$\tilde{m}_1(t) = -\int_t^\infty e^{-(A-S_1K_1)^T(t-s)}K_1S_2P_2x(s)ds,$$

where, with $A_{cl} := A - S_1P_1 - S_2P_2$, $x(t) = e^{A_{cl}s}x_0$ or, stated differently,

$$\dot{x} = A_{cl}x(t), \quad x(0) = x_0.$$

Furthermore, $\tilde{x}$ satisfies the differential equation

$$\dot{\tilde{x}}(t) = (A - S_1K_1)\tilde{x}(t) - S_2P_2x(t) - S_1\tilde{m}_1(t), \quad \tilde{x}_0 = x_0. \tag{7.11.31}$$

Notice that we can also rewrite $\tilde{m}_1(t)$ as the solution of the next differential equation:

$$\dot{\tilde{m}}_1(t) = -(A - S_1K_1)^T\tilde{m}_1(t) + K_1S_2P_2x(t), \quad \tilde{m}_1(0) = -\int_0^\infty e^{(A-S_1K_1)^Ts}K_1S_2P_2x(s)ds. \tag{7.11.32}$$

Next we show that with

$$m_1(t) := (P_1 - K_1)x(t) \quad \text{and} \quad \hat{x}(t) := x(t) \tag{7.11.33}$$

both the differential equations (7.11.31) and (7.11.32) are satisfied. As a consequence we then immediately have that $\tilde{u}_1(t) = u_1^*(t)$.

From expression (7.11.33) we have that

$$\dot{m}_1(t) = (P_1 - K_1)\dot{x}(t)$$
$$= (P_1 - K_1)(A - S_1P_1 - S_2P_2)x(t)$$
$$= (-A^TP_1 - Q_1 - K_1(A - S_1P_1 - S_2P_2))x(t)$$

(using equation (7.4.5)).
Consequently,

$$\dot{m}_1(t) + (A - S_1K_1)^Tm_1(t) - K_1S_2P_2x(t) =$$
$$(-A^TP_1 - Q_1 - K_1(A - S_1P_1 - S_2P_2) + (A - S_1K_1)^T(P_1 - K_1) - K_1S_2P_2)x(t) =$$
$$(-Q_1 - K_1A - A^TK_1 + K_1S_1K_1)x(t) = 0.$$

Furthermore, by partial integration, we see that the initial state $\tilde{m}_1(0)$ satisfies

$$
\begin{aligned}
\tilde{m}_1(0) &= -\int_0^\infty e^{(A-S_1K_1)^T s} K_1 S_2 P_2 x(s) ds \\
&= \int_0^\infty e^{(A-S_1K_1)^T s} K_1 A_{cl} x(s) ds - \int_0^\infty e^{(A-S_1K_1)^T s} (A - S_1 K_1)^T P_1 x(s) ds \\
&\quad + \int_0^\infty e^{(A-S_1K_1)^T s} (A^T P_1 - K_1 A) x(s) ds \\
&= -K_1 x_0 + P_1 x_0 - \int_0^\infty e^{(A-S_1K_1)^T s} (K_1 A + (A - S_1 K_1)^T K_1) x(s) ds \\
&\quad + \int_0^\infty e^{(A-S_1K_1)^T s} (A^T P_1 + P_1 (A - S_1 P_1 - S_2 P_2)) x(s) ds \\
&= (P_1 - K_1) x_0,
\end{aligned}
$$

where for the last equality we used equations (7.4.5) and (7.4.7), respectively. So, $m_1(t)$ satisfies equation (7.11.32).

On the other hand we have

$$
\dot{\hat{x}}(t) - (A - S_1 K_1)\hat{x}(t) + S_2 P_2 x(t) + S_1 m_1(t) =
$$
$$
((A - S_1 P_1 - S_2 P_2) - (A - S_1 K_1) + S_2 P_2 + S_1 (P_1 - K_1)) x(t) = 0.
$$

Thus, both m_1 and x satisfy the differential equations (7.11.31) and (7.11.32). Since the solution of these differential equations is uniquely defined we conclude that $\tilde{x}(.) = x(.)$ and $\tilde{m}_1(.) = (P_1 - K_1) x(.)$, from which it follows directly that $\tilde{u}_1(.) = u_1^*(.)$. $\square$

Proof of Theorem 7.16

In the proof of this theorem we need a result about observable and detectable systems, respectively. These results are stated in the next two separate lemmas.

Lemma 7.37

Consider the system

$$
\dot{x}(t) = Ax(t), \quad y(t) = Cx(t).
$$

Assume that (C, A) is detectable. Then whenever $y(t) = Cx(t) \to 0$, if $t \to \infty$, it then also follows that $x(t) \to 0$ if $t \to \infty$.

Proof of Lemma 7.37

Without loss of generality (Anderson and Moore, 1989), assume that the system is given by

$$
\begin{bmatrix} \dot{x}_1 \\ \dot{x}_2 \end{bmatrix} = \begin{bmatrix} A_{11} & 0 \\ A_{21} & A_{22} \end{bmatrix} \begin{bmatrix} x_1 \\ x_2 \end{bmatrix}; \quad y(t) = [C_1, \ 0] \begin{bmatrix} x_1 \\ x_2 \end{bmatrix},
$$

where the pair (C_1, A_{11}) is observable and $\sigma(A_{22}) \subset \mathbb{C}^-$.

Next, consider

$$\int_t^{t+1} y^T(s)y(s)ds = x_1^T(t)\int_t^{t+1} e^{A_{11}^T(s-t)}C_1^T C_1 e^{A_{11}(s-t)}ds x_1(t).$$

Since (C_1, A_{11}) is observable, the matrix (in literature known as the observability gramian)

$$W := \int_t^{t+1} e^{A_{11}^T(s-t)}C_1^T C_1 e^{A_{11}(s-t)}ds$$

is a full rank positive definite matrix. Moreover, it is easily verified that W is constant and does not depend on the time t.

Now, since $y(t)$ converges to zero it also follows that

$$\int_t^{t+1} y^T(s)y(s)ds = x_1^T(t)Wx_1(t)$$

converges to zero. However, since $W > 0$ we then conclude that $x_1(t)$ also converges to zero; but this implies that $x(t)$ converges to zero.   □

Lemma 7.38

Assume there exists an initial state $x_0 \neq 0$ such that

$$x(t) = e^{-A^T t}x_0 \to 0 \quad \text{if} \quad t \to \infty \quad \text{and} \quad B^T x(t) = 0.$$

Then (B^T, A^T) is not detectable.

Proof of Lemma 7.38

By assumption $x_0 \in S$, where S is the stable subspace of matrix $-A^T$. According to the definition S satisfies $-A^T S = \Lambda S$, where $\sigma(\Lambda) \subset \mathbb{C}^-$. Elementary rewriting shows that $A^T S = -\Lambda S$. Therefore it follows that x_0 belongs to the unstable subspace of matrix A^T.

So, x_0 is an initial state which does not converge to zero in the system $\dot{x}(t) = A^T x(t)$, $x(0) = x_0$, whereas the observed state $B^T x(t)$ is identically zero by assumption. So this initial state cannot be observed. Therefore, by definition, (B^T, A^T) is not detectable.   □

Next, we proceed with the proof of the theorem.

'$\Rightarrow$ **part**' That the Riccati equations (7.4.7) must have a stabilizing solution follows directly from Theorem 7.11.

Assume that matrix M has an s-dimensional stable graph subspace S, with $s > n$. Let $\{b_1, \ldots, b_s\}$ be a basis for S. Denote $d_i := [I, 0, 0]b_i$ and assume (without loss of generality) that $\text{Span}\,[d_1, \ldots, d_n] = \mathbb{R}^n$. Then $d_{n+1} = \mu_1 d_1 + \cdots + \mu_n d_n$ for some μ_i, $i = 1, \ldots, n$. Furthermore, let $x_0 = \alpha_1 d_1 + \cdots + \alpha_n d_n$. Then also for arbitrary $\lambda \in [0, 1]$,

$$\begin{aligned}
x_0 &= \lambda(\alpha_1 d_1 + \cdots + \alpha_n d_n) + (1 - \lambda)(d_{n+1} - \mu_1 d_1 - \cdots - \mu_n d_n) \\
&= [I, 0, 0]\{\lambda(\alpha_1 b_1 + \cdots + \alpha_n b_n) + (1 - \lambda)(b_{n+1} - \mu_1 b_1 - \cdots - \mu_n b_n)\} \\
&= [I, 0, 0]\{(\lambda\alpha_1 - (1 - \lambda)\mu_1)b_1 + \cdots + (\lambda\alpha_n - (1 - \lambda)\mu_n)b_n + (1 - \lambda)b_{n+1}\}.
\end{aligned}$$

Next consider

$$v_\lambda := (\lambda\alpha_1 - (1 - \lambda)\mu_1)b_1 + \cdots + (\lambda\alpha_n - (1 - \lambda)\mu_n)b_n + (1 - \lambda)b_{n+1}.$$

Notice that $v_{\lambda_1} \neq v_{\lambda_2}$ whenever $\lambda_1 \neq \lambda_2$.

According to Theorem 7.11 all solutions $v^T(t) = [x^T, \psi_1^T, \psi_2^T]$ of

$$\dot{v}(t) = Mv(t), \quad v(0) = v_\lambda,$$

then induce open-loop Nash equilibrium strategies

$$u_{i,\lambda} := -R_i^{-1}B_i^T\psi_{i,\lambda}(t), \quad i = 1, 2.$$

Since by assumption for every initial state there is a unique equilibrium strategy it follows on the one hand that the state trajectory $x_\lambda(t)$ induced by these equilibrium strategies coincides for all λ and, on the other hand, that

$$B_i^T\psi_{i,\lambda_1}(t) = B_i^T\psi_{i,\lambda_2}(t), \quad \forall\lambda_1, \lambda_2 \in [0, 1]. \tag{7.11.34}$$

Since $\dot{\psi}_{i,\lambda} = -Q_ix_\lambda(t) - A^T\psi_{i,\lambda}$ it follows that

$$\dot{\psi}_{i,\lambda_1} - \dot{\psi}_{i,\lambda_2} = -A^T(\psi_{i,\lambda_1} - \psi_{i,\lambda_2}),$$
$$B_i^T(\psi_{i,\lambda_1}(t) - \psi_{i,\lambda_2}(t)) = 0.$$

Notice that both $\psi_{i,\lambda_1}(t)$ and $\psi_{i,\lambda_2}(t)$ converge to zero. Furthermore, since $v_{\lambda_1} \neq v_{\lambda_2}$ whenever $\lambda_1 \neq \lambda_2$, $\{b_1, \ldots, b_{n+1}\}$ are linearly independent and $\text{Span}[d_1, \ldots, d_n] = \mathbb{R}^n$, it can easily be verified that at least for one i, $\psi_{i,\lambda_1}(0) \neq \psi_{i,\lambda_2}(0)$, for some λ_1 and λ_2.

By Lemma 7.38 (B_i^T, A^T) is not then detectable. So (see Corollary 3.23), (A, B_i) is not stabilizable, but this violates our basic assumption. So, our assumption that $s > n$ must have been wrong and we conclude that matrix M has an n-dimensional stable graph subspace and that the dimension of the subspace corresponding to non-stable eigenvalues is $2n$. By Theorem 7.10 the set of Riccati equations (7.4.5) and (7.4.6) then has a strongly stabilizing solution.

'$\Leftarrow$ **part**' Since by assumption the stable subspace, E^s, is a graph subspace we know that every initial state, x_0, can be written uniquely as a combination of the first n entries of the basis vectors in E^s. Consequently, with every x_0 there corresponds a unique ψ_1 and ψ_2 for which the solution of the differential equation $\dot{z}(t) = Mz(t)$, with $z_0^T = [x_0^T, \psi_1^T, \psi_2^T]$, converges to zero. So, according to Theorem 7.11, for every x_0 there is a Nash equilibrium. On the other hand, we have from the proof of Theorem 7.11 that all Nash equilibrium actions (u_1^*, u_2^*) satisfy

$$u_i^*(t) = -R_i^{-1}B_i^T\psi_i(t), \quad i = 1, 2,$$

where $\psi_i(t)$ satisfy the differential equation

$$\begin{bmatrix} \dot{x}(t) \\ \dot{\psi}_1(t) \\ \dot{\psi}_2(t) \end{bmatrix} = M \begin{bmatrix} x(t) \\ \psi_1(t) \\ \psi_2(t) \end{bmatrix}, \quad \text{with } x(0) = x_0.$$

Now, consider the system

$$\dot{z}(t) = Mz(t); \quad y(t) = Cz(t),$$

where

$$C := \begin{bmatrix} I & 0 & 0 \\ 0 & -R_1^{-1}B_1 & 0 \\ 0 & 0 & -R_2^{-1}B_2 \end{bmatrix}.$$

Since (A, B_i), $i = 1, 2$, is stabilizable, it is easily verified that the pair (C, M) is detectable. Consequently, due to our assumption that $x(t)$ and $u_i^*(t)$, $i = 1, 2$, converge to zero, we have from Lemma 7.37 that $[x^T(t), \psi_1^T(t), \psi_2^T(t)]$ converges to zero. Therefore, $[x^T(0), \psi_1^T(0), \psi_2^T(0)]$ has to belong to the stable subspace of M. However, as we argued above, for every x_0 there is exactly one vector $\psi_1(0)$ and vector $\psi_2(0)$ such that $[x^T(0), \psi_1^T(0), \psi_2^T(0)] \in E^s$. So we conclude that for every x_0 there exists exactly one Nash equilibrium.

Finally, by Theorem 7.13, notice that the game has an equilibrium for every initial state given by equation (7.4.8). Since for every initial state the equilibrium actions are uniquely determined, it follows that the equilibrium actions u_i^*, $i = 1, 2$, have to coincide with equation (7.4.8). $\qquad\square$

Proof of Theorem 7.25

From the proof of Theorem 7.4 we recall that by definition (see expression (7.11.15))

$$P_i(t) = G_i(T - t)H^{-1}(T - t), \tag{7.11.35}$$

where

$$H(t) = [I \quad 0 \quad 0]e^{-Mt} \begin{bmatrix} I \\ Q_{1T} \\ Q_{2T} \end{bmatrix}.$$

Since, (see equation (7.11.13))

$$\psi_1(0) = [0 \quad I \quad 0]e^{-MT} \begin{bmatrix} I \\ Q_{1T} \\ Q_{2T} \end{bmatrix} H^{-1}(T)x(0) = P_1(0, T)x(0),$$

for arbitrarily x_0,

$$P_1(0, T) = [0 \quad I \quad 0]e^{-MT} \begin{bmatrix} I \\ Q_{1T} \\ Q_{2T} \end{bmatrix} \left([I \quad 0 \quad 0]e^{-MT} \begin{bmatrix} I \\ Q_{1T} \\ Q_{2T} \end{bmatrix} \right)^{-1}. \tag{7.11.36}$$

In a similar way, it follows that

$$P_2(0,T) = [0 \quad 0 \quad I]e^{-MT}\begin{bmatrix} I \\ Q_{1T} \\ Q_{2T} \end{bmatrix}\left([I \quad 0 \quad 0]e^{-MT}\begin{bmatrix} I \\ Q_{1T} \\ Q_{2T} \end{bmatrix}\right)^{-1}. \qquad (7.11.37)$$

Now, choose $Y_1 := \left\{\begin{bmatrix} X_0 \\ X_1 \\ X_2 \end{bmatrix}\right\}$ as a basis for V_1 and let Y_2 be a basis for V_2. Then, the columns of matrix $V := [Y_1 \quad Y_2]$ form a basis for $\mathbb{R}^{3n}$. Moreover, because M is dichotomically separable, X_0 is invertible and there exist matrices J_1, J_2 such that

$$M = V\begin{bmatrix} J_1 & 0 \\ 0 & J_2 \end{bmatrix}V^{-1},$$

where $\sigma(J_i) = \sigma(M \mid V_i)$, $i = 1, 2$.

Using this, we can rewrite $P_1(0,T)$ and $P_2(0,T)$ in equations (7.11.36) and (7.11.37) as $\tilde{G}_i(T)\tilde{H}^{-1}(T)$, $i = 1, 2$, where

$$\tilde{G}_1(T) = [0 \quad I \quad 0]Ve^{\rho T}\begin{bmatrix} e^{-J_1 T} & 0 \\ 0 & e^{-J_2 T} \end{bmatrix}V^{-1}\begin{bmatrix} I \\ Q_{1T} \\ Q_{2T} \end{bmatrix},$$

$$\tilde{G}_2(T) = [0 \quad 0 \quad I]Ve^{\rho T}\begin{bmatrix} e^{-J_1 T} & 0 \\ 0 & e^{-J_2 T} \end{bmatrix}V^{-1}\begin{bmatrix} I \\ Q_{1T} \\ Q_{2T} \end{bmatrix},$$

$$\tilde{H}(T) = [I \quad 0 \quad 0]Ve^{\rho T}\begin{bmatrix} e^{-J_1 T} & 0 \\ 0 & e^{-J_2 T} \end{bmatrix}V^{-1}\begin{bmatrix} I \\ Q_{1T} \\ Q_{2T} \end{bmatrix}.$$

Here $\rho \in \mathbb{R}$ is any real number which separates the two spectra J_1 and J_2. That is, ρ satisfies $\lambda_i < \rho < \mu_j$ for all $\lambda_i \in \sigma(M \mid V_1)$ and all $\mu_j \in \sigma(M \mid V_2)$.

Next, consider $\begin{bmatrix} T_1 \\ T_2 \\ T_3 \end{bmatrix} := V^{-1}\begin{bmatrix} I \\ Q_{1T} \\ Q_{2T} \end{bmatrix}$. Since, by assumption, the direct sum of $\begin{bmatrix} I \\ Q_{1T} \\ Q_{2T} \end{bmatrix}$ and V_2 is $\mathbb{R}^{3n}$ there exist Z_i, $i = 1, 2, 3$, with Z_1 invertible, such that

$$\begin{bmatrix} I \\ Q_{1T} \\ Q_{2T} \end{bmatrix} = V\begin{bmatrix} Z_1 \\ Z_2 \\ Z_3 \end{bmatrix}.$$

Consequently,

$$T_1 = [I \quad 0 \quad 0]V^{-1}\begin{bmatrix} I \\ Q_{1T} \\ Q_{2T} \end{bmatrix} = [I \quad 0 \quad 0]V^{-1}V\begin{bmatrix} Z_1 \\ Z_2 \\ Z_3 \end{bmatrix} = Z_1,$$

is invertible. Therefore, we can rewrite $\tilde{H}(T)$ as

$$\tilde{H}(T) = e^{\rho T} [I \quad 0 \quad 0][Y_1 \quad Y_2] \begin{bmatrix} e^{-J_1 T} & 0 \\ 0 & e^{-J_2 T} \end{bmatrix} \begin{bmatrix} T_1 \\ T_2 \\ T_3 \end{bmatrix}$$

$$= e^{\rho T} [X_0 \quad U \quad W] \begin{bmatrix} e^{-J_1 T} & 0 \\ 0 & e^{-J_2 T} \end{bmatrix} \begin{bmatrix} T_1 \\ T_2 \\ T_3 \end{bmatrix}$$

$$= e^{\rho T} \left(X_0 e^{-J_1 T} T_1 + [U \quad W] e^{-J_2 T} \begin{bmatrix} T_2 \\ T_3 \end{bmatrix} \right),$$

for some matrices U, W. Since X_0 is invertible, matrix $X_0 e^{-J_1 T} T_1$ is invertible too. So,

$$\tilde{H}^{-1}(T) = \left(e^{\rho T} X_0 e^{-J_1 T} T_1 \left(I + (X_0 e^{-J_1 T} T_1)^{-1} [U \quad W] e^{-J_2 T} \begin{bmatrix} T_2 \\ T_3 \end{bmatrix} \right) \right)^{-1}$$

$$= \left(I + (X_0 e^{-J_1 T} T_1)^{-1} [U \quad W] e^{-J_2 T} \begin{bmatrix} T_2 \\ T_3 \end{bmatrix} \right)^{-1} e^{-\rho T} T_1^{-1} e^{J_1 T} X_0^{-1}$$

$$= \left(I + T_1^{-1} e^{(J_1 - \rho I) T} X_0^{-1} [U \quad W] e^{(\rho I - J_2) T} \begin{bmatrix} T_2 \\ T_3 \end{bmatrix} \right)^{-1} T_1^{-1} e^{(J_1 - \rho I) T} X_0^{-1}.$$

From this it is clear that both $T_1^{-1} e^{(J_1 - \rho I) T} X_0^{-1}$ and $[U \quad W] e^{(\rho I - J_2) T} \begin{bmatrix} T_2 \\ T_3 \end{bmatrix}$ converge to zero if $T \to \infty$. As a result, $\tilde{H}^{-1}(T)$ remains bounded if $T \to \infty$.

Furthermore, it follows that $\tilde{G}_1(T) = e^{\rho T} (X_1 e^{-J_1 T} T_1 + U_1 e^{-J_2 T} W_1)$ and $\tilde{G}_2(T) = e^{\rho T} (X_2 e^{-J_1 T} T_1 + U_2 e^{-J_2 T} W_2)$, for some matrices U_i, W_i, $i = 1, 2$.

Next, consider $\tilde{G}_1(T) - X_1 X_0^{-1} \tilde{H}(T)$. Simple calculations show that this matrix can be rewritten as

$$e^{\rho T} \left(U_1 e^{-J_2 T} W_1 - X_1 X_0^{-1} [U \quad W] e^{-J_2 T} \begin{bmatrix} T_2 \\ T_3 \end{bmatrix} \right). \tag{7.11.38}$$

As $e^{\rho T} e^{-J_2 T}$ converges to zero for $T \to \infty$, it is obvious now that $\tilde{G}_1(T) - X_1 X_0^{-1} \tilde{H}(T)$ converges to zero for $T \to \infty$. Similarly it can also be shown that $\tilde{G}_2(T) - X_2 X_0^{-1} \tilde{H}(T)$ converges to zero for $T \to \infty$. The final conclusion that $P_1(0, T) \to X_1 X_0^{-1}$, and $P_2(0, T) \to X_2 X_0^{-1}$, then follows by recalling the fact that $\tilde{H}^{-1}(T)$ remains bounded for $T \to \infty$.

In a similar way one can show that the solutions $K_i(0, T)$ of both the Riccati equations (7.6.5) also converge if $T \to \infty$. Since H_i is dichotomically separable, by Lemma 7.23, H_i has no eigenvalues on the imaginary axis. Consequently, the converged solutions stabilize the corresponding closed-loop systems. So we conclude (see Lemma 7.24) that both the algebraic Riccati equations (7.6.7) have a stabilizing solution K_i, $i = 1, 2$, respectively, and that $K_i(0, T) \to K_i$, if $T \to \infty$.   □

8

Non-cooperative feedback information games

8.1 Introduction

As in the previous chapter, we continue to study the game where the evolution of the state variable is described by the linear differential equation

$$\dot{x}(t) = Ax(t) + B_1 u_1(t) + \cdots + B_N u_N(t), \ x(0) = x_0, \tag{8.1.1}$$

and each player has a quadratic cost function, they like to minimize, given by:

$$J_i(u_1, \ldots, u_N) = \int_0^T \{x^T(t)Q_i x(t) + \sum_{j=1}^N u_j^T(t)R_{ij}u_j(t)\}dt + x^T(T)Q_{iT}x(T), \ i = 1, \ldots, N.$$

$$\tag{8.1.2}$$

Again all matrices in the cost functions J_i are assumed to be symmetric, and R_{ii}, $i = 1, \ldots, N$, are positive definite. The model and the objective functions are assumed to be common knowledge. Throughout this section we will use the shorthand notation:

$$S_i := B_i R_{ii}^{-1} B_i^T \quad \text{and} \quad S_{ij} := B_i R_{ii}^{-1} R_{ji} R_{ii}^{-1} B_i^T, \quad \text{for } i \neq j.$$

In the previous chapter the open-loop Nash equilibria of this game were studied. The open-loop Nash equilibria were defined as the Nash equilibria which result if the strategy spaces Γ_i, $i = 1, \ldots, N$, are

$$\Gamma_i = \{u_i(.) \mid u_i(.) = f_i(t, x_0) \quad \text{and} \quad (u_1(.), \ldots, u_N(.)) \in \mathcal{U}\}, \quad i = 1, \ldots, N.$$

LQ Dynamic Optimization and Differential Games   J. Engwerda
© 2005 John Wiley & Sons, Ltd.

A disadvantage of the open-loop equilibrium concept is that if, due to whatever circumstances, the actual state of the game at some point in time t_1 differs from the state implied by the equilibrium actions $u_i^*(0, t_1)$, $i = 1, \ldots, N$, in general the rest of the a priori designed equilibrium actions $u_i^*(t_1, T)$, $i = 1, \ldots, N$, will not be equilibrium actions any more if the players might reconsider their actions at t_1 for the rest of the planning horizon. That is, consider the subgame that consists of just the final part $[t_1, T]$ of the original game, starting at time t_1 at some arbitrary state x_{t_1}. Then the open-loop equilibrium actions $u_i^*(t_1, T)$, $i = 1, \ldots, N$, will only be a set of equilibrium actions for this subgame if x_{t_1} coincides with the state of the game $x^*(t_1)$ at time t_1 that is attained by using the equilibrium actions $u_i^*(0, t_1)$, $i = 1, \ldots, N$, during the time period $[0, t_1)$. We will formalize the above considerations somewhat. To that end we first introduce the notion of a truncated game (or subgame).

Definition 8.1

Consider the game defined by eqautions by (8.1.1) and (8.1.2). Denote this game by $\Gamma(0, x_0)$. Then, the **truncated game** (or **subgame**) $\Gamma(t_1, x_{t_1})$ of $\Gamma(0, x_0)$ is the game defined by:

$$\dot{x}(t) = Ax(t) + B_1 u_1(t) + \cdots + B_N u_N(t), \quad x(t_1) = x_{t_1}, \tag{8.1.3}$$

with, for each player, the quadratic cost function given by:

$$J_i(u_1, \ldots, u_N) = \int_{t_1}^{T} \{x^T(t) Q_i x(t) + \sum_{j=1}^{N} u_j^T(t) R_{ij} u_j(t)\} dt + x^T(T) Q_{iT} x(T), \quad i = 1, \ldots, N.$$
$$\tag{8.1.4} \square$$

Using this notion of a truncated game, we can then introduce the concepts of strong and weak time consistency, respectively. These notions formalize the above stated considerations.

Definition 8.2

A set of equilibrium actions $u_i^*(0, T)$ $i = 1, \ldots, N$, is called:

(a) **weakly time consistent** if the actions $u_i^*(t_1, T)$, $i = 1, \ldots, N$, constitute a set of equilibrium actions for the truncated game $\Gamma(t_1, x^*(t_1))$, where $x^*(t_1) = x(0, t_1, u_1^*(0, t_1))$, $\ldots, u_N^*(0, t_1))$, for every $t_1 \in (0, T)$;

(b) **strongly time consistent** or **subgame perfect** if, for every $t_1 \in (0, T)$, the actions $u_i^*(t_1, T)$, $i = 1, \ldots, N$, constitute a set of equilibrium actions for the truncated game $\Gamma(t_1, x_{t_1})$, where $x_{t_1} \in \mathbb{R}^n$ is an arbitrarily chosen state which is reachable from some initial state at $t = 0$. $\square$

So, weak time consistency means that the continuation of the equilibrium solution remains an equilibrium solution along the equilibrium path for all $t_1 \in (0, T)$, whereas strong time consistency means that the continuation of the equilibrium solution also remains an equilibrium solution for all initial conditions x_{t_1} (that can be attained at t_1 from some initial state at $t = 0$) off the equilibrium path. Obviously, if an equilibrium solution is strongly time consistent it is also weakly time consistent.

Example 8.1

Consider the two person scalar game with $a = -3$, $q_i = b_i = 2$, $r_{ii} = 1$, $i = 1, 2$ and $r_{ij} = 0$, $i \neq j$.

Then the unique infinite-planning horizon open-loop Nash equilibrium strategies are (see Theorem 7.29):

$$u_i^*(t) = -\frac{1}{2}x^*(t), \quad i = 1, 2,$$

where $x^*(t) = e^{-5t}x_0$. Now, if $x(t_1) \neq e^{-5t_1}x_0$, the equilibrium actions for the truncated game $\Gamma(t_1, x_{t_1})$ are

$$\tilde{u}_i(t) = -\frac{1}{2}\tilde{x}(t), \quad i = 1, 2,$$

where $\tilde{x}(t) = e^{-5(t-t_1)}x_{t_1}$. Obviously, $\tilde{u}(.) \neq u_i^*(.)$. So, the infinite-planning horizon open-loop Nash equilibrium strategies are not strongly time consistent. On the other hand it is trivially verified that if $x_{t_1} = e^{-5t_1}x_0$, the actions $u_i^*(t)$, $i = 1, 2$, are the open-loop Nash equilibrium actions for the truncated game $\Gamma(t_1, x_{t_1})$ as well. So, the infinite-planning horizon open-loop Nash equilibrium strategies are weakly time consistent. $\square$

It can easily be verified that for our linear quadratic differential game the open-loop Nash equilibrium actions are weakly time consistent. The following corollary formalizes this result.

Corollary 8.1

Consider the differential game (8.1.1) and (8.1.2). Any open-loop Nash equilibrium is weakly time consistent. $\square$

It is often argued that weak time consistency is a minimal requirement for the credibility of an equilibrium solution. That is, if the advertised equilibrium action of, say, player one is not weakly time consistent, player one would have an incentive to deviate from this action during the course of the game. For the other players, knowing this, it is therefore rational to incorporate this defection of player one into their own actions, which would lead to a different equilibrium solution. On the other hand, the property that the equilibrium solution does not have to be adapted by the players during the course of the game, although the system evolves not completely as expected beforehand, is generally regarded as a very nice property. Since the open-loop Nash equilibria in general do not have this property, the question arises whether there exist strategy spaces Γ_i such that if we look for Nash equilibria within these spaces, the equilibrium solutions do satisfy this strong time consistency property. In this chapter we will introduce such strategy spaces for both the finite-planning and infinite-planning horizon game. The corresponding equilibrium strategies will be derived and convergence properties discussed.

8.2 Finite-planning horizon

As argued in the introduction we look for strategy spaces such that within this class of feasible actions the non-cooperative Nash equilibrium solution has the strong time consistency property. Given this requirement a natural choice of the strategy space is the set of so-called Markov functions. That is, the set of functions where each function depends only on the current state of the system and time. Therefore, we consider

$$\Gamma_i^{fb} := \{u_i(0, T) \mid u_i(t) = f_i(t, x(t)) \text{ and } (u_1(.), \dots, u_N(.)) \in \mathcal{U}\}, \ i = 1, \dots, N. \quad (8.2.1)$$

From a conceptual point of view, the restriction to this class of control functions can be justified by the assumption that the players participate in a game where they only have access to the current state of the system.

By just restricting to this class of feasible actions it is hard to expect that an arbitrary Nash equilibrium will satisfy the strong time consistency property. One way to achieve this property is to consider a subclass of all Nash equilibria of this game which satisfy some additional requirements. We call this subclass the feedback Nash equilibria and they satisfy the following requirements.

Definition 8.3

The set of control actions $u_i^*(t) = \gamma_i^*(t, x(t))$ constitute a **feedback Nash equilibrium** solution if these strategies provide a Nash equilibrium for the truncated game $\Gamma(t_1, x_{t_1})$, for all $t_1 \in [0, T)$, and $x_{t_1} \in \mathbb{R}^n$ that are reachable from some initial state at time $t = 0$. $\qquad\square$

Since by definition the equilibrium actions are a function of the current state of the system, they can be interpreted as policy rules (Reinganum and Stokey (1985)). They require no precommitment of the players, and hence are also applicable if players are not 'credible'.

For notational simplicity we again confine ourselves to the two-player case from now on.

As in Section 4.4 we next introduce $J_i^*(t, x)$ as the minimum cost to go for player i if he evaluates his cost J at time t starting in the initial state x and knowing the action of his opponent. Using Theorem 4.10 an elementary evaluation of Definition 8.3 yields the following theorem.

Theorem 8.2

Assume $u_i^*(.)$, $i = 1, 2$, provide a feedback Nash equilibrium and let $x^*(t)$ be the corresponding closed-loop state trajectory. Let $V_i(t, x) := J_i^*(t, x)$ and assume that both partial derivatives of V_i exist and, moreover, $\frac{\partial V_i}{\partial x}$ is continuous and $\frac{d}{dt} V_i(t, x(t))$

exist. Then, for $t_0 \leq t \leq T$, $V_i : [0,T] \times \mathbb{R}^n \to \mathbb{R}$, $i = 1,2$, satisfy the partial differential equations:

$$-\frac{\partial V_1(t,x_1)}{\partial t} = \min_{u_1 \in U_1} \left[\frac{\partial V_1(t,x_1)}{\partial x_1} (Ax_1 + B_1 u_1 + B_2 u_2^*) + x_1^T Q_1 x_1 + u_1^T R_{11} u_1 + u_2^{*T} R_{12} u_2^* \right],$$

(8.2.2)

$$V_1(T,x_1) = x_1^T(T) Q_{1T} x_1(T)$$

(8.2.3)

$$-\frac{\partial V_2(t,x_2)}{\partial t} = \min_{u_2 \in U_2} \left[\frac{\partial V_2(t,x_1)}{\partial x_2} (Ax_2 + B_1 u_1^* + B_2 u_2) + x_2^T Q_2 x_2 + u_1^{*T} R_{21} u_1^* + u_2^T R_2 u_2 \right],$$

(8.2.4)

$$V_2(T,x_2) = x_2^T(T) Q_{2T} x_2(T),$$

(8.2.5)

where x_1 and x_2 are the solutions of the differential equations

$$\dot{x}_1 = Ax_1 + B_1 u_1 + B_2 u_2^*, \quad x_1(0) = x(0)$$
$$\dot{x}_2 = Ax_2 + B_1 u_1^* + B_2 u_2, \quad x_2(0) = x(0),$$

respectively.

If there exists V_i, $i = 1,2$, with the above mentioned properties such that they satisfy the set of partial differential equations (8.2.2)–(8.2.4) and

$$u_1^* = \arg\min_{u_1 \in U_1} \left[\frac{\partial V_1(t,x_1)}{\partial x_1} (Ax_1 + B_1 u_1 + B_2 u_2^*) + x_1^T Q_1 x_1 + u_1^T R_{11} u_1 + u_2^{*T} R_{12} u_2^* \right], \text{ and}$$

$$u_2^* = \arg\min_{u_2 \in U_2} \left[\frac{\partial V_2(t,x_2)}{\partial x_2} (Ax_2 + B_1 u_1^* + B_2 u_2) + x_2^T Q_2 x_2 + u_1^{*T} R_{21} u_1^* + u_2^T R_2 u_2 \right],$$

then $u_i^*(.)$, $i = 1,2$, provide a feedback Nash equilibrium. This equilibrium is strongly time consistent and the minimum costs for player i are $J_i^* = V_i(t_0, x_0)$. ☐

Since the system we are considering is linear, it is often argued that the equilibrium actions should be a linear function of the state too. This argument implies that we should consider either a refinement of the feedback Nash equilibrium concept or strategy spaces that only contain functions of the above-mentioned type. The first option amounts to considering only those feedback Nash equilibria which permit a linear feedback synthesis as being relevant. For the second option one has to consider the strategy spaces defined by

$$\Gamma_i^{lfb} := \{u_i(0,T) | u_i(t) = F_i(t) x(t)\},$$

where $F_i(.)$ is a piecewise continuous function, $i = 1,2$ and consider Nash equilibrium actions (u_1^*, u_2^*) within the strategy space $\Gamma_1^{lfb} \times \Gamma_2^{lfb}$.

It turns out that both equilibrium concepts yield the same characterization of these equilibria for the linear quadratic differential game, which will be presented in Theorem 8.5. Therefore, we will define just one equilibrium concept here and leave the formulation and proof of the other concept as an exercise for the reader.

Definition 8.4

The set of control actions $u_i^*(t) = F_i^*(t)x(t)$ constitute a **linear feedback Nash equilibrium** solution if both

$$J_1(u_1^*, u_2^*) \leq J_1(u_1, u_2^*) \quad \text{and} \quad J_2(u_1^*, u_2^*) \leq J_1(u_1^*, u_2),$$

for all $u_i \in \Gamma_i^{lfb}$.   □

Note

In the sequel, with some abuse of notation, sometimes the pair $(F_1^*(t), F_2^*(t))$ will be called a linear feedback Nash equilibrium.   □

In the same way as for open-loop Nash equilibria, it turns out that linear feedback Nash equilibria can be explicitly determined by solving a set of coupled Riccati equations.

Theorem 8.3

The two-player linear quadratic differential game (8.1.1) and (8.1.2) has, for every initial state, a linear feedback Nash equilibrium if and only if the following set of coupled Riccati differential equations has a set of symmetric solutions K_1, K_2 on $[0, T]$

$$\dot{K}_1(t) = -(A - S_2 K_2(t))^T K_1(t) - K_1(t)(A - S_2 K_2(t)) + K_1(t) S_1 K_1(t)$$
$$- Q_1 - K_2(t) S_{21} K_2(t),$$
$$K_1(T) = Q_{1T} \tag{8.2.6}$$

$$\dot{K}_2(t) = -(A - S_1 K_1(t))^T K_2(t) - K_2(t)(A - S_1 K_1(t)) + K_2(t) S_2 K_2(t)$$
$$- Q_2 - K_1(t) S_{12} K_1(t),$$
$$K_2(T) = Q_{2T}. \tag{8.2.7}$$

Moreover, in that case there is a unique equilibrium. The equilibrium actions are

$$u_i^*(t) = -R_i^{-1} B_i^T K_i(t) x(t), \quad i = 1, 2.$$

The cost incurred by player i is $x_0^T K_i(0) x_0$, $i = 1, 2$.

Proof

Assume $u_i^*(t) = F_i^*(t)x(t)$, $t \in [0, T]$, $i = 1, 2$, is a set of linear feedback equilibrium actions. Then, according the definition of a linear feedback equilibrium, the following linear quadratic regulator problem has as a solution $u_1^*(t) = F_1^*(t)x_1(t)$, for all x_0.

$$\min \int_0^T \{x_1^T(s)(Q_1 + F_2^{*T}(s) R_{12} F_2^*(s))x_1(s) + u_1^T(s) R_1 u_1(s)\}ds + x_1^T(T) Q_{1T} x_1(T),$$

subject to the system

$$\dot{x}_1(t) = (A + B_2 F_2^*(t))x_1(t) + B_1 u_1(t), \quad x_1(0) = x_0.$$

According to Theorem 5.1 (see also the Note following Corollary 5.10) this regulator problem has a solution if and only if the Riccati differential equation

$$\dot{K}_1(t) = -(A + B_2 F_2^*(t))^T K_1(t) - K_1(t)(A + B_2 F_2^*(t)) + K_1(t)S_1 K_1(t)$$
$$- (Q_1 + F_2^{*T}(t)R_{12}F_2^*(t)),$$
$$K_1(T) = Q_{1T} \tag{8.2.8}$$

has a symmetric solution $K_1(.)$ on $[0, T]$. Moreover, the solution for this optimization problem is unique, and is given by

$$u_1^*(t) = -R_1^{-1}B_1^T K_1(t)x_1(t).$$

So, we conclude that $F_1^*(t) = -R_1^{-1}B_1^T K_1(t)$. Similarly, it follows by definition that $u_2^*(t) = F_2^*(t)x_2(t)$ solves the problem

$$\min \int_0^T \{x_2^T(s)(Q_2 + K_1(s)S_{12}K_1(s))x_2(s) + u_2^T(s)R_2 u_2(s)\}ds + x_2^T(T)Q_{2T}x_2(T),$$

subject to the system

$$\dot{x}_2(t) = (A - S_1 K_1(t))x_2(t) + B_2 u_2(t), \quad x_2(0) = x_0.$$

According to Theorem 5.1 this regulator problem has a solution if and only if the Riccati differential equation

$$\dot{K}_2(t) = -(A - S_1 K_1(t))^T K_2(t) - K_2(t)(A - S_1 K_1(t)) + K_2(t)S_2 K_2(t) - Q_2, \quad K_2(T) = Q_{2T}$$

has a symmetric solution $K_2(.)$ on $[0, T]$. Moreover, the solution for the optimization problem is unique, and is given by

$$u_2^*(t) = -R_2^{-1}B_2^T K_2(t)x_2(t).$$

Therefore, $F_2^*(t)$ must coincide with $-R_2^{-1}B_2^T K_2(t)$. Substituting this result into equation (8.2.8) then shows that it is necessary for the set of coupled Riccati equations (8.2.6) and (8.2.7) to have a set of symmetric solutions $(K_1(.), K_2(.))$ on $[0, T]$.

The converse statement, that if the set of coupled Riccati differential equations (8.2.6) and (8.2.7) has a symmetric solution the strategies

$$u_i^*(t) = -R_i^{-1}B_i^T K_i(t)x(t)$$

constitute an equilibrium solution, follows directly from Theorem 8.2 by considering $V_i(t, x) = x^T(t)K_i(t)x(t)$. That there is only one equilibrium follows from the first part of the proof. Finally, by Theorem 8.2, the cost incurred by player i is $x_0^T K_i(0)x_0$. ☐

Unfortunately, up to now it is unclear whether there also exist feedback Nash equilibria which are nonlinear functions of the state. That is, it is unclear whether Theorem 8.3 still applies if we enlarge the strategy space to the set of nonlinear feedback Nash equilibria.

Example 8.2 illustrates Theorem 8.3. Moreover, the example shows the important role that information plays in games. It gives an example of a game where for every planning horizon a linear feedback equilibrium exists, whereas an open-loop equilibrium does not exist. Furthermore, by a slight change in the parameter values it shows that situations exist where the game has neither a linear feedback equilibrium nor an open-loop equilibrium.

Example 8.2

Consider the scalar zero-sum dynamic game:

$$\dot{x}(t) = \sqrt{2}u_1(t) - u_2(t), \quad x(0) = x_0,$$

where player one likes to minimize

$$J_1(u_1, u_2) = \int_0^T \{u_1^2(t) - u_2^2(t)\}dt + \alpha x^2(T),$$

whereas player two likes to maximize this performance criterion. So, $J_2 = -J_1$. With $a = 0$, $b_1 = \sqrt{2}$, $b_2 = -1$, $q_i = 0$, $r_{ii} = 1$ and $r_{ij} = -1$, $i \neq j = 1, 2$ according to Theorem 8.3 there exists a linear feedback Nash equilibrium if and only if the following set of coupled differential equations

$$\dot{k}_1(t) = 2k_1^2(t) + 2k_1(t)k_2(t) + k_2^2(t), \ k_1(T) = \alpha, \tag{8.2.9}$$

$$\dot{k}_2(t) = 2k_1^2(t) + 4k_1(t)k_2(t) + k_2^2(t), \ k_1(T) = -\alpha, \tag{8.2.10}$$

has a solution on $[0, 3]$. Adding equation (8.2.9) to (8.2.10) shows that $k_1 + k_2$ satisfies the differential equation

$$\frac{d(k_1 + k_2)(t)}{dt} = 4k_1(k_1 + k_2) + 2k_2(k_1 + k_2), \ (k_1 + k_2)(0) = 0.$$

Obviously, $(k_1 + k_2)(.) = 0$ satisfies this differential equation. Since the solution for this differential equation is uniquely determined it follows that $k_1(t) = -k_2(t)$. Substitution of this into equation (8.2.9) shows that the set of equations (8.2.9) and (8.2.10) has a solution on $[0, T]$ if and only if the following differential has a solution on $[0, T]$

$$\dot{k}_1(t) = k_1^2(t), \quad k_1(T) = \alpha.$$

The solution for this differential equation is

$$k_1(t) = \frac{1}{T + \frac{1}{\alpha} - t}.$$

If $\alpha > 0$, we see that $k_1(t)$ exists on $[0, T]$. Therefore, if $\alpha > 0$, the game has a linear feedback Nash equilibrium for every planning horizon T. The equilibrium actions are

$$u_1^*(t) = -\sqrt{2}k_1(t)x(t) = -\frac{\sqrt{2}}{T + \frac{1}{\alpha} - t}x(t),$$

and

$$u_2^*(t) = -k_1(t)x(t) = -\frac{1}{T + \frac{1}{\alpha} - t}x(t).$$

The cost for player one is $k_1(0)x_0^2 = \frac{1}{T + \frac{1}{\alpha}}x_0^2$ and the cost for player two is $-k_1(0)x_0^2 = \frac{-1}{T + \frac{1}{\alpha}}x_0^2$.

Next, consider the case that $\alpha < 0$. Then, $k_1(t)$ does not exist on $[0, T]$. It has a finite escape time at $t = T + \frac{1}{\alpha}$. So the game has no linear feedback Nash equilibrium in that case, whatever the length of the chosen planning horizon is.

Now consider the open-loop Nash equilibria for this game. According to Proposition 7.6 this game has, for all initial states, a unique open-loop Nash equilibrium for all $T \in [0, t_1)$ if and only if the following differential equations have a solution on $[0, t_1)$

$$\dot{p}(t) = p^2(t), \quad p(T) = \alpha \tag{8.2.11}$$

and

$$\dot{k}_1(t) = 2k_1^2(t), \quad k_1(T) = \alpha \tag{8.2.12}$$
$$\dot{k}_2(t) = k_2^2(t), \quad k_2(T) = -\alpha. \tag{8.2.13}$$

So a necessary condition for the existence of an open-loop Nash equilibrium, for all x_0 and all $T \in [0, t_1)$, is that the differential equations (8.2.13) and (8.2.11) have a solution on $[0, t_1)$. However, as we saw from the analysis of the feedback case, one of these differential games always has no solution (assuming $\alpha \neq 0$). So, whatever the sign of α is, the game has no open-loop equilibrium on $[0, t_1)$. $\quad\square$

Example 8.3

Reconsider the finite-planning horizon version of the dynamic duopoly game with sticky prices (see Examples 3.24 and 7.5).

$$\dot{p}(t) = s_1\left(\frac{c}{s_1} - \frac{s_2}{s_1}(v_1 + v_2) - p(t)\right), \quad p(0) = p_0,$$

where $p(t)$ denotes the price of the product, v_i is the output rate of company i and s_1 the speed at which the price p converges to the price $\tilde{p}$ dictated by the inverse demand function $\tilde{p} = \frac{c}{s_1} - \frac{s_2}{s_1}(v_1 + v_2)$; Also, the profit functions are given by:

$$J_i(v_1, v_2) = \int_0^T \{p(t)v_i(t) - c_i v_i(t) - \frac{1}{2}v_i^2(t)\}dt.$$

All parameters c, s_1, s_2, p_0 and c_i are positive. Assume that we look for Nash equilibria within the class of affine functions of the price variable, i.e. we consider strategy spaces defined by

$$\Gamma_i^{aff} := \{u_i(0, T) \mid u_i(t) = F_i(t)p(t) + g_i(t), \text{ with } F_i(.), g_i(.) \text{ piecewise continuous functions}\}.$$

We can proceed then in the same way as in Example 7.5. Introducing as state variable $x(t) := [p(t)\ 1]^T$ and $u_i(t) := v_i(t) + [-1\ c_i]x(t)$, the above maximization problems can be rewritten as

$$\min_{u_i} J_i(u_1, u_2) = \frac{1}{2}\int_0^T \left\{ x^T(t)\begin{bmatrix} -1 & c_i \\ c_i & -c_i^2 \end{bmatrix}x(t) + u_i^2(t) \right\}dt$$

subject to

$$\dot{x}(t) = \begin{bmatrix} -(s_1 + 2s_2) & c + s_2(c_1 + c_2) \\ 0 & 0 \end{bmatrix}x(t) + \begin{bmatrix} -s_2 \\ 0 \end{bmatrix}u_1 + \begin{bmatrix} -s_2 \\ 0 \end{bmatrix}u_2.$$

According to Theorem 8.3 the game has, for every initial state, a unique Nash equilibrium within $\Gamma_1^{aff} \times \Gamma_2^{aff}$ if and only if the following set of coupled Riccati differential equations has a set of symmetric solutions $(K_1(.), K_2(.))$ on $[0, T]$

$$\dot{K}_1(t) = -(A - S_2K_2(t))^T K_1(t) - K_1(t)(A - S_2K_2(t)) + K_1(t)S_1K_1(t) - Q_1, \ K_1(T) = 0$$
$$\dot{K}_2(t) = -(A - S_1K_1(t))^T K_2(t) - K_2(t)(A - S_1K_1(t)) + K_2(t)S_2K_2(t) - Q_2, \ K_2(T) = 0.$$

Here

$$A := \begin{bmatrix} -(s_1 + 2s_2) & c + s_2(c_1 + c_2) \\ 0 & 0 \end{bmatrix}, \ S_i := \begin{bmatrix} 2s_2^2 & 0 \\ 0 & 0 \end{bmatrix} \text{ and } Q_i := \frac{1}{2}\begin{bmatrix} -1 & c_i \\ c_i & -c_i^2 \end{bmatrix}.$$

Moreover, the equilibrium actions are:

$$v_i = 2[s_2 \ 0] K_i x(t) - [-1 \ c_i]x(t).$$

As an example consider the symmetric case, i.e. $c_1 = c_2$. By considering the differential equation for $V(t) := K_1(t) - K_2(t)$, it is easily verified that in this case $K_1(t) = K_2(t)$. Denoting

$$K(t) =: \begin{bmatrix} k_1(t) & k_2(t) \\ k_2(t) & k_3(t) \end{bmatrix},$$

we obtain the following set of differential equations:

$$\dot{k}_1(t) = 2(s_1 + 2s_2)k_1(t) + 6s_2^2 k_1^2(t) + \frac{1}{2}, \quad k_1(T) = 0;$$

$$\dot{k}_2(t) = (s_1 + 2s_2 + 6s_2^2 k_1(t))k_2(t) - (c + 2s_2c_1)k_1(t) - \frac{1}{2}c_1, \quad k_2(T) = 0;$$

$$\dot{k}_3(t) = -2(c + 2s_2c_1)k_2(t) + 6s_2^2 k_2^2(t) + \frac{1}{2}c_1^2, \quad k_3(T) = 0.$$

Taking a closer look at the structure of this control problem one observes that to find its solution it is sufficient to solve the system of differential equations in k_1 and k_2. In fact, the differential equation in k_1 can be solved explicitly. With $w := 2\sqrt{s_1^2 + 4s_1s_2 + s_2^2}$, the solution is given by

$$k_1(t) = \frac{2(s_1 + 2s_2) + w}{6s_2^2}\left(-\frac{1}{2} + \frac{w}{2(s_1 + 2s_2) + w + (-2(s_1 + 2s_2) + w)e^{w(t-T)}}\right).$$

Next, a linear time-varying differential equation for k_2 results whose solution cannot, in general, be determined explicitly. Fortunately, since $k_1(.)$ is a continuous function, we know from the general theory (see equation (3.1.6)) that $k_2(.)$ exists. So, usually using numerical techniques to find an explicit expression, the equilibrium actions are

$$v_i(t) = (2s_2k_1(t) + 1)p(t) + 2s_2k_2(t) - c_1,$$

where $p(t)$ satisfies the differential equation

$$\dot{p}(t) = -(s_1 + 2s_2 + 4s_2^2k_1(t))p(t) + c + 2s_2c_1 - 4s_2^2k_2(t), \quad p(0) = p_0.$$

In Figure 8.1 the equilibrium price path and action trajectories are plotted for the parameter values we chose in Example 7.5. That is, $p_0 = 2$, $c_1 = c_2 = s_1 = s_2 = 1$, $c = 5$

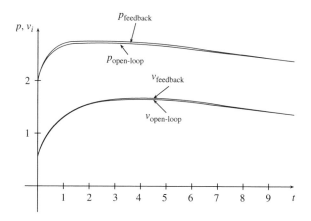

Figure 8.1   Some optimal state and control trajectories for Example 8.3

and $T = 10$. These solutions are obtained by solving the above set of differential equations numerically for $(k_1(.), k_2(.), p(.))$. To show the difference from the results obtained in Example 7.13 those numerical results are also included. We see that the steady-state values of both trajectories are slightly larger for the feedback paths than those for the open-loop paths.   ☐

We conclude this section by considering the zero-sum differential game.

Theorem 8.4 (Zero-sum differential game)

Consider the differential game described by

$$\dot{x}(t) = Ax(t) + B_1u_1(t) + B_2u_2(t), \quad x(0) = x_0, \tag{8.2.14}$$

with, for player one, the quadratic cost function:

$$J_1(u_1, u_2) = \int_0^T \{x^T(t)Qx(t) + u_1^T(t)R_1u_1(t) - u_2^T(t)R_2u_2(t)\}dt + x^T(T)Q_Tx(T),$$

and, for player two, the opposite objective function:

$$J_2(u_1, u_2) = -J_1(u_1, u_2)$$

where the matrices Q and Q_T are symmetric and R_i, $i = 1, 2$, are positive definite.

This zero-sum linear quadratic differential game has a linear feedback Nash equilibrium for every initial state if and only if the following Riccati differential equation has a symmetric solution K on $[0, T]$

$$\dot{K}(t) = -A^T K(t) - K(t)A + K(t)(S_1 - S_2)K(t) - Q, \quad K(T) = Q_T. \qquad (8.2.15)$$

Moreover, if equation (8.2.15) has a solution the game has a unique equilibrium. The equilibrium actions are

$$u_1^*(t) = -R_1^{-1}B_1^T K(t)x(t) \quad \text{and} \quad u_2^*(t) = R_2^{-1}B_2^T K(t)x(t).$$

The cost incurred by player one is $x_0^T K(0)x_0$ and by player two $-x_0^T K(0)x_0$.

Proof

According to Theorem 8.3 this game has a linear feedback Nash equilibrium if and only if the following two coupled differential equations have a symmetric solution $K_i(.)$, $i = 1, 2$, on $[0, T]$

$$\dot{K}_1(t) = -(A - S_2K_2(t))^T K_1(t) - K_1(t)(A - S_2K_2(t)) + K_1(t)S_1K_1(t) - Q + K_2(t)S_2K_2(t),$$
$$K_1(T) = Q_T \qquad (8.2.16)$$

$$\dot{K}_2(t) = -(A - S_1K_1(t))^T K_2(t) - K_2(t)(A - S_1K_1(t)) + K_2(t)S_2K_2(t) + Q + K_1(t)S_1K_1(t),$$
$$K_2(T) = -Q_T. \qquad (8.2.17)$$

Adding equation (8.2.16) to (8.2.17) gives the following differential equation in $(K_1 + K_2)(t)$

$$(K_1 + K_2) = -(K_1 + K_2)(t)(A - S_1K_1(t) - S_2K_2(t)) - (A - S_1K_1(t) - S_2K_2(t))^T(K_1 + K_2)(t),$$
$$= (K_1 + K_2)(T)\,0.$$

Obviously, $(K_1 + K_2)(.) = 0$ satisfies this differential equation. Since the solution to this differential equation is unique, we conclude that $K_1(t) = -K_2(t)$. Substitution of this into equation (8.2.16) then shows that equations (8.2.16) and (8.2.17) have a solution on $[0, T]$ if and only if (8.2.15) has a solution on $[0, T]$.

The corresponding equilibrium strategies and cost then follow directly from Theorem 8.3. $\qquad \square$

Note

Recall from Proposition 7.6 that this game has a unique open-loop Nash equilibrium for all $T \in [0, t_1)$ if and only if in addition to equation (8.2.15) the following two Riccati differential equations have a solution on $[0, t_1)$

$$\dot{K}_1(t) = -A^T K_1(t) - K_1(t)A + K_1(t)S_1 K_1(t) - Q, \quad K_1(T) = Q_T \qquad (8.2.18)$$
$$\dot{K}_2(t) = -A^T K_2(t) - K_2(t)A + K_2(t)S_2 K_2(t) + Q, \quad K_2(T) = -Q_T. \qquad (8.2.19)$$

If this open-loop equilibrium exists, the equilibrium strategies coincide with the linear feedback equilibrium strategies. ☐

8.3 Infinite-planning horizon

As in the open-loop case we consider in this section the minimization of the performance criterion

$$J_i(x_0, u_1, u_2) = \lim_{T \to \infty} J_i(x_0, u_1, u_2, T) \qquad (8.3.1)$$

for player i, $i = 1, 2$, where

$$J_i(x_0, u_1, u_2, T) = \int_0^T \left\{ x^T(t) Q_i x(t) + u_i^T(t) R_{ii} u_i(t) + u_j(t) R_{ij} u_j(t) \right\} dt, \quad j \neq i,$$

subject to the dynamical system

$$\dot{x}(t) = Ax(t) + B_1 u_1(t) + B_2 u_2(t), \quad x(0) = x_0. \qquad (8.3.2)$$

Here Q_i and R_{ij}, $i, j = 1, 2$, are symmetric and R_{ii}, $i = 1, 2$, is positive definite.

From a conceptual point of view it is difficult to generalize the notion of feedback Nash equilibrium which we introduced in the previous section for an infinite horizon. On the other hand it is clear that, maybe under some additional regularity conditions, the notion of linear feedback Nash equilibrium can be generalized straightforwardly if we are restricted to stationary policy rules. So it seems reasonable to study Nash equilibria within the class of linear time-invariant state feedback policy rules. Therefore we shall limit our set of permitted controls to the constant linear feedback strategies. That is, to $u_i = F_i x$, with $F_i \in \mathbb{R}^{m_i \times n}$, $i = 1, 2$, and where (F_1, F_2) belongs to the set

$$\mathcal{F} := \{ F = (F_1, F_2) \mid A + B_1 F_1 + B_2 F_2 \text{ is stable} \}.$$

The stabilization constraint is imposed to ensure the finiteness of the infinite-horizon cost integrals that we will consider. This assumption can also be justified from the supposition that one is studying a perturbed system which is temporarily out of equilibrium. In that case it is reasonable to expect that the state of the system remains close to the origin. Obviously the stabilization constraint is a bit unwieldy since it introduces dependence between the strategy spaces of the players. So it presupposes that there is at least the possibility of some coordination between both players. This coordination assumption seems to be more stringent in this case than for the equilibrium concepts we introduced previously. However, the stabilization constraint can be justified from the supposition that both players have a first priority in stabilizing the system. Whether this coordination actually takes place, depends on the outcome of the game. Only if the players have the impression that their actions are such that the system becomes unstable, will they coordinate their actions in order to realize this meta-objective and adapt their actions accordingly. Probably for most games the equilibria without this stabilization constraint coincide with the equilibria of the game if one does consider this additional stabilization constraint. That is, the stabilization constraint will not be active in most cases, but there are games where it does play a role. We will give an example of this at the end of this section.

To make sure that our problem setting makes sense, we assume throughout this chapter that the set $\mathcal{F}$ is non-empty. A necessary and sufficient condition for this to hold is that the matrix pair $(A, [B_1, B_2])$ is stabilizable.

Summarizing, we define the concept of a linear feedback Nash equilibrium on an infinite-planning horizon as follows.

Definition 8.5

$(F_1^*, F_2^*) \in \mathcal{F}$ is called a **stationary linear feedback Nash equilibrium** if the following inequalities hold

$$J_1(x_0, F_1^*, F_2^*) \leq J_1(x_0, F_1, F_2^*) \quad \text{and} \quad J_2(x_0, F_1^*, F_2^*) \leq J_2(x_0, F_1^*, F_2)$$

for each x_0 and for each state feedback matrix F_i, $i = 1, 2$ such that (F_1^*, F_2) and $(F_1, F_2^*) \in \mathcal{F}$. □

Unless otherwise stated, the phrases 'stationary' and 'linear' in the notion of stationary linear feedback Nash equilibrium are dropped in this section. This is because it is clear from the context here which equilibrium concept we are dealing with.

Next, consider the set of coupled algebraic Riccati equations

$$0 = -(A - S_2K_2)^T K_1 - K_1(A - S_2K_2) + K_1S_1K_1 - Q_1 - K_2S_{21}K_2, \quad (8.3.3)$$
$$0 = -(A - S_1K_1)^T K_2 - K_2(A - S_1K_1) + K_2S_2K_2 - Q_2 - K_1S_{12}K_1. \quad (8.3.4)$$

Theorem 8.5 below states that feedback Nash equilibria are completely characterized by symmetric **stabilizing solutions** of equations (8.3.3) and (8.3.4). That is, by symmetric solutions (K_1, K_2) for which the closed-loop system matrix $A - S_1K_1 - S_2K_2$ is stable.

Theorem 8.5

Let (K_1, K_2) be a symmetric stabilizing solution of equations (8.3.3) and (8.3.4) and define $F_i^* := -R_{ii}^{-1}B_i^T K_i$ for $i = 1, 2$. Then (F_1^*, F_2^*) is a feedback Nash equilibrium. Moreover, the cost incurred by player i by playing this equilibrium action is $x_0^T K_i x_0$, $i = 1, 2$. Conversely, if (F_1^*, F_2^*) is a feedback Nash equilibrium, there exists a symmetric stabilizing solution (K_1, K_2) of equations (8.3.3) and (8.3.4) such that $F_i^* = -R_{ii}^{-1}B_i^T K_i$.

Proof

Let (K_1, K_2) be a stabilizing solution of equations (8.3.3) and (8.3.4) and define $F_2^* := -R_{22}^{-1} B_2^T K_2$. Next, consider the minimization by player one of the cost function

$$J_1(x_0, u_1, F_2^*) = \int_0^\infty \left\{ x^T (Q_1 + F_2^{*T} R_{12} F_2^*) x + u_1^T R_{11} u_1 \right\} dt,$$

subject to the system $\dot{x}(t) = (A - S_2 K_2) x(t) + B_1 u_1(t)$, $x(0) = x_0$.

By assumption (see equations (8.3.3)), the equation

$$0 = -(A - S_2 K_2)^T X - X(A - S_2 K_2) + X S_1 X - Q_1 - K_2 S_{21} K_2$$

has a stabilizing solution $X = K_1$. So, according to Theorem 5.14, the above minimization problem has a solution. The solution is given by $u_1^* = -R_{11}^{-1} B_1^T X$ and the corresponding minimum cost by $x_0^T X x_0$. Since $X = K_1$ it follows that $F_1 = F_1^*$ solves the minimization problem. That is,

$$J_1(x_0, F_1^*, F_2^*) \leq J_1(x_0, F_1, F_2^*).$$

In a similar way it can be shown that the corresponding minimization problem for player two is solved by F_2^*, which proves the first part of the claim.

Next, assume that $(F_1^*, F_2^*) \in \mathcal{F}$ is a feedback Nash equilibrium. Then, by definition,

$$J_1(x_0, F_1^*, F_2^*) \leq J_1(x_0, F_1, F_2^*) \quad \text{and} \quad J_2(x_0, F_1^*, F_2^*) \leq J_2(x_0, F_1^*, F_2)$$

for all x_0 and for all state feedback matrices F_i, $i = 1, 2$, such that (F_1^*, F_2) and $(F_1, F_2^*) \in \mathcal{F}$. Hence, according to Theorem 5.14, there exist real symmetric matrices K_i, $i = 1, 2$, satisfying the set of equations

$$0 = -(Q_1 + F_2^{*T} R_{12} F_2^*) - (A + B_2 F_2^*)^T K_1 - K_1 (A + B_2 F_2^*) + K_1 S_1 K_1 \tag{8.3.5}$$

$$0 = -(Q_2 + F_1^{*T} R_{21} F_1^*) - (A + B_1 F_1^*)^T K_2 - K_2 (A + B_1 F_1^*) + K_2 S_2 K_2 \tag{8.3.6}$$

such that both $A + B_2 F_2^* - S_1 K_1$ and $A + B_1 F_1^* - S_2 K_2$ are stable.

Theorem 5.22 implies, moreover, that $F_i^* = -R_{ii}^{-1} B_i^T K_i$ and $J_i(x_0, F_1^*, F_2^*) = x_0^T K_i x_0$. Replacing F_2^* in equation (8.3.5) by $-R_{22}^{-1} B_2^T K_2$ and F_1^* in equation (8.3.6) by $-R_{11}^{-1} B_1^T K_1$ shows that (K_1, K_2) satisfies the coupled set of algebraic Riccati equations (8.3.3) and (8.3.4). Furthermore, replacing F_1^* in $A + B_1 F_1^* - S_2 K_2$ by $-R_{11}^{-1} B_1^T K_1$ shows that matrix $A - S_1 K_1 - S_2 K_2$ is stable, which completes the proof. $\qquad\square$

Example 8.4

Reconsider the government debt stabilization game from section 7.8.1. In this game it is assumed that government debt accumulation, $\dot{d}$, is the sum of interest payments on government debt, $rd(t)$, and primary fiscal deficits, $f(t)$, minus the seignorage, $m(t)$:

$$\dot{d}(t) = rd(t) + f(t) - m(t), \quad d(0) = d_0.$$

The objective of the fiscal authority is to minimize a sum of time profiles of the primary fiscal deficit, base-money growth and government debt:

$$L^F = \int_0^\infty \{(f(t) - \bar{f})^2 + \eta(m(t) - \bar{m})^2 + \lambda(d(t) - \bar{d})^2\}e^{-\delta t}dt,$$

whereas the monetary authorities set the growth of base money so as to minimize the loss function:

$$L^M = \int_0^\infty \{(m(t) - \bar{m})^2 + \kappa(d(t) - \bar{d})^2\}e^{-\delta t}dt.$$

Here all variables denoted with a bar are assumed to be fixed targets which are given a priori.

Introducing $x_1(t) := (d(t) - \bar{d})e^{-\frac{1}{2}\delta t}$, $x_2(t) := (r\bar{d} + \bar{f} - \bar{m})e^{-\frac{1}{2}\delta t}$, $u_1(t) := (f(t) - \bar{f})e^{-\frac{1}{2}\delta t}$ and $u_2(t) := (m(t) - \bar{m})e^{-\frac{1}{2}\delta t}$ we showed that the above game can be rewritten in our standard notation as

$$\min_{u_1} L^F = \int_0^\infty \{\lambda x_1^2(t) + u_1^2(t) + \eta u_2^2(t)\}dt$$

and

$$\min_{u_2} L^M = \int_0^\infty \{\kappa x_1^2(t) + u_2^2(t)\}dt$$

subject to

$$\dot{x}(t) = Ax(t) + B_1 u_1(t) + B_2 u_2(t)$$

with:

$$A = \begin{bmatrix} r - \frac{1}{2}\delta & 1 \\ 0 & -\frac{1}{2}\delta \end{bmatrix}, \quad B_1 = \begin{bmatrix} 1 \\ 0 \end{bmatrix}, \quad B_2 = \begin{bmatrix} -1 \\ 0 \end{bmatrix}, \quad Q_1 = \begin{bmatrix} \lambda & 0 \\ 0 & 0 \end{bmatrix}, \quad Q_2 = \begin{bmatrix} \kappa & 0 \\ 0 & 0 \end{bmatrix},$$

$R_{11} = 1, R_{22} = 1, R_{12} = \eta$ and $R_{21} = 0$.

By Theorem 8.5 this game has a set of feedback Nash equilibria $u_i^* = F_i^* x(t)$ within the set of constant feedback matrices $\mathcal{F}$ if and only if the set of algebraic Riccati equations (8.3.3) and (8.3.4) has a set of stabilizing solutions (K_1, K_2). Or, stated in terms of the original problem formulation, the game has a set of feedback Nash equilibria $(f(t), m(t)) = (f_{11}d(t) + f_{12}, f_{21}d(t) + f_{22})$ within the class of stabilizing affine functions of the government debt

$$\Gamma_1^{aff} \times \Gamma_2^{aff} := \left\{(f_{11}, f_{21}) \mid r - \frac{1}{2}\delta + f_{11} - f_{21} < 0\right\}$$

if and only if equations (8.3.3) and (8.3.4) has a set of stabilizing solutions. With

$$S_i := \begin{bmatrix} 1 & 0 \\ 0 & 0 \end{bmatrix}, \ i = 1, 2, \ S_{12} = 0, \ S_{21} = \begin{bmatrix} \eta & 0 \\ 0 & 0 \end{bmatrix},$$

and

$$K_1 := \begin{bmatrix} l_1 & l_2 \\ l_2 & l_3 \end{bmatrix}, \ K_2 := \begin{bmatrix} k_1 & k_2 \\ k_2 & k_3 \end{bmatrix},$$

the equations reduce to the following six expressions

$$2l_1 \left(r - \frac{1}{2}\delta - k_1 \right) + l_1^2 - \lambda - \eta k_1^2 = 0 \tag{8.3.7}$$

$$k_1^2 - 2k_1 \left(r - \frac{1}{2}\delta - l_1 \right) - \kappa = 0 \tag{8.3.8}$$

$$(-r + \delta + k_1 + l_1)l_2 + (l_1 - \eta k_1)k_2 = l_1 \tag{8.3.9}$$

$$k_1 l_2 + (-r + \delta + k_1 + l_1)k_2 = k_1 \tag{8.3.10}$$

$$\frac{1}{\delta}(2l_2(1 - k_2) - l_2^2 + \eta k_2^2) = l_3 \tag{8.3.11}$$

$$\frac{1}{\delta}(2k_2(1 - l_2) - k_2^2) = k_3. \tag{8.3.12}$$

From equation (8.3.8) it follows, using the fact that the closed-loop system must be stable, that

$$k_1 = r - \frac{1}{2}\delta - l_1 + \sqrt{\left(r - \frac{1}{2}\delta - l_1 \right)^2 + \kappa}.$$

Using this result, one can then solve the rest of the equations by first substituting, for example, this result into equation (8.3.7). This yields an, in principle, fourth-order polynomial equation in l_1 that has to be solved. Once the solution of this equation has been found (l_2, k_2) are obtained from equations (8.3.9) and (8.3.10) as

$$\begin{bmatrix} l_2 \\ k_2 \end{bmatrix} = \begin{bmatrix} -r + \delta + k_1 + l_1 & l_1 - \eta k_1 \\ k_1 & -r + \delta + k_1 + l_1 \end{bmatrix}^{-1} \begin{bmatrix} l_1 \\ k_1 \end{bmatrix},$$

whereas l_3 and k_3 are then directly obtained from equations (8.3.11) and (8.3.12), respectively.

The equilibrium actions are then

$$u_1^*(t) = -[l_1 \ l_2]x(t) \quad \text{and} \quad u_2^*(t) = [k_1 \ k_2]x(t).$$

Or, stated in the original variables,

$$f(t) = -l_1 d(t) + (l_1 - l_2 r)\bar{d} + (1 - l_2)\bar{f} - l_2 \bar{m}$$

and

$$m(t) = k_1 d(t) - (k_1 - k_2 r)\bar{d} + k_2 \bar{f} + (1 - k_2)\bar{m},$$

where

$$d(t) = \alpha e^{(r - l_1 - k_1)t} + \bar{d} + \frac{(1 - l_2 - k_2)(r\bar{d} + \bar{f} - \bar{m})}{l_1 + k_1 - r},$$

with $\alpha = d_0 - \bar{d} - \frac{(1 - l_2 - k_2)(r\bar{d} + \bar{f} - \bar{m})}{l_1 + k_1 - r}$.

In particular we see that the debt accumulates over time to the constant level

$$\bar{d} + \frac{(1 - l_2 - k_2)(r\bar{d} + \bar{f} - \bar{m})}{l_1 + k_1 - r}.$$

With $\delta = 0.04$, $r = 0.06$, $\eta = 0.2$, $\lambda = \kappa = 1$ the above equations (8.3.7)–(8.3.12) have the unique stabilizing solution

$$K_1 = \begin{bmatrix} 0.6295 & 0.4082 \\ 0.4082 & 10.8101 \end{bmatrix} \text{ and } K_2 = \begin{bmatrix} 0.5713 & 0.2863 \\ 0.2863 & 6.4230 \end{bmatrix}.$$

Consequently, the convergence rate of the government debt towards its long-term equilibrium, $r - l_1 - k_1$, is -1.1408. Moreover, the long-term equilibrium debt is $\bar{d} + \frac{0.3055}{1.1408}(r\bar{d} + \bar{f} - \bar{m})$.

Comparing these numbers with the open-loop outcomes (see Section 7.8.1 and, in particular, equation (7.8.6)) we see that the corresponding convergence speed of the government debt towards its long-term equilibrium, $\frac{1}{2}\delta - l$, is -1.3948 and the long-term equilibrium debt is $\bar{d} - \frac{0.0139}{1.3948}(r\bar{d} + \bar{f} - \bar{m})$. Assuming that the gap between the desired financing, $\bar{f} + r\bar{d}$, and desired monetary accomodation, $\bar{m}$, is positive we see that for this case the long-term equilibrium debt is below the government debt target $\bar{d}$ under the open-loop scenario, whereas under the feedback scenario it is above this target value. Furthermore, we see that the debt converges more rapidly towards its equilibrium value in the open-loop scenario. □

Example 8.5

In this example we reconsider the model on duopolistic competition with sticky prices that was analyzed in Section 7.8.2. We now look for the feedback Nash equilibria of the game defined by the revenue functions

$$J_i(u_1, u_2) = \int_0^\infty e^{-rt}\{p(t)u_i(t) - c_i u_i(t) - \frac{1}{2}u_i^2(t)\}dt, \tag{8.3.13}$$

subject to the dynamic constraint

$$\dot{p}(t) = s\{a - (u_1(t) + u_2(t)) - p(t)\}, \quad p(0) = p_0. \tag{8.3.14}$$

For the interpretation of the variable we refer to section 7.8.2.

The strategy spaces from which the players choose their actions is assumed to be given by the set of stabilizing affine functions of the price p, i.e.

$$\Gamma_1^{aff} \times \Gamma_2^{aff} := \left\{ (u_1, u_2) \mid u_i(t) = f_{ii}p(t) + g_i, \text{ with } -s - sf_{11} - sf_{22} < \frac{1}{2}r \right\}.$$

To determine the feedback Nash equilibrium actions for this game (8.3.13) and (8.3.14) we first reformulate it into our standard framework. Recall from section 7.8.2 that by introducing the variables

$$x^T(t) := e^{-\frac{1}{2}rt}[p(t) \quad 1] \text{ and } v_i(t) := e^{-\frac{1}{2}rt}u_i(t) + [-1 \quad c]x(t), \ i = 1, 2, \quad (8.3.15)$$

the problem can be rewrittten as the minimization of

$$J_i := \int_0^\infty \{x^T(t)Q_ix(t) + v_i^T(t)R_iv_i(t)\}dt, \ i = 1, 2,$$

subject to the dynamics

$$\dot{x}(t) = Ax(t) + B_1v_1(t) + B_2v_2(t), \ x^T(0) = [p_0 \quad 1].$$

Here

$$A = \begin{bmatrix} -\frac{1}{2}r - 3s & (a + 2c)s \\ 0 & -\frac{1}{2}r \end{bmatrix}, \ B_i = \begin{bmatrix} -s \\ 0 \end{bmatrix}, \ Q_i = \begin{bmatrix} -\frac{1}{2} & \frac{1}{2}c \\ \frac{1}{2}c & -\frac{1}{2}c^2 \end{bmatrix} \text{ and } R_i = \frac{1}{2}.$$

According to Theorem 8.5 this game has a feedback Nash equilibrium if and only if equations (8.3.3) and (8.3.4) have a set of stabilizing solutions. With

$$S_i := \begin{bmatrix} 2s^2 & 0 \\ 0 & 0 \end{bmatrix}, \ i = 1, 2; \ S_{12} = S_{21} = 0$$

and

$$K_1 := \begin{bmatrix} l_1 & l_2 \\ l_2 & l_3 \end{bmatrix}, K_2 := \begin{bmatrix} k_1 & k_2 \\ k_2 & k_3 \end{bmatrix},$$

the equations reduce to the following six expressions

$$-2l_1\left(-\frac{1}{2}r - 3s - 2s^2k_1\right) + 2s^2l_1^2 + \frac{1}{2} = 0 \tag{8.3.16}$$

$$2s^2k_1^2 - 2k_1\left(-\frac{1}{2}r - 3s - 2s^2l_1\right) + \frac{1}{2} = 0 \tag{8.3.17}$$

$$(r + 3s + 2s^2k_1 + 2s^2l_1)l_2 + 2s^2l_1k_2 = \frac{1}{2}c + (a + 2c)sl_1 \tag{8.3.18}$$

$$2s^2k_1l_2 + (r + 3s + 2s^2k_1 + 2s^2l_1)k_2 = \frac{1}{2}c + (a + 2c)sk_1 \tag{8.3.19}$$

$$\frac{1}{r}(2l_2((a + 2c)s - 2s^2k_2) - 2s^2l_2^2 - \frac{1}{2}c^2) = l_3 \tag{8.3.20}$$

$$\frac{1}{r}(2k_2((a + 2c)s - 2s^2l_2) - 2s^2k_2^2 - \frac{1}{2}c^2 = k_3. \tag{8.3.21}$$

The stability requirement, that matrix $A - S_1 K_1 - S_2 K_2$ must be stable, reduces to

$$\frac{1}{2} r + 3s + 2s^2 (l_1 + k_1) > 0. \tag{8.3.22}$$

We can solve this set of equations in the same way as in Example 8.4. First, solve (k_1, l_1) from equations (8.3.16) and (8.3.17). Then (k_2, l_2) are determined by the set of linear equations (8.3.18) and (8.3.19). Finally, equations (8.3.20) and (8.3.21) then give us l_3 and k_3, respectively.

Subtracting equation (8.3.17) from equation (8.3.16) gives

$$(l_1 - k_1)(r + 6s + 2s^2 (l_1 + k_1)) = 0.$$

If in this equation the term $r + 6s + 2s^2 (l_1 + k_1) = 0$, we have a contradiction with the stability requirement (8.3.22). So, it follows that $l_1 = k_1$. As a consequence, the equations (8.3.18) and (8.3.19) become symmetric in (l_2, k_2), and it is straightforwardly verified that $l_2 = k_2$. Finally, by a direct inspection of equations (8.3.20) and (8.3.21), we also obtain that $l_3 = k_3$. Or, stated differently, $K_1 = K_2$.

Using this result equations (8.3.16) and (8.3.18) can be rewritten as

$$6s^2 k_1^2 + (r + 6s)k_1 + \frac{1}{2} = 0, \tag{8.3.23}$$

and

$$k_2 = \frac{\frac{1}{2} c + (a + 2c)sk_1}{r + 3s + 6s^2 k_1}, \tag{8.3.24}$$

respectively.

From the stability condition (8.3.22) it next follows that the appropriate solution k_1 in equation (8.3.23) above is

$$k_1 = \frac{-(6s + r) + \sqrt{(r + 6s)^2 - 12s^2}}{12s^2}. \tag{8.3.25}$$

The equilibrium actions now follow directly from expression (8.3.15). That is

$$u_i(t) = (2sk_1 + 1)p(t) + 2sk_2 - c, \quad i = 1, 2, \tag{8.3.26}$$

with k_1 given by equation (8.3.25) and k_2 by equation (8.3.24).

The resulting dynamics of the equilibrium price path are

$$\dot{p}(t) = (-4s^2 k_1 - 3s)p(t) + s(a - 4sk_2 + 2c),$$

or, stated differently,

$$p(t) = \alpha e^{(-4s^2 k_1 - 3s)t} + \frac{a - 4sk_2 + 2c}{4sk_1 + 3},$$

where $\alpha = p_0 - \frac{a - 4sk_2 + 2c}{4sk_1 + 3}$. In particular we infer from this that the feedback equilibrium price converges to the stationary value

$$\bar{p} := \frac{a - 4sk_2 + 2c}{4sk_1 + 3}.$$

In the same way as for the open-loop case, we next study the behavior of the stationary equilibrium price if the price adjustment parameter, s, increases. Elementary calculation shows that

$$\lim_{s \to \infty} sk_1 = \frac{-6 + \sqrt{24}}{12} \quad \text{and} \quad \lim_{s \to \infty} sk_2 = \left(\frac{1}{6} - \frac{1}{\sqrt{24}}\right)a + \left(\frac{2}{6} - \frac{1}{\sqrt{24}}\right)c.$$

Consequently,

$$\lim_{s \to \infty} \bar{p} = \theta a + (1 - \theta)c, \quad \text{where} \quad \theta = \frac{1 + \sqrt{6}}{3 + 2\sqrt{6}}.$$

Comparing these numbers with the open-loop and static case, we see that the feedback price resembles the static equilibrium price more.

We conclude this example by a comparison of the convergence speed of the open-loop and feedback equilibrium prices towards their stationary values. The convergence speed of the feedback price is

$$r^{fb} := s - \frac{1}{3}\left(r - \sqrt{(r + 6s)^2 - 12s^2}\right),$$

whereas that of the open-loop price is

$$r^{ol} := -\frac{1}{2}r + \sqrt{\left(\frac{1}{2}r + 3s\right)^2 - 2s^2}.$$

We will now show that

$$r^{fb} < r^{ol}.$$

Notice that the inequality can be rewritten as

$$s + \frac{1}{6}r < \sqrt{\left(\frac{1}{2}r + 3s\right)^2 - 2s^2} - \frac{1}{3}\sqrt{(r + 6s)^2 - 12s^2}$$

where the right-hand side of this inequality is positive. By squaring both sides of this inequality and rearranging terms the question reduces to whether the following inequality holds

$$\frac{2}{3}\sqrt{\left(\frac{1}{2}r + 3s\right)^2 - 2s^2}\sqrt{(r + 6s)^2 - 12s^2} < \frac{1}{3}r^2 + 4rs + 8\frac{2}{3}s^2.$$

Squaring both sides of this inequality again and comparing terms then show that indeed this inequality holds, which proves the claim.   □

Next we consider the results from Theorem 8.5 for the special case of the zero-sum game.

Corollary 8.6

Consider the zero-sum differential game described by

$$\dot{x}(t) = Ax(t) + B_1 u_1(t) + B_2 u_2(t), \quad x(0) = x_0,$$

with, for player one, the quadratic cost function:

$$J_1(u_1, u_2) = \int_0^\infty \{x^T(t)Qx(t) + u_1^T(t)R_1 u_1(t) - u_2^T(t)R_2 u_2(t)\}dt,$$

and, for player two, the opposite objective function

$$J_2(u_1, u_2) = -J_1(u_1, u_2),$$

where the matrices Q and R_i, $i = 1, 2$, are symmetric and R_i, $i = 1, 2$, are positive definite.

This zero-sum linear quadratic differential game has for every initial state a feedback Nash equilibrium if and only if the following Riccati equation has a symmetric solution K such that matrix $A - S_1 K + S_2 K$ is stable

$$-A^T K - KA + K(S_1 - S_2)K - Q = 0.$$

Moreover, in that case there is a unique equilibrium. The equilibrium actions are then

$$u_1^*(t) = -R_1^{-1} B_1^T K x(t) \quad \text{and} \quad u_2^*(t) = R_2^{-1} B_2^T K x(t).$$

The cost incurred by player one is $x_0^T K x_0$ and by player two $-x_0^T K x_0$.

Proof

By Theorem 8.5 this zero-sum game has an equilibrium if and only if the set of algebraic Riccati equations

$$0 = -(A - S_2 K_2)^T K_1 - K_1(A - S_2 K_2) + K_1 S_1 K_1 - Q + K_2 S_2 K_2, \tag{8.3.27}$$
$$0 = -(A - S_1 K_1)^T K_2 - K_2(A - S_1 K_1) + K_2 S_2 K_2 + Q + K_1 S_1 K_1, \tag{8.3.28}$$

has a stabilizing set of solutions (K_1^*, K_2^*). Adding equations (8.3.27) and (8.3.28) yields the Lyapunov equation in $K_1 + K_2$:

$$-(K_1 + K_2)(A - S_1 K_1 - S_2 K_2) - (A - S_1 K_1 - S_2 K_2)^T (K_1 + K_2) = 0.$$

Since, by assumption, matrix $A - S_1 K_1^* - S_2 K_2^*$ is stable, the equation has a unique solution (see Corollary 2.32 or Example 2.16). Obviously, $K_1 + K_2 = 0$ satisfies this Lyapunov equation. Thus, $K_1^* = -K_2^*$. Substitution of $K_1 = -K_2$ into equations (8.3.27) and (8.3.28) shows that both equations reduce to the equation

$$-A^T K - KA + K(S_1 - S_2)K - Q = 0. \qquad (8.3.29)$$

So, this equation (8.3.29) has a solution $K = K_1^*$ such that $A - S_1 K + S_2 K$ is stable; but equation (8.3.29) is an algebraic Riccati equation which, according to Theorem 2.33, has at most one stabilizing solution. Therefore, K_1^* is uniquely determined and there is exactly one equilibrium.

Conversely, assume that the algebraic Riccati equation (8.3.29) has a stabilizing solution K_1. Introducing $K_2 := -K_1$ then shows that the set of coupled algebraic Riccati equations (8.3.27) and (8.3.28) has a stabilizing solution. From this we immediately conclude (see Theorem 8.5) that the game has a feedback Nash equilibrium. ☐

Notice the difference between this Corollary 8.6 and its open-loop counterpart as formulated in Proposition 7.20. The result stated above holds under less stringent conditions. First, the uncontrolled system does not necessarily have to be stable and, secondly, no additional algebraic Riccati equations appear other than equation (8.3.29) in the existence conditions.

As in the open-loop case, there are examples illustrating the fact that the game can have either no, a finite number or even an infinite number of feedback Nash equilibria. Examples will be provided in the following sections, when we have some numerical algorithms available to deal with the problem to find all zeros of the set of nonlinear coupled Riccati equations (8.3.3) and (8.3.4). Here, we present an example illustrating the fact that the set of algebraic Riccati equations may have an infinite number of stabilizing solutions.

Example 8.6

Consider

$$A = B_i = R_{ii} = \begin{bmatrix} 1 & 0 \\ 0 & 1 \end{bmatrix}, \; i = 1, 2; \quad \text{and} \quad Q_i = R_{ij} = \begin{bmatrix} 0 & 0 \\ 0 & 0 \end{bmatrix}, \; i \neq j = 1, 2.$$

Then straightforward calculations show that for $0 \leq a \leq 2$

$$K_1 = \begin{bmatrix} a & \sqrt{-a^2 + 2a} \\ \sqrt{-a^2 + 2a} & 2 - a \end{bmatrix} \quad \text{and} \quad K_2 = \begin{bmatrix} 2 - a & -\sqrt{-a^2 + 2a} \\ -\sqrt{-a^2 + 2a} & a \end{bmatrix}$$

solve the set of coupled algebraic Riccati equations. Furthermore,

$$A - S_1 K_1 - S_2 K_2 = \begin{bmatrix} -1 & 0 \\ 0 & -1 \end{bmatrix},$$

which, obviously, is a stable matrix. So, there exist an infinite number of feedback Nash equilibrium strategies, all yielding the same closed-loop system. It is easily verified, using

Algorithm 7.1, that this game has two open-loop Nash equilibria which permit a feedback synthesis. ☐

The restriction that the feedback matrices have to belong to the set $\mathcal{F}$ is essential. Indeed, there exist feedback Nash equilibria in which a player can improve unilaterally by choosing a feedback matrix for which the closed-loop system is unstable. The following example is from Mageirou (1976).

Example 8.7

Consider the scalar zero-sum dynamic game

$$\dot{x}(t) = x(t) + u_1(t) + u_2(t), \quad x(0) = x_0 \neq 0,$$

with performance criterion

$$J = \int_0^\infty \{x^2(t) + u_1^2(t) - 2u_2^2(t)\}dt,$$

which player one likes to minimize and player two to maximize. Then, according to Corollary 8.6 this game has a stationary feedback Nash equilibrium if and only if the algebraic Riccati equation

$$\frac{1}{2}k^2 - 2k - 1 = 0$$

has a stabilizing solution, i.e. a solution k such that $1 - s_1k + s_2k < 0$. It is easily verified that $k = 2 + \sqrt{6}$ is the unique stabilizing solution of this equation. So, the equilibrium actions for this game are

$$u_1^*(t) = -(2 + \sqrt{6})x(t) \quad \text{and} \quad u_2^*(t) = -\frac{1}{2}(2 + \sqrt{6})x(t),$$

whereas the corresponding cost for player one is $(2 + \sqrt{6})x_0^2$ and for player two $-(2 + \sqrt{6})x_0^2$.

However, if player two uses the action $u_2^*(t)$ and player one is allowed to choose an action outside the class of stabilizing feedback strategies, things change drastically. For, if player two uses the action $u_2^*(t)$, player one is confronted with the problem to minimize

$$J = \int_0^\infty \{(-4 - 2\sqrt{6})x^2(t) + u_1^2(t)\}dt,$$

subject to the system

$$\dot{x}(t) = \frac{4 + \sqrt{6}}{2}x(t) + u_1(t), \quad x(0) = x_0.$$

Obviously, if $x_0 \neq 0$, player one can obtain an arbitrarily small cost by playing $u_1(.) = 0$. So, the action $u_1^*(.)$ is not an equilibrium action anymore, if we permit player one to choose his actions from a different action set. ☐

8.4 Two-player scalar case

The previous section showed that all infinite-planning horizon feedback Nash equilibria can be found by solving a set of coupled algebraic Riccati equations. Solving the system (8.3.3) and (8.3.4) is in general a difficult problem. To gain some intuition for the solution set we next consider the scalar two-player game, where players are not interested in the control actions pursued by the other player. In that case it is possible to derive some analytic results. In particular it will be shown that in this game never more than three equilibria occur. Furthermore a complete characterization of parameters which give rise to either zero, one, two, or three equilibria will be given.

So, the object of study in this section is the next game:

$$J_i(x_0, u_1, u_2) = \int_0^\infty \{q_i x^2(t) + r_i u_i^2\} dt, \quad i = 1, 2, \tag{8.4.1}$$

subject to the dynamical system

$$\dot{x}(t) = ax(t) + b_1 u_1(t) + b_2 u_2(t), \quad x(0) = x_0. \tag{8.4.2}$$

The algebraic Riccati equations which provide the key to finding the feedback Nash equilibria for this game (see Theorem 8.5) are obtained from equations (8.3.3) and (8.3.4) by substitution of $R_{21} = R_{12} = 0$, $A = a$, $B_i = b_i$, $Q_i = q_i$, $R_{ii} = r_i$ and $s_i := b_i^2/r_i$, $i = 1, 2$, into these equations. By Theorem 8.5 a pair of control actions $f_i^* := -\frac{b_i}{r_i} k_i$, $i = 1, 2$, then constitute a feedback Nash equilibrium if and only if the following equations have a solution $x_i = k_i$, $i = 1, 2$,

$$s_1 x_1^2 + 2s_2 x_1 x_2 - 2ax_1 - q_1 = 0 \tag{8.4.3}$$

$$s_2 x_2^2 + 2s_1 x_1 x_2 - 2ax_2 - q_2 = 0 \tag{8.4.4}$$

$$a - s_1 x_1 - s_2 x_2 < 0. \tag{8.4.5}$$

Geometrically, the equations (8.4.3) and (8.4.4) represent two hyperbolas in the (x_1, x_2) plane, whereas the inequality (8.4.5) divides this plane into a 'stable' and an 'anti-stable' region. So, all feedback Nash equilibria are obtained as the intersection points of both hyperbolas in the 'stable' region. In Examples 8.8 and 8.9 below we illustrate two different situations that can occur.

Example 8.8

Consider $a = b_i = r_i = 1$, $i = 1, 2$, $q_1 = \frac{1}{4}$ and $q_2 = \frac{1}{5}$. Then the hyperbolas describing equations (8.4.3) and (8.4.4) are

$$x_2 = 1 - \frac{1}{2}x_1 + \frac{1}{8x_1}, \quad \text{and} \quad x_1 = 1 - \frac{1}{2}x_2 + \frac{1}{10x_2},$$

respectively. Both the hyperbolas, as well as the 'stability-separating' line $x_2 = 1 - x_1$, are plotted in Figure 8.2. From the plot we see that both hyperbolas have three intersection points in the stable region. So, the game has three feedback Nash equilibria. □

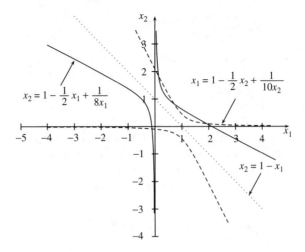

Figure 8.2   A game with three feedback Nash equilibria: $a = b_i = r_i = 1$, $q_1 = \frac{1}{4}$, $q_2 = \frac{1}{5}$

Example 8.9

Consider the same parameters as in Example 8.8 except a, which we now choose to be equal to zero. Then the hyperbolas describing equations (8.4.3) and (8.4.4) are

$$x_2 = -\frac{1}{2}x_1 + \frac{1}{8x_1}, \quad \text{and} \quad x_1 = -\frac{1}{2}x_2 + \frac{1}{10x_2},$$

respectively. Both hyperbolas, as well as the 'stability-separating' line $x_2 = -x_1$, are plotted in Figure 8.3. We see that both hyperbolas have two intersection points, from which one is located in the stable region. So, this game has one feedback Nash equilibrium.   □

Next we will analyze the various situations that can occur in detail (see also van den Broek (2001)) where the various situations that can occur are exhaustively discussed in

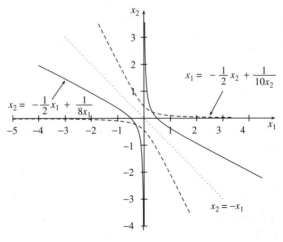

Figure 8.3   A game with one feedback Nash equilibrium: $a = 0$, $b_i = r_i = 1$, $q_1 = \frac{1}{4}$, $q_2 = \frac{1}{5}$

geometric terms). Before discussing the general case, we first consider some 'degenerate' cases. Proposition 8.7 summarizes these cases. These results can easily be verified.

Proposition 8.7

Consider the equations (8.4.3)–(8.4.5). Assume that $s_1 = 0$.

1. If $s_2 \neq 0$, there exists a solution $(x_1, x_2) \in \mathbb{R}^2$ of equations (8.4.3)–(8.4.5) if and only if $a^2 + s_2 q_2 > 0$. If this condition holds the solution is unique and is equal to $(q_1/(2\sqrt{a^2 + s_2 q_2}), (a + \sqrt{a^2 + s_2 q_2})/s_2)$.

2. If $s_2 = 0$, there exists a solution if and only if $a < 0$. If this condition holds, the solution is unique and is given by $(-q_1/(2a), -q_2/(2a))$. □

Given this result we assume from now on that $s_i \neq 0$, $i = 1, 2$. Next, introduce $y_i := s_i x_i$ and $\sigma_i := s_i q_i$, $i = 1, 2$. Without loss of generality assume henceforth that $\sigma_1 \geq \sigma_2$. Then, multiplication of equation (8.4.3) by s_1 and equation (8.4.4) by s_2 shows that equations (8.4.3)–(8.4.5) have a solution (x_1, x_2) if and only if the equations

$$y_i^2 - 2y_3 y_i + \sigma_i = 0, \quad i = 1, 2, \tag{8.4.6}$$

$$y_3 := -a + y_1 + y_2 > 0, \tag{8.4.7}$$

have a solution $(y_1, y_2) \in \mathbb{R}^2$. More precisely, there is a one-to-one correspondence between both solution sets. The following result shows that the system of equations (8.4.6) and (8.4.7) can equivalently be formulated as a set of four equations in one unknown, again with a one-to-one correspondence between solution sets.

Lemma 8.8

The system of equations (8.4.6) and (8.4.7) has a solution if and only if there exist $t_1, t_2 \in \{-1, 1\}$ such that the equation

$$y_3 + t_1 \sqrt{y_3^2 - \sigma_1} + t_2 \sqrt{y_3^2 - \sigma_2} = a \tag{8.4.8}$$

has a solution $y_3 > 0$ satisfying $y_3^2 \geq \sigma_1$. □

Next define for all $x > 0$, satisfying $x^2 \geq \sigma_1$, the functions;

$$f_1(x) = x - \sqrt{x^2 - \sigma_1} - \sqrt{x^2 - \sigma_2} \tag{8.4.9}$$

$$f_2(x) = x + \sqrt{x^2 - \sigma_1} - \sqrt{x^2 - \sigma_2} \tag{8.4.10}$$

$$f_3(x) = x - \sqrt{x^2 - \sigma_1} + \sqrt{x^2 - \sigma_2} \tag{8.4.11}$$

$$f_4(x) = x + \sqrt{x^2 - \sigma_1} + \sqrt{x^2 - \sigma_2}. \tag{8.4.12}$$

Notice that each possible choice of a pair (t_1, t_2) in the left-hand side of equation (8.4.8) corresponds to an f_i. So, finding all equilibria is equivalent to determining all intersection points of the graphs $f_i(x)$ with the level a. Using elementary analysis one can determine

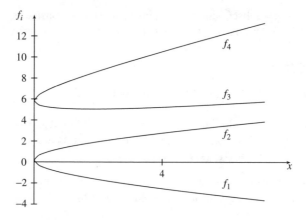

Figure 8.4   The curves f_i for $\sigma_1 = 9$ and $\sigma_2 = 1$

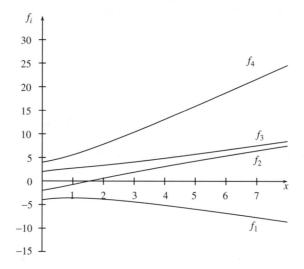

Figure 8.5   The curves f_i for $\sigma_1 = -1$ and $\sigma_2 = -9$

the graphs of these functions. The number of different equilibria that can occur can then be analyzed by plotting all four functions in one graph (see, for example, Figures 8.4 and 8.5). Lemma 8.9 summarizes some essential properties of the functions f_i. It turns out that we have to distinguish two cases: $\sigma_1 > 0$ and $\sigma_1 < 0$.

Lemma 8.9

Let $\sigma_1 > \sigma_2$. Then,

1. $f_1(x) \leq f_2(x) \leq f_3(x) \leq f_4(x)$.

2. If $\sigma_1 > 0$:

   (a) $f_i(\sqrt{\sigma_1}) = f_{i+1}(\sqrt{\sigma_1}), \quad i = 1, 3$;

   (b) f_1 is strictly decreasing and both f_2 and f_4 are strictly increasing;

   (c) f_3 has a unique minimum at $x^* > \sqrt{\sigma_1}$.

3. If $\sigma_1 < 0$:

   (a) f_2 and f_3 are strictly increasing;

   (b) f_1 has exactly one global maximum at $x^* > 0$ and $\lim_{x \to \infty} f_1(x) = -\infty$;

   (c) f_4 has exactly one global minimum at $x^* < 0$ and $\lim_{x \to \infty} f_1(x) = \infty$, so f_4 is strictly increasing for $x > 0$ as well;

   (d) $\max f_1(x) < f_2(0)$.

Proof

The proofs of most parts are rather elementary and are therefore skipped. The uniqueness result in 3(b) and (c) is due to the concavity (convexity) of the functions. Claim 3(d) follows from the fact that $f_1(x) - f_2(0) = (x - \sqrt{x^2 - \sigma_1}) + (\sqrt{-\sigma_2} - \sqrt{x^2 - \sigma_2}) - \sqrt{-\sigma_1} < 0$. Furthermore, since f_1 is concave and $f_1'(0) > 0$, it is clear that f_1 has a unique maximum. $\square$

Figures 8.4 and 8.5 show both situations which can occur. Completing the analysis by an elementary elaboration of the cases $\sigma_1 = \sigma_2 > 0$ and $\sigma_1 = \sigma_2 < 0$, we arrive at the following result.

Theorem 8.10

Consider the differential game (8.4.1) and (8.4.2), with $\sigma_i = \frac{b_i^2 q_i}{r_i}$, $i = 1, 2$. Assume, without loss of generality, that $\sigma_1 \geq \sigma_2$. Moreover, let $f_i(x)$, $i = 1, \ldots, 4$, be defined as in equations (8.4.9)–(8.4.12).

1(a) If $\sigma_1 > 0$ and $\sigma_1 > \sigma_2$, the game has

   - one equilibrium if $-\infty < a < \min f_3(x)$,
   - two equilibria if $a = \min f_3(x)$,
   - three equilibria if $a > \min f_3(x)$.

1(b) If $\sigma_1 = \sigma_2 > 0$ the game has

   - one equilibrium if $a \leq \sqrt{\sigma_1}$,
   - three equilibria if $a > \sqrt{\sigma_1}$.

2(a) If $\sigma_1 < 0$ and $\sigma_1 > \sigma_2$, the game has

   - no equilibrium if $\max f_1(x) < a \leq \sqrt{-\sigma_1} - \sqrt{-\sigma_2}$,
   - one equilibrium if either
     - (i) $a = \max f_1(x)$,
     - (ii) $a \leq -\sqrt{-\sigma_1} - \sqrt{-\sigma_2}$, or
     - (iii) $\sqrt{-\sigma_1} - \sqrt{-\sigma_2} < a \leq \sqrt{-\sigma_1} + \sqrt{-\sigma_2}$,
   - two equilibria if either
     - (i) $-\sqrt{-\sigma_1} - \sqrt{-\sigma_2} < a < \max f_1(x)$, or
     - (ii) $-\sqrt{-\sigma_1} + \sqrt{-\sigma_2} < a \leq \sqrt{-\sigma_1} + \sqrt{-\sigma_2}$,
   - three equilibria if $a > \sqrt{-\sigma_1} + \sqrt{-\sigma_2}$.

2(b) If $\sigma_1 = \sigma_2 < 0$ the game has

- no equilibrium if $-\sqrt{-3\sigma_1} < a \leq 0$,
- one equilibrium if $a \leq -\sqrt{-3\sigma_1}$,
- two equilibria if $0 < a \leq 2\sqrt{-\sigma_1}$,
- three equilibria if $a > 2\sqrt{-\sigma_1}$. □

Example 8.10

1. Take $a = 3$, $b_i = r_i = 1$, $i = 1, 2$, $q_1 = 9$ and $q_2 = 1$. Then, $\sigma_1 = 9 > 1 = \sigma_2$. Furthermore,

$$f_3(x) - 3 = x - 3 - \sqrt{x^2 - 9} + \sqrt{x^2 - 1} > 0, \text{ if } x \geq 3.$$

So, $\min_{x \geq 3} f_3(x) > 3$, and therefore

$$-\infty < a < \min f_3(x).$$

According to Theorem 8.10, part 1(a), the game has one feedback Nash equilibrium. In Figure 8.4 one can also see this graphically.

2. Consider $a = 3$, $b_i = r_i = 1$, $i = 1, 2$, $q_1 = -1$ and $q_2 = -9$. Then, $\sigma_1 = -1 > -9 = \sigma_2$. Furthermore,

$$-\sqrt{-\sigma_1} + \sqrt{-\sigma_2} = 2 < 3 \leq 4 = \sqrt{-\sigma_1} + \sqrt{-\sigma_2}.$$

So by Theorem 8.10, part 2(a), the game has two feedback Nash equilibria. By considering Figure 8.5 one can also see this graphically. □

Note

Theorem 8.10 shows that only if both q_i and a are negative might the game have no equilibrium. For all other parameter values, the number of equilibria is always at least one and at most three. In the same way as in the proof of Lemma 8.9 one can show that, if $\sigma_1 < 0$, the maximum of $f_1(x)$ is smaller than $-\sqrt{-\sigma_2}$. So, we conclude from part 2 of Theorem 8.10 that the interval of parameters a for which there is no equilibrium grows if σ_1 becomes more negative. □

The next example is included to give the reader some intuition for the fact that sometimes there are games for which no feedback equilibrium exists.

Example 8.11

Consider the following model for the planning of marketing expenditures.

$$\min_{u_i} \int_0^\infty \{-q_i x^2(t) + r_i u_i^2(t)\} dt, \tag{8.4.13}$$

subject to the dynamic constraint

$$\dot{x}(t) = -ax(t) + u_1(t) + u_2(t), \quad x(0) = x_0. \tag{8.4.14}$$

Here all parameters are positive. The variables and parameters are: x is the brand equity, u_i are the marketing efforts by company i, a is the forgetting factor by the consumers of the product. So, equation (8.4.14) models the evolution of the brand equity over time. It is assumed that consumers forget about the product if the companies do not actively merchandise. Equation (8.4.13) models the suppositions that the revenues of company i increase quadratically with the brand reputation of the product, whereas the cost of marketing increase quadratically with u_i. The fact that we are looking for strategies which ultimately have the effect that the brand equity will converge to zero is motified by the consideration that due to innovations the product will not sell anymore as time passes. So it does not make sense to keep a good reputation for the brand forever.

Next we will try to provide some intuition for the fact that there is in general a small bounded interval of a-parameters for which there exists no equilibrium. To that end assume that q_i is much larger than r_i and a is close to zero. That is, the companies benefit a lot from a good reputation of the brand, whereas the consumers have a good memory w.r.t. the brand. Now consider for a fixed strategy $\bar{f}_2 x(t)$ the corresponding cost, J_1, for player one. Simple calculations show that

$$J_1(f_1, \bar{f}_2) = -\frac{-q_1 + f_1^2 r_1}{2(-a + f_1 + \bar{f}_2)}. \tag{8.4.15}$$

Since company two has to take into account the stabilization constraint and it seems plausible that both companies will try to improve the brand equity with their marketing efforts, assume that $-a + \bar{f}_2 =: \epsilon$ is negative. Because a is close to zero and f_2 is positive ϵ will be a negative number also close to zero. Then company one can obtain arbitrarily large gains by taking f_1 close to $-\epsilon$ (in fact f_1 should be a little bit smaller than $-\epsilon$ in order to meet the stability requirement). The gains for company one are positive, since with f_1 small and q_1 (relative to r_1) large $-q_1 + f_1^2 r_1$ will be negative. However, similarly it follows that once company one has chosen some strategy company two can obtain a much better profit by choosing a slightly different action. So, in this way we see that actually a set of equilibrium actions will never occur that will stabilize the system.

Notice that if the forgetting factor, a, becomes larger the feedback parameters, f_i, can be chosen larger. As a consequence the numerator in equation (8.4.15) becomes positive. Consequently arbitrarily large gains are not possible anymore and equilibria will occur.

So, in conclusion, in this model the situation where consumers have a large brand loyalty seems, from a marketing point of view for the companies, to raise a problem since they know that in the long-term there will be no market for the product anymore. Notice that this phenomenon disappears if the companies discount their future profits (with a 'large enough' factor) in this model. □

8.5   Computational aspects

Section 8.3 showed that in order to find the stationary linear feedback Nash equilibria of the linear quadratic differential game (8.3.1) and (8.3.2), one has to find all stabilizing solutions of the set of coupled algebraic Riccati equations (ARE) (8.3.3) and (8.3.4).

Moreover, the examples illustrated that the number of equilibria may vary between zero and infinity.

For the multivariable case a number of algorithms have been proposed for calculating a solution of the involved algebraic Riccati equations (8.3.3) and (8.3.4) (for example, Krikelis and Rekasius (1971), Tabak (1975), Papavassilopoulos, Medanic and Cruz (1979) and Li and Gajic (1994)). What all these algorithms have in common is that whenever they converge they only provide one solution of the ARE. Obviously, particularly when there is no additional information that a certain type of equilibrium point is preferred, or the number of equilibria is unknown, one would like to have an overview of all possible equilibria.

Papavassilopoulos and Olsder (1984) considered a geometric approach for calculating the feedback Nash ARE similar to the approach used in Chapter 7 to calculate the solutions for the algebraic Riccati equations associated with the open-loop Nash equilibria. Their approach requires the finding of subspaces which simultaneously satisfy some invariance properties. However, up to now, it is not known how these subspaces can be found.

Two other interesting methods that have been proposed in the past for finding all isolated solutions to a system of polynomial constraints over real numbers are interval methods (for example, Van Hentenryck, McAllester and Kapur (1997)) and continuation methods (for example, Morgan (1987) and Verschelde, Verlinden and Cools (1994)). Continuation methods have been shown to be effective for problems for which the total degree of the constraints is not too high, since the number of paths explored depends on the estimation of the number of solutions. Interval methods are generally robust but used to be slow. The recent approach taken in Van Hentenryck, McAllester and Kapur (1997), however, seems to overcome this bottleneck and be rather efficient. One important point that has to be managed in using the interval method is, however, the choice of the initial interval that contains all equilibria. Moreover, it is unclear how this method will perform if there are an infinite number of feedback Nash solutions. Numerical experience using the above methods to find feedback Nash equilibria is, unfortunately, still lacking at this moment.

In this section we will present an eigenvalue-based algorithm which gives us all equilibrium points if we are dealing with the scalar case. If there are only a few players, this algorithm is very efficient and can easily be implemented using, for example, MATLAB. Compared with the interval method, a possible disadvantage of this eigenvalue approach is that the matrix from which we have to determine the eigenvalues grows exponentially with the number, N, of players considered. However, since this matrix is rather sparse, computational efficiency might be considerably increased if N becomes large.

Obviously, the above mentioned numerical problems are highly relevant if one wants to calculate the equilibria for a game. However, since we believe that these topics are somewhat outside the scope of this book, we will not elaborate on these issues further. We stick to just presenting the basic algorithm for the scalar case together with a formal proof of it.

8.5.1 Preliminaries

The introduction and understanding of the eigenvalue algorithm requires first some technicalities to be set out. This will be the subject of this subsection.

For simplicity reasons, as in section 8.4 (see equations (8.4.1) and (8.4.2)), the starting point will be to consider the scalar game where players have no direct interest in each other's control actions and that $b_i \neq 0$. From the general N-player result (see the notes at the end of this chapter) it follows that this N-player game has an equilibrium if and only if, with $s_i := \frac{b_i^2}{r_i}$, the following equations have a solution $x_i = k_i$, $i = 1, \ldots, N$:

$$\left(a - \sum_{j=1}^{N} x_j s_j\right) x_i + x_i \left(a - \sum_{j=1}^{N} s_j x_j\right) + q_i + x_i s_i x_i = 0, \quad i = 1, \ldots, N, \quad (8.5.1)$$

$$a - \sum_{j=1}^{N} s_j x_j < 0. \quad (8.5.2)$$

Notice that this result is in line with the two-player result as outlined in equations (8.4.3)–(8.4.4).

Next, for notational convenience, introduce the variables:

$$\sigma_i := s_i q_i, \quad \sigma_{max} = \max_i \sigma_i, \quad y_i := s_i x_i, \quad i = 1, \ldots, N, \quad \text{and} \quad y_{N+1} := -a_{cl} := -\left(a - \sum_{j=1}^{N} y_j\right)$$

and assume, without loss of generality, that $\sigma_1 \geq \cdots \geq \sigma_N$. Multiplying equation (8.5.1) by s_i, these equations can then be rewritten as

$$y_i^2 - 2y_{N+1} y_i + \sigma_i = 0, \quad i = 1, \ldots, N. \quad (8.5.3)$$

In other words, the equilibria of the game are obtained by determining all real solutions y_i, $i = 1, \ldots, N$, with $y_{N+1} > 0$, of the above N quadratic equations and the equation

$$y_{N+1} = -a + \sum_{j=1}^{N} y_j. \quad (8.5.4)$$

In section 8.5.2 a numerical algorithm will be provided to find all the solutions of these equations. In this subsection some preliminary analytic properties are derived.

First we consider the generalization of Lemma 8.8 to the N-player case. Obviously, the solutions of equation (8.5.3) are

$$y_i = y_{N+1} + \sqrt{y_{N+1}^2 - \sigma_i} \quad \text{and} \quad y_i = y_{N+1} - \sqrt{y_{N+1}^2 - \sigma_i}, \quad i = 1, \ldots, N.$$

Substitution of this into equation (8.5.4) shows the following lemma.

Lemma 8.11

1. The set of equations (8.5.3) and (8.5.4) has a solution if and only if there exist $t_i \in \{-1, 1\}$, $i = 1, \ldots, N$, such that the equation

$$(N - 1)y_{N+1} + t_1 \sqrt{y_{N+1}^2 - \sigma_1} + \cdots + t_N \sqrt{y_{N+1}^2 - \sigma_N} = a \quad (8.5.5)$$

has a solution y_{N+1}. In fact all solutions of equations (8.5.3) and (8.5.4) are obtained by considering all possible sequences $(t_1, \ldots, t_N)$ in equation (8.5.5).

2. The game has a feedback Nash equilibrium if and only if there exist $t_i \in \{-1, 1\}$, $i = 1, \ldots, N$, such that equation (8.5.5) has a solution $y_{N+1} > 0$ with $y_{N+1}^2 \geq \sigma_1$.   □

For the two-player case in section 8.4 the functions $f_i(x)$ were introduced in (8.4.9)–(8.4.12) to study the solution set of equations (8.4.6) and (8.4.7) in detail. Analogously the functions f_i^N are next defined to study the solution set of equations (8.5.3) and (8.5.4) in some more detail.

$$f_i^{n+1}(x) := f_i^n(x) + x - \sqrt{x^2 - \sigma_{n+1}}, \quad i = 1, \ldots, 2^n, \tag{8.5.6}$$

$$f_{i+2^n}^{n+1}(x) := f_i^n(x) + x + \sqrt{x^2 - \sigma_{n+1}}, \quad i = 1, \ldots, 2^n, \tag{8.5.7}$$

with

$$f_1^1(x) := -\sqrt{x^2 - \sigma_1} \quad \text{and} \quad f_2^1(x) := \sqrt{x^2 - \sigma_1}. \tag{8.5.8}$$

Each function f_i^N, $i = 1, \ldots, 2^N - 1$, corresponds to a function obtained from the left-hand side of equation (8.5.5) by making a specific choice of t_j, $j = 1, \ldots, N$, and substituting x for y_{N+1}. From Lemma 8.11 it is obvious then that equations (8.5.3) and (8.5.4) have a solution if and only if $f_i^N(x) = a$ has a solution for some $i \in \{1, \ldots, 2^N\}$. Or, stated differently, equations (8.5.3) and (8.5.4) have a solution if and only if the following function has a root

$$\prod_{i=1}^{2^N} (f_i^N(x) - a) = 0. \tag{8.5.9}$$

Introducing

$$f(x) := \prod_{i=1}^{2^N} (f_i^N(x) - a),$$

we obtain the following theorem.

Theorem 8.12

y_i is a solution of equations (8.5.3) and (8.5.4) if and only if y_{N+1} is a zero of $f(x)$ and there exist $t_i \in \{-1, 1\}$, such that $y_i = y_{N+1} + t_i \sqrt{y_{N+1}^2 - \sigma_i}$. Moreover, $f(x)$ is a polynomial of degree 2^N.   □

The first part of the theorem follows directly from the above reasoning. The proof that $f(x)$ is a polynomial can be found in the Appendix at the end of this chapter.

Theorem 8.14, below, presents some properties concerning the number of feedback Nash equilibria, e, of the game. In particular it relates this number of equilibria to the sum of the algebraic multiplicities, m, of all roots $\bar{x} > 0$ of equation (8.5.9) that additionally have the property that $\bar{x}^2 \geq \sigma_1$. The proof of this theorem uses the following preliminary result.

Lemma 8.13

Assume $f(\sqrt{|\sigma_1|}) \neq 0$. Then, for each feedback Nash equilibrium yielding $\bar{x} := -a_{cl}$ there is exactly one function $f_i^N(x) - a$ that has a zero at $\bar{x}$.

Proof

Assume that y_i, $i = 1, \ldots, N$, is a feedback Nash equilibrium. Then, there is an $\bar{x}$ such that $f_i^N(\bar{x}) = a$. Or, stated differently, there is a sequence $t_i \in \{-1, 1\}$, $i = 1, \ldots, N$, such that $f_i^N(\bar{x}) = a$, having the additional property that $y_i = \bar{x} + t_i \sqrt{\bar{x}^2 - \sigma_i}$, $i = 1, \ldots, N$. Now, also assume that $f_j^N(\bar{x}) = a$ for some $j \neq i$. This implies the existence of a sequence $s_i \in \{-1, 1\}$, $i = 1, \ldots, N$, such that $f_j^N(\bar{x}) = a$, having the additional property that $y_i = \bar{x} + s_i \sqrt{\bar{x}^2 - \sigma_i}$, $i = 1, \ldots, N$. So, $y_i = \bar{x} + s_i \sqrt{\bar{x}^2 - \sigma_i} = \bar{x} + t_i \sqrt{\bar{x}^2 - \sigma_i}$. Since by assumption $\bar{x}^2 - \sigma_i > \sigma_1 - \sigma_i \geq 0$, it follows necessarily that $s_i = t_i$. ☐

Theorem 8.14

Let e denote the number of feedback Nash equilibria of the game and m denote the sum of the algebraic multiplicities of all roots $\bar{x} > 0$ of equation (8.5.9) that additionally have the property that $\bar{x}^2 \geq \sigma_1$. Then:

1. $e \leq m$;

2. assume $f(\sqrt{|\sigma_1|}) \neq 0$, then $e = m$ if and only if $\frac{df_i^N}{dx}(\bar{x}) \neq 0$, whenever $f_i^N(\bar{x}) = 0$. ☐

Notice that the leading coefficient (i.e. the coefficient of x^{2N}) in $f(x)$ is negative. Moreover, if $\sigma_1 \geq 0$ it follows straightforwardly from the definition of $f(x)$ that $f(\sqrt{\sigma_1}) > 0$. Similarly it follows that if $\sigma_1 < 0$, $f(-\sqrt{-\sigma_1}) > 0$. Theorem 8.14 then immediately provides the next conclusion.

Corollary 8.15

The N-player scalar game has at most $2^N - 1$ feedback Nash equilibria. ☐

8.5.2 A scalar numerical algorithm: the two-player case

In this subsection a numerical algorithm is developed to calculate all feedback Nash equilibria of the two-player scalar game (8.4.1) and (8.4.2). Section 8.5.3 generalizes this result for the corresponding N-player game.

Let p_1, p_2 be a (possibly complex) solution of equations (8.5.3) and (8.5.4). Denote the negative of the resulting closed-loop system parameter by

$$\lambda := -a + p_1 + p_2. \tag{8.5.10}$$

Then, by equations (8.5.3) and (8.5.4), p_1 and p_2 satisfy

$$p_1^2 - 2\lambda p_1 + \sigma_1 = 0, \tag{8.5.11}$$

and

$$p_2^2 - 2\lambda p_2 + \sigma_2 = 0. \tag{8.5.12}$$

From the definition of λ and equation (8.5.11), respectively, then it follows that

$$
\begin{aligned}
p_1\lambda &= -p_1 a + p_1^2 + p_1 p_2 \\
&= -p_1 a + 2\lambda p_1 - \sigma_1 + p_1 p_2.
\end{aligned}
$$

Some elementary rewriting then shows that

$$p_1\lambda = \sigma_1 + ap_1 - p_1 p_2. \tag{8.5.13}$$

In a similar way, using the definition of λ and equation (8.5.12), respectively, one obtains

$$p_2\lambda = \sigma_2 + ap_2 - p_1 p_2. \tag{8.5.14}$$

Finally, using the definition of λ and both equations (8.5.11) and (8.5.12), respectively, we have

$$
\begin{aligned}
p_1 p_2\lambda &= -p_1 p_2 a + p_1^2 p_2 + p_1 p_2^2 \\
&= -p_1 p_2 a + 4\lambda p_1 p_2 - \sigma_1 p_2 - \sigma_2 p_1.
\end{aligned}
$$

Which gives

$$p_1 p_2\lambda = 1/3(\sigma_2 p_1 + \sigma_1 p_2 + ap_1 p_2). \tag{8.5.15}$$

So, denoting

$$
\tilde{M} := \begin{bmatrix} -a & 1 & 1 & 0 \\ \sigma_1 & a & 0 & -1 \\ \sigma_2 & 0 & a & -1 \\ 0 & 1/3\sigma_2 & 1/3\sigma_1 & 1/3a \end{bmatrix} \tag{8.5.16}
$$

we conclude from equations (8.5.10), (8.5.13), (8.5.14) and (8.5.15) that every solution p_1, p_2 of equations (8.5.3) and (8.5.4) satisfies the equation

$$
\tilde{M} \begin{bmatrix} 1 \\ p_1 \\ p_2 \\ p_1 p_2 \end{bmatrix} = \lambda \begin{bmatrix} 1 \\ p_1 \\ p_2 \\ p_1 p_2 \end{bmatrix}. \tag{8.5.17}
$$

This observation leads to the following theorem.

Theorem 8.16

1. Assume that (k_1, k_2) is a feedback Nash equilibrium strategy. Then the negative of the corresponding closed-loop system parameter $\lambda := -a + \sum_{i=1}^{2} s_i k_i > 0$ is an eigenvalue of the matrix

$$
M := \begin{bmatrix} -a & s_1 & s_2 & 0 \\ q_1 & a & 0 & -s_2 \\ q_2 & 0 & a & -s_1 \\ 0 & \frac{1}{3}q_2 & \frac{1}{3}q_1 & \frac{1}{3}a \end{bmatrix}. \tag{8.5.18}
$$

Furthermore, $[1, k_1, k_2, k_1 k_2]^T$ is a corresponding eigenvector and $\lambda^2 \geq \sigma_{max}$.

2. Assume that $[1, k_1, k_2, k_3]^T$ is an eigenvector corresponding to a positive eigenvalue λ of M, satisfying $\lambda^2 \geq \sigma_{max}$, and that the eigenspace corresponding to λ has dimension one. Then, (k_1, k_2) is a feedback Nash equilibrium.

Proof

1. The correctness of this statement follows directly from Lemma 8.11 part 2, the above arguments and by substituting the parameters $\sigma_i = s_i q_i$ and $p_i = s_i k_i$, $i = 1, 2$, into equation (8.5.17).

2. Notice that, without loss of generality, we may consider matrix $\tilde{M}$ instead of M.

Let $p(x)$ be the characteristic polynomial of $\tilde{M}$. Obviously, $p(x)$ has degree four. Let p_1, p_2 be an arbitrary (possibly complex) solution to equations (8.5.3) and (8.5.4). From equation (8.5.17) it then follows that $\lambda := -a + p_1 + p_2$ is an eigenvalue of $\tilde{M}$ or, stated differently, λ is a root of $p(x)$.

On the other hand it follows from the previous section that λ is a root of

$$
f(x) = (f_1(x) - a)(f_2(x) - a)(f_3(x) - a)(f_4(x) - a),
$$

where $f_i := f_i^2$ are defined by equations (8.4.9)–(8.4.12). Since all roots of $f(x)$ yield solutions of equations (8.5.3) and (8.5.4) and the degree of $f(x)$ is also four (see Theorem 8.12), it follows that $f(x) = \alpha p(x)$ for some scalar α.

Now, assume $\lambda \geq \sigma_{max}$ is an eigenvalue of $\tilde{M}$ and $v_1 := [1, k_1, k_2, k_3]^T$ a corresponding eigenvector. Then, since $f(x) = \alpha p(x)$, λ must also be a root of $f(x)$. So, according to Lemma 8.11 the game has an equilibrium. Consequently, by part 1 of this theorem, $\tilde{M}$ has a real eigenvector corresponding with λ of the form $v_2 := [1, p_1, p_2, p_1 p_2]^T$. Since the dimension of the eigenspace corresponding to λ is one-dimensional it follows that $v_1 = v_2$, which concludes the proof. □

From Lemma 8.11 part 1, Theorem 8.14 part 1 and Theorem 8.16 the following numerical algorithm then results.

Algorithm 8.1

This algorithm calculates all feedback Nash equilibria of the linear quadratic differential game (8.4.1) and (8.4.2).

Step 1 Calculate matrix M in equation (8.5.18) and $\sigma := \max_i \frac{b_i^2 q_i}{r_i}$.

Step 2 Calculate the eigenstructure (λ_i, m_i), $i = 1, \ldots, k$, of M, where λ_i are the eigenvalues and m_i the corresponding algebraic multiplicities.

Step 3 For $i = 1, \ldots, k$ repeat the following steps.

    3.1. If (i) $\lambda_i \in \mathbb{R}$, (ii) $\lambda_i > 0$ and (iii) $\lambda_i^2 \geq \sigma$ then proceed with Step 3.2 of the algorithm. Otherwise, return to Step 3.

    3.2. If $m_i = 1$ then carry out the following.

        3.2.1. Calculate an eigenvector v corresponding with λ_i of M. Denote the entries of v by $[v_0, v_1, v_2, \ldots]^T$. Calculate $k_j := \frac{v_j}{v_0}$ and $f_j := -\frac{b_j k_j}{r_j}$. Then, $(f_1, \ldots, f_N)$ is a feedback Nash equilibrium and $J_j = k_j x_0^2$, $j = 1, \ldots, N$. Return to Step 3.

    If $m_i > 1$ then carry out the following.

    3.2.2. Calculate $\sigma_i := \frac{b_i^2 q_i}{r_i}$.

    3.2.3. For all 2^N sequences $(t_1, \ldots, t_N)$, $t_k \in \{-1, 1\}$,

        (i) calculate

$$y_j := \lambda_i + t_j \sqrt{\lambda_i^2 - \sigma_j}, \ j = 1, \ldots, N;$$

        (ii) if $\lambda_i = -a + \sum_{j=1,\ldots,N} y_j$ then calculate $k_j := \frac{y_j r_j}{b_j^2}$ and $f_j := -\frac{b_j k_j}{r_j}$; then $(f_1, \ldots, f_N)$ is a feedback Nash equilibrium and $J_j = k_j x_0^2$, $j = 1, \ldots, N$.

Step 4 End of the algoritm.   ☐

Example 8.12

Reconsider Example 7.9 where, for $A = 3$, $B_i = Q_i = 2$ and $R_i = 1$, $i = 1, 2$, we calculated the open-loop Nash equilibrium for an infinite planning horizon. To calculate the feedback Nash equilibria for this game, according to Algorithm 8.1, we first have to determine the eigenstructure of matrix

$$M := \begin{bmatrix} -3 & 4 & 4 & 0 \\ 2 & 3 & 0 & -4 \\ 2 & 0 & 3 & -4 \\ 0 & \frac{2}{3} & \frac{2}{3} & 1 \end{bmatrix}.$$

Using MATLAB, we find the eigenvalues $\{-4.8297, 2.8297, 3, 3\}$. Since both the square of 2.8297 and 3 are larger than $\sigma := 8$, we have to process Step 3 of the algorithm for both these eigenvalues.

First, consider the eigenvalue 2.8297. A corresponding eigenvector is

$$[v_0, v_1, v_2, v_3] := [-0.6532, -0.4760, -0.4760, -0.3468].$$

So,

$$k_1 := \frac{v_1}{v_0} = 0.7287 \quad \text{and} \quad k_2 := \frac{v_2}{v_0} = k_1.$$

This gives the symmetric feedback Nash equilibrium actions

$$u_i = -\frac{b_i k_i}{r_i} x(t) = -1.4574 x(t).$$

The corresponding closed-loop system and cost are

$$\dot{x}(t) = -2.8297 x(t), \quad x(0) = x_0; \quad \text{and} \quad J_i = 0.7287 x_0^2, \quad i = 1, 2,$$

respectively.

Next consider the eigenvalue 3. This eigenvalue has an algebraic multiplicity 2. So we have to proceed with Step 3.2.3 of the algorithm to calculate the equilibria associated with this eigenvalue.

For the sequence $(t_1, t_2) := (1, -1)$ we obtain in Step 3.2.3(i) $y_1 = 4$ and $y_2 = 2$, respectively. These numbers satisfy the equality under Step 3.2.3(ii). Therefore, with

$$k_1 := \frac{y_1 r_1}{b_1^2} = 1 \quad \text{and} \quad k_2 := \frac{y_2 r_2}{b_2^2} = \frac{1}{2},$$

the corresponding equilibrium actions are

$$u_1 = -\frac{b_1 k_1}{r_1} x(t) = -2x(t) \quad \text{and} \quad u_2 = -\frac{b_2 k_2}{r_2} x(t) = -x(t).$$

The resulting closed-loop system and cost in this case are

$$\dot{x}(t) = -3x(t), \quad x(0) = x_0; \quad J_1 = x_0^2 \quad \text{and} \quad J_2 = \frac{1}{2} x_0^2.$$

In a similar way we obtain for the sequence $(-1, 1)$ the 'reversed' solution

$$k_1 = \frac{1}{2}; \quad k_2 = 1; \quad f_1 = -1; \quad \text{and} \quad f_2 = -2.$$

respectively, which gives rise to the closed-loop and cost

$$\dot{x}(t) = -3x(t), \quad x(0) = x_0; \quad J_1 = \frac{1}{2} x_0^2 \quad \text{and} \quad J_2 = x_0^2.$$

Finally, it is easily verified that the numbers y_j implied by both the sequences $(1, 1)$ and $(-1, -1)$ do not satisfy the equality mentioned under Step 3.2.3(ii). So the game has three feedback Nash equilibria.

Comparing the symmetric equilibrium with the open-loop equilibrium that permits a feedback synthesis we see that both players incur a higher cost since the pursued actions are more moderate. In the non-symmetric feedback equilibria one player is better off and the other player worse off (in terms of cost) than in the open-loop equilibrium that permits a feedback synthesis. By an inspection of the cost functions of all other open-loop Nash equilibria (see Example 7.17) it is seen that for every initial state x_0 there always exists an open-loop Nash equilibrium in which the costs for one player are lower and for the other player are higher than in any of the feedback equilibria. However, for all feedback equilibria the convergence rate of the closed-loop system is larger in the open-loop case than in the feedback case. ☐

Notes

1. Generically the multiplicity of eigenvalues of a matrix is one. Of course, our matrix M has some additional structure so we cannot directly use this result here. However, simulation results suggest that the imposed structure of matrix M does not have an impact on this generic result. So usually the algoritm skips parts 3.2.2 and 3.2.3. Obviously, it would be nice if one could always derive all equilibria from the eigenstructure of matrix M. In principle, one has to search the eigenspace corresponding to the appropriate eigenvalue for points that satisfy (if $N = 2$) an additional quadratic equation. From a numerical point of view it seems that by proceeding in this direction will not lead (at least for small N) to a faster procedure than the straightforward approach of screening all potential candidate solutions we take here. However, for large dimensions, this might be an issue for further investigation.

2. All real eigenvalues which are smaller than σ_{max} either correspond to partial complex solutions of (ARE) or to solutions which yield an anti-stable closed-loop system.

3. Notice that two different equilibria which give the same closed-loop system $-\lambda$ correspond to two independent eigenvectors of M. Therefore, one might expect that in general the number of equilibria giving a closed-loop system $-\lambda$ equals at most the geometric multiplicity of λ (see also Example 8.13 below). However, to prove this statement does not seem to be straightforward.

4. In Papavassilopoulos and Olsder (1984) it was shown that solving the set of (ARE) equations is equivalent to finding a 3×1 matrix H satisfying

$$A_S^N H = L_1 H x_1 + L_2 H y_1 + L_3 H z_1,$$

for some scalars x_1, y_1 and z_1, where

$$A_S^N = \begin{bmatrix} a & -s_1 & -s_2 \\ 0 & 0 & -s_1 \\ 0 & -s_2 & 0 \\ -q_1 & -a & 0 \\ -q_2 & 0 & -a \end{bmatrix}, L_1 = \begin{bmatrix} 1 & 0 & 0 \\ 0 & 0 & 0 \\ 0 & 0 & 0 \\ 0 & 1 & 0 \\ 0 & 0 & 1 \end{bmatrix}, L_2 = \begin{bmatrix} 0 & 0 & 0 \\ 1 & 0 & 0 \\ 0 & 0 & 0 \\ 0 & 0 & 0 \\ 0 & 1 & 0 \end{bmatrix}, \text{ and } L_3 = \begin{bmatrix} 0 & 0 & 0 \\ 0 & 0 & 0 \\ 1 & 0 & 0 \\ 0 & 0 & 1 \\ 0 & 0 & 0 \end{bmatrix}.$$

As already noted in the introduction to this section it is not clear how one can solve this set of equations. Given an appropriate eigenvalue λ and corresponding eigenvector of matrix M, $v = [v_1 \ v_2 \ v_3 \ v_4]^T$, it is however clear that with $H = [v_1 \ v_2 \ v_3]^T$, $x_1 = -\lambda$, $y_1 = -\frac{s_1 v_3}{v_1}$ and $z_1 = -\frac{s_2 v_2}{v_1}$ the above equation is satisfied. ☐

The first part of Example 8.13 illustrates that in general in Theorem 8.14 a strict inequality holds. Furthermore, it also illustrates the somewhat extraordinary role σ_1 plays in the analysis. The second part of Example 8.13 shows that in the two-player case three equilibria may occur. The third part shows that situations also exist where no equilibrium exists.

Example 8.13

1. Consider $b_1 = b_2 = 1$, $r_1 = r_2 = 1$, $q_1 = q_2 = 4$ and $a = 2$. Then the characteristic polynomial of M is $(\lambda + \frac{10}{3})(\lambda - 2)^3$. The geometric multiplicity of the eigenvalue 2 is 2. From the algorithm we find that there is only one equilibrium: $(f_1, f_2) = (-2, -2)$.

2. Consider $b_1 = b_2 = r_1 = r_2 = q_1 = 1$, $q_2 = 2$ and $a = 4$. Then, M has three different positive eigenvalues which each correspond with an equilibrium.

3. Consider $b_1 = b_2 = r_1 = r_2 = 1$, $q_1 = -1$, $q_2 = -9$ and $a = -3$. Then, M has two negative and two complex eigenvalues. So, an equilibrium does not exist. $\qquad\square$

8.5.3 The *N*-player scalar case

This section considers the extension of Algorithm 8.1 for the general N-player case. It will be shown that the numerical procedure to determine all feedback Nash equilibria for the N-player scalar game coincides with Algorithm 8.1. All that has to be adapted is the calculation of matrix M.

So, let p_i, $i = 1, \ldots, N$, be a solution of equations (8.5.3) and (8.5.4). Denoting again the negative of the resulting closed-loop system parameter by

$$\lambda := -a + \sum_i p_i, \qquad (8.5.19)$$

it is obvious from equations (8.5.3) and (8.5.4) that p_i satisfy

$$p_i^2 - 2\lambda p_i + \sigma_i = 0, \quad i = 1, \ldots, N. \qquad (8.5.20)$$

Next we derive for each index set Ω, with elements from the set $\{1, \ldots, N\}$, a linear equation (linear in terms of products of p_i variables (Πp_i)). This gives us in addition to equation (8.5.19) another $2^N - 1$ linear equations, which determine our matrix M.

For didactical reasons we will first outline the situation if Ω contains only one number. For this particular case we have, using the definition of λ and equation (8.5.20), respectively

$$p_j \lambda = p_j\left(-a + \sum_{i=1}^N p_i\right) = -ap_j + p_j^2 + p_j \sum_{i \neq j} p_i$$

$$= -ap_j + 2\lambda p_j - \sigma_j + p_j \sum_{i \neq j} p_i.$$

Some elementary rewriting of this equation then gives that

$$p_j\lambda = \sigma_j + ap_j - p_j\sum_{i\neq j}p_i, \quad j = 1,\ldots,N. \tag{8.5.21}$$

Next consider the general case $\prod_{j\in\Omega}p_j\lambda$. For notational convenience we use the notation Ω_{-i} to denote the set of all numbers that are in Ω except number i. Then,

$$\begin{aligned}
\prod_{j\in\Omega}p_j\lambda &= \prod_{j\in\Omega}p_j(-a+\sum_{i=1}^{N}p_i)\\
&= -a\prod_{j\in\Omega}p_j + \prod_{j\in\Omega}p_j\sum_{i\in\Omega}p_i + \prod_{j\in\Omega}p_j\sum_{i\notin\Omega}p_i\\
&= -a\prod_{j\in\Omega}p_j + \sum_{i\in\Omega}\prod_{j\in\Omega}p_jp_i + \sum_{i\notin\Omega}\prod_{j\in\Omega}p_jp_i\\
&= -a\prod_{j\in\Omega}p_j + \sum_{i\in\Omega}p_i^2\prod_{j\in\Omega_{-i}}p_j + \sum_{i\notin\Omega}\prod_{j\in\Omega}p_jp_i\\
&= -a\prod_{j\in\Omega}p_j + \sum_{i\in\Omega}(2\lambda p_i - \sigma_i)\prod_{j\in\Omega_{-i}}p_j + \sum_{i\notin\Omega}\prod_{j\in\Omega}p_jp_i\\
&= -a\prod_{j\in\Omega}p_j + 2\lambda\sum_{i\in\Omega}\prod_{j\in\Omega}p_j - \sum_{i\in\Omega}\sigma_i\prod_{j\in\Omega_{-i}}p_j + \sum_{i\notin\Omega}\prod_{j\in\Omega}p_jp_i.
\end{aligned}$$

Denoting the number of elements in Ω by n_Ω, we conclude that

$$\prod_{j\in\Omega}p_j\lambda = \frac{1}{2n_\Omega - 1}\left\{a\prod_{j\in\Omega}p_j + \sum_{i\in\Omega}\sigma_i\prod_{j\in\Omega_{-i}}p_j - \sum_{i\notin\Omega}\prod_{j\in\Omega}p_jp_i\right\}. \tag{8.5.22}$$

Equations (8.5.19) and (8.5.22) determine the matrix $\tilde{M}$. That is, introducing

$$p := \left[1, p_1, \ldots, p_N, p_1p_2, \ldots, p_{N-1}p_N, \ldots, \prod_{i=1}^{N}p_i\right]^T$$

we have that $\tilde{M}p = \lambda p$. Since $p_i = s_ik_i$ and $\sigma_i = s_iq_i$, matrix M is then easily obtained from $\tilde{M}$ by rewriting p as $p = Dk$, where $k := [1, k_1, \ldots, k_N, k_1k_2, \ldots, k_{N-1}k_N, \ldots, \Pi_{i=1}^{N}k_i]^T$ and D is a diagonal matrix defined by $D := \text{diag}\{1, s_1, \ldots, s_N, s_1s_2, \ldots, s_{N-1}s_N, \ldots, \Pi_{i=1}^{N}s_i\}$. Obviously, $M = D^{-1}\tilde{M}D$. Example 8.14, below, works out the case for $N = 3$.

Example 8.14

Consider the three-player case. With $p := [1, p_1, p_2, p_3, p_1p_2, p_1p_3, p_2p_3, p_1p_2p_3]^T$,

$$D = \text{diag}\{1, s_1, s_2, s_3, s_1s_2, s_1s_3, s_2s_3, s_1s_2s_3\}$$

and

$$
\tilde{M} =
\begin{bmatrix}
-a & 1 & 1 & 1 & 0 & 0 & 0 & 0 \\
\sigma_1 & a & 0 & 0 & -1 & -1 & 0 & 0 \\
\sigma_2 & 0 & a & 0 & -1 & 0 & -1 & 0 \\
\sigma_3 & 0 & 0 & a & 0 & -1 & -1 & 0 \\
0 & \frac{1}{3}\sigma_2 & \frac{1}{3}\sigma_1 & 0 & \frac{1}{3}a & 0 & 0 & -\frac{1}{3} \\
0 & \frac{1}{3}\sigma_3 & 0 & \frac{1}{3}\sigma_1 & 0 & \frac{1}{3}a & 0 & -\frac{1}{3} \\
0 & 0 & \frac{1}{3}\sigma_3 & \frac{1}{3}\sigma_2 & 0 & 0 & \frac{1}{3}a & -\frac{1}{3} \\
0 & 0 & 0 & 0 & \frac{1}{5}\sigma_3 & \frac{1}{5}\sigma_2 & \frac{1}{5}\sigma_1 & \frac{1}{5}a
\end{bmatrix}.
$$

This gives

$$
M =
\begin{bmatrix}
-a & s_1 & s_2 & s_3 & 0 & 0 & 0 & 0 \\
q_1 & a & 0 & 0 & -s_2 & -s_3 & 0 & 0 \\
q_2 & 0 & a & 0 & -s_1 & 0 & -s_3 & 0 \\
q_3 & 0 & 0 & a & 0 & -s_1 & -s_2 & 0 \\
0 & \frac{1}{3}q_2 & \frac{1}{3}q_1 & 0 & \frac{1}{3}a & 0 & 0 & -\frac{1}{3}s_3 \\
0 & \frac{1}{3}q_3 & 0 & \frac{1}{3}q_1 & 0 & \frac{1}{3}a & 0 & -\frac{1}{3}s_2 \\
0 & 0 & \frac{1}{3}q_3 & \frac{1}{3}q_2 & 0 & 0 & \frac{1}{3}a & -\frac{1}{3}s_1 \\
0 & 0 & 0 & 0 & \frac{1}{5}q_3 & \frac{1}{5}q_2 & \frac{1}{5}q_1 & \frac{1}{5}a
\end{bmatrix}.
$$

Using this matrix M in Algorithm 8.1 one can determine all feedback Nash equilibria of the scalar three-player linear quadratic differential game. $\qquad\square$

Example 8.15

Consider macroeconomic policy design in an EMU consisting of two blocks of countries that share a common central bank. Assume that the competitiveness, $s(t)$, (measured by the differences in prices in both countries) satisfies the differential equation

$$
\dot{s}(t) = as(t) - b_1 f_1(t) + b_2 f_2(t) + b_E i_E(t), \quad s(0) = s_0.
$$

Here s_0 measures the initial disequilibrium in intra-EMU competitiveness, $f_i(t)$ is the national fiscal deficit of country i, $i = 1, 2$, and $i_E(t)$ is the common interest rate set by the central bank.

The aim of the fiscal authorities is to use their fiscal policy instrument such that the following loss function is minimized

$$
J_i := \int_0^\infty \{q_i s^2(t) + r\, i f_i^2(t)\}dt, \quad i = 1, 2,
$$

whereas the central bank is confronted with the minimization of

$$
J_E := \int_0^\infty \{q_E s^2(t) + r_E i_E^2(t)\}dt,
$$

where for the sake of simplicity it is assumed that the long-term equilibrium level of the interest rate is zero.

Let $q_1 = 2$, $r_1 = 1$, $q_2 = 2$, $r_2 = 2$, $q_E = 1$ and $r_E = 3$. These numbers reflect the fact that country 1 has a preference for stabilizing the price differential, country 2 is indifferent between fiscal stabilization and price differential stabilization, and the central bank is primarily interested in targeting the interest rate at its equilibrium value. Assume $a = -1$, $b_1 = b_2 = 1$ and $b_E = \frac{1}{2}$. Then, $s_1 := 1$, $s_2 := \frac{1}{2}$ and $s_3 := \frac{1}{12}$. Using these numbers

$$
M = \begin{bmatrix}
1 & 1 & \frac{1}{2} & \frac{1}{12} & 0 & 0 & 0 & 0 \\
2 & -1 & 0 & 0 & \frac{-1}{2} & \frac{-1}{12} & 0 & 0 \\
2 & 0 & -1 & 0 & -1 & 0 & \frac{-1}{12} & 0 \\
1 & 0 & 0 & -1 & 0 & -1 & \frac{-1}{2} & 0 \\
0 & \frac{2}{3} & \frac{2}{3} & 0 & \frac{-1}{3} & 0 & 0 & \frac{-1}{36} \\
0 & \frac{1}{3} & 0 & \frac{2}{3} & 0 & \frac{-1}{3} & 0 & \frac{-1}{6} \\
0 & 0 & \frac{1}{3} & \frac{2}{3} & 0 & 0 & \frac{-1}{3} & \frac{-1}{3} \\
0 & 0 & 0 & 0 & \frac{1}{5} & \frac{2}{5} & \frac{2}{5} & \frac{-1}{5}
\end{bmatrix}.
$$

The eigenvalues of M are

$$\{1.9225, \ -1.5374, \ -0.6112 \pm 0.9899i, \ -0.8171 \pm 0.4462i, \ -0.3643 \pm 0.1261i\}.$$

So, the game has a unique feedback Nash equilibrium which can be calculated from an eigenvector $v =: [v_0, v_1, \ldots, v_7]^T$ corresponding with the eigenvalue 1.9225. MATLAB gives

$$v^T = [0.7170, \ 0.4447, \ 0.4023, \ 0.1875, \ 0.2495, \ 0.1163, \ 0.1052, \ 0.0653].$$

According to Algorithm 8.1 the equilibrium strategies are then

$$f_1^*(t) = k_1 s(t), \quad f_2^*(t) = -\frac{1}{2} k_2 s(t), \quad \text{and} \quad i_E^*(t) = -\frac{1}{6} k_E s(t),$$

where

$$k_1 = \frac{v_1}{v_0} = 0.6202, \quad k_2 = \frac{v_2}{v_0} = 0.5611, \quad k_E = \frac{v_3}{v_0} = 0.2616.$$

The resulting closed-loop system and cost are

$$\dot{s}(t) = -1.9225 s(t) \quad \text{and} \quad J_i = k_i s_0^2, \quad i = 1, 2, E.$$

Notice that country 1 is indeed much more active in using its fiscal instrument to reduce price differences than country 2. Moreover, we see that the central bank is indeed rather cautious in using its interest rate as a policy instrument to stabilize an initial price differential between both countries.

Using the three-player extension of Algorithm 7.1 the open-loop Nash equilibria for this game can be calculated either directly from Theorem 7.33 or from the eigenstructure of the matrix

$$M_0 := \begin{bmatrix} -1 & -1 & \frac{-1}{2} & \frac{-1}{12} \\ -2 & 1 & 0 & 0 \\ -2 & 0 & 1 & 0 \\ -1 & 0 & 0 & 1 \end{bmatrix}.$$

The resulting unique open-loop equilibrium actions are

$$f_1^*(t) = p_1 s(t), \quad f_2^*(t) = -\frac{1}{2} p_2 s(t), \quad \text{and} \quad i_E^*(t) = -\frac{1}{6} p_E s(t),$$

with

$$p_1 = p_2 = 0.6621 \quad \text{and} \quad p_E = 0.3310.$$

The resulting closed-loop system and cost are

$$\dot{s}(t) = -2.0207 s(t), \quad J_1 = 1.2067 s_0^2, \quad J_2 = 1.0982 s_0^2 \quad \text{and} \quad J_3 = 0.4994 s_0^2.$$

Comparing both equilibrium strategies we see that, within the open-loop framework, any initial price differential converges to zero more quickly. Moreover, all players use more active control policies in the open-loop case. However, the difference in the amount of control used by all players becomes less pronounced. Finally, we see that these more active policies ultimately imply almost a doubling of the cost for all the involved players.

□

8.6   Convergence results for the two-player scalar case

This section studies the convergence of the equilibrium strategies for a special scalar two-player finite-planning horizon game. The game that will be studied reads as follows:

$$J_i(x_0, u_1, u_2) = \int_0^T \{q_i x^2(t) + r_i u_i^2\} dt + q_{iT} x^2(T), \quad i = 1, 2, \tag{8.6.1}$$

subject to the dynamical system

$$\dot{x}(t) = ax(t) + b_1 u_1(t) + b_2 u_2(t), \quad x(0) = x_0. \tag{8.6.2}$$

It will be assumed throughout this section that q_i, q_{iT} and $s_i = \frac{b_i^2}{r_i}$, $i = 1, 2$, are strictly positive. First, we show that this game (8.6.1) and (8.6.2) always has a unique linear feedback Nash equilibrium. Then, the behavior of the equilibrium actions is analyzed if the planning horizon T expands. It will be shown that if these actions converge, the converged actions depend critically on the imposed final weighting factors q_{iT}.

By Theorem 8.2 this game (8.6.1) and (8.6.2) has a feedback equilibrium if and only if the following set of coupled Riccati differential equations

$$\dot{k}_1(t) = -2ak_1(t) + s_1k_1^2(t) + 2s_2k_1(t)k_2(t) - q_1, \quad k_1(T) = q_{1T}, \qquad (8.6.3)$$
$$\dot{k}_2(t) = -2ak_2(t) + s_2k_2^2(t) + 2s_1k_1(t)k_2(t) - q_2, \quad k_2(T) = q_{2T}, \qquad (8.6.4)$$

has a unique solution $(k_1(.), k_2(.))$ on $[0, T]$. Since we need to study convergence properties of the solution later on, it is more convenient to rewrite this terminal value problem for $(k_1(t), k_2(t))$ as an initial value problem. To that end introduce the new variable

$$m_i(t) := k_i(T - t), \quad i = 1, 2.$$

The derivative of this variable satisfies

$$\frac{dm_i(t)}{dt} = \frac{dk_i(T - t)}{dt} = -\frac{dk_i(s)}{ds}, \quad i = 1, 2.$$

Using this, it is easily verified that the set of coupled Riccati differential equations (8.6.3) and (8.6.4) has a solution if and only if the next initial value problem has a solution on $[0, T]$

$$\dot{m}_1(t) = 2am_1(t) - s_1m_1^2(t) - 2s_2m_1(t)m_2(t) + q_1, \quad m_1(0) = q_{1T}, \qquad (8.6.5)$$
$$\dot{m}_2(t) = 2am_2(t) - s_2m_2^2(t) - 2s_1m_1(t)m_2(t) + q_2, \quad m_2(0) = q_{2T}. \qquad (8.6.6)$$

Introducing the variables

$$\sigma_i := s_iq_i \quad \text{and} \quad \kappa_i := s_im_i, \quad i = 1, 2,$$

both equations can be rewritten as

$$\dot{\kappa}_1(t) = 2a\kappa_1(t) - \kappa_1^2(t) - 2\kappa_1(t)\kappa_2(t) + \sigma_1, \quad \kappa_1(0) = q_{1T}, \qquad (8.6.7)$$
$$\dot{\kappa}_2(t) = 2a\kappa_2(t) - \kappa_2^2(t) - 2\kappa_1(t)\kappa_2(t) + \sigma_2, \quad \kappa_2(0) = q_{2T}. \qquad (8.6.8)$$

The study of planar quadratic systems in general is a very complicated topic, as can be seen, for example, in the survey papers by Coppel (1996) and Reyn (1987). For example, the famous 16th Hilbert problem, to determine the maximal number of limit cycles, H_d, for dth degree polynomial planar systems, is still unsolved even for quadratic systems ($d = 2$). Hence, in general one can expect complicated dependence on the parameters for the quadratic system (8.6.7) and (8.6.8). Reyn (1987), for example, finds 101 topologically different global phase portraits for a six-parameter family of quadratic systems. Below some of the characteristics of the quadratic system (8.6.7) and (8.6.8) that one is typically interested in will be addressed. The most urgent question at this moment is whether this system has a solution on $[0, \infty)$. The next result is shown in the Appendix at the end of this chapter.

Theorem 8.17

Assume that q_i, q_{iT} and s_i, $i = 1, 2$, are strictly positive. Then the set of coupled Riccati equations (8.6.7) and (8.6.8) has a solution on $[0, \infty)$. Moreover, the solution remains bounded. $\qquad\square$

To construct the phase portrait of the planar system (8.6.7) and (8.6.8). the question arises whether this system has periodic solutions. According to Bendixson's theorem (Theorem 3.18) with

$$f_i(\kappa_1, \kappa_2) := 2a\kappa_i - \kappa_i^2 - 2\kappa_1\kappa_2 + \sigma_i, \quad i = 1, 2, \tag{8.6.9}$$

the planar system (8.6.7) and (8.6.8) has no periodic solutions on a set E if

$$\frac{\partial f_1}{\partial \kappa_1} + \frac{\partial f_2}{\partial \kappa_2}$$

is not identically zero and does not change sign on this set E. Here,

$$\frac{\partial f_1}{\partial \kappa_1} + \frac{\partial f_2}{\partial \kappa_2} = 4(a - \kappa_1 - \kappa_2).$$

Hence, $\frac{\partial f_1}{\partial \kappa_1} + \frac{\partial f_2}{\partial \kappa_2} = 0$ on the line

$$a - \kappa_1 - \kappa_2 = 0. \tag{8.6.10}$$

So, for the sets $E_1 = \{(\kappa_1, \kappa_2) \mid a - \kappa_1 - \kappa_2 > 0\}$ and $E_2 = \{(\kappa_1, \kappa_2) \mid a - \kappa_1 - \kappa_2 < 0\}$ there are no periodic solutions which lie entirely in these regions. Furthermore it follows that a periodic solution of the quadratic system (8.6.7) and (8.6.8) has to cross the line (8.6.10) at least twice. However, on the line (8.6.10)

$$\dot{\kappa}_1 = \sigma_1 + \kappa_1^2 > 0 \quad \text{and} \quad \dot{\kappa}_2 = \sigma_2 + \kappa_2^2 > 0.$$

Hence any solution of equations (8.6.7) and (8.6.8) can cross the line (8.6.10) once at the most. So, the planar system (8.6.7) and (8.6.8) has no periodic solutions.

Corollary 8.18

Assume that q_i, q_{iT} and s_i, $i = 1, 2$, are strictly positive. Then

1. the game (8.6.1) and (8.6.2) has for all $T < \infty$ a feedback Nash equilibrium;

2. $\lim_{T \to \infty} k_i(0, T)$, $i = 1, 2$, in equations (8.6.3) and (8.6.4) exists and satisfies the corresponding set of algebraic Riccati equations. $\qquad\square$

Next, consider the equilibria of the planar system (8.6.7) and (8.6.8). These equilibria coincide with the solutions of the feedback Nash algebraic Riccati equations

$$2a\kappa_1 - \kappa_1^2 - 2\kappa_1\kappa_2 + \sigma_1 = 0, \tag{8.6.11}$$

$$2a\kappa_2 - \kappa_2^2 - 2\kappa_1\kappa_2 + \sigma_2 = 0. \tag{8.6.12}$$

Recall from Section 8.4 that, if $\sigma_i > 0$, the infinite-planning horizon game has either one, two or three feedback Nash equilibria (see Theorem 8.10). Therefore one would expect a similar result here. There are, however, two differences compared with the analysis of Section 8.4. The first point is that in Section 8.4 we did not pay attention to the location of the feedback Nash equilibria. For our phase plane analysis of the planar system (8.6.7) and (8.6.8) we are, however, only interested in the positive quadrant. The second point is that in Section 8.4 the inter Section points of both hyperbolas that satisfy some additional constraint were studied. The constraint plays no role here. Therefore, in principle, there is the possibility that there are some additional points satisfying both equations but not the constraint. Therefore both equations will be reconsidered in some more detail.

Assuming that $\kappa_i > 0$, $i = 1, 2$, equations (8.6.7) and (8.6.8) can be rewritten as

$$\kappa_2 = a - \frac{1}{2}\kappa_1 + \frac{\sigma_1}{2\kappa_1} \quad \text{and} \quad \kappa_1 = a - \frac{1}{2}\kappa_2 + \frac{\sigma_2}{2\kappa_2}.$$

From this it is clear that hyperbola (8.6.11) has the asymptotes $\kappa_1 = 0$ and $\kappa_2 = a - \frac{1}{2}\kappa_1$ and hyperbola (8.6.12) has the asymptotes $\kappa_2 = 0$ and $\kappa_2 = 2a - \kappa_1$. Furthermore, hyperbola (8.6.11) intersects the positive κ_1-axis at the point where $\kappa_1 = a + \sqrt{a^2 + \sigma_1}$, and hyperbola (8.6.12) intersects the positive κ_2-axis in the points where $\kappa_2 = a + \sqrt{a^2 + \sigma_2}$. From this it is easily seen that the equilibrium points of the planar system (8.6.7) and (8.6.8) which are located in the positive quadrant are located in the region

$$G_1 := (0, a + \sqrt{a^2 + \sigma_1}) \times (0, a + \sqrt{a^2 + \sigma_2}).$$

Lemma 8.19

The equilibrium points $(\kappa_1^*, \kappa_2^*) \geq (0,0)$ of the planar system (8.6.7) and (8.6.8) are located in the region G_1. Moreover, G_1 contains at least one equilibrium point.

Proof

The region G_1 lies entirely in the positive quadrant. Hyperbola (8.6.7) intersects the κ_1-axis at the point where $\kappa_1 = a + \sqrt{a^2 + \sigma_1}$, and hence any equilibrium point in the first quadrant has to be located to the left of $\kappa_1 = a + \sqrt{a^2 + \sigma_1}$. Similarly, any critical point in the first quadrant has to be located below the line $\kappa_2 = a + \sqrt{a^2 + \sigma_2}$. Hence, any equilibrium point in the positive quadrant has to be located in G_1. Furthermore, it is easily seen that hyperbola (8.6.7) enters G_1 at the point $(0, a + \sqrt{a^2 + \sigma_2})$, and leaves G_1 through the line $\kappa_1 = a + \sqrt{a^2 + \sigma_1}$. Hyperbola (8.6.8) enters G_1 through the line $\kappa_2 = a + \sqrt{a^2 + \sigma_2}$ and leaves G_1 at the point $(a + \sqrt{a^2 + \sigma_1}, 0)$. Necessarily, since both hyperbolas (8.6.7) and (8.6.8) are continuous on G_1 they have to intersect at least once in G_1. $\qquad\square$

Lemma 8.20

Every equilibrium point (κ_1^*, κ_2^*) in G_1 corresponds to a stationary feedback Nash equilibrium.

Proof

Let $S := (\kappa_1^*, \kappa_2^*)$ be an equilibrium point in G_1. Because S is located on the hyperbola (8.6.7), we know S is located above the asymptote $\kappa_2 = a - \frac{1}{2}\kappa_1$, and thus

$$\kappa_2^* > a - \frac{1}{2}\kappa_1^*.$$

Hence,

$$a - \kappa_1^* - \kappa_2^* < a - \kappa_1^* - a + \frac{1}{2}\kappa_1* = -\frac{1}{2}\kappa_1^* < 0. \tag{8.6.13}$$

Since $\kappa_i = s_i k_i$, the claim is now obvious. $\qquad\square$

Theorem 8.10 shows that the number of feedback Nash equilibria can vary between one and three. Actually, Example 8.13, part 2, demonstrates that there exist games, within the restricted class of games we consider here, which have three positive feedback Nash equilibria. So, the following corollary is now clear.

Corollary 8.21

The planar system (8.6.7) and (8.6.8) has either one, two or three equilibrium points in the positive quadrant. $\qquad\square$

Next, we consider the local behavior of the solution curves near each equilibrium. It turns out that almost all different situations which can occur in our planar system can be analyzed using the phase-plane approach we developed in Chapter 3. There are only two situations which require some additional theory. They deal with the case that the linearized system at the equilibrium point has an eigenvalue on the imaginary axis (so the eigenvalue has a zero real part). Equilibria which have this property are called **nonhyperbolic critical points** of a planar system. In Perko (2001), for example, one can find a complete treatment of the different types of local behavior of trajectories which can occur near nonhyperbolic equilibria. From these types of local behavior there are two types of behavior which we have not yet treated, but which can also occur in our planar system. A complete treatment of all the local behaviors which can occur at a nonhyperbolic equilibrium is somewhat beyond the scope of this book. Therefore, we will just present an example of a case containing a saddle-node, one type of behavior which can also occur in our planar system (8.6.7) and (8.6.8). The second type of behavior which can also occur will not be covered. This is because, on the one hand, its analysis requires some additional theory in order to gain a clear impression of the local phase diagram. On the other hand, its local behavior corresponds to a stable node, a type of behavior which we are already familiar with.

Example 8.16

Consider the planar system (8.6.7) and (8.6.8) with $a = 3\frac{1}{2}$, $\sigma_1 = 8$ and $\sigma_2 = 6\frac{3}{4}$. That is

$$\dot{\kappa}_1(t) = 7\kappa_1(t) - \kappa_1^2(t) - 2\kappa_1(t)\kappa_2(t) + 8, \quad \kappa_1(0) = q_{1T},$$

$$\dot{\kappa}_2(t) = 7\kappa_2(t) - \kappa_2^2(t) - 2\kappa_1(t)\kappa_2(t) + 6\frac{3}{4}, \quad \kappa_2(T) = q_{2T}.$$

Then, by simple substitution, it is seen that $(\kappa_1^*, \kappa_2^*) = (2, 4\frac{1}{2})$ is an equilibrium point of this planar system.

To investigate the local behavior of the solution curves near this equilibrium, consider the derivative of

$$f_1(\kappa_1, \kappa_2) := 7\kappa_1(t) - \kappa_1^2(t) - 2\kappa_1(t)\kappa_2(t) + 8$$

and

$$f_2(\kappa_1, \kappa_2) := 7\kappa_2(t) - \kappa_2^2(t) - 2\kappa_1(t)\kappa_2(t) + 6\frac{3}{4}$$

at the equilibrium point $(2, 4\frac{1}{2})$. Straightforward differentiation shows that the derivative of $f := (f_1, f_2)$ at the equilibrium point is

$$f'\left(2, 4\frac{1}{2}\right) = \begin{bmatrix} 7 - 2\kappa_1^* - 2\kappa_2^* & -2\kappa_1^* \\ -2\kappa_2^* & 7 - 2\kappa_1^* - 2\kappa_2^* \end{bmatrix} = \begin{bmatrix} -6 & -4 \\ -9 & -6 \end{bmatrix}.$$

The eigenvalues of $f'(2, 4\frac{1}{2})$ are 0 and -12. The corresponding eigenvectors are $e_1 := \begin{bmatrix} -2 \\ 3 \end{bmatrix}$ and $e_2 := \begin{bmatrix} 2 \\ 3 \end{bmatrix}$, respectively. Since $f'(2, 4\frac{1}{2})$ has an eigenvalue zero, this case does not satisfy the assumptions of the Stable Manifold theorem, (Theorem 3.15). $(2, 4\frac{1}{2})$ is a nonhyperbolic critical point of the planar system. To analyze the local behavior of trajectories of our planar system near this equilibrium point we will follow the approach outlined in Perko (2001).

In that approach the first step is to rewrite our planar system near the equilibrium point $(2, 4\frac{1}{2})$ as a set of differential equations of the following type

$$\dot{x}(t) = p_2(x(t), y(t))$$
$$\dot{y}(t) = y + q_2(x(t), y(t)),$$

where p_2 and q_2 are analytic in the neighborhood of the origin and have Taylor expansions that begin with second-degree terms in x and y. To that purpose we first rewrite our planar system, using the transformation

$$\mu_1(t) = \kappa_1(t) - 2 \quad \text{and} \quad \mu_2(t) = \kappa_2(t) - 4\frac{1}{2},$$

as

$$\begin{bmatrix} \dot{\mu}_1(t) \\ \dot{\mu}_2(t) \end{bmatrix} = \begin{bmatrix} -6 & -4 \\ -9 & -6 \end{bmatrix} \begin{bmatrix} \mu_1(t) \\ \mu_2(t) \end{bmatrix} - \begin{bmatrix} \mu_1^2(t) + 2\mu_1(t)\mu_2(t) \\ \mu_2^2(t) + 2\mu_1(t)\mu_2(t) \end{bmatrix}.$$

Let $S := [e_1 \ e_2]$. Then, $f'(2, 4\frac{1}{2}) = SDS^{-1}$, where D is a diagonal matrix containing the eigenvalues 0 and -12 on its diagonal. Next, rewrite the above system with

$$\begin{bmatrix} \tau_1 \\ \tau_2 \end{bmatrix} := S^{-1} \begin{bmatrix} \mu_1 \\ \mu_2 \end{bmatrix} \quad \text{and, thus,} \quad \begin{bmatrix} \mu_1 \\ \mu_2 \end{bmatrix} = S \begin{bmatrix} \tau_1 \\ \tau_2 \end{bmatrix}$$

as

$$\begin{bmatrix} \dot{\tau}_1 \\ \dot{\tau}_2 \end{bmatrix} = D \begin{bmatrix} \tau_1 \\ \tau_2 \end{bmatrix} - S^{-1} \begin{bmatrix} (-2\tau_1 + 2\tau_2)^2 + 2(-2\tau_1 + 2\tau_2)(3\tau_1 + 3\tau_2) \\ (3\tau_1 + 3\tau_2)^2 + 2(-2\tau_1 + 2\tau_2)(3\tau_1 + 3\tau_2) \end{bmatrix}.$$

Reversing the time-axis we obtain, with $\rho_i(-t) := \tau_i(t)$, $i = 1, 2$,

$$\begin{bmatrix} \dot{\rho}_1 \\ \dot{\rho}_2 \end{bmatrix} = -D \begin{bmatrix} \rho_1 \\ \rho_2 \end{bmatrix} + S^{-1} \begin{bmatrix} (-2\rho_1 + 2\rho_2)^2 + 2(-2\rho_1 + 2\rho_2)(3\rho_1 + 3\rho_2) \\ (3\rho_1 + 3\rho_2)^2 + 2(-2\rho_1 + 2\rho_2)(3\rho_1 + 3\rho_2) \end{bmatrix}.$$

Finally, introducing $y := \tau_2^{1/12}$ and $x := \tau_1$ we obtain the equivalent description of our planar system

$$\dot{y}(t) = \frac{1}{12} \tau_2^{-11/12}(t) \dot{\tau}_2(t)$$

$$= \tau_2^{1/12} + \tau_2^{-11/12}[0, \ 1]S^{-1} \begin{bmatrix} (-2\tau_1 + 2\tau_2)^2 + 2(-2\tau_1 + 2\tau_2)(3\tau_1 + 3\tau_2) \\ (3\tau_1 + 3\tau_2)^2 + 2(-2\tau_1 + 2\tau_2)(3\tau_1 + 3\tau_2) \end{bmatrix}$$

$$= y + y^{-11}[0, \ 1]S^{-1} \begin{bmatrix} (-2x + 2y^{12})^2 + 2(-2x + 2y^{12})(3x + 3y^{12}) \\ (3x + 3y^{12})^2 + 2(-2x + 2y^{12})(3x + 3y^{12}) \end{bmatrix}$$

and

$$\dot{x} = [1, \ 0]S^{-1} \begin{bmatrix} (-2x + 2y^{12})^2 + 2(-2x + 2y^{12})(3x + 3y^{12}) \\ (3x + 3y^{12})^2 + 2(-2x + 2y^{12})(3x + 3y^{12}) \end{bmatrix}.$$

The second step is to rewrite y as a function $\phi(x)$ from the equation $\dot{y} = 0$. That is, we have to solve y as a function of x from the equation

$$y + y^{-11}[0, \ 1]S^{-1} \begin{bmatrix} (-2x + 2y^{12})^2 + 2(-2x + 2y^{12})(3x + 3y^{12}) \\ (3x + 3y^{12})^2 + 2(-2x + 2y^{12})(3x + 3y^{12}) \end{bmatrix} = 0.$$

Multiplying the above equation by y^{11} and introducing $\tau := y^{12}$, we can rewrite the equation as

$$\tau + [0, \ 1]S^{-1} \begin{bmatrix} (-2x + 2\tau)^2 + 2(-2x + 2\tau)(3x + 3\tau) \\ (3x + 3\tau)^2 + 2(-2x + 2\tau)(3x + 3\tau) \end{bmatrix} = 0.$$

This is a simple quadratic equation in τ. Solving the equation gives

$$\tau = \frac{1}{15}\left\{ -(x + 1) + \sqrt{(x + 1)^2 + 75x^2} \right\}.$$

The third step is to substitute this result into the differential equation we have for $x(t)$ and, next, make a Taylor expansion of the resulting right-hand side of this differential equation. So, we have to make a Taylor series expansion of

$$[1, \ 0]S^{-1} \begin{bmatrix} (-2x + 2\tau(x))^2 + 2(-2x + 2\tau(x))(3x + 3\tau(x)) \\ (3x + 3\tau(x))^2 + 2(-2x + 2\tau(x))(3x + 3\tau(x)) \end{bmatrix} = \frac{3}{2}x^2 + 5x\tau(x) - \frac{1}{2}\tau^2(x)$$

around $x = 0$.

Elementary differentiation shows that $\tau'(0) = 0$. Consequently

$$\frac{3}{2}x^2 + 5x\tau(x) - \frac{1}{2}\tau^2(x) = \frac{3}{2}x^2 + \text{ higher-order terms.}$$

The fourth step is to determine the first nonvanishing term in this Taylor series expansion. In this case this term is $\frac{3}{2}x^2$. The order of this first term is 2, and thus even. According to Theorem 2.11.1 in Perko (2001), the equilibrium $(2, 4\frac{1}{2})$ is then a saddle-node. It can easily be verified (see, for example, Theorem 8.22 below) that this planar system has still one other equilibrium point which is a stable node. This then produces the phase diagram as plotted in Figure 8.6. $\qquad\square$

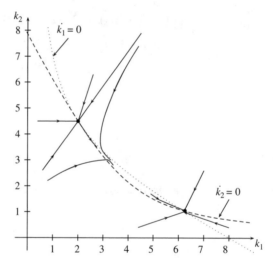

Figure 8.6   Phase diagram of hyperbolas (8.6.7) and (8.6.8) with $a = 3\frac{1}{2}$, $\sigma_1 = 8$ and $\sigma_2 = 6\frac{3}{4}$

To analyze the general case (8.6.7) and (8.6.8), we consider the derivative of $(f_1(\kappa_1, \kappa_2), f_2(\kappa_1, \kappa_2))$ at the equilibrium point (κ_1^*, κ_2^*). From equation (8.6.9)

$$f' = \begin{bmatrix} 2(a - \kappa_1 - \kappa_2) & -2\kappa_1 \\ -2\kappa_2 & 2(a - \kappa_1 - \kappa_2) \end{bmatrix}.$$

The eigenvalues of f' are

$$\lambda_{1,2} = 2(a - \kappa_1 - \kappa_2) \pm 2\sqrt{\kappa_1\kappa_2}.$$

From this, some elementary analysis gives the following theorem (see the Appendix at the end of this chapter).

Theorem 8.22

If the planar system (8.6.7) and (8.6.8) has

- one equilibrium, then this equilibrium is a stable node;
- two equilibria, then one of them is a stable node and one is a saddle-node;
- three equilibria, then two of them are stable nodes and one is a saddle.   □

   It is now possible to give a complete picture of the different convergence behavior which can occur if the planning horizon expands. The different global behavior of the trajectories that may occur is sketched in Figures 8.7(a)–(c). We see that this behavior

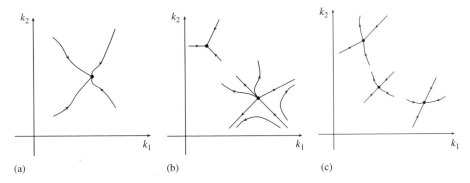

(a)   (b)   (c)

Figure 8.7   Global phase portrait of equations (8.6.7) and (8.6.8): (a) ARE has one positive stabilizing solution; (b) ARE has two positive stabilizing solutions (c) ARE has three positive stabilizing solutions

depends crucially on the number of solutions that the set of algebraic Riccati equations has. If there is only one solution for these coupled equations then, independent of the scrap value q_{iT}, $i = 1, 2$, the solutions of the Riccati differential equations (8.6.7) and (8.6.8) converge to this equilibrium point (cf. Figure 8.7(a)). If there are two or more solutions of the set of algebraic Riccati equations in the positive quadrant the final outcome of the equilibrium, if the planning horizon expands, depends crucially on the scrap values. In that case the solutions converge to either one of these solutions. This convergence behavior is sketched in Figures 8.7(b) and (c), respectively. If there are three equilibria, the central equilibrium is a saddle-point. So, convergence towards this equilibrium will only occur with rare combinations of the scrap values.

Example 8.17

Reconsider Example 8.8, where we showed that for $a = b_i = r_i = 1$, $i = 1, 2$, $q_1 = \frac{1}{4}$ and $q_2 = \frac{1}{5}$, the game has three equilibria. Using Algorithm 8.1, one can calculate the corresponding solutions of the algebraic Riccati equations. They are:

$$(p_1^l, p_2^l) := (0.1387, 1.8317), (p_1^m, p_2^m) := (0.7125, 0.8192) \text{ and } (p_1^r, p_2^r) := (1.9250, 0.1024),$$

respectively.

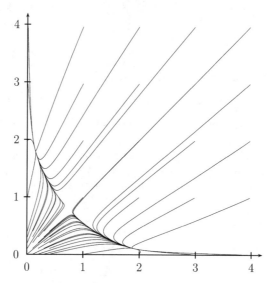

Figure 8.8   Global phase portrait of Example 8.17

Figure 8.8 illustrates the solutions $(p_1(0), p_2(0))$ for different scrap values and different planning horizons. The scrap values q_{1T} and q_{2T} both varied over the interval $[0, 4]$. The planning horizon T varied from 0 to 10. We clearly observe the phenomenon that the choice of the scrap value plays an important role in the evolution of the equilibrium if the planning horizon expands. A modest change in the scrap values may have as a consequence that convergence towards the infinite-planning horizon equilibrium (p_1^l, p_2^l) is replaced by a convergence towards the equilibrium (p_1^r, p_2^r). We also see that it is almost impossible to trace a scrap value that will lead us to the equilibrium (p_1^m, p_2^m).   □

8.7   Notes and references

For this chapter in particular the papers by van den Broek, Engwerda and Schumacher (2003a), Engwerda (2000a, b, 2003) and Weeren, Schumacher and Engwerda (1999) have been consulted.

Theorems 8.3 and 8.5 can be straightforwardly generalized for the N-player game (8.1.1) and (8.1.2). One just has to replace the equations (8.2.6) and (8.2.7) and (8.3.3) and (8.3.4) by

$$\dot{K}_i(t) = -\left(A - \sum_{j\neq i}^{N} S_j K_j\right)^T K_i - K_i\left(A - \sum_{j\neq i}^{N} S_j K_j\right) + K_i S_i K_i - Q_i - \sum_{j\neq i}^{N} K_j S_{ij} K_j$$

$$K_i(T) = Q_{iT}, \; i = 1, \ldots, N,$$

and

$$0 = -\left(A - \sum_{j\neq i}^{N} S_j K_j\right)^T K_i - K_i\left(A - \sum_{j\neq i}^{N} S_j K_j\right) + K_i S_i K_i - Q_i - \sum_{j\neq i}^{N} K_j S_{ij} K_j,$$

where $A - \sum_{i=1}^{N} S_i K_i$ is stable, respectively.

Finally, we should stress once again the crucial role played by our assumption about the information the players have on the game for the results presented in this chapter. The Example 8.7 from Mageirou (1976) clearly demonstrates that if one leaves the pre-supposed information framework, then the Nash equilibrium property of a set of strategies might be lost. Details and references concerning Nash equilibria under different feedback information patterns can be found in Başar and Olsder (1999). In this context the model on duopolistic competition studied by Tsutsui and Mino (1990) should also be mentioned. They study the linear quadratic differential game we considered, for example, in Example 8.3. However, this is under the restriction that the used control functions should be positive functions of the price $p \geq 0$. So on the one hand they restrict the domain of the state space and on the other hand they consider strategy spaces consisting of the set of nonlinear positive functions in p, which clearly is not a subset of the set of affine functions we considered in this chapter. As a consequence the equilibrium actions in their model become nonlinear functions of the price too.

8.8 Exercises

1. Reconsider the differential games from Exercise 7.1.

   (a) Determine the differential equations that have to be solvable in order that the game has, for every initial state, a linear feedback Nash equilibrium.

   (b) Determine numerically which of the differential games has, for every initial state, a linear feedback Nash equilibrium. In case an equilibrium exists, compute the equilibrium actions and involved cost. Compare your answers with those of Exercise 7.1.

2. Reconsider the zero-sum differential game from Exercise 7.3.

   (a) Determine the differential equation that has to be solvable in order that the game has, for every initial state, a linear feedback Nash equilibrium.

   (b) Determine the equilibrium actions and involved cost in item (a) in case $T = 1$.

   (c) Does there exist a finite time T such that the game does not have, for every initial state, a linear feedback Nash equilibrium?

   (d) Consider a scrap value of $-200x^2(1)$ in part (a). Does the game have a linear feedback Nash equilibrium for every initial state? Does there exist for all $T < 1$ an open-loop Nash equilibrium for every initial state?

3. Reconsider the fiscal stabilization differential game from Exercise 7.19.

Consider the class of affine functions of the state variable $s(t)$, i.e.

$$\Gamma_i^{aff} := \{f_i(0,T)|f_i(t) = F_i(t)s(t) + g_i(t), \text{ with } F_i(.), \ g_i(.)$$
$$\text{piecewise continuous functions, } i = 1,2\}.$$

Answer the same questions as in Exercise 7.19(a)–(d) but now w.r.t. the class of control functions Γ_i^{aff}.

4. Reconsider the government debt stabilization game from Section 7.8.1. There we assumed that government debt accumulation, $\dot{d}$, is the sum of interest payments on government debt, $rd(t)$, and primary fiscal deficits, $f(t)$, minus the seignorage, $m(t)$:

$$\dot{d}(t) = rd(t) + f(t) - m(t), \quad d(0) = d_0.$$

Assume that the objective of the fiscal authority is to minimize

$$L^F = \int_0^T \{(f(t) - \bar{f})^2 + \eta(m(t) - \bar{m})^2 + \lambda(d(t) - \bar{d})^2\}e^{-\delta t}dt,$$

whereas the monetary authorities want to minimize the loss function:

$$L^M = \int_0^T \{(m(t) - \bar{m})^2 + \kappa(d(t) - \bar{d})^2\}e^{-\delta t}dt.$$

Here all variables denoted with a bar are assumed to be fixed targets which are given a priori. Consider the class of affine functions of the state variable $s(t)$, i.e.

$$\Gamma_F^{aff} := \{f(0,T)|f(t) = F(t)d(t) + g(t), \text{ with } F(.), \ g(.) \text{ piecewise continuous functions}\},$$
$$\Gamma_M^{aff} := \{m(0,T)|m(t) = M(t)d(t) + h(t), \text{ with } M(.), \ h(.) \text{ piecewise continuous functions}\}.$$

(a) Let $x_1(t) := (d(t) - \bar{d})e^{-\frac{1}{2}\delta t}, x_2(t) := (r\bar{d} + \bar{f} - \bar{m})e^{-\frac{1}{2}\delta t}, u_1(t) := (f(t) - \bar{f})e^{-\frac{1}{2}\delta t}$ and $u_2(t) := (m(t) - \bar{m})e^{-\frac{1}{2}\delta t}$.

Show that the above game can be rewritten in our standard notation as

$$\min_{u_1} L^F = \int_0^T \{\lambda x_1^2(t) + u_1^2(t) + \eta u_2^2(t)\}dt$$

and

$$\min_{u_2} L^M = \int_0^T \{\kappa x_1^2(t) + u_2^2(t)\}dt$$

subject to

$$\dot{x}(t) = Ax(t) + B_1 u_1(t) + B_2(t)u_2(t)$$

with : $A = \begin{bmatrix} r - \frac{1}{2}\delta & 1 \\ 0 & -\frac{1}{2}\delta \end{bmatrix}$, $B_1 = \begin{bmatrix} 1 \\ 0 \end{bmatrix}$, $B_2 = \begin{bmatrix} -1 \\ 0 \end{bmatrix}$, $Q_1 = \begin{bmatrix} \lambda & 0 \\ 0 & 0 \end{bmatrix}$, $Q_2 = \begin{bmatrix} \kappa & 0 \\ 0 & 0 \end{bmatrix}$,

$R_{11} = 1, R_{22} = 1, R_{12} = \eta$ and $R_{21} = 0$.

(b) Present both necessary and sufficient conditions under which this game has a Nash equilibrium within $\Gamma_F \times \Gamma_M$ for every initial state.

(c) Determine the Nash equilibrium actions in (a) and the incurred cost for the players (you are not asked to solve the involved differential equations explicitly).

(d) Let $\eta = \frac{1}{4}$, $\lambda = \kappa = \frac{1}{2}$, $r = 0.02$ and $\delta = 0.05$. Plot (numerically) the corresponding Nash equilibrium actions if $d_0 = 1$. Determine also the corresponding cost for the players.

(e) Answer the same question as in part (d) if $r = 0.04$ and $r = 0.06$, respectively. Compare the answers and formulate your expectations about the effect of r on this model.

5. Reconsider Exercise 7.4. That is, consider the zero-sum game (in terms of country 1)

$$\min_{u_1} J := \int_0^\infty \{x^2(t) + ru_1^2(t) - u_2^2(t)\}dt,$$

subject to

$$\dot{x}(t) = ax(t) + u_1(t) + u_2(t), \quad x(0) = x_0.$$

(a) Consider $r = 1$. Show that this game has a feedback Nash equilibrium if and only if $a < 0$.

(b) Determine the equilibrium actions and cost for both countries if $a < 0$ and $r = 1$.

(c) Assume that both countries differ in the sense that country 1 has an industry which pollutes less than that of country 2. Justify how this fact can be incorporated into this model by choosing a value of r different from one.

(d) Determine the equilibrium actions and cost in part (c) and discuss the impact of a more polluting industry on the outcome of the game. In particular, discuss the existence of a feedback Nash equilibrium and compare this result with the case $r = 1$. What can you say about the consequences of a higher value of the parameter a on the outcome of the game?

6. Consider an industry which is constantly lobbying the government to exploit natural resources and an environmental party which is lobbying to avoid this exploitation in order to save nature. Let $c(t)$ denote the success rate of the industry lobby, u_i the efforts of the industry lobbyists and u_e the efforts of the

environmental lobbyists. The problem is formalized as follows:

$$\min_{u_e} \int_0^\infty e^{-rt}\{c(t) - \rho u_i^2(t) + u_e^2(t)\}dt$$

subject to $\dot{c}(t) = -\delta c(t) + u_i(t) - u_e(t)$, $c(0) = c_0$. The industry lobby likes to maximize the above utility function. Here $\rho < 1$ indicates the fact that industry has much more money to spend on the lobby than the environmental party has, and therefore the normalized impact of their control efforts is larger.

(a) Reformulate the problem into the standard framework. Define the class of affine control functions that is appropriate in this setting if we want to consider feedback Nash equilibria for this game.

(b) Show that the problem does not have a feedback Nash equilibrium within the class of stabilizing affine control functions.

7. Consider the two-player scalar linear quadratic differential game (8.4.1) and (8.4.2) with $b_i = r_i = 1$, $i = 1, 2$, $q_1 = 8$ and $q_2 = 6\frac{3}{4}$.

(a) Use Algorithm 8.1 to calculate all feedback Nash equilibria of this game if (i) $a = -1$, (ii) $a = 3\frac{1}{2}$ and (iii) $a = 4$, respectively.

(b) Show that the function $f(x) = x - \sqrt{x^2 - 8} + \sqrt{x^2 - 6\frac{3}{4}}$ has a minimum at $x = 3$.

(c) Verify whether your answers in part (a) are in line with Theorem 8.10.

8. Consider the two-player scalar linear quadratic differential game (8.4.1) and (8.4.2) with $b_i = r_i = 1$, $i = 1, 2$, $q_1 = -5$ and $q_2 = -32$.

(a) Use Algorithm 8.1 to calculate all feedback Nash equilibria of this game if (i) $a = -10$, (ii) $a = -5$, (iii) $a = 0$ (iv) $a = 5$ and (v) $a = 10$, respectively.

(b) Show that the function $f(x) = x - \sqrt{x^2 + 5} - \sqrt{x^2 + 32}$ has a maximum at $x = 2$.

(c) Verify whether your answers in part (a) are in line with Theorem 8.10.

9. Consider the two-player scalar linear quadratic differential game (8.4.1) and (8.4.2). Let $\sigma_i = \frac{b_i^2}{r_i} = 1$, $i = 1, 2$, and assume that $\sigma := \sigma_1 = \sigma_2$.

(a) Let $a_1 := a - \sqrt{a^2 + 3\sigma}$ and $a_2 := a + \sqrt{a^2 + 3\sigma}$. Show that in equation (8.5.16) matrix $\tilde{M} = SJS^{-1}$, with

$$S := \begin{bmatrix} 0 & 1 & -\frac{a_2}{\sigma} & -\frac{a_1}{\sigma} \\ -1 & 2a & 1 & 1 \\ 1 & 0 & 1 & 1 \\ 0 & \sigma & \frac{a_1}{3} & \frac{a_2}{3} \end{bmatrix} \quad \text{and} \quad J := \begin{bmatrix} a & 0 & 0 & 0 \\ 0 & a & 0 & 0 \\ 0 & 0 & \frac{-a+\sqrt{a^2+3\sigma}}{3} & 0 \\ 0 & 0 & 0 & \frac{-a-\sqrt{a^2+3\sigma}}{3} \end{bmatrix}.$$

(b) Use Algorithm 8.1 to calculate all feedback Nash equilibria of this game if $a \le 0$. Distinguish between the cases (i) $a^2 + 3\sigma < 0$, (ii) $a^2 + 4\sigma \ge 0$ and (iii) $a^2 + 4\sigma < 0$, $a^2 + 3\sigma \ge 0$.

(c) Use Algorithm 8.1 to calculate all feedback Nash equilibria of this game if $a > 0$ and $a^2 = \sigma > 0$.

(d) Use Algorithm 8.1 to calculate all feedback Nash equilibria of this game if $\sigma > 0$.

(e) Verify whether your results from parts (b) and (d) are in line with Theorem 8.10.

10. The following algorithm is proposed by Li and Gajic (1994) to find a stabilizing solution of the set of coupled algebraic Riccati equations (8.3.3) and (8.3.4).

1. Initialization: determine the stabilizing solution K_1^0 of

$$K_1^0 A + A^T K_1^0 + Q_1 - K_1^0 S_1 K_1^0 = 0.$$

Next determine the stabilizing solution K_2^0 of the Riccati equation

$$K_2^0 (A - S_1 K_1^0) + (A - S_1 K_1^0)^T K_2^0 + Q_2 + K_1^0 S_{12} K_1^0 - K_2^0 S_2 K_2^0 = 0.$$

2. Let $i := 0$. Repeat the next iterations until the matrices K_j^i, $j = 1, 2$, below have converged. Here $A_{cl}^i := A - S_1 K_1^i - S_2 K_2^i$, and K_j^{i+1}, $j = 1, 2$, are the solutions of the Lyapunov equations

$$A_{cl}^{i^T} K_1^{i+1} + K_1^{i+1} A_{cl}^i = -(Q_1 + K_1^i S_1 K_1^i + K_2^i S_{21} K_2^i)$$

$$A_{cl}^{i^T} K_2^{i+1} + K_2^{i+1} A_{cl}^i = -(Q_2 + K_2^i S_2 K_2^i + K_1^i S_{12} K_1^i),$$

respectively, for $i = 0, 1, \ldots$.

(a) Implement this algorithm.

(b) Use this algorithm to calculate the solutions for the game considered in Exercise 7.

(c) Use this algorithm to calculate a feedback Nash equilibrium for the game (using the standard notation) with $A = \begin{bmatrix} -2 & 4 \\ 1 & -4 \end{bmatrix}$, $B_1 = \begin{bmatrix} 1 \\ 0 \end{bmatrix}$, $B_2 = \begin{bmatrix} 0 \\ 1 \end{bmatrix}$, $Q_1 = \begin{bmatrix} 2 & 1 \\ 1 & 1 \end{bmatrix}$, $Q_2 = \begin{bmatrix} 1 & 1 \\ 1 & 2 \end{bmatrix}$ and $R_{11} = R_{22} = 1$.

11. Consider the next numerical algorithm to calculate a stabilizing solution of the set of coupled algebraic Riccati equations (8.3.3) and (8.3.4).

1. Initialization: determine (K_1^0, K_2^0) such that $A - S_1 K_1^0 - S_2 K_2^0$ is stable.

2. Let $i := 0$. Repeat the next iterations until the matrices K_j^i, $j = 1, 2$, below have converged. Here K_j^{i+1}, $j = 1, 2$, are the stabilizing solutions of the algebraic Riccati equations

$$0 = -(A - S_2 K_2^i)^T K_1^{i+1} - K_1^{i+1}(A - S_2 K_2^i) + K_1^{i+1} S_1 K_1^{i+1} - Q_1 - K_2^i S_{21} K_2^i,$$

$$0 = -(A - S_1 K_1^{i+1})^T K_2^{i+1} - K_2^{i+1}(A - S_1 K_1^{i+1}) + K_2^{i+1} S_2 K_2^{i+1} - Q_2 - K_1^{i+1} S_{12} K_1^{i+1},$$

respectively.

(a) Implement this algorithm and calculate the solutions for the exercises studied in Exercise 10.

(b) Discuss the (dis)advantages of the algorithm in this exercise and Exercise 10.

12. Consider the differential game (see also Exercise 7)

$$\min_{u_1} \int_0^T \{8x^2(t) + u_1^2(t)\}dt + q_{1T}x^2(T)$$

$$\min_{u_2} \int_0^T \left\{6\frac{3}{4}x^2(t) + u_2^2(t)\right\}dt + q_{2T}x^2(T)$$

subject to $\dot{x}(t) = ax(t) + u_1(t) + u_2(t)$, $x(0) = x_0$.

(a) Show that this game has a feedback Nash equilibrium.

(b) Consider $a = -1$. Assume that $u_i^*(t, T, x_0, q_{1T}, q_{2T})$, $i = 1, 2$, are the equilibrium actions. Determine $\lim_{T\to\infty} u_i^*(t, T, x_0, q_{1T}, q_{2T})$, $i = 1, 2$.

(c) Consider $a = 4$. Determine numerically $\lim_{T\to\infty} u_i^*(t, T, x_0, 5, 0)$ and $\lim_{T\to\infty} u_i^*(t, T, x_0, 0, 5)$, $i = 1, 2$.

(d) Consider $a = 3.5$. Determine numerically the phase diagram of the set of coupled Riccati differential equations corresponding to this game.

13. Consider two fishermen who fish at a lake. Let s be the number of fish in the lake. Assume that the price $p(t)$ the fishermen get for their fish is fixed, i.e. $p(t) = p$. The growth of the fish stock in the lake is described by

$$\dot{s}(t) = \beta s(t) - u_1(t) - u_2(t), \quad s(0) = s_0 > 0.$$

The fishermen consider the next optimization problem

$$\min_{u_i \in \mathcal{F}^{aff}} \int_0^\infty e^{-rt}\{-pu_i + \gamma_i u_i^2\}dt, \quad i = 1, 2,$$

where

$$\mathcal{F}^{aff} := \left\{(u_1, u_2)|u_i(t) = f_{ii}s(t) + g_i, \quad \text{with } \beta - f_{11} - f_{22} < \frac{1}{2}r\right\}.$$

In this formulation all constants, $r, \alpha_i, \beta, \gamma_i$ and v_i, are positive.

(a) Show that with $\hat{x}^T(t) := e^{-\frac{1}{2}rt}[s(t)\ 1]$ and $\hat{u}_i(t) := e^{-\frac{1}{2}rt}(u_i - \frac{p}{2\gamma_i})$, the optimization problem can be rewritten as

$$\min_{\hat{u}_i \in \mathcal{F}} \int_0^\infty \{\hat{x}^T(t)\begin{bmatrix} 0 & 0 \\ 0 & \frac{-p^2}{4\gamma_i} \end{bmatrix}\hat{x}(t) + \gamma_i \hat{u}_i^2(t)\}dt, \quad i = 1, 2,$$

subject to the dynamics

$$\dot{x}(t) = \begin{bmatrix} \beta - \frac{1}{2}r & \frac{-p}{2\gamma_1} + \frac{-p}{2\gamma_2} \\ 0 & -\frac{1}{2}r \end{bmatrix} \hat{x}(t) + \begin{bmatrix} -1 \\ 0 \end{bmatrix} \hat{u}_1(t) + \begin{bmatrix} -1 \\ 0 \end{bmatrix} \hat{u}_2(t), \ \hat{x}(0) = \begin{bmatrix} s_0 \\ 1 \end{bmatrix}.$$

(b) Denote $K_1 =: \begin{bmatrix} x_1 & x_2 \\ x_2 & x_3 \end{bmatrix}$, $K_2 =: \begin{bmatrix} z_1 & z_2 \\ z_2 & z_3 \end{bmatrix}$ and $A := \begin{bmatrix} \beta - \frac{1}{2}r & \frac{-p}{2\gamma_1} + \frac{-p}{2\gamma_2} \\ 0 & -\frac{1}{2}r \end{bmatrix}$ in

equations (8.3.3) and (8.3.4). Show that this game has a feedback Nash equilibrium if and only if the following six equations have a solution x_i, z_i, $i = 1, 2, 3$, such that $\beta - \frac{1}{2}r - \frac{x_1}{\gamma_1} - \frac{z_1}{\gamma_2} < 0$.

$$\left(-2\beta + r + \frac{2}{\gamma_2}z_1 \right) x_1 + \frac{1}{\gamma_1}x_1^2 = 0 \tag{8.8.1}$$

$$\left(\frac{p}{2\gamma_1} + \frac{p}{2\gamma_2} + \frac{1}{\gamma_2}z_2 \right) x_1 + \left(r - \beta + \frac{1}{\gamma_2}z_1 \right) x_2 + \frac{1}{\gamma_1}x_1 x_2 = 0 \tag{8.8.2}$$

$$\frac{p^2}{4\gamma_1} + \left(\frac{p}{\gamma_1} + \frac{p}{\gamma_2} + \frac{2}{\gamma_2}z_2 \right) x_2 + \frac{1}{\gamma_1}x_2^2 + rx_3 = 0 \tag{8.8.3}$$

$$\left(-2\beta + r + \frac{2}{\gamma_1}x_1 \right) z_1 + \frac{1}{\gamma_2}z_1^2 = 0 \tag{8.8.4}$$

$$\left(\frac{p}{2\gamma_1} + \frac{p}{2\gamma_2} + \frac{1}{\gamma_1}x_2 \right) z_1 + \left(r - \beta + \frac{1}{\gamma_1}x_1 \right) z_2 + \frac{1}{\gamma_2}z_1 z_2 = 0 \tag{8.8.5}$$

$$\frac{p^2}{4\gamma_2} + \left(\frac{p}{\gamma_1} + \frac{p}{\gamma_2} + \frac{2}{\gamma_1}x_2 \right) z_2 + \frac{1}{\gamma_2}z_2^2 + rz_3 = 0. \tag{8.8.6}$$

(c) Consider the equations (8.8.1) and (8.8.4). Show that either (i) $(x_1, z_1) = (0, 0)$, (ii) $(x_1, z_1) = (0, \gamma_2(2\beta - r))$, (iii) $(x_1, z_1) = (\gamma_1(2\beta - r), 0)$ or (iv) $(x_1, z_1) = \frac{2\beta - r}{3}(\gamma_1, \gamma_2)$.

(d) Assume $\beta \neq r$. Show that case (i) provides a set of equilibrium actions $\hat{u}_i^*(t) = \frac{p}{2\gamma_i}$ if and only if $2\beta - r < 0$.

(e) Let $c := -\frac{p(\gamma_1 + \gamma_2)(2\beta - r)}{2\beta\gamma_1\gamma_2}$. Show that cases (ii)–(iv) yield the equilibrium actions

$$(u_1^*(t), u_2^*(t)) = \left(\frac{p}{2\gamma_1}, (2\beta - r)s(t) + c + \frac{p}{2\gamma_2} \right),$$

$$(u_1^*(t), u_2^*(t)) = \left((2\beta - r)s(t) + c + \frac{p}{2\gamma_1}, \frac{p}{2\gamma_2} \right),$$

$$(u_1^*(t), u_2^*(t)) = \left(\frac{2\beta - r}{3}s(t) + \frac{c}{3} + \frac{p}{2\gamma_1}, \frac{2\beta - r}{3}s(t) + \frac{c}{3} + \frac{p}{2\gamma_2} \right),$$

respectively, if and only if $2\beta - r > 0$.

(f) Calculate the profits of both fishermen in case (i). The same question in case (ii) if $p = 1$ and $\gamma_1 = \gamma_2$; and if $p = \gamma_1 = s_0 = \beta = 1$, $r = \frac{3}{2}$ and $\gamma_2 = 2$. Conclude.

14. Reconsider the game on fiscal stabilization policies between two countries (see Exercise 7.19). That is,

$$\min_{f_i} \int_0^\infty e^{-\delta t} \{s^2(t) + r_i f_i^2(t)\} dt, \quad i = 1, 2.$$

subject to the differential equation

$$\dot{s}(t) = -as(t) + f_1(t) - f_2(t), \quad s(0) = s_0,$$

where all parameters are positive numbers.

(a) Show that the problem can be reformulated into the standard framework as

$$\min_{u_i} \int_0^\infty \{x^2(t) + r_i u_i^2(t)\} dt, \quad i = 1, 2.$$

subject to the differential equation

$$\dot{x}(t) = -\left(a + \frac{1}{2}\delta\right) x(t) + u_1(t) - u_2(t), \quad x(0) = s_0.$$

(b) Let $\tilde{s} := \frac{1}{\alpha r_1} + \frac{1}{(1-\alpha) r_2}$, $\tilde{a} := -(a + \frac{1}{2}\delta)$ and $k := \frac{\tilde{a} + \sqrt{\tilde{a}^2 + \tilde{s}}}{\tilde{s}}$. Show that the Pareto frontier is parameterized by

$$(J_1^*, J_2^*) = s_0^2 \left(\frac{1 + \frac{k^2}{\alpha^2 r_1}}{2\sqrt{\tilde{a}^2 + \tilde{s}}}, \frac{1 + \frac{k^2}{(1-\alpha)^2 r_2}}{2\sqrt{\tilde{a}^2 + \tilde{s}}} \right), \quad \alpha \in (0, 1).$$

(c) Let $\sigma := \frac{1}{r_1} + \frac{1}{r_2}$ and $p := \frac{\tilde{a} + \sqrt{\tilde{a}^2 + \sigma}}{\sigma}$. Show that the game has a unique open-loop Nash equilibrium and that the associated costs are

$$(J_1^*, J_2^*) = s_0^2 \left(\frac{1 + \frac{p^2}{r_1}}{\sqrt{\tilde{a}^2 + \sigma}}, \frac{1 + \frac{p^2}{r_2}}{\sqrt{\tilde{a}^2 + \sigma}} \right).$$

(d) Show that the game has a unique feedback Nash equilibrium.

(e) Consider the case $r := r_1 = r_2$. Let $s := \frac{1}{r}$. Show that the costs associated with the feedback Nash equilibrium are

$$(J_1^*, J_2^*) = s_0^2 \left(\frac{1}{-\tilde{a} + \sqrt{\tilde{a}^2 + 3s}}, \frac{1}{\sqrt{-\tilde{a} + \tilde{a}^2 + 3s}} \right).$$

(f) Show that, in case $r_1 = r_2$, the open-loop equilibrium costs are always larger than the feedback equilibrium costs.

(g) Choose $\tilde{a} = -1$ and $r = r_1 = r_2$. To assess the effect of r on the distance between the Pareto frontier, the open-loop and the feedback cost, calculate numerically for $r = 1/4, r = 1/2, r = 1, r = 2$ and $r = 4$, respectively, these items. Calculate the maximal distance of the Pareto frontier towards the open-loop and feedback cost, within the negotiation area. Conclude.

(h) Perform a similar exercise as in part (g) but now w.r.t. the parameter $\tilde{a}$.

8.9 Appendix

Proof of Theorem 8.12

Let

$$f(x) := \prod_{i=1}^{2^N} (f_i^N - a).$$

To show that $f(x)$ is a polynomial of degree 2^N we first consider the two-player case. Let

$$a_0 := x - a \quad \text{and} \quad a_i := \sqrt{x^2 - \sigma_i}, \ i = 1, 2.$$

Then $f(x)$ has the following algebraic structure

$$f(a_0, a_1, a_2) := (a_0 - a_1 - a_2)(a_0 + a_1 - a_2)(a_0 - a_1 + a_2)(a_0 + a_1 + a_2). \quad (8.9.1)$$

The structure of f for the general N-player case is similar and is omitted in order to avoid unnecessary cumbersome notation. It is easily verified that both

$$f(-a_0, a_1, \ldots, a_N) = (-1)^{2^N} f(a_0, \ldots, a_N) = f(a_0, \ldots, a_N) \text{ and}$$
$$f(a_0, \ldots, -a_i, \ldots, a_N) = f(a_0, \ldots, a_i, \ldots, a_N), \ i = 1, \ldots, N.$$

From this it follows rather straightforwardly that all entries a_i in f appear quadratically. For, assume that f has a term in which, for example, a_0 has an odd exponent. Next collect all terms of f containing odd exponents in a_0. As a consequence $f = a_0 g(a_0, \ldots, a_N) + h(a_0, \ldots, a_N)$, where a_0 appears with an even exponent in all terms of both g and h. Since $f(-a_0, a_1, \ldots, a_N) = f(a_0, a_1, \ldots, a_N)$ we conclude immediately from this that g must be zero. So, $f(a_0, \ldots, a_N)$ is a sum of terms, in which each term can be written as $\prod_{i=0}^{N} a_i^{2k_i}$ for some nonnegative integers k_i satisfying $\sum_{i=0}^{N} 2k_i = 2^N$.

Since $a_0 = (N-1)x - a$ and $a_i = \sqrt{x^2 - \sigma_i}, \ i = 1, \ldots, N, f(x)$ must necessarily be a polynomial of degree 2^N. □

Proof of Theorem 8.14

1. We will split this part of the proof into two cases: (i) case $e = 1$ and (ii) case $e > 1$. Since for general $e > 1$ the proof is similar to the case $e = 2$, the proof will be given only for this case.

(i) $e = 1$ In this case the proof follows directly from Theorem 8.12.

(ii) $e = 2$ This case will be analyzed in three steps.

(a) First consider the case that the game has two equilibria which yield a different closed-loop system $\bar{x}_i$, $i = 1, 2$. Without loss of generality assume that $f_1(\bar{x}_1) - a = 0$ and $f_2(\bar{x}_2) - a = 0$. This implies that $f(x)$ has two different roots $\bar{x}_i$, $i = 1, 2$. That is, $f(x) = (x - \bar{x}_1)(x - \bar{x}_2)p(x)$, where $p(x)$ is a polynomial of degree 2^{N-1}. So, the claim is now obvious in this case.

(b) Next consider the case that the game has two equilibria which yield the same closed-loop system $\bar{x} \neq \sqrt{|\sigma_1|}$. Then, a Taylor expansion of $f(x)$ at $\bar{x}$ is

$$f(x) = f(\bar{x}) + \sum_{i=1}^{2^N} (x - \bar{x})^i \frac{d^i f}{dx^i}(\bar{x}). \tag{8.9.2}$$

Since $f(x) = \prod_{i=1}^{2^N}(f_i^N - a)$ and $\bar{x} \neq \sqrt{|\sigma_1|}$ it is easily verified that all $f_i^N(x)$ are differentiable of any order at $\bar{x}$. Since for at least two different functions $f_i^N(.), f_i^N(\bar{x}) - a = 0$, it is easily shown that $f(\bar{x}) = \frac{df}{dx}(\bar{x}) = 0$. So, $f(x)$ has a factor $(x - \bar{x})^2$, which shows the correctness of the claim for this case.

(c) Finally, consider the case that $\sigma_1 \geq 0$ and the game has two equilibria which yield the same closed-loop system $\sqrt{\sigma_1}$. Introduce the functions g_i^N which coincide with f_i^N, $i = 1, \dots, 2^N$, except for the term $\sqrt{x^2 - \sigma_1}$ which is dropped, formal: $g_{2i-1}^N := f_{2i-1}^N + \sqrt{x^2 - \sigma_1}$, $i = 1, \dots, 2^{N-1}$ and $g_{2i}^N := f_{2i}^N - \sqrt{x^2 - \sigma_1}$, $i = 1, \dots, 2^{N-1}$.

Then it is easily verified that $g_i^N(x)$ is differentiable; $g_i^N(x) = g_{i+1}^N(x)$, $i = 1, 3, 5, \dots$; and there exist two different numbers $\bar{i}$ and $\bar{j}$ for which

$$g_i^N(\sqrt{\sigma_1}) - a = 0. \tag{8.9.3}$$

Similar to the proof of Theorem 8.12 it follows that $\prod_{i=1}^{2^{N-1}}(g_{2i-1}^N - a)$ is a polynomial of degree 2^{N-1}. According to part (b) of this theorem then $\prod_{i=1}^{2^{N-1}}(g_{2i-1}^N - a) = (x - \sqrt{\sigma_1})^2 h(x)$, where $h(x)$ is a polynomial of degree $2^{N-1} - 2$. Next, rewrite $f(x)$ as follows:

$$f(x) = \prod_{i=1}^{2^N}(f_i^N - a)$$

$$= \prod_{i=1}^{2^{N-1}}\left(g_{2i-1}^N - a - \sqrt{x^2 - \sigma_1}\right) \prod_{i=1}^{2^{N-1}}\left(g_{2i}^N - a + \sqrt{x^2 - \sigma_1}\right)$$

$$= \prod_{i=1}^{2^{N-1}}((g_{2i-1}^N - a)^2 - (x^2 - \sigma_1))$$

$$= (x - \sqrt{\sigma_1})^4 h^2(x) - (x^2 - \sigma_1) \sum_{i=1}^{2^N-1} \prod_{j \neq i}(g_{2j-1}^N - a)^2 + (x^2 - \sigma_1)^2 h_2(x),$$

where $h_2(x)$ is a linear combination of terms containing products of $(g_{2i-1}^N - a)^2$ and $x^2 - \sigma_1$. Or, stated differently,

$$f(x) - (x - \sqrt{\sigma_1})^4 h^2(x) = (x^2 - \sigma_1)\left(-\sum_{i=1}^{2^N-1}\prod_{j\neq i}(g_{2j-1}^N - a)^2 + (x^2 - \sigma_1)h_2(x)\right).$$

Since the left-hand side of this equation is a polynomial, the right-hand side must be a polynomial as well. Furthermore, by equation (8.9.3), for each i, $\prod_{j\neq i}(g_{2j-1}^N(\sqrt{\sigma_1}) - a)^2 = 0$. Therefore, the polynomial on the right-hand side must have an additional factor $x - \sqrt{\sigma_1}$, which concludes this part of the proof.

2. To prove this part of the theorem first recall from Lemma 8.13 that under the stated assumption for each feedback Nash solution with $\bar{x} := -a_{cl}$, there is exactly one function $f_i^N(x) - a$ that has $\bar{x}$ as a zero. Furthermore, recall that all these functions are infinitely many times differentiable. First consider the case that there are k feedback Nash solutions for which $-a_{cl} = \bar{x}$. Without loss of generality assume that $f_i^N(\bar{x}) - a = 0$, $i = 1, \ldots, k$. Then,

$$\frac{df}{dx}(x) = \sum_{i=1}^{2^N} \frac{df_i^N}{dx}(x)\prod_{j\neq i}(f_j^N(x) - a). \tag{8.9.4}$$

So, if $k = 1$, $\frac{df}{dx}(\bar{x}) = \frac{df_1^N}{dx}(\bar{x})\prod_{j\neq 1}(f_j^N(\bar{x}) - a)$. In that case, according to Lemma 8.13, all functions $f_j^N(\bar{x}) - a \neq 0$, $(j \neq 1)$. Therefore, $\frac{df}{dx}(\bar{x}) \neq 0$ if and only if $\frac{df_1^N}{dx}(\bar{x}) \neq 0$. By making a Taylor expansion of $f(x)$ at $\bar{x}$ (see equation (8.9.2)), it follows then that $f(x) = (x - \bar{x})p(x)$, with $p(\bar{x}) \neq 0$ iff $\frac{df_1^N}{dx}(\bar{x}) \neq 0$. If $k > 1$, by (8.9.4), $\frac{df}{dx}(\bar{x}) = 0$. By differentiating (8.9.4) once again we see that if $k = 2$, $\frac{d^2f}{dx^2}(\bar{x}) = 2\frac{df_1^N}{dx}(\bar{x})\frac{df_2^N}{dx}(\bar{x})\prod_{i=3}^{2^N}(f_i^N(\bar{x}) - a)$, which differs from zero iff $\frac{df_i^N}{dx}(\bar{x}) \neq 0$, $i = 1, 2$. In the same way as for the case $k = 1$ it then follows that $f(x) = (x - \bar{x})^2 p(x)$, with $p(\bar{x}) \neq 0$ if and only if $\frac{df_i^N}{dx}(\bar{x}) \neq 0$, $i = 1, 2$. Furthermore, if $k > 2$ it is clear that $\frac{d^2f}{dx^2}(\bar{x}) = 0$.

In a similar way it can be shown that if k is an arbitrary number smaller than 2^N necessarily $\frac{d^i f}{dx^i}(\bar{x}) = 0$, $i = 1, \ldots, k - 1$, and that $\frac{d^k f}{dx^k}(\bar{x}) = 0$ if and only if $\frac{df_i^N}{dx}(\bar{x}) = 0$, $i = 1, \ldots, k$. So, since the number of different feedback Nash solutions yielding the same closed-loop system a_{cl} coincides with the algebraic multiplicity the root $-a_{cl}$ has in $f(x)$, the statement in the theorem follows straightforwardly. $\square$

Proof of Theorem 8.17

On the axis $\kappa_1 = 0$ we find that $\dot{\kappa}_1 = \sigma_1 > 0$ and similarly on the axis $\kappa_2 = 0$ we find that $\dot{\kappa}_2 > 0$. Since $\kappa_i(0) > 0$, $i = 1, 2$, it follows therefore that $\kappa_i(t) \geq 0$ for all $t > 0$. Next assume that, without loss of generality, $\kappa_1(t)$ has a finite escape time at some final point in time $t = t_1$. Since $\kappa_1(t) > 0$, necessarily then $\lim_{t\to t_1}\kappa_1(t) = \infty$. In particular this implies that also $\lim_{t\to t_1}\dot{\kappa}_1(t) = \infty$; but according to equation (8.6.7)

$$\dot{\kappa}_1(t) = -(\kappa_1(t) - a)^2 - 2\kappa_1(t)\kappa_2(t) + \sigma_1 + a^2.$$

Since $\kappa_2(t) \geq 0$ it follows that, if $\kappa_1(t)$ becomes arbitrarily large, the derivative of $\kappa_1(t)$ becomes arbitrarily negative. This contradicts our previous conclusion that $\lim_{t \to t_1} \dot{\kappa}_1(t) = \infty$. So, κ_i, $i = 1, 2$, has no finite escape time. Consequently the set of coupled differential equations has a solution for all $t > 0$.

To see that the solution remains bounded, it is convenient to introduce polar coordinates (r, θ) and to rewrite the system (8.6.5) and (8.6.6) in polar coordinates. If we let

$$r^2(t) := \kappa_1^2(t) + \kappa_2^2(t) \quad \text{and} \quad \tan \theta(t) := \frac{\kappa_2(t)}{\kappa_1(t)}$$

differentiation of both sides of the first equation w.r.t. time t immediately gives

$$\dot{r}(t) = \frac{\kappa_1(t)\dot{\kappa}_1(t) + \kappa_2(t)\dot{\kappa}_2(t)}{r(t)}. \tag{8.9.5}$$

On the other hand, differentiating both sides of the second equation w.r.t. time t gives

$$\frac{d \tan \theta(t)}{dt} = \frac{1}{\cos^2(\theta(t))} \dot{\theta}(t)$$

$$= \frac{\dot{\kappa}_2(t)\kappa_1(t) - \dot{\kappa}_1(t)\kappa_2(t)}{\kappa_1^2(t)}$$

From this we conclude that

$$\dot{\theta}(t) = \frac{\dot{\kappa}_2(t)\kappa_1(t) - \dot{\kappa}_1(t)\kappa_2(t)}{r^2(t)} \frac{\kappa_1^2(t) + \kappa_2^2(t)}{\kappa_1^2(t)} \cos^2(\theta)$$

$$= \frac{\dot{\kappa}_2(t)\kappa_1(t) - \dot{\kappa}_1(t)\kappa_2(t)}{r^2(t)} (1 + \tan^2(\theta)) \cos^2(\theta)$$

$$= \frac{\kappa_1(t)\dot{\kappa}_2(t) - \kappa_2(t)\dot{\kappa}_1(t)}{r^2(t)}.$$

Rewriting the system (8.6.7) and (8.6.8) in polar coordinates gives for $r > 0$

$$\dot{r}(t) = \frac{(2a - \kappa_1(t) - \kappa_2(t))(\kappa_1^2(t) + \kappa_2^2(t)) - \kappa_1\kappa_2(\kappa_1(t) + \kappa_2(t)) + \sigma_1\kappa_1(t) + \sigma_2\kappa_2(t)}{r(t)}$$

$$\tag{8.9.6}$$

$$\dot{\theta}(t) = \frac{-\kappa_2(t)\kappa_1^2(t) + \kappa_1(t)\kappa_2^2(t) + \sigma_2\kappa_1(t) - \sigma_1\kappa_2(t)}{r^2(t)}. \tag{8.9.7}$$

Next, consider the right-hand side of equation (8.9.6) in more detail. Since $\frac{\kappa_i}{r} < 1$, $i = 1, 2$, and $(\kappa_1 + \kappa_2)^2 \geq r^2$, this side can be estimated by

$$\frac{(2a - \kappa_1(t) - \kappa_2(t))(\kappa_1^2(t) + \kappa_2^2(t)) - \kappa_1\kappa_2(\kappa_1(t) + \kappa_2(t)) + \sigma_1\kappa_1(t) + \sigma_2\kappa_2(t)}{r(t)}$$

$$\leq (2a - \kappa_1(t) - \kappa_2(t))r(t) + \sigma_1 + \sigma_2$$

$$\leq 2ar(t) - r^2(t) + \sigma_1 + \sigma_2.$$

So,

$$\dot{r}(t) \leq -(r(t) - a)^2 + \sigma_1 + \sigma_2 + a^2. \tag{8.9.8}$$

From this it is clear that, whenever $r(t) \geq a + \sqrt{\sigma_1 + \sigma_2 + a^2}$, $r(t)$ will decrease. Therefore, $r(t)$ is always bounded by the maximum value of $r(0)$ and $a + \sqrt{\sigma_1 + \sigma_2 + a^2}$, which proves the claim that every solution remains bounded. □

Proof of Theorem 8.22

From Lemma 8.8 it is clear that the equilibria of the planar system (8.6.7) and (8.6.8) are obtained as the solutions of the equations $f_i(\kappa_3) = a$, $i = 1, \ldots, 4$ (where we assumed without loss of generality that $\sigma_1 \geq \sigma_2$). We analyze the eigenvalues at each solution separately. To that end first notice that with $\kappa_3 = -(a - \kappa_1 - \kappa_2)$

$$\lambda_{1,2} < 0 \quad \text{if and only if} \quad \kappa_1 \kappa_2 < \kappa_3^2 \tag{8.9.9}$$

and

$$\lambda_1 < 0, \quad \lambda_2 > 0 \quad \text{if and only if} \quad \kappa_1 \kappa_2 > \kappa_3^2 \tag{8.9.10}$$

at the equilibrium point (κ_1, κ_2).

Now, first consider the case that $a = f_1(\kappa_3)$ has a solution κ_3^*. Then,

$$\kappa_1^* = \kappa_3^* - \sqrt{\kappa_3^{*2} - \sigma_1} \quad \text{and} \quad \kappa_2^* = \kappa_3^* - \sqrt{\kappa_3^{*2} - \sigma_2}.$$

Obviously, unless $\kappa_3^{*2} = \sigma_1 = \sigma_2$, $\kappa_i^* \leq \kappa_3^*$, $i = 1, 2$, where at least one of the inequalities is strict. Therefore also $\kappa_1^* \kappa_2^* < \kappa_3^{*2}$. So, from expression (8.9.9) we conclude that this equilibrium is a stable node.

Next, consider the case that $a = f_2(\kappa_3)$ has a solution κ_3^*. Then,

$$\kappa_1^* = \kappa_3^* + \sqrt{\kappa_3^{*2} - \sigma_1} \quad \text{and} \quad \kappa_2^* = \kappa_3^* - \sqrt{\kappa_3^{*2} - \sigma_2}.$$

So,

$$\kappa_1^* \kappa_2^* = \kappa_3^{*2} + \kappa_3^* \left(\sqrt{\kappa_3^{*2} - \sigma_1} - \sqrt{\kappa_3^{*2} - \sigma_2} \right) - \sqrt{\kappa_3^{*2} - \sigma_1} \sqrt{\kappa_3^{*2} - \sigma_2}$$
$$< \kappa_3^{*2},$$

unless $\kappa_3^{*2} = \sigma_1 = \sigma_2$. Therefore, it follows from expression (8.9.9) again that this equilibrium is a stable node. The case that the equation $a = f_4(\kappa_3)$ has a solution κ_3^* can be analyzed in the same way as the first case we considered except we now obtain the opposite conclusion that $\kappa_1^* \kappa_2^* > \kappa_3^{*2}$ at the equilibrium point. So, according to expression (8.9.10) this equilibrium is a saddle.

Finally, consider the case that the equation $a = f_3(\kappa_3)$ has a solution κ_3^*. Then,

$$\kappa_1^* = \kappa_3^* - \sqrt{\kappa_3^{*^2} - \sigma_1} \quad \text{and} \quad \kappa_2^* = \kappa_3^* + \sqrt{\kappa_3^{*^2} - \sigma_2}.$$

So,

$$\kappa_1^* \kappa_2^* = \kappa_3^{*^2} + \kappa_3^* \left(\sqrt{\kappa_3^{*^2} - \sigma_2} - \sqrt{\kappa_3^{*^2} - \sigma_1} \right) - \sqrt{\kappa_3^{*^2} - \sigma_1} \sqrt{\kappa_3^{*^2} - \sigma_2}. \quad (8.9.11)$$

To analyze this case we consider the derivative of $f_3(\kappa_3)$. Since $f_3(\kappa_3) = \kappa_3 - \sqrt{\kappa_3^2 - \sigma_1} + \sqrt{\kappa_3^2 - \sigma_2}$,

$$f'(\kappa_3) = 1 - \frac{\kappa_3}{\sqrt{\kappa_3^2 - \sigma_1}} + \frac{\kappa_3}{\sqrt{\kappa_3^2 - \sigma_2}}$$

$$= \frac{1}{\sqrt{\kappa_3^2 - \sigma_1} \sqrt{\kappa_3^2 - \sigma_2}} \left\{ \sqrt{\kappa_3^2 - \sigma_1} \sqrt{\kappa_3^2 - \sigma_2} - \left(\sqrt{\kappa_3^2 - \sigma_2} - \sqrt{\kappa_3^2 - \sigma_1} \right) \right\}.$$

From Lemma 8.9 we recall that $f_3(\kappa_3)$ has a unique minimum at some point $\kappa_3 = \bar{\kappa}_3$. Consequently, $f'_3(\kappa_3) < 0$ for all $\kappa_3 < \bar{\kappa}_3$ and $f'_3(\kappa_3) > 0$ for all $\kappa_3 > \bar{\kappa}_3$. Therefore if the equation $a = f_3(\kappa_3)$ has two solutions, necessarily at the smallest solution $f'_3(\kappa_3^*) < 0$. From $f'_3(\kappa_3)$ we conclude that this implies that at this equilibrium point then necessarily

$$\sqrt{\kappa_3^{*^2} - \sigma_1} \sqrt{\kappa_3^{*^2} - \sigma_2} - \left(\sqrt{\kappa_3^{*^2} - \sigma_2} - \sqrt{\kappa_3^{*^2} - \sigma_1} \right) < 0.$$

However, this implies (see expression (8.9.11)) that at this equilibrium point

$$\kappa_1^* \kappa_2^* > \kappa_3^{*^2}$$

So, this equilibrium point is a saddle.

In a similar way it is shown that if the equation $a = f_3(\kappa_3)$ has a solution to the right of $\bar{\kappa}_3$, this equilibrium point is a stable node.

This leaves the next two cases for inspection:

(i) $\kappa_3^{*^2} = \sigma_1 = \sigma_2$, and

(ii) $a = \min f_3(\kappa)$.

Both cases can be analyzed in a similar way to Example 8.16. It turns out that the first equilibrium (i) is a stable node, whereas the second equilibrium (ii) is a saddle-node. From this the claims made in the theorem are then straightforwardly obtained. □

9

Uncertain non-cooperative feedback information games

Dynamic game theory brings together three features that are key to many situations in economics, ecology and elsewhere: optimizing behavior, the presence of multiple agents and enduring consequences of decisions. In this chapter we add a fourth aspect, namely robustness with respect to variability in the environment. In our formulation of dynamic games, so far, we specified a set of differential equations including input functions that are controlled by the players, and players are assumed to optimize a criterion over time. The dynamic model is supposed to be an exact representation of the environment in which the players act; optimization takes place with no regard for possible deviations. It can safely be assumed, however, that in reality agents follow a different strategy. If an accurate model can be formed at all, it will in general be complicated and difficult to handle. Moreover, it may be unwise to optimize on the basis of a model which is too detailed, in view of possible changes in dynamics that may take place in the course of time and that may be hard to predict. It makes more sense for agents to work on the basis of a relatively simple model and to look for strategies that are robust with respect to deviations between the model and reality. In an economics context, the importance of incorporating aversion to specification uncertainty has been stressed for instance by Hansen, Sargent and Tallarini (1999).

In control theory, an extensive theory of robust design is already in place (see Başar (2003) for a recent survey). We use this background to arrive at suitable ways of describing aversion to model risk in a dynamic game context. We assume linear dynamics and quadratic cost functions. These assumptions are reasonable for situations of dynamic quasi-equilibrium, where no large excursions of the state vector are to be expected (see Section 5.1).

Following a pattern that has become standard in control theory two approaches will be considered. The first one is based on a stochastic approach. This approach assumes that the dynamics of the system are corrupted by a standard Wiener process (white noise). Basic assumptions will be that the players have access to the current value of the state of

LQ Dynamic Optimization and Differential Games   J. Engwerda
© 2005 John Wiley & Sons, Ltd

the system and that the positive definite covariance matrix does not depend on the state of the system. Basically it turns out that under these assumptions the feedback Nash equilibria also constitute an equilibrium in such an uncertain environment. In the second approach, a malevolent disturbance input is introduced which is used in the modeling of aversion to specification uncertainty. That is, it is assumed that the dynamics of the system is corrupted by a deterministic noise component, and that each player has his own expectation about this noise. This is modeled by adapting, for each player, their cost function accordingly. The players cope with this uncertainty by considering a worst-case scenario. Consequently in this approach the equilibria of the game, in general, depend on the worst-case scenario expectations about the noise of the players.

This chapter, basically, restricts the analysis to the infinite-planning horizon case. Furthermore only the feedback information structure is considered. For some results dealing with an open-loop information structure see for example, Section 7.4 of this book, Başar and Bernhard (1995) and Kun (2001).

9.1　Stochastic approach

In this Section we assume that the state of the system is generated by a linear noisy system,

$$\dot{x}(t) = Ax(t) + B_1 u_1(t) + \cdots + B_N u_N(t) + w(t). \tag{9.1.1}$$

The noise w is white, Gaussian, of zero mean and has covariance $V(t)\delta(t - \tau)$, where $V(.) > 0$ is continuous. The initial state at time $t = 0$, x_0, is a Gaussian random variable of mean m_0 and covariance P_0. This random variable is independent of w. The strategy spaces considered by the players are assumed to be the set of linear feedback actions Γ_i^{lfb}, as defined in Section 8.2. For the finite-planning horizon we consider the cost functions

$$J_i := \mathrm{E}\left\{ \int_0^T x^T(t)\left(Q_i + \sum_{j=1}^N F_j(t)^T(t) R_{ij} F_j(t) \right) x(t) dt + x^T(T) Q_{iT} x(T) \right\}, \quad i = 1, \ldots, N, \tag{9.1.2}$$

where the expectation operation is taken with respect to the statistics of $w(t)$, $t \in [0, T]$.

First, consider the one-player case. Completely analogous to the proof Davis (1977) gives of his Theorem 5.2.5, we have the following result.

Theorem 9.1

Assume that the Riccati differential equation

$$\dot{K}(t) = -A^T K(t) - K(t)A + K(t)SK(t) - Q, \quad K(T) = Q_T,$$

has a symmetric solution on $[0, T]$. Then the control

$$u^*(t) = -R^{-1} B^T K(t) x(t)$$

is optimal for the stochastic one-player control problem

$$\min_{u(t)=F(t)x(t)} \mathrm{E}\left\{ \int_0^T x^T(t)(Q + F^T(t)RF(t))x(t)dt + x^T(T)Q_Tx(T) \right\}, \qquad (9.1.3)$$

subject to the system

$$\dot{x}(t) = Ax(t) + Bu(t) + w(t).$$

Moreover, the cost is

$$\int_0^T \mathrm{tr}\{V^T(s)K(s)V(s)\}ds + m_0^T K(0)m_0 + \mathrm{tr}\{P_0K(0)\}. \qquad \square$$

Using this result one can straightforwardly derive the next result for the multi-player case.

Theorem 9.2

The set of linear feedback Nash equilibrium actions (see Theorem 8.3) also provides a Nash equilibrium for the stochastic differential game defined by (9.1.1) and (9.1.2). Moreover, the costs involved with these equilibrium actions are

$$J_i^* := \int_0^T \mathrm{tr}\{V^T(s)K_i(s)V(s)\}ds + m_0^T K_i(0)m_0 + \mathrm{tr}\{P_0K_i(0)\}.$$

Proof ($N = 2$)

Assume that the set of 'feedback' Riccati differential equations

$$\dot{K}_1(t) = -(A - S_2K_2(t))^T K_1(t) - K_1(t)(A - S_2K_2(t)) + K_1(t)S_1K_1(t) - Q_1 - K_2(t)S_{21}K_2(t),$$
$$K_1(T) = Q_{1T} \qquad (9.1.4)$$
$$\dot{K}_2(t) = -(A - S_1K_1(t))^T K_2(t) - K_2(t)(A - S_1K_1(t)) + K_2(t)S_2K_2(t) - Q_2 - K_1(t)S_{12}K_1(t),$$
$$K_2(T) = Q_{2T}. \qquad (9.1.5)$$

has a solution (K_1, K_2). Then, by Theorem 9.1 and equation (9.1.4), with

$$u_2^*(t) := -R_2^{-1}K_2(t)x(t),$$

the stochastic optimization problem

$$\min_{u = F_1(t)x(t)} J_1(u_1, u_2^*)$$

subject to the system

$$\dot{x}(t) = (A - S_2K_2(t))x(t) + B_1u_1(t) + w(t)$$

has the solution

$$u_1^*(t) := -R_1^{-1}K_1(t)x(t).$$

That is,

$$J_1(u_1^*, u_2^*) \le J_1(u_1, u_2^*).$$

Similarly it follows that

$$J_2(u_1^*, u_2^*) \le J_2(u_1^*, u_2(t)).$$

Which proves the claim.   □

Next consider the infinite-planning horizon case. In that case, at least when $Q > 0$, the costs of the stochastic regulator problem (9.1.3) become unbounded if $T \to \infty$ (for example, Davis (1977)). This is also clear intuitively, as the system is constantly perturbed by the noise w. For that reason we have to consider different cost functions for the players. Instead of minimizing the total cost, we will consider the minimization of the average cost per unit time:

$$L_i(V, u_1, \ldots, u_N) = \lim_{T \to \infty} \frac{1}{T} \mathrm{E} \left\{ \int_0^T \left(x^T Q_i x + \sum_{j=1}^N u_j^T R_{ij} u_j \right) dt \right\} \qquad (9.1.6)$$

Again, assume that the players have full access to the current state of the system and that the control actions that are used by the players are the constant linear feedback strategies. That is,

$$u_i(t) = F_i x(t), \quad \text{with } F_i \in \mathbb{R}^{m_i \times n}, \quad i = 1, \ldots, N,$$

and where $(F_1, \ldots, F_N)$ belong to the set

$$\mathcal{F} = \left\{ F = (F_1, \ldots, F_N) \mid A + \sum_{i=1}^N B_i F_i \text{ is stable} \right\}.$$

Moreover, assume that the covariance matrix $V(t)$ does not depend on the time, t, any more.

Given this context, we consider the next equilibrium concept.

Definition 9.1

An N-tuple $\hat{F} = (\hat{F}_1, \ldots, \hat{F}_N) \in \mathcal{F}_N$ is called a **stochastic variance-independent feedback Nash equilibrium** if, for all i, the following inequality

$$L_i(V, \hat{F}) \le L_i(V, \hat{F}_{-i}(\alpha))$$

holds for each $V \in \mathcal{V}$ and for each state feedback matrix α such that $\hat{F}_{-i}(\alpha) \in \mathcal{F}$. Here $\mathcal{V}$ is the set of all real positive semi-definite $n \times n$ matrices.   □

Here $F_{-i}(\alpha) := (F_1, \ldots, F_{i-1}, \alpha, F_{i+1}, \ldots, F_N)$. As a preliminary to the general N-player case the one-player case is first treated again.

To analyse this case we introduce the function

$$\varphi : \mathcal{F} \to \mathbb{R}^{n \times n} \text{ defined by } \varphi(F) = P, \tag{9.1.7}$$

where P is the unique solution of the Lyapunov equation

$$(A + BF)^T P + P(A + BF) = -(Q + F^T RF). \tag{9.1.8}$$

In the derivation of the one-player result, the following result will be used.

Lemma 9.3

For each $V \in \mathcal{V}$ and $F \in \mathcal{F}$

$$L(V, F) = \text{tr}(V\varphi(F)), \tag{9.1.9}$$

where φ is defined by expression (9.1.7).

Proof

Let $S \in \mathcal{S}$ and $F \in \mathcal{F}$. Analogously to, for example, Anderson and Moore (1989) one can show that $L(V, F)$ can be written as $L(V, F) = \text{tr}(W(Q + F^T RF))$, where W is the unique solution of the Lyapunov equation

$$(A + BF)W + W(A + BF)^T = -V.$$

Denote $P = \varphi(F)$. Multiplying the Lyapunov equation (9.1.8) by W produces

$$W(Q + F^T RF) = -W(A + BF)^T P - WP(A + BF).$$

Hence, it is easily seen that $L(V, F) = \text{tr}(VP)$. $\qquad\square$

Theorem 9.4

The one-player stochastic optimization problem

$$\min_{F \in \mathcal{F}} L(V, F) := \min_{F \in \mathcal{F}} \lim_{T \to \infty} \frac{1}{T} \text{E} \left\{ \int_0^T x^T(t)(Q + F^T RF)x(t)dt \right\},$$

subject to the system

$$\dot{x}(t) = (A + BF)x(t) + w(t),$$

has a solution independent of the covariance matrix V of w if and only if the algebraic Riccati equation

$$Q + A^T X + XA - XSX = 0 \tag{9.1.10}$$

has a stabilizing solution. If this condition holds, the solution is uniquely given by $\hat{F} :=$ $-R^{-1}B^T X_s$ with X_s the stabilizing solution of the ARE. Moreover, $L(V, \hat{F}) = \text{tr}(VX_s)$.

Proof

'$\Rightarrow$ **part**' Choose $\hat{F}$ as the feedback matrix that solves the stochastic optimization problem. Let $x_0 \in \mathbb{R}^n$ and $F \in \mathcal{F}$ be arbitrary. Define the matrix $V := x_0 x_0^T$. Clearly, $V \in \mathcal{V}$, which implies that $L(V, \hat{F}) \le L(V, F)$. Now, let

$$J(x_0, F) := \int_0^\infty x^T(t)(Q + F^T RF)x(t)dt = x_0^T \varphi(F)x_0. \qquad (9.1.11)$$

Then, using equation (9.1.9),

$$J(x_0, \hat{F}) = x_0^T \varphi(\hat{F})x_0 = \text{tr}(V\varphi(\hat{F})) = L(V, \hat{F}) \le L(V, F) = J(x_0, F).$$

So, $\hat{F}$ solves the regular linear optimal control problem (5.4.1) and (5.4.2). According to Theorem 5.14 this implies that (9.1.10) has a stabilizing solution X_s. Moreover, $\hat{F} = -R^{-1}B^T X_s$.

'$\Leftarrow$ **part**' Choose $\hat{F} = -R^{-1}B^T X_s$, where X_s is the stabilizing solution of equation (9.1.10). Let $V \in \mathcal{V}$ and $F \in \mathcal{F}$ be arbitrary. Since V is positive semi-definite there exists a matrix Y such that $V = YY^T$. Denote the ith column of Y by y_i. From equation (9.1.9) it follows that

$$L(V, F) = \text{tr}(V\varphi(F)) = \text{tr}(Y^T \varphi(F)Y) = \sum_{i=1}^n y_i^T \varphi(F)y_i = \sum_{i=1}^n J(y_i, F),$$

with $J(y_i, F)$ as defined in equation (9.1.11). For each $i = 1, \ldots, n$, $J(y_i, \hat{F}) \le J(y_i, F)$. Hence

$$L(V, \hat{F}) = \sum_{i=1}^n J(y_i, \hat{F}) \le \sum_{i=1}^n J(y_i, F) = L(V, F). \qquad \square$$

As an immediate consequence of this theorem we then obtain for the general N-player case the following result.

Theorem 9.5

Let $(X_1, \ldots, X_N)$ be a stabilizing solution of the algebraic Riccati equations associated with the feedback Nash equilibria (see, for example, equations (8.3.3) and (8.3.4))

$$0 = -\left(A - \sum_{j \ne i}^N S_j X_j\right)^T X_i - X_i\left(A - \sum_{j \ne i}^N S_j X_j\right) + X_i S_i X_i - Q_i - \sum_{j \ne i}^N X_j S_{ij} X_j, \qquad (9.1.12)$$

and define $\hat{F}_i := -R_{ii}^{-1}B_i^T X_i$ for $i = 1, \ldots, N$. Then $\hat{F} := (\hat{F}_1, \ldots, \hat{F}_N)$ is a stochastic variance-independent feedback Nash equilibrium. Conversely, if $(\hat{F}_1, \ldots, \hat{F}_N)$ is a stochastic variance-independent feedback Nash equilibrium, there exists a stabilizing solution $(X_1, \ldots, X_N)$ of equation (9.1.12) such that $\hat{F}_i = -R_{ii}^{-1} B_i^T X_i$. Moreover, $L_i(V, \hat{F}) = \mathrm{tr}(V X_i)$. $\qquad\square$

Theorem 9.5 shows that the linear feedback Nash equilibrium actions from Chapter 8 are also equilibrium actions for the stochastic games studied in this Section. Obviously, the corresponding costs differ. In particular, notice that in the finite-planning horizon case the additional cost which players expect to incur using this stochastic framework compared to the noise-free case is

$$\int_0^T \mathrm{tr}\{V^T(s)K_i(s)V(s)\}ds + \mathrm{tr}\{P_0 K_i(0)\}.$$

The infinite-planning horizon costs are more difficult to compare, since in the stochastic framework the average cost is used as a performance criterion rather than the total cost.

9.2 Deterministic approach: introduction

Our second approach to deal with uncertainty in the game assumes that the system is corrupted by some deterministic input. The dynamic model now reads:

$$\dot{x}(t) = Ax(t) + \sum_{i=1}^N B_i u_i(t) + Ew(t), \quad x(0) = x_0. \tag{9.2.1}$$

Here $w \in L_2^q(0, \infty)$ is a q-dimensional disturbance vector affecting the system and E is a constant matrix.

As in the stochastic case, we assume in this deterministic approach that all players know the current state of the system at any point in time. Moreover, it is assumed that the players use stabilizing constant linear feedback strategies. That is, only controls $u_i(.)$ of the type $u_i(t) = F_i x(t)$ are considered, with $F_i \in \mathbb{R}^{m_i \times n}$, and where $(F_1, \ldots, F_N)$ belongs to the set

$$\mathcal{F} := \left\{ F = (F_1, \ldots, F_N) \mid A + \sum_{i=1}^N B_i F_i \text{ is stable} \right\}.$$

So the information structure and strategy spaces are similar to those considered in Section 9.1 and Chapter 8.

The above stabilization constraint is imposed to ensure the finiteness of the infinite-horizon cost integrals that will be considered; also, this assumption helps to justify our basic supposition that the state vector remains close to the origin. As already mentioned in Chapter 8, the constraint is a bit unwieldy since it introduces dependence between the strategy spaces of the players. However, we will now focus on equilibria in which the inequalities that ensure the stability property are inactive constraints. It will be a standing assumption that the set $\mathcal{F}$ is non-empty; a necessary and sufficient condition for this to hold is that the matrix pair $(A, [B_1 \cdots B_N])$ is stabilizable. Given that we work with an infinite horizon, restraining the players to constant feedback strategies seems reasonable; to prescribe linearity may also seem natural in the linear quadratic context that we assume, although there is no way to exclude a priori equilibria in nonlinear feedback strategies. Questions regarding the existence of such equilibria are beyond the scope of this book.

We now come to the formulation of the objective functions of the players. The starting point is the usual quadratic criterion which assigns to player i the cost function

$$J_i := \int_0^\infty \left\{ x(t)^T Q_i x(t) + \sum_{j=1}^N u_j(t)^T R_{ij} u_j(t) \right\} dt. \tag{9.2.2}$$

Here, Q_i is symmetric and R_{ii} is positive definite for all $i = 1, \ldots, N$. Due to the assumption that the players use constant linear feedbacks, the criterion in equation (9.2.2) may be rewritten as

$$J_i := \int_0^\infty \left\{ x^T \left(Q_i + \sum_{j=1}^N F_j^T R_{ij} F_j \right) x \right\} dt \tag{9.2.3}$$

where F_i is the feedback chosen by player i. Written in the above form, the criterion may be looked at as a function of the initial condition x_0 and the state feedbacks F_i.

The description of the players' objectives given above needs to be modified in order to express a desire for robustness. To that end, we modify the criterion (9.2.3) to

$$\bar{J}_i^{SC}(F_1, \ldots, F_N, x_0) := \sup_{w \in L_2^q(0,\infty)} J_i(F_1, \ldots, F_N, w, x_0) \tag{9.2.4}$$

where

$$J_i(F_1, \ldots, F_N, w, x_0) := \int_0^\infty \left\{ x^T \left(Q_i + \sum_{j=1}^N F_j^T R_{ij} F_j \right) x - w^T V_i w \right\} dt. \tag{9.2.5}$$

The weighting matrix V_i is symmetric and positive definite for all $i = 1, \ldots, N$. Because it occurs with a minus sign in equation (9.2.5), this matrix constrains the disturbance vector w in an indirect way so that it can be used to describe the aversion to model risk of player i. Specifically, if the quantity $w^T V_i w$ is large for a vector $w \in R^q$, this means that player i does not expect large deviations of the nominal dynamics in the direction of Ew. Furthermore, the larger player i chooses V_i, the closer the worst case signal player i can be confronted with in this model will approach the zero input signal (that is: $w(.) = 0$).

In line with the nomenclature used in control theory literature we will call this the 'soft-constrained' formulation. Note that, since we do not assume positive definiteness of the state weighting matrix, this development' even in the one-player case, extends the standard results that may be found for instance in Francis (1987), Başar and Bernhard (1995), Lancaster and Rodman (1995), Zhou, Doyle and Glover (1996), Başar and Olsder (1999) and Başar (2003).

The equilibrium concept that will be used throughout this chapter is based on the adjusted cost functions (9.2.4). A formal definition is given below.

Definition 9.2

An N-tuple $\overline{F} = (\overline{F}_1, \ldots, \overline{F}_N) \in \mathcal{F}$ is called a **soft-constrained Nash equilibrium** if for each $i = 1, \ldots, N$ the following inequality

$$\overline{J}_i^{SC}(\overline{F}, x_0) \leq \overline{J}_i^{SC}(\overline{F}_{-i}(F), x_0)$$

holds for all $x_0 \in \mathbb{R}^n$ and for all $F \in \mathbb{R}^{m_i \times n}$ that satisfy $\overline{F}_{-i}(F) \in \mathcal{F}$.   □

The remainder of this chapter is organized as follows. In Section 9.3 the one-player case is discussed. The results obtained for that particular case are the basis for the derivation of results for the general N-player case in Section 9.5. To obtain a better understanding of the results in the one-player case, we elaborate the scalar case in Section 9.4 in detail. Section 9.6 considers a worked example. Section 9.8 deals with a stochastic interpretation of the soft-constrained feedback Nash equilibrium concept based on the well-known connection between the H_∞ control problem and the risk sensitive linear exponential quadratic gaussian (LEQG) control problem (for example, Başar (2003), Başar and Bernhard (1995), Glover and Doyle (1988), Jacobson (1973) and Pratt (1964)). For this purpose we use some results from Runolfsson (1994). A numerical computation scheme for the N-player scalar case, similar to the algorithm that was developed in Section 8.5 to calculate the feedback Nash equilibria, is provided in Section 9.7.

9.3   The one-player case

This Section studies the one-player case. That is, we consider

$$\dot{x} = (A + BF)x + Ew, \quad x(0) = x_0, \tag{9.3.1}$$

with (A, B) stabilizable, $F \in \mathcal{F}$ and

$$J(F, w, x_0) = \int_0^\infty \{x^T(Q + F^T RF)x - w^T Vw\}dt. \tag{9.3.2}$$

The matrices Q, R and V are symmetric, $R > 0$ and $V > 0$. The problem is to determine for each $x_0 \in \mathbb{R}^n$ the value

$$\inf_{F \in \mathcal{F}} \sup_{w \in L_2^q(0, \infty)} J(F, w, x_0). \tag{9.3.3}$$

Furthermore, if this infimum is finite, we like to know whether there is a feedback matrix $\overline{F} \in \mathcal{F}$ that achieves the infimum, and to determine all matrices that have this property. This soft-constrained differential game can also be interpreted as a model for a situation where the controller designer is minimizing the criterion (9.3.2) by choosing an appropriate $F \in \mathcal{F}$, while the uncertainty is maximizing the same criterion by choosing an appropriate $w \in L_2^q(0, \infty)$.

A necessary condition for the expression (9.3.3) to be finite is that the supremum

$$\sup_{w \in L_2^q(0, \infty)} J(F, w, x_0)$$

is finite for at least one $F \in \mathcal{F}$. However, this condition is not sufficient. It may happen that the infimum in (9.3.3) becomes arbitrarily small. Item 3 in the Note following Lemma 9.6, below, provides an example illustrating this point.

Lemma 9.6 gives necessary and sufficient conditions for the supremum in (9.3.3) to attain a finite value for a given stabilizing feedback matrix F. This result will be used later on in Theorem 9.8, which provides a sufficient condition under which the soft-constrained differential game associated with (9.3.1) and (9.3.2) has a saddle point. For notational convenience let

$$\|x\|_M := x^T M x,$$

where M is a square matrix.

Lemma 9.6

Let A be stable. Consider the system

$$\dot{x} = Ax + Ew \tag{9.3.4}$$

and the cost function

$$\phi(w, x_0) := \int_0^\infty \{x^T Q x - w^T V w\} dt, \quad x(0) = x_0,$$

with $Q = Q^T$ and $V > 0$. Let $M := EV^{-1}E^T$. The following conditions are equivalent.

(i) For each $x_0 \in \mathbb{R}^n$ there exists a $\bar{w} \in L_2^q(0, \infty)$ such that $\phi(w, x_0) \leq \phi(\bar{w}, x_0)$.

(ii) The Hamiltonian matrix

$$H := \begin{bmatrix} A & M \\ -Q & -A^T \end{bmatrix}$$

has no eigenvalues on the imaginary axis.

(iii) The algebraic Riccati equation

$$Q + A^T X + XA + XMX = 0 \tag{9.3.5}$$

has a stabilizing solution (i.e. $\sigma(A + MX) \subset \mathbb{C}^-$).

If these conditions hold, the maximum of $\phi(w, x_0)$ is uniquely attained by

$$\bar{w}(t) := V^{-1} E^T X e^{(A+MX)t} x_0$$

where X is the stabilizing solution of (9.3.5). Furthermore $\phi(\bar{w}, x_0) = x_0^T X x_0$.

Proof

We will show the following implications: (i) $\Rightarrow$ (ii) $\Rightarrow$ (iii) $\Rightarrow$ (i). The second part of the lemma follows from the proof that (iii) implies (i).

(i) $\Rightarrow$ **(ii)** : Denote the state trajectory corresponding to $\bar{w}$ by $\bar{x}$. Then according to the maximum principle (see Theorem 4.6) there exists a costate variable p such that

$$\dot{\bar{x}} = A\bar{x} + E\bar{w}, \quad \bar{x}(0) = x_0$$
$$\dot{p} = -Q\bar{x} - A^T p$$
$$\bar{w}(t) = \arg \max_{w \in \mathbb{R}^q} (\bar{x}^T Q\bar{x} - w^T V w + 2p^T (A\bar{x} + Ew)).$$

A completion of squares shows that

$$-w^T V w + 2p^T E w = -(w - V^{-1} E^T p)^T V (w - V^{-1} E^T p) - p^T M p.$$

Since $V > 0$, it follows that $\bar{w} = V^{-1} E^T p$. Hence

$$\begin{bmatrix} \dot{\bar{x}} \\ \dot{p} \end{bmatrix} = \begin{bmatrix} A & M \\ -Q & -A^T \end{bmatrix} \begin{bmatrix} \bar{x} \\ p \end{bmatrix} = H \begin{bmatrix} \bar{x} \\ p \end{bmatrix}, \quad \bar{x}(0) = x_0.$$

Since $\bar{w} \in L_2^q(0, \infty)$ and A is stable, $\bar{x}(t) \to 0$ for $t \to \infty$ for all $x_0 \in \mathbb{R}^n$. This shows that the eigenspace corresponding to the eigenvalues in the open left-half plane has at least dimension n. Since H is a Hamiltonian matrix this implies that H has no eigenvalues on the imaginary axis.

(ii) $\Rightarrow$ **(iii)** : This implication follows from Theorem 2.37.

(iii) $\Rightarrow$ **(i)** : Let $w \in L_2^q(0, \infty)$ and x be generated by equation (9.3.4). Since A is stable, $x(t) \to 0$ for $t \to \infty$. A completion of the squares then gives that

$$\phi(w, x_0) = \int_0^\infty \left\{ x^T Q x - w^T V w + \frac{d}{dt} x^T X x - \frac{d}{dt} x^T X x \right\} dt$$
$$= x_0^T X x_0 - \int_0^\infty \| w - V^{-1} E^T X x \|_V^2 dt.$$

Hence $\phi(w, x_0) \le x_0^T X x_0$ and equality holds if and only if $w = V^{-1} E^T X x$. Substituting this in equation (9.3.4) shows that $\phi(\cdot, x_0)$ is uniquely maximized by $\bar{w}$. $\qquad\square$

Notes

1. Notice that Lemma 9.6 does not imply that if the Hamiltonian matrix H has eigenvalues on the imaginary axis, the cost will be unbounded. Consider, for example, $a = -1$, $q = r = e = v = 1$. Then $X = 1$ is the unique (though not stabilizing) solution of equation (9.3.5). A completion of squares (see proof above) shows that $\phi(w, x_0) \le x_0^2$. Furthermore, it is easily verified that with $w = (1 - \epsilon)x$, for an arbitrarily small positive ϵ, we can approach arbitrarily close to this cost.

2. If there exists a $\bar{F} \in \mathcal{F}$ such that $\sup_{w \in L_2^q(0, \infty)} J(F, w, x_0)$ is finite, this does not imply that there is an open neighborhood of $\bar{F} \in \mathcal{F}$ for which the supremum is also finite. This fact can be demonstrated using the result part (i). Take, for example, $a = -\frac{1}{2}$, $b = \frac{1}{2}$, $\bar{f} = -1$, $q = r = e = v = 1$. Then according to part (i) the cost will be bounded.

However, for every $\epsilon > 0$, $\sup\limits_{w \in L_2^q(0,\infty)} J((\bar{f} + \epsilon), w, x_0)$ is infinite. This can be seen by choosing, for example, for a fixed ϵ, $w(t) = \left(1 - \frac{1}{2}\epsilon - \mu\right)x(t)$, where $\mu > 0$ is an arbitrarily small number.

3. Since we did not assume that the state weighting Q in equation (9.3.2) is nonnegative definite, it may well happen that the value of the expression in (9.3.3) is $-\infty$. As an example, consider the scalar case with $E = 0$, $A = -1$, $B = R = V = 1$ and $Q = -2$. Then $w = 0$ gives the supremum for every $f < 1$ in expression (9.3.3). Consequently

for all $f < 1$, $\sup\limits_{w \in L_2^q(0,\infty)} J(f, w, x_0) = \dfrac{-2 + f^2}{2(1 - f)}x_0^2$. From this it is obvious that by choosing f close to 1, the costs become arbitrarily small. $\qquad\square$

Following on from the result of Lemma 9.6 part (ii) we define for each $F \in \mathcal{F}$ the Hamiltonian matrix

$$H_F := \begin{bmatrix} A + BF & M \\ -Q - F^T RF & -(A + BF)^T \end{bmatrix} \qquad (9.3.6)$$

and introduce the set

$$\bar{\mathcal{F}} := \{F \in \mathcal{F} \mid H_F \text{ has no eigenvalues on the imaginary axis}\}.$$

Lemma 9.7 provides a convenient expression for the objective function of the game.

Lemma 9.7

Consider equations (9.3.1) and (9.3.2) with $F \in \mathcal{F}$ and $w \in L_2^q(0,\infty)$. Let X be an arbitrary symmetric matrix; then

$$J(F, w, x_0) = x_0^T X x_0 + \int_0^\infty \{x^T(t)(Q + A^T X + XA - XSX + XMX)x(t)$$
$$+ \|(F + R^{-1}B^T X)x(t)\|_R^2 - \|w(t) - V^{-1}E^T Xx(t)\|_V^2\}dt \qquad (9.3.7)$$

where $S := BR^{-1}B^T$ and $M := EV^{-1}E^T$.

Proof

Since $F \in \mathcal{F}$ and $w \in L_2^q(0,\infty)$, $x(t) \to 0$ for $t \to \infty$. Thus

$$J(F, w, x_0) = \int_0^\infty \left\{x^T(Q + F^T RF)x - w^T Vw + \frac{d}{dt}x^T Xx - \frac{d}{dt}x^T Xx\right\}dt$$
$$= x_0^T X x_0 + \int_0^\infty \{x^T(Q + A^T X + XA)x + x^T F^T RFx$$
$$+ 2x^T F^T B^T Xx - w^T Vw + 2w^T E^T Xx\}dt.$$

Hence, the two completions of the squares

$$x^T F^T RFx + 2x^T F^T B^T Xx = \|(F + R^{-1}B^T X)x\|_R^2 - x^T XSXx$$

and

$$-w^T V w + 2 w^T E^T X x = -\|w - V^{-1} E^T X x\|_V^2 + x^T X M X x$$

show that equation (9.3.7) holds.   □

In particular, Lemma 9.7 shows that if X satisfies the algebraic Riccati equation (9.3.9) below, an optimal choice for the minimizing player is $-R^{-1} B^T X$, which is an admissible choice if X is the stabilizing solution of this equation. If the maximizing player were to be restricted to choose linear state feedback matrices as well, his optimal choice would be the state feedback matrix $V^{-1} E^T X$. The following theorem shows that under the open-loop information structure, the optimal choice for the maximizing player, given that the minimizing player chooses $-R^{-1} B^T X$, can indeed be obtained from the feedback law $x \to V^{-1} E^T X x$. This theorem provides a set of sufficient conditions for a saddlepoint solution to exist. Consequently, it also generates a solution for problem (9.3.3).

To justify the conditions in the theorem, consider the scalar case for the moment, without going into too much detail. We replace the upper case symbols for matrices with their lower case equivalents to emphasize that these matrices are now just real numbers. Under the assumption that the conditions of Lemma 9.6 are satisfied, the equation (cf. equation (9.3.5))

$$m x^2 + 2(a + bf)x + q + f^2 r = 0 \tag{9.3.8}$$

holds for each f, and $\sup\limits_{w \in L_2^q(0,\infty)} J(f, w, x_0) = x(f) x_0^2$. In particular, the minimizing $\bar{f}$ satisfies $x'(\bar{f}) = 0$. Differentiation of equation (9.3.8) with respect to f shows that $\bar{f} = -bx/r$. Substitution of this relationship into equation (9.3.8) gives that x should be a stabilizing solution of $(m - s)x^2 + 2ax + q = 0$ (see equation (9.3.9)). On the other hand, a direct inspection of $x(f)$ shows that if the boundary of the set of potential feasible feedbacks (i.e. the f for which $a + bf = 0$) does not belong to our set $\bar{\mathcal{F}}$, $x(f)$ always attains a minimum. So the condition $-a/b \notin \bar{\mathcal{F}}$ is enough to conclude that $\bar{f}$ yields a minimum. By a direct evaluation of equation (9.3.6) it is easily verified that this condition is equivalent to $a^2 + qs > 0$. This requirement is the scalar version of condition (9.3.10) below. The proof of this theorem is given in the Appendix at the end of this chapter.

Theorem 9.8

Consider equations (9.3.1) and (9.3.2) and let the matrices S and M be defined as in Lemma 9.7. Assume that the algebraic Riccati equation

$$Q + A^T X + X A - X S X + X M X = 0 \tag{9.3.9}$$

has a stabilizing solution X (i.e. $A - S X + M X$ is stable) and that additionally $A - S X$ is stable. Furthermore, assume that there exists a real symmetric $n \times n$ matrix Y that satisfies the matrix inequality

$$Q + A^T Y + Y A - Y S Y \geq 0. \tag{9.3.10}$$

Define $\bar{F} := -R^{-1}B^T X$ and $\bar{w}(t) := V^{-1}E^T X e^{(A-SX+MX)t}x_0$. Then the matrix $\bar{F}$ belongs to $\mathcal{F}$, the function $\bar{w}$ is in $L_2^q(0, \infty)$, and for all $F \in \mathcal{F}$ and $w \in L_2^q(0, \infty)$

$$J(\bar{F}, w, x_0) \leq J(\bar{F}, \bar{w}, x_0) \leq J(F, \bar{w}, x_0).$$

Moreover, $J(\bar{F}, \bar{w}, x_0) = x_0^T X x_0$.   □

Notice that if $Q \geq 0$, condition (9.3.10) is trivially satisfied by choosing $Y = 0$. Corollary 9.9 summarizes the consequences of Theorem 9.8 for the problem (9.3.3) posed at the beginning of this Section.

Corollary 9.9

Let the assumptions of Theorem 9.8 hold and let X, $\bar{F}$ and $\bar{w}$ be as in that theorem. Then

$$\min_{F \in \mathcal{F}} \sup_{w \in L_2^q(0,\infty)} J(F, w, x_0) = \max_{w \in L_2^q(0,\infty)} J(\bar{F}, w, x_0) = x_0^T X x_0$$

and

$$\max_{w \in L_2^q(0,\infty)} \inf_{F \in \mathcal{F}} J(F, w, x_0) = \min_{F \in \mathcal{F}} J(F, \bar{w}, x_0) = x_0^T X x_0.$$ □

Example 9.1

In Example 5.5 we considered the minimization of

$$J := \int_0^\infty \{x^T(t)x(t) + 2u^2(t)\}dt,$$

subject to the system

$$\dot{x}(t) = \begin{bmatrix} 1 & 0 \\ 0 & -1 \end{bmatrix} x(t) + \begin{bmatrix} 1 \\ 0 \end{bmatrix} u(t), \quad x(0) = \begin{bmatrix} 1 \\ 2 \end{bmatrix}.$$

Assume that the input channel used by the control designer is corrupted by noise. The control designer includes this information by considering the adapted optimization problem

$$\min_{F \in \mathcal{F}} \sup_{w \in L_2(0,\infty)} \int_0^\infty \{x^T(t)x(t) + 2u^2(t) - vw^2(t)\}dt,$$

subject to the system

$$\dot{x}(t) = \begin{bmatrix} 1 & 0 \\ 0 & -1 \end{bmatrix} x(t) + \begin{bmatrix} 1 \\ 0 \end{bmatrix} u(t) + \begin{bmatrix} 1 \\ 0 \end{bmatrix} w(t), \quad x(0) = \begin{bmatrix} 1 \\ 2 \end{bmatrix}.$$

Here v is some parameter expressing the magnitude of the noise expected by the control designer. We will assume that $v > 2$. This assumption can be justified, for example, if the control designer knows that the input signal dominates the noise.

Since (using the notation of Theorem 9.8) matrix $Q = I \geq 0$, then according to Theorem 9.8, this problem has a solution $\hat{F} \in \mathcal{F}$ if the following algebraic Riccati equation

$$\begin{bmatrix} 1 & 0 \\ 0 & 1 \end{bmatrix} + \begin{bmatrix} 1 & 0 \\ 0 & -1 \end{bmatrix} X + X \begin{bmatrix} 1 & 0 \\ 0 & -1 \end{bmatrix} - X \begin{bmatrix} \frac{1}{2} & 0 \\ 0 & 0 \end{bmatrix} X + X \begin{bmatrix} \frac{1}{v} & 0 \\ 0 & 0 \end{bmatrix} X = 0,$$

has a solution X such that both

$$\begin{bmatrix} 1 & 0 \\ 0 & -1 \end{bmatrix} - \begin{bmatrix} \frac{1}{2} - \frac{1}{v} & 0 \\ 0 & 0 \end{bmatrix} X \text{ and } \begin{bmatrix} 1 & 0 \\ 0 & -1 \end{bmatrix} - \begin{bmatrix} \frac{1}{2} & 0 \\ 0 & 0 \end{bmatrix} X$$

are stable.

Simple calculations show that

$$X = \begin{bmatrix} x_1 & 0 \\ 0 & \frac{1}{2} \end{bmatrix}, \text{ with } x_1 = \frac{1 + \sqrt{\frac{3}{2} - \frac{1}{v}}}{\frac{1}{2} - \frac{1}{v}}$$

is the solution of this Riccati equation which satisfies both the above mentioned stability requirements. Therefore, the resulting optimal control and cost are

$$u^*(t) = -\frac{1}{2}[x_1, \ 0]x(t), \quad \text{and } J^* = 2 + x_1,$$

respectively. Notice that, if $v \to \infty$, the optimal control and cost converge to those of the 'noise-free' case. Furthermore, we see that the input gain x_1 is a monotonically decreasing function of v on $(2, \infty)$, with $\lim_{v \downarrow 2} x_1(v) = \infty$. This implies that the less the control designer expects that the input signal can be discerned from the noise, the more control efforts will be used to accomplish the system stabilization goal. □

Example 9.2

Consider a salesperson who can sell any amount of their products at a prespecified price p. This salesperson has an initial stock q_0 of some goods sell. The costs of selling, warehousing and advertising are a quadratic function of the amount of sold goods. Some of these goods are, unfortunately, returned. The cost associated with returned goods is also assumed to be a quadratic function of this amount. The salesperson uses the following stylized model

$$\dot{q}(t) = -u(t) + w(t), \quad q(0) = q_0,$$

where $q(t)$ is the stock of the goods, $u(t)$ the number of products sold and $w(t)$ the number of goods returned. The optimization problem is formulated as follows

$$\max_{u \in \mathcal{F}^{aff}} \inf_{w \in L_2(0,\infty)} \int_0^\infty e^{-rt}\{pq(t) - u^2(t) + vw^2(t)\}dt,$$

where

$$\mathcal{F}^{aff} := \left\{ u(.) \mid u(t) = f_1 q(t) + f_2, \quad \text{with } f_1 < \frac{1}{2}r \right\}.$$

In this model $v > 1$ is a measure of the proportion of the goods that will be returned and $r > 0$ is a discount factor. Now, let $x^T(t) := [q(t) \; 1]$. Then the optimization problem can be rewritten as

$$\min_u \sup_{w \in L_2(0,\infty)} \int_0^\infty e^{-rt} \left\{ x^T(t) \begin{bmatrix} 0 & -\frac{1}{2}p \\ -\frac{1}{2}p & 0 \end{bmatrix} x(t) + u^2(t) - vw^2(t) \right\} dt,$$

subject to the dynamics

$$\dot{x}(t) = \begin{bmatrix} 0 & 0 \\ 0 & 0 \end{bmatrix} x(t) + \begin{bmatrix} -1 \\ 0 \end{bmatrix} u(t) + \begin{bmatrix} 1 \\ 0 \end{bmatrix} w(t).$$

Next, introducing $\tilde{x}(t) := e^{-\frac{1}{2}rt} x(t)$, $\tilde{u}(t) := e^{-\frac{1}{2}rt} u(t)$ and $\tilde{w}(t) := e^{-\frac{1}{2}rt} w(t)$, the optimization problem can be rewritten in standard form as

$$\min_{\tilde{u} \in \mathcal{F}} \sup_{\tilde{w} \in L_2(0,\infty)} \int_0^\infty \left\{ \tilde{x}^T(t) \begin{bmatrix} 0 & -\frac{1}{2}p \\ -\frac{1}{2}p & 0 \end{bmatrix} \tilde{x}(t) + \tilde{u}^2(t) - v\tilde{w}^2(t) \right\} dt,$$

subject to the dynamics

$$\dot{\tilde{x}}(t) = \begin{bmatrix} -\frac{1}{2}r & 0 \\ 0 & -\frac{1}{2}r \end{bmatrix} \tilde{x} + \begin{bmatrix} -1 \\ 0 \end{bmatrix} \tilde{u}(t) + \begin{bmatrix} 1 \\ 0 \end{bmatrix} \tilde{w}(t).$$

According to Theorem 9.8 this problem has a solution if the algebraic Riccati equation

$$\begin{bmatrix} 0 & -\frac{1}{2}p \\ -\frac{1}{2}p & 0 \end{bmatrix} - rX + \left(-1 + \frac{1}{v} \right) X \begin{bmatrix} 1 & 0 \\ 0 & 0 \end{bmatrix} X = 0$$

has a solution X such that both $-\frac{1}{2}I - \begin{bmatrix} 1 & 0 \\ 0 & 0 \end{bmatrix} X$ and $-\frac{1}{2}I + \left(-1 + \frac{1}{v} \right) \begin{bmatrix} 1 & 0 \\ 0 & 0 \end{bmatrix} X$ are stable, and the matrix inequality

$$\begin{bmatrix} 0 & -\frac{1}{2}p \\ -\frac{1}{2}p & 0 \end{bmatrix} - rY - Y \begin{bmatrix} 1 & 0 \\ 0 & 0 \end{bmatrix} Y \geq 0$$

has a symmetric solution Y.

It is easily verified that the latter inequality is satisfied if we choose, for example,

$Y = \begin{bmatrix} \frac{-1}{2}r & \frac{-2p}{r} \\ \frac{-2p}{r} & \frac{-5p^2}{r^3} \end{bmatrix}$. With $X := \begin{bmatrix} x_1 & x_2 \\ x_2 & x_3 \end{bmatrix}$, the algebraic Riccati equation has an appropriate solution if and only if the following three equations

$$-rx_1 + \left(-1 + \frac{1}{v}\right)x_1^2 = 0$$

$$-\frac{1}{2}p - rx_2 + \left(-1 + \frac{1}{v}\right)x_1x_2 = 0$$

$$-rx_3 + \left(-1 + \frac{1}{v}\right)x_2^2 = 0.$$

have a solution (x_1, x_2, x_3) satisfying $-\frac{1}{2}r - x_1 < 0$ and $-\frac{1}{2}r + \left(-1 + \frac{1}{v}\right)x_1 < 0$. Direct substitution shows that

$$X = \begin{bmatrix} 0 & \frac{-p}{2r} \\ \frac{-p}{2r} & \left(-1 + \frac{1}{v}\right)\frac{p^2}{4r^3} \end{bmatrix}$$

satisfies the (in)equalities.

So, according to Theorem 9.8, the optimal control is

$$\tilde{u}(t) = \frac{p}{2r}e^{-\frac{1}{2}rt}.$$

Or, stated in the original model parameters,

$$u(t) = \frac{p}{2r}.$$

So, within the class of affine strategies, the optimal policy is constantly to sell a fixed number $\frac{p}{2r}$ of goods. Notice that, in particular, this quantity does not depend on the stock of goods and the parameters modeling the consequences of expected returned goods. □

We conclude this Section by considering the question under which conditions the assumption that the algebraic Riccati equation (9.3.9) has a symmetric solution such that both $A + MX - SX$ and $A - SX$ are stable is also a necessary condition in order to conclude that (9.3.3) is finite. Theorem 9.10, below, shows that this condition must hold if the infimum in (9.3.3) is achieved at some $\bar{F} \in \bar{\mathcal{F}}$. The proof of this theorem is provided in the Appendix at the end of this chapter.

Theorem 9.10

Assume there exists an $\bar{F} \in \bar{\mathcal{F}}$ such that for each $x_0 \in \mathbb{R}^n$

$$\min_{F \in \bar{\mathcal{F}}} \sup_{w \in L_2^q(0,\infty)} J(F, w, x_0) = \max_{w \in L_2^q(0,\infty)} J(\bar{F}, w, x_0).$$

Then the algebraic Riccati equation (9.3.9) has a stabilizing solution X. Furthermore, the matrix $A - SX$ is stable and $\bar{F} = -R^{-1}B^TX$. □

A direct consequence of this result and the discussion after Theorem 9.8 is the following corollary.

Corollary 9.11

Let $Q \geq 0$. Then, there exists an $\bar{F} \in \bar{\mathcal{F}}$ such that for each $x_0 \in \mathbb{R}^n$

$$\min_{F \in \bar{\mathcal{F}}} \sup_{w \in L_2^q(0,\infty)} J(F,w,x_0) = \max_{w \in L_2^q(0,\infty)} J(\bar{F},w,x_0)$$

if and only if the algebraic Riccati equation

$$Q + A^T X + XA - XSX + XMX = 0$$

has a solution X such that both $A - SX + MX$ and $A - SX$ are stable.
Furthermore,

$$\bar{F} := -R^{-1}B^T X \quad \text{and} \quad \bar{w}(t) := V^{-1}E^T Xe^{(A-SX+MX)t}x_0$$

are the worst-case optimal feedback and disturbance, respectively.
Finally, $J(\bar{F},\bar{w},x_0) = x_0^T X x_0$. $\qquad\qquad\qquad\square$

9.4 The one-player scalar case

This Section elaborates on the scalar case of the one-player problem (9.3.1) and (9.3.2). That is, we look for conditions on the parameters under which

$$\bar{J} := \inf_{f \in \mathcal{F}} \sup_{w \in L_2^q(0,\infty)} J(f,w,x_0) \tag{9.4.1}$$

exists. Furthermore, the role condition (9.3.10) plays in Theorem 9.8 will be explained.
    We begin with necessary and sufficient conditions under which the supremum in the inner optimization problem of (9.4.1) takes a finite value.

Lemma 9.12

Let $f \in \mathcal{F}$ be fixed. Then $\sup\limits_{w \in L_2^q(0,\infty)} J(f,w,x_0)$ is finite if and only if

$$g(f) := (a + bf)^2 - m(q + f^2 r) \geq 0.$$

Furthermore, if this supremum is finite its value is $-\frac{1}{2}(q + f^2 r)x_0^2/(a + bf)$ if $e = 0$ and is $-(a + bf + \sqrt{g(f)})x_0^2/m$ otherwise.

Proof

If $e = 0$, the supremum is achieved at $w = 0$ and so it is finite for any $f \in \mathcal{F}$ also in this case $m = e^2/v = 0$ and so trivially $g(f) \geq 0$. If $e \neq 0$, the pair $(a + bf, e)$ is controllable. According to the Note following Example 5.5 (or using elementary analysis) the supremum is finite if and only if the algebraic Riccati equation

$$mx^2 + 2(a + bf)x + q + f^2 r = 0 \tag{9.4.2}$$

has a real solution. Furthermore, the value of the supremum is $x_0^2 x_s$, where x_s is the smallest solution of equation (9.4.2). From this, the above statement follows directly.   □

Next, we consider the outer minimization in (9.4.1). From Lemma 9.12 it is clear that the case $e = 0$ is a special one. Therefore, this case is analyzed separately.

Theorem 9.13

Suppose $e = 0$ in the scalar version of equations (9.3.1) and (9.3.2). Let $t := a^2 + sq$. Then

1. if $b \neq 0$ and

   (i) $t > 0$, $\bar{J} = (qs + (a + \sqrt{t})^2)/2s\sqrt{t}$ and $\bar{f} = -(a + t)/b$,

   (ii) $t = 0$, $\bar{J} = a/s$ and the infimum in problem (9.4.1) is not achieved (actually, the infimum is attained at $f = -a/b$),

   (iii) $t < 0$, $\bar{J} = -\infty$;

2. If $b = 0$, then $\bar{J}$ exists if and only if $a < 0$. Furthermore, $\bar{J} = -q/(2a)$ and $\bar{f} = 0$.

Proof

All statements follow by an elementary analysis of the function $\mathcal{F} \ni f \mapsto -\frac{1}{2}(q + f^2 r)/(a + bf)$ (see Lemma 9.12). If $t > 0$ this function has a unique minimum at $\bar{f}$; if $t = 0$ its graph is a line; if $t < 0$ it is a monotonic function that has a vertical asymptote at $f = -a/b$.   □

Next, consider the case $e \neq 0$ or, equivalently, $m \neq 0$. Let

$$\bar{\mathcal{F}}_e := \{f \in \mathcal{F} \mid g(f) = (a + bf)^2 - m(q + f^2 r) \geq 0\}. \qquad (9.4.3)$$

By Lemma 9.12, $\bar{\mathcal{F}}_e$ is the set of all stabilizing feedback 'matrices' for which $\sup_{w \in L_2^q(0,\infty)} J(f, w, x_0)$ is finite. Moreover according to this Lemma 9.12 for each $f \in \bar{\mathcal{F}}_e$ this supremum equals $x_0^2 x_s(f)$, where $x_s(f)$ is given by

$$x_s(f) = -\frac{a + bf + \sqrt{g(f)}}{m}.$$

So solving problem (9.4.1) is now reduced to looking for the infimum of $x_s(f)$ over all f in the set $\bar{\mathcal{F}}_e$. Obviously to find this infimum the domain $\bar{\mathcal{F}}_e$ first has to be fixed. Lemma 9.14, below, characterizes this domain in geometric terms.

Lemma 9.14

Assume $e \neq 0$. The set $\bar{\mathcal{F}}_e$ defined in equation (9.4.3) is either empty, a single point, a half-line, a bounded interval, or the union of a half-line and a bounded interval.

Proof

Define $\mathcal{G} := \{f \mid (a+bf)^2 - m(q+f^2r) \geq 0\}$. Then $\bar{\mathcal{F}}_e = \mathcal{G} \cap \mathcal{F}$. Note that $\mathcal{F}$ is an open half-line. To determine $\mathcal{G}$, consider the graph of $g(f) := (a+bf)^2 - m(q+f^2r), f \in \mathbb{R}$. If g is concave, $\mathcal{G}$ is a closed interval (possibly empty) or just a single point. So $\bar{\mathcal{F}}_e$ is also a possibly empty interval or single point. If g is convex, $\mathcal{G}$ consists of either the whole real line or the union of two closed half-lines. From this the other possibilities mentioned in the lemma are easily established.   □

Notice that whenever $\bar{\mathcal{F}}_e$ consists of more than one single point, one can use differentiation arguments to investigate the finiteness of equation (9.4.1). However, from Lemma 9.14 it follows that $\bar{\mathcal{F}}_e$ may either be empty or consist of just one single point. In these cases differentiation arguments make no sense. Therefore, these two cases are now elaborated.

Theorem 9.15

Let $e \neq 0$. Then

1. $\bar{\mathcal{F}}_e = \emptyset$ if and only if $s < m$ and either (i) $a^2 + q(s-m) < 0$ or (ii) $a^2 + q(s-m) \geq 0$, $a \geq 0$ and $a^2 + qs \geq 0$ – in this case, $\bar{J} = \infty$;

2. $\bar{\mathcal{F}}_e$ consists of only one point if and only if simultaneously $s - m < 0$, $a^2 + q(s-m) = 0$ and $-ma/(s-m) < 0$ hold. Then, $\bar{J} = a/(s-m)$ and $\bar{f} = -ab/(r(s-m))$.

Proof

1. $\bar{\mathcal{F}}_e = \emptyset$ if and only if (see Lemma 9.14) either $\mathcal{G}$ is empty, or the inter Section of $\mathcal{F}$ with $\mathcal{G}$ (with $\mathcal{G}$ a bounded interval) is empty. The first case occurs if both $s - m < 0$ and $a^2 + q(s-m) < 0$. The second case occurs if $s - m < 0$, $a^2 + q(s-m) \geq 0$ and (assume without loss of generality $b > 0$) $-a/b \leq -(ab/r + \sqrt{m/r}\sqrt{a^2 + q(s-m)})/(s-m)$. This holds if and only if $a \geq 0$ and $a^2 + qs \geq 0$.

2. $\bar{\mathcal{F}}_e$ consists of only one point if and only if $g(f) = 0$ has exact one solution in $\mathcal{F}$. Elementary calculations then show the stated result.   □

Note

It is easily verified that the conditions of Theorem 9.15 are not met if the equation (9.3.9),

$$q + 2ax - sx^2 + mx^2 = 0,$$

has a stabilizing solution x^* (i.e. $a + (m-s)x^* < 0$) for which additionally $a - sx^* < 0$. That is, under assumption (9.3.9) of Theorem 9.8 the set $\bar{\mathcal{F}}_e$ is either a half-line, a bounded interval or the union of both these two sets (see Lemma 9.14).   □

Next consider the case $\mathcal{G} = \mathbb{R}$. Or, in terms of the system parameters, $s > m$ and $a^2 + q(s-m) \leq 0$. It can be shown that under these conditions the derivative of $x_s(f)$ is negative. So, the infimum is finite, but is attained at the boundary of $\mathcal{F}$.

With this, the analysis of the case that the parameters satisfy $a^2 + q(s - m) \leq 0$ is almost complete. The only case that has not been dealt with yet is $s = m$. Obviously, if $s = m$, $a^2 + q(s - m) \leq 0$ if and only if $a = 0$. It is then easily verified that $x'_s(f) < 0$ again. So, in this case as well a finite infimum is attained at the boundary of $\mathcal{F}$.

Finally, consider the case that $a^2 + q(s - m) > 0$. Elementary calculations show that in that case the derivative of $x_s(f)$ has a unique zero f^*. This zero coincides with $-(b/r)x^*$, where x^* is the smallest solution of the algebraic Riccati equation (9.3.9). Furthermore, by differentiating equation (9.4.2) twice it is easily seen that $x''_s(f^*) < 0$, so $x_s(f)$ has a minimum at f^*. Moreover, $g(f^*) = (a - sx^* + mx^*)^2 \geq 0$. So, $f^* \in \mathcal{G}$. If additionally $a - sx^* \in \mathcal{F}$, then $x_s(f)$ has a minimum in $\bar{\mathcal{F}}_e$ which, moreover, is a global minimum if, for example, $\bar{\mathcal{F}}_e$ is connected. On the other hand, it is clear that if $a - sx^* \notin \mathcal{F}$ the infimum value is again attained at the boundary of $\mathcal{F}$. The following example illustrates the case in which $\bar{\mathcal{F}}_e$ is not connected.

Example 9.3

Let $a = 5$, $b = 1$, $m = 1$, $r = \frac{1}{9}$ and $q = -3$. Then, $\bar{\mathcal{F}}_e = (-\infty, -6) \cup (-5\frac{1}{4}, -5)$. Moreover, $J(-5, \bar{w}, x_0) = -\frac{1}{3}\sqrt{2}x_0^2$, $f^* = -6\frac{3}{4}$ and $J(f^*, \bar{w}, x_0) = \frac{3}{4}x_0^2$. In this case the infimum is not achieved. Note that if $f = -5$ the worst-case action (from the player's point of view) the disturbance can take is to stabilize the system since the player's aim is to maximize the revenues x (subject to the constraint that the undisturbed closed-loop system must be stable). ☐

The following lemma gives conditions, in terms of the problem parameters, under which a nonempty set $\bar{\mathcal{F}}_e$ is not connected.

Lemma 9.16

Assume that $\bar{\mathcal{F}}_e \neq \emptyset$. Then $\bar{\mathcal{F}}_e$ is not connected if and only if the following four conditions are satisfied

(i) $s - m > 0$

(ii) $a^2 + q(s - m) \geq 0$

(iii) $a^2 + qs < 0$

(iv) $a > 0$.

Proof

If g is concave (see proof Lemma 9.14), the set $\bar{\mathcal{F}}_e$ is an interval and is thus connected. It is easily verified that this situation occurs if and only if $s - m \leq 0$. Next consider the case that g is convex. If g has no zeros it is obvious that $\bar{\mathcal{F}}_e$ is connected. This occurs if and only if $a^2 + (s - m)q < 0$. Otherwise, $\mathcal{G} = (-\infty, a_0) \cup (a_1, \infty)$. Then, $\bar{\mathcal{F}}_e$ is connected if and only if (assume without loss of generality $b > 0$) $-a/b \leq -(ab + \sqrt{a^2b^2 - (a^2 - mq)(b^2 - mr)})/(r(s - m))$. This condition holds if and only if either $a \leq 0$ or $a^2 + qs \geq 0$. ☐

If $\bar{\mathcal{F}}_e$ is not connected, $J(f, \bar{w}(f), x_0)$ does not have a global minimum since

$$J(-a/b, \bar{w}(-a/b), x_0) = -\left(\sqrt{-m(q + a^2/s)/m}\right)x_0^2 < 0$$

$$< \left(a + \sqrt{a^2 + q(s - m)}\right)/(s - m)x_0^2 = J(f^*, \bar{w}(f^*), x_0).$$

Actually one can show that $x_s(f)$ again attains an infimum at the boundary of the stabilization constraint interval $-a/b$. So we conclude the following theorem.

Theorem 9.17

Consider the scalar version of the one-player game (9.3.1) and (9.3.2). Assume that the set $\bar{\mathcal{F}}_e$ defined in equation (9.4.3) has more than one element and that $e \neq 0$. Then

1. the one-player game has a solution if and only if either one of the four conditions in Lemma 9.16 is violated and equation (9.3.9) has a stabilizing solution x^* for which additionally $a - sx^*$ is stable – in that case the optimal feedback and worst-case disturbance are $\bar{f} := -\frac{b}{r}x^*$ and $\bar{w}(t) := \frac{e}{v}x^*e^{(a-sx^*+mx^*)t}x_0$;

2. if the conditions in part 1 are not met, the infimum in equation (9.4.1) is attained at the boundary of the stabilization constraint interval $-\frac{a}{b}$ and $\bar{J} = J\left(-\frac{a}{b}, \bar{w}\left(-\frac{a}{b}\right), x_0\right)$.   □

Note

The assumption in Theorem 9.8 that there exists a number y such that $q + 2ay - sy^2 \geq 0$ is equivalent to the assumption that $a^2 + qs \geq 0$. According to Lemma 9.16 this condition (together with assumption (9.3.9), see the Note following Theorem 9.15) is enough to conclude that in the scalar case the set $\bar{\mathcal{F}}_e$ will be connected. So for the scalar case, Theorem 9.8 is a special case of Theorem 9.17, part 1.   □

Example 9.4

Reconsider Example 7.8 (see also Example 3.25). The problem that will now be addressed is to find

$$\inf_{g \in \mathcal{F}} \sup_{w \in L_2} L_\lambda(g, w) \tag{9.4.4}$$

where

$$L_\lambda(g, w) := \int_0^\infty \{x^2(t) + \phi g^2(t)\}dt - \lambda \int_0^\infty w^2(t)dt.$$

subject to

$$\dot{x}(t) = -\alpha\delta x(t) + \delta g(t) + w(t), \quad x(0) = x_0. \tag{9.4.5}$$

By either Theorem 9.17, part 1, or Corollary 9.11, this problem has a solution if and only if

$$1 - 2\alpha\delta x - \left(\frac{\delta^2}{\phi} - \frac{1}{\lambda}\right)x^2 = 0 \tag{9.4.6}$$

has a solution $\bar{p}_\lambda$ such that both $-\alpha\delta - \left(\frac{\delta^2}{\phi} - \frac{1}{\lambda}\right)\bar{p}_\lambda < 0$ and $-\alpha\delta - \frac{\delta^2}{\phi}\bar{p}_\lambda < 0$.

According to Example 7.8

$$\bar{p}_\lambda = \frac{1}{\alpha\delta + \sqrt{\alpha^2\delta^2 + \frac{\delta^2}{\phi} - \frac{1}{\lambda}}} > 0.$$

From this it is easily verified that the problem will have a solution if and only if

$$\alpha^2\delta^2 + \left(\frac{\delta^2}{\phi} - \frac{1}{\lambda}\right) > 0.$$

That is, for all $\lambda > \hat{\lambda} := \frac{1}{\alpha^2\delta^2 + \frac{\delta^2}{\phi}}$ the optimal worst-case control is

$$g^*(t) = -\frac{\delta}{\phi}\bar{p}_\lambda x(t),$$

where $x(t)$ is the solution of the differential equation

$$\dot{x}(s) = \left(-\alpha\delta - \frac{\delta^2}{\phi}\bar{p}_\lambda\right)x(s) + \tilde{w}(s), \quad x(0) = x_0,$$

at time t and $\tilde{w}$ is the realization of the disturbance up to time t. For all $\lambda \leq \hat{\lambda}$ there exists no worst-case optimal control.

Recall that the evolution of the state under the optimal open-loop worst-case control is

$$\dot{x}(s) = -\alpha\delta x(s) - \frac{\delta^2}{\phi}\bar{p}_\lambda e^{\left(-\alpha\delta - \left(\frac{\delta^2}{\phi} - \frac{1}{\lambda}\right)\bar{p}_\lambda\right)s}x_0 + \tilde{w}(s), \quad x(0) = x_0.$$

Comparing both closed-loop systems we see that in the feedback information case, in general the system converges faster to zero. This is because on the one hand $\alpha\delta + \frac{\delta^2}{\phi}\bar{p}_\lambda > \alpha\delta$ and on the other hand the feedback system does not contain an additional external term, that can usually be viewed as an additional disturbance.

Furthermore, the conclusion in the open-loop case that the smaller the government chooses λ the more active control policy it will use carries over to this case. Moreover, we see that the set of risk-attitude parameters for which a solution exists is in this feedback information case larger than that for the open-loop case. From an intuitive point this is something that one would expect. In the feedback case the players can react to the disturbance during the evolution of the game and adapt their action accordingly, something which is not possible in the open-loop case. So, wilder disturbances (from the governments point of view) can be dealt with. Notice that the closer one chooses λ to $\hat{\lambda}$ the better the estimate obtained for the performance: with $x_0 = 0$,

$$\int_0^\infty \{x^2(t) + \phi g^2(t)\}dt \leq \lambda \int_0^\infty w^2(t)dt, \quad \forall w \in L_2.$$

However, although this estimate becomes better the reverse is that the system becomes almost unstable under the worst-case disturbance scenario if λ is close to $\hat{\lambda}$. □

9.5　The two-player case

Next we study the two-player case. That is, we consider

$$\dot{x}(t) = (A + B_1 F_1 + B_2 F_2)x(t) + Ew(t), \quad x(0) = x_0, \qquad (9.5.1)$$

with $(A, [B_1 \ B_2])$ stabilizable, $(F_1, F_2) \in \mathcal{F}$ and

$$J_i(F_1, F_2, w, x_0) = \int_0^\infty \{x^T(t)(Q_i + F_1^T R_{i1} F_1 + F_2^T R_{i2} F_2)x(t) - w^T(t)V_i w(t)\}dt. \quad (9.5.2)$$

Here the matrices Q_i, R_{ij} and V_i are symmetric, $R_{ii} > 0$, $V_i > 0$, and

$$\mathcal{F} := \{(F_1, F_2) | A + B_1 F_1 + B_2 F_2 \text{ is stable}\}.$$

For this game we want to determine all soft-constrained Nash equilibria. That is, to find all $(\bar{F}_1, \bar{F}_2) \in \mathcal{F}$ such that

$$\sup_{w \in L_2^q(0,\infty)} J_1(\bar{F}_1, \bar{F}_2, w, x_0) \leq \sup_{w \in L_2^q(0,\infty)} J_1(F_1, \bar{F}_2, w, x_0), \quad \text{for all } (F_1, \bar{F}_2) \in \mathcal{F} \quad (9.5.3)$$

and

$$\sup_{w \in L_2^q(0,\infty)} J_2(\bar{F}_1, \bar{F}_2, w, x_0) \leq \sup_{w \in L_2^q(0,\infty)} J_2(\bar{F}_1, F_2, w, x_0), \quad \text{for all } (\bar{F}_1, F_2) \in \mathcal{F} \quad (9.5.4)$$

for all $x_0 \in \mathbb{R}^m$.

From Corollary 9.9, a sufficient condition for the existence of a soft-constrained feedback Nash equilibrium follows in a straightforward way. Using, throughout this Section, the shorthand notation

$$S_i := B_i R_{ii}^{-1} B_i^T, \quad S_{ij} := B_i R_{ii}^{-1} R_{ji} R_{ii}^{-1} B_i^T, \quad i \neq j, \quad \text{and} \quad M_i := E V_i^{-1} E^T,$$

we have the following result.

Theorem 9.18

Consider the differential game defined by expressions (9.5.1)–(9.5.4). Assume there exist real symmetric $n \times n$ matrices X_i, $i = 1, 2$, and real symmetric $n \times n$ matrices Y_i, $i = 1, 2$, such that

$$- (A - S_2 X_2)^T X_1 - X_1 (A - S_2 X_2) + X_1 S_1 X_1 - Q_1 - X_2 S_{21} X_2 - X_1 M_1 X_1 = 0, \quad (9.5.5)$$

$$- (A - S_1 X_1)^T X_2 - X_2 (A - S_1 X_1) + X_2 S_2 X_2 - Q_2 - X_1 S_{12} X_1 - X_2 M_2 X_2 = 0, \quad (9.5.6)$$

$$A - S_1 X_1 - S_2 X_2 + M_1 X_1 \text{ and } A - S_1 X_1 - S_2 X_2 + M_2 X_2 \text{ are stable}, \quad (9.5.7)$$

$$A - S_1 X_1 - S_2 X_2 \text{ is stable}, \quad (9.5.8)$$

$$- (A - S_2 X_2)^T Y_1 - Y_1 (A - S_2 X_2) + Y_1 S_1 Y_1 - Q_1 - X_2 S_{21} X_2 \leq 0, \quad (9.5.9)$$

$$- (A - S_1 X_1)^T Y_2 - Y_2 (A - S_1 X_1) + Y_2 S_2 Y_2 - Q_2 - X_1 S_{12} X_1 \leq 0. \quad (9.5.10)$$

Define $\overline{F} = (\overline{F}_1, \overline{F}_2)$ by

$$\overline{F}_i := -R_{ii}^{-1} B_i^T X_i, \quad i = 1, 2.$$

Then $\overline{F} \in \mathcal{F}$, and $\overline{F}$ is a soft-constrained Nash equilibrium. Furthermore, the worst-case signal $\overline{w}_i$ from player i's perspective is

$$\overline{w}(t) = V_i^{-1} E^T X_i e^{(A - S_1 X_1 - S_2 X_2 + M_i X_i) t} x_0.$$

Moreover the costs for player i under their worst-case expectations are

$$\overline{J}_i^{SC} (\overline{F}_1, \overline{F}_2, x_0) = x_0^T X_i x_0, \quad i = 1, 2.$$

Conversely, if $(\overline{F}_1, \overline{F}_2)$ is a soft-constrained Nash equilibrium, the equations (9.5.5)–(9.5.8) have a set of real symmetric solutions (X_1, X_2).

Proof

The assumption (9.5.8) immediately implies that $\overline{F} \in \mathcal{F}$. Let $x_0 \in \mathbb{R}^n$ and $1 \leq i \leq 2$. Let the functional ϕ be defined by

$$\phi : \{F \in \mathbb{R}^{m_i \times n} \mid \overline{F}_{-i}(F) \in \mathcal{F}\} \to \mathbb{R}, \quad \phi(F) = \overline{J}_i^{SC} (\overline{F}_{-i}(F), x_0).$$

We need to show that this function is minimal at $F = \overline{F}_i$. We have

$$\phi(F) = \sup_{w \in L_2^q (0, \infty)} \int_0^\infty \left(x^T \left(Q_i + \sum_{j \neq i}^2 X_j S_{ij} X_j + F^T R_{ii} F \right) x - w^T V_i w \right) dt$$

where x follows from

$$\dot{x} = \left(A - \sum_{j \neq i}^2 S_j X_j + B_i F \right) x + Ew, \quad x(0) = x_0.$$

This function ϕ coincides with the function J, as defined in Theorem 9.8, with A replaced by $A - \sum\limits_{j \neq i}^{2} S_j X_j$, $B := B_i$, $Q := Q_i + \sum\limits_{j \neq i}^{2} X_j S_{ij} X_j$, $R := R_{ii}$, $V = V_i$, and the same values for E and x_0. It is easily seen that the conditions (9.5.5)–(9.5.10) guarantee that the conditions of Theorem 9.8 are satisfied with $X := X_i$ and $Y := Y_i$. So, according to Theorem 9.8, the function ϕ is minimal at $F = -R_{ii}^{-1} B_i^T X_i = \bar{F}_i$ and $w = \bar{w}_i$, and the minimal value is $x_0^T X_i x_0$.

The converse statement follows directly from Theorem 9.10. Assume that $(\bar{F}_1, \bar{F}_2)$ is a soft-constrained Nash equilibrium. Then both the optimization problems (9.5.3) and (9.5.4) have a solution. So, by Theorem 9.10, the algebraic Riccati equation

$$Q_1 + (A + B_2 \bar{F}_2)^T X_1 + X_1 (A + B_2 \bar{F}_2) - X_1 S_1 X_1 + X_1 M_1 X_1 = 0$$

has a solution X_1 such that both $A - S_1 X_1 + B_2 \bar{F}_2$ and $A + B_2 \bar{F}_2 + (M_1 - S_1) X_1$ are stable. Furthermore, according to this theorem, $\bar{F}_1 = -R^{-1} B_1^T X_1$.

Similarly it also follows that the algebraic Riccati equation

$$Q_2 + (A + B_1 \bar{F}_1)^T X_2 + X_2 (A + B_1 \bar{F}_1) - X_2 S_2 X_2 + X_2 M_2 X_2 = 0$$

has a solution X_2 such that both $A + B_1 \bar{F}_1 - S_2 X_2$ and $A + B_1 \bar{F}_1 + (M_2 - S_2) X_2$ are stable, whereas $\bar{F}_2 = -R_2^{-1} B_2^T X_2$. Substitution of $\bar{F}_i = -R_i^{-1} B_i^T X_i$, $i = 1, 2$, into both these algebraic Riccati equations then shows the result. $\qquad \square$

Corollary 9.19

If $Q_i \geq 0$, $i = 1, 2$, and $S_{ij} \geq 0$, $i, j = 1, 2$, the matrix inequalities (9.5.9)–(9.5.10) are trivially satisfied with $Y_i = 0$, $i = 1, 2$. So, under these conditions the differential game defined by (9.5.1)–(9.5.4) has a soft-constrained Nash equilibrium if and only if the equations (9.5.5)–(9.5.8) have a set of real symmetric $n \times n$ matrices X_i, $i = 1, 2$. $\qquad \square$

Example 9.5

Consider the differential game where two players have to solve a regulator problem defined by

$$\min_{u_1 = F_1 x} \int_0^\infty \left\{ x^T(t) \begin{bmatrix} 7 & 13 \\ 13 & 40 \end{bmatrix} x(t) + u_1^T \begin{bmatrix} 1 & -1 \\ -1 & 2 \end{bmatrix} u_1(t) \right\} dt \quad \text{and}$$

$$\min_{u_2 = F_2 x} \int_0^\infty \left\{ x^T(t) \begin{bmatrix} 18 & -3 \\ -3 & 41 \end{bmatrix} x(t) + u_2^T \begin{bmatrix} 1 & 0 \\ 0 & \frac{1}{2} \end{bmatrix} u_2(t) \right\} dt,$$

where $(F_1, F_2) \in \mathcal{F}$, subject to the dynamics

$$\dot{x}(t) = \begin{bmatrix} -1 & 0 \\ 0 & -2 \end{bmatrix} x(t) + u_1(t) + u_2(t); \quad x(0) = x_0.$$

Then, with

$$K_1 := \begin{bmatrix} 1 & 1 \\ 1 & 2 \end{bmatrix} \quad \text{and} \quad K_2 := \begin{bmatrix} 2 & -1 \\ -1 & 3 \end{bmatrix}, \tag{9.5.11}$$

K_i, $i = 1, 2$, is a stabilizing set of solutions of the algebraic Riccati equations (8.3.3) and (8.3.4). So, according to Theorem 8.5,

$$u_1^* = -\begin{bmatrix} 3 & 4 \\ 2 & 3 \end{bmatrix} x(t)$$

$$u_2^* = -\begin{bmatrix} 2 & -1 \\ -2 & 6 \end{bmatrix} x(t)$$

constitute a set of feedback Nash equilibrium actions for this regulator game.

Next assume that the dynamics are subject to noise, which is modeled as follows

$$\dot{y}(t) = \begin{bmatrix} -1 & 0 \\ 0 & -2 \end{bmatrix} y(t) + u_1(t) + u_2(t) + w(t); \quad x(0) = x_0,$$

where $w(.) \in L_2^2$ is some deterministic noise. The players deal with this uncertainty by considering the following worst-case cost functions

$$J_1 := \int_0^\infty \left\{ x^T(t) \begin{bmatrix} 7 & 13 \\ 13 & 40 \end{bmatrix} x(t) + u_1^T \begin{bmatrix} 1 & -1 \\ -1 & 2 \end{bmatrix} u_1(t) - \frac{3}{656} w^T(t) \begin{bmatrix} 93 & 66 \\ 66 & 68 \end{bmatrix} w(t) \right\} dt \quad \text{and}$$

$$J_2 := \int_0^\infty \left\{ x^T(t) \begin{bmatrix} 18 & -3 \\ -3 & 41 \end{bmatrix} x(t) + u_2^T \begin{bmatrix} 1 & 0 \\ 0 & \frac{1}{2} \end{bmatrix} u_2(t) - \frac{3}{4382} w^T(t) \begin{bmatrix} 730 & -530 \\ -530 & 835 \end{bmatrix} w(t) \right\} dt.$$

Notice that since, approximately, $V_1 = \begin{bmatrix} 0.5569 & 0.3952 \\ 0.3952 & 0.3772 \end{bmatrix}$ and $V_2 = \begin{bmatrix} 0.4998 & -0.3628 \\ -0.3628 & 0.5717 \end{bmatrix}$ both players expect an impact of the noise that could be rather severe. Then,

$$X_1 := \frac{3}{2} \begin{bmatrix} 1 & 1 \\ 1 & 2 \end{bmatrix} \quad \text{and} \quad X_2 := \frac{3}{2} \begin{bmatrix} 2 & -1 \\ -1 & 3 \end{bmatrix}, \tag{9.5.12}$$

satisfy equations (9.5.5)–(9.5.8). Moreover, since $Q_i > 0$, the inequalities (9.5.9)–(9.5.10) are satisfied with $Y_i = 0$, $i = 1, 2$. Therefore, according to Theorem 9.18,

$$u_1^* = -\frac{3}{2} \begin{bmatrix} 3 & 4 \\ 2 & 3 \end{bmatrix} x(t)$$

$$u_2^* = -\frac{3}{2} \begin{bmatrix} 2 & -1 \\ -2 & 6 \end{bmatrix} x(t)$$

constitute a soft-constrained Nash equilibrium.

Comparing expressions (9.5.11) and (9.5.12) we see that the consideration of the above adapted cost function results in a worst-case cost increase of 50% for both players. Furthermore, both players use more control to realize their primary regulation goal.

More in general, it follows that if we consider in the above formulation for $d > 1$

$$V_1 = \frac{d^2}{(2d+3)(30d+37)(d-1)} \begin{bmatrix} 17d+21, & 12d+15 \\ 12d+15, & 12d+16 \end{bmatrix} \quad \text{and}$$

$$V_2 = \frac{5d^2}{(406d^2+1096d+729)(d-1)} \begin{bmatrix} 26d+34, & -(20d+23) \\ -(20d+23), & 31d+37 \end{bmatrix},$$

then

$$X_1 = d * \begin{bmatrix} 1 & 1 \\ 1 & 2 \end{bmatrix} \quad \text{and} \quad X_2 := d \begin{bmatrix} 2 & -1 \\ -1 & 3 \end{bmatrix},$$

satisfy equations (9.5.5) and (9.5.6). Since

$$A - S_1 X_1 - S_2 X_2 + M_1 X_1 = \frac{1}{d} \begin{bmatrix} -(5d^2+1), & -(15d^2+2d-14) \\ 5d^2+d-6, & 13d^2+3d-27 \end{bmatrix}, \quad \text{and}$$

$$A - S_1 X_1 - S_2 X_2 + M_2 X_2 = \frac{1}{5d} \begin{bmatrix} 17d^2+4d-51, & 14d^2+3d-32 \\ 14d^2-2d-12, & 13d^2+11d-79 \end{bmatrix},$$

an inspection of the eigenvalues of both matrices shows that both these matrices only have stable eigenvalues if $d < \frac{-3+\sqrt{905}}{16} =: d_0$. Moreover, it is easily verified that the closed-loop system (9.5.8) is always stable.

So, for $1 < d < d_0$, $V_i(d)$, $i = 1, 2$, represents the noise-weighting matrices that correspond with a worst-case equilibrium cost for both players that is a factor d of the 'noise-free' regulator equilibrium cost. For example, the noise-weighting matrices that correspond with a worst-case cost that is 10% above the noise-free cost are obtained by taking $d = 1.1$, which gives

$$V_1 = \begin{bmatrix} 1.3197 & 0.9374 \\ 0.9374 & 0.9707 \end{bmatrix} \quad \text{and} \quad V_2 = \begin{bmatrix} 1.5612 & -1.1223 \\ -1.1223 & 1.7732 \end{bmatrix},$$

whereas $d = 1.01$ gives

$$V_1 = \begin{bmatrix} 11.5251 & 8.1887 \\ 8.1887 & 8.4906 \end{bmatrix} \quad \text{and} \quad V_2 = \begin{bmatrix} 13.6595 & -9.7924 \\ -9.7924 & 15.4843 \end{bmatrix}.$$

From this we observe that for large V_i, $i = 1, 2$, there is almost no effect on the equilibrium cost and control actions for both players, compared with the noise-free case. Conversely, if these matrices V_i are 'small', a small change in V_i has a large effect on both the cost and control actions. □

9.6 A fishery management game

This Section illustrates some consequences of taking deterministic noise into account by means of a simple fishery management problem. Consider two fishermen who fish in a lake. Let s be the number of fish in the lake. Assume that the price $p(t)$ the fishermen get for their fish is fixed, i.e.

$$p(t) = p.$$

The growth of the fish stock in the lake is described by

$$\dot{s}(t) = \beta s(t) - u_1(t) - u_2(t) - w(t), \quad s(0) = s_0 > 0$$

where w is a factor which has a negative impact on the growth of the fish stock (e.g. water pollution, weather, birds, local fishermen, etc.). Both fishermen have their own expectations about the consequences of these negative influences on the fish growth and cope with this by considering the following optimization problem

$$J_i := \min_{u_i \in \mathcal{F}^{aff}} \sup_{w \in L_2} \int_0^\infty e^{-rt} \{-pu_i(t) + \gamma_i u_i^2(t) - v_i w^2(t)\} dt, \quad i = 1, 2,$$

where

$$\mathcal{F}^{aff} := \left\{ (u_1, u_2) \mid u_i(t) = f_{ii} s(t) + g_i, \quad \text{with} \quad \beta - f_{11} - f_{22} < \frac{1}{2} r \right\}.$$

In this formulation all constants, $r, \alpha_i, \beta, \gamma_i$ and v_i, are positive. The term $\gamma_i u_i^2$ models the cost involved for fisherman i in catching an amount u_i of fish. We will assume that $v_i > \gamma_i$, $i = 1, 2$. That is, each fisherman does not expect that a situation will occur where the deterministic cost will be larger than his normal cost of operation. Notice that, since in this formulation the involved cost for the fishermen depends quadratically on the amount of fish they catch, catching large amounts of fish is not profitable for them. This observation might model the fact that catching a large amount of fish is, from a practical point of view, impossible for them. This might be due to either technical restrictions and/ or the fact that there is not an abundant amount of fish in the lake. That is, catching a lot more fish requires much more advanced technology the cost of which rises quadratically.
Introducing $x^T(t) := [s(t)\ 1]$, the optimization problem can be rewritten as

$$\min_{u_i \in \mathcal{F}^{aff}} \sup_{w \in L_2(0,\infty)} \int_0^\infty e^{-rt} \left\{ [x^T(t)\ u_i^T(t)] \begin{bmatrix} 0 & 0 & 0 \\ 0 & 0 & -\frac{1}{2}p \\ 0 & -\frac{1}{2}p & \gamma_i \end{bmatrix} \begin{bmatrix} x(t) \\ u_i(t) \end{bmatrix} - v_i w^2(t) \right\} dt, \quad i = 1, 2,$$

subject to the dynamics

$$\dot{x}(t) = \begin{bmatrix} \beta & 0 \\ 0 & 0 \end{bmatrix} x(t) + \begin{bmatrix} -1 \\ 0 \end{bmatrix} u_1(t) + \begin{bmatrix} -1 \\ 0 \end{bmatrix} u_2(t) + \begin{bmatrix} 1 \\ 0 \end{bmatrix} w(t), \quad x(0) = \begin{bmatrix} s_0 \\ 1 \end{bmatrix}.$$

Using the transformation (see Section 3.6, part III)

$$\tilde{u}_i := u_i - \frac{p}{2\gamma_i}, \quad i = 1, 2,$$

the optimization problem can be rewritten as

$$\min_{\tilde{u}_i \in \mathcal{F}^{aff}} \sup_{w \in L_2(0,\infty)} \int_0^\infty e^{-rt} \{\tilde{x}^T(t) \begin{bmatrix} 0 & 0 \\ 0 & \frac{-p^2}{4\gamma_i} \end{bmatrix} \tilde{x}(t) + \gamma_i \tilde{u}_i^2(t) - v_i w^2(t)\} dt, \quad i = 1, 2,$$

subject to the dynamics

$$\dot{\tilde{x}}(t) = \begin{bmatrix} \beta & \frac{-p}{2\gamma_1} + \frac{-p}{2\gamma_2} \\ 0 & 0 \end{bmatrix} \tilde{x}(t) + \begin{bmatrix} -1 \\ 0 \end{bmatrix} \tilde{u}_1(t) + \begin{bmatrix} -1 \\ 0 \end{bmatrix} \tilde{u}_2(t) + \begin{bmatrix} -1 \\ 0 \end{bmatrix} w(t), \tilde{x}(0) = \begin{bmatrix} s_0 \\ 1 \end{bmatrix}.$$

With, $\hat{x}(t) := e^{-\frac{1}{2}rt}\tilde{x}(t)$, $\hat{u}_i(t) := e^{-\frac{1}{2}rt}\tilde{u}_i(t)$, $i = 1, 2$, and $\hat{w}(t) := e^{-\frac{1}{2}rt}w(t)$, we obtain the standard formulation

$$\min_{\hat{u}_i \in \mathcal{F}} \sup_{\hat{w} \in L_2(0,\infty)} \int_0^\infty \left\{ \hat{x}^T(t) \begin{bmatrix} 0 & 0 \\ 0 & \frac{-p^2}{4\gamma_i} \end{bmatrix} \hat{x}(t) + \gamma_i \hat{u}_i^2(t) - v_i \hat{w}^2(t) \right\} dt, \quad i = 1, 2,$$

subject to the dynamics

$$\dot{\hat{x}}(t) = \begin{bmatrix} \beta - \frac{1}{2}r & \frac{-p}{2\gamma_1} + \frac{-p}{2\gamma_2} \\ 0 & -\frac{1}{2}r \end{bmatrix} \hat{x}(t) + \begin{bmatrix} -1 \\ 0 \end{bmatrix} \hat{u}_1(t) + \begin{bmatrix} -1 \\ 0 \end{bmatrix} \hat{u}_2(t) + \begin{bmatrix} -1 \\ 0 \end{bmatrix} \hat{w}(t), \hat{x}(0) = \begin{bmatrix} s_0 \\ 1 \end{bmatrix}.$$

According to Theorem 9.18 the soft-constrained Nash equilibria for this game are obtained as

$$\hat{u}_i(t) = \begin{bmatrix} \frac{1}{\gamma_i} & 0 \end{bmatrix} X_i \hat{x}(t),$$

where with

$$A := \begin{bmatrix} \beta - \frac{1}{2}r & \frac{-p}{2\gamma_1} + \frac{-p}{2\gamma_2} \\ 0 & -\frac{1}{2}r \end{bmatrix}, \quad S_i := \begin{bmatrix} \frac{1}{\gamma_i} & 0 \\ 0 & 0 \end{bmatrix}, \quad S_{ij} = 0, \quad M_i := \begin{bmatrix} \frac{1}{v_i} & 0 \\ 0 & 0 \end{bmatrix} \text{ and } Q_i := \begin{bmatrix} 0 & 0 \\ 0 & \frac{-p^2}{4\gamma_i} \end{bmatrix}$$

(X_1, X_2) solve equations (9.5.5) and (9.5.6) and satisfy the conditions (9.5.7) and (9.5.8). Notice that with

$$Y_i := \begin{bmatrix} 0 & 0 \\ 0 & \frac{-p^2}{4r\gamma_i} \end{bmatrix}, \quad i = 1, 2,$$

the inequalities (9.5.9) and (9.5.10) are satisfied.

In case the discount factor, r, is more than two times larger than the exogenous growth rate, β, of the fish population we see by straightforward substitution that

$$X_i := \begin{bmatrix} 0 & 0 \\ 0 & \frac{-p^2}{4r\gamma_i} \end{bmatrix}, \quad i = 1, 2, \tag{9.6.1}$$

satisfy the equations (9.5.5)–(9.5.8). So one soft-constrained Nash equilibrium, in that case, is provided by

$$u_i^*(t) = \frac{p}{2\gamma_i}.$$
(9.6.2)

That is, irrespective of the growth of the fish population, the fishermen catch a constant amount of fish each time. This amount is completely determined by their cost function and the price of the fish.

To see whether different equilibria exist, we elaborate on the equations (9.5.5) and (9.5.6). To that end introduce

$$X_1 =: \begin{bmatrix} x_1 & x_2 \\ x_2 & x_3 \end{bmatrix} \quad \text{and} \quad X_2 =: \begin{bmatrix} z_1 & z_2 \\ z_2 & z_3 \end{bmatrix}.$$

Then, the equations (9.5.5) and (9.5.6) can be rewritten as

$$\left(-2\beta + r + \frac{2}{\gamma_2}z_1\right)x_1 + \left(\frac{1}{\gamma_1} - \frac{1}{v_1}\right)x_1^2 = 0$$
(9.6.3)

$$\left(\frac{p}{2\gamma_1} + \frac{p}{2\gamma_2} + \frac{1}{\gamma_2}z_2\right)x_1 + \left(r - \beta + \frac{1}{\gamma_2}z_1\right)x_2 + \left(\frac{1}{\gamma_1} - \frac{1}{v_1}\right)x_1 x_2 = 0$$
(9.6.4)

$$\frac{p^2}{4\gamma_1} + \left(\frac{p}{\gamma_1} + \frac{p}{\gamma_2} + \frac{2}{\gamma_2}z_2\right)x_2 + \left(\frac{1}{\gamma_1} - \frac{1}{v_1}\right)x_2^2 + rx_3 = 0$$
(9.6.5)

$$\left(-2\beta + r + \frac{2}{\gamma_1}x_1\right)z_1 + \left(\frac{1}{\gamma_2} - \frac{1}{v_2}\right)z_1^2 = 0$$
(9.6.6)

$$\left(\frac{p}{2\gamma_1} + \frac{p}{2\gamma_2} + \frac{1}{\gamma_1}x_2\right)z_1 + \left(r - \beta + \frac{1}{\gamma_1}x_1\right)z_2 + \left(\frac{1}{\gamma_2} - \frac{1}{v_2}\right)z_1 z_2 = 0$$
(9.6.7)

$$\frac{p^2}{4\gamma_2} + \left(\frac{p}{\gamma_1} + \frac{p}{\gamma_2} + \frac{2}{\gamma_1}x_2\right)z_2 + \left(\frac{1}{\gamma_2} - \frac{1}{v_2}\right)z_2^2 + rz_3 = 0.$$
(9.6.8)

From the first equation (9.6.3) it follows that either

$$\text{(i) } x_1 = 0 \quad \text{or (ii) } \left(\frac{1}{\gamma_1} - \frac{1}{v_1}\right)x_1 + \frac{2}{\gamma_2}z_1 = 2\beta - r.$$

In case (i), $x_1 = 0$, equations (9.6.4) and (9.6.6) yield that $x_2 = 0$ (under the assumptions that $\beta \neq r$ and $\beta + \frac{\gamma_2}{v_2}(\beta - r) \neq 0$). Equation (9.6.5) then shows that necessarily $x_3 = -\frac{p^2}{4r\gamma_1}$. Equations (9.6.6)–(9.6.8) then give that either X_2 is as reported in equation (9.6.1) or

$$X_2 = \begin{bmatrix} \frac{\gamma_2 v_2(2\beta - r)}{v_2 - \gamma_2} & \bar{x}_2 \\ \bar{x}_2 & -\frac{p^2}{4r\gamma_2} - \frac{p}{2\beta}\left(\frac{1}{\gamma_1} + \frac{1}{\gamma_2}\right)\bar{x}_2 \end{bmatrix},$$
(9.6.9)

where $\bar{x}_2 = -\frac{v_2}{\gamma_1(v_2 - \gamma_2)}\frac{p(\gamma_1 + \gamma_2)(2\beta - r)}{2\beta}$

Similarly, a lengthy analysis of case (ii) shows that besides the solutions (9.6.1) and (9.6.9) this set of equations still has (given our parametric assumptions) two other solutions. Introducing, for notational convenience,

$$c := -\frac{p(\gamma_1 + \gamma_2)(2\beta - r)}{2\beta\gamma_1\gamma_2} \quad \text{and} \quad d := 3v_1v_2 + \gamma_1v_2 + \gamma_2v_1 - \gamma_1\gamma_2,$$

these solutions are

$$X_1 := \begin{bmatrix} \frac{\gamma_1 v_1 (2\beta - r)}{v_1 - \gamma_1} & \bar{y}_2 \\ \bar{y}_2 & -\frac{p^2}{4r\gamma_1} - \frac{p}{2\beta}\left(\frac{1}{\gamma_1} + \frac{1}{\gamma_2}\right)\bar{y}_2 \end{bmatrix}, \quad X_2 = \begin{bmatrix} 0 & 0 \\ 0 & -\frac{p^2}{4r\gamma_2} \end{bmatrix}, \quad (9.6.10)$$

where $\bar{y}_2 = \frac{v_1\gamma_1}{(v_1 - \gamma_1)}c$; and

$$X_1 := \begin{bmatrix} \frac{\gamma_1 v_1 (\gamma_2 + v_2)(2\beta - r)}{d} & \bar{z}_2 \\ \bar{z}_2 & -\frac{p^2}{4r\gamma_1} - \frac{p}{2\beta}\left(\frac{1}{\gamma_1} + \frac{1}{\gamma_2}\right)\bar{z}_2 \end{bmatrix},$$

$$X_2 := \begin{bmatrix} \frac{\gamma_2 v_2 (\gamma_1 + v_1)(2\beta - r)}{d} & \tilde{z}_2 \\ \tilde{z}_2 & -\frac{p^2}{4r\gamma_2} - \frac{p}{2\beta}\left(\frac{1}{\gamma_1} + \frac{1}{\gamma_2}\right)\tilde{z}_2 \end{bmatrix}, \quad (9.6.11)$$

where $\bar{z}_2 = \frac{v_1\gamma_1(v_2 + \gamma_2)}{d}c$ and $\tilde{z}_2 = \frac{v_2\gamma_2(v_1 + \gamma_1)}{d}c$.

From expressions (9.6.9)–(9.6.11) the following potential equilibrium actions result

$$(u_1^*(t), u_2^*(t)) = \left(\frac{p}{2\gamma_1}, \frac{v_2}{v_2 - \gamma_2}\left((2\beta - r)s(t) + c\right) + \frac{p}{2\gamma_2}\right); \quad (9.6.12)$$

$$(u_1^*(t), u_2^*(t)) = \left(\frac{v_1}{v_1 - \gamma_1}\left((2\beta - r)s(t) + c\right) + \frac{p}{2\gamma_1}, \frac{p}{2\gamma_2}\right); \quad (9.6.13)$$

$$(u_1^*(t), u_2^*(t)) = \left(\frac{v_1(\gamma_2 + v_2)}{d}\left((2\beta - r)s(t) + c\right) + \frac{p}{2\gamma_1}, \frac{v_2(\gamma_1 + v_1)}{d}\right.$$

$$\left. \times \left((2\beta - r)s(t) + c\right) + \frac{p}{2\gamma_2}\right), \quad (9.6.14)$$

respectively.

To see whether they actually can arise as equilibria we have to verify whether there are parametric conditions such that the stability constraints (9.5.7) and (9.5.8) are met. Straightforward calculations show that for each of these three equilibria (9.6.12)–(9.6.14) these stability conditions are satisfied if and only if $2\beta - r > 0$. As an example consider the equilibrium strategy (9.6.14). Using this strategy,

$$A - S_1 X_1 - S_2 X_2 = \begin{bmatrix} -\frac{(2\beta - r)(v_1 + \gamma_1)(v_2 + \gamma_2)}{2d} & h_1 \\ 0 & -\frac{1}{2}r \end{bmatrix},$$

whereas

$$A - S_1 X_1 - S_2 X_2 + M_i X_i = \begin{bmatrix} -\frac{(2\beta - r)(v_i - \gamma_i)(v_j + \gamma_j)}{2d} & h_2 \\ 0 & -\frac{1}{2}r \end{bmatrix}, \quad i,j = 1,2, \ j \neq i.$$

Here, h_i, $i = 1, 2, 3$, are some parameters which are not important for the stability analysis. The claim then follows directly by considering the first entry of all these matrices. All these entries are negative if and only if $2\beta - r > 0$.

From this analysis it follows that if $r > 2\beta$ the game has a unique equilibrium (9.6.2) which is characterized by the fishing of a fixed amount of fish by both fishermen. Due to the large discounting rate the players seem to be indifferent to their noise expectations. This is because the fixed amount of fish they catch is independent of these expectations. This equilibrium results in a situation where, under the assumptions that the initial fish stock s_0 is larger than $\frac{p}{2\beta}\left(\frac{1}{\gamma_1} + \frac{1}{\gamma_2}\right)$ and the deterministic negative impact factor w is not too large, the amount of fish will grow steadily with a factor β. Notice that the expected worst-case revenues (i.e. $-J_i^*$) of fisherman i are $-\hat{x}^T(0)X_i\hat{x}(0) = \frac{p^2}{4r\gamma_i}$, $i = 1, 2$. This coincides with the actual revenues obtained using these actions, as measured by

$$-\int_0^\infty e^{-rt}\{-pu_i(t) + \gamma_i u_i^2(t)\}dt, \quad i = 1, 2.$$

If $r < 2\beta$ a different situation occurs. Then, three different equilibria occur. Two equilibria correspond to a situation where one fisherman fishes a fixed amount of fish, whereas the amount of fish the other fisherman catches consists of a fixed amount (that might be negative, which can be interpreted as that the fisherman plants some fish), and an additional amount that depends on both the fishstock and his expectations about the deterministic disturbance. In the third equilibrium both fishermen catch an amount of fish that depends on the fishstock additional to some fixed (possibly negative) amount.

If $g_i := v_i(r - \beta) - \beta\gamma_i < 0$ and the external factors w are modest, the fish stock will converge to some fixed amount in the first two equilibria (9.6.12) and (9.6.13), respectively. This amount depends on the actual realization of the external factor w. In case $g_i > 0$, on the other hand, the fish stock will grow steadily with a growth factor g provided $s_0 > \frac{p(\gamma_1 + \gamma_2)}{2\beta\gamma_1\gamma_2(v_i - \gamma_i)}$, $i = 1, 2$ (assuming again that the external factors are not too unwieldy). The expected worst-case revenues of one of the fishermen, i, in these two equilibria are $\frac{p^2}{4r\gamma_i}$ (which coincide again with his actual obtained revenues), whereas the revenues under the worst-case realization of the noise for fisherman j are

$$-J_j^* = -\hat{x}_0^T\left(\frac{\gamma_j}{v_j - \gamma_j}H_{1,j} + H_{2,j}\right)\hat{x}_0, \tag{9.6.15}$$

where

$$H_{1,j} = \begin{bmatrix} \gamma_j(2\beta - r) & c\gamma_j \\ c\gamma_j & -\frac{pc\gamma_j}{2\beta}\left(\frac{1}{\gamma_1} + \frac{1}{\gamma_2}\right) \end{bmatrix} \quad \text{and} \quad H_{2,j} = \begin{bmatrix} \gamma_j(2\beta - r) & c\gamma_j \\ c\gamma_j & -\frac{p^2}{4r\gamma_j} - \frac{pc\gamma_j}{2\beta}\left(\frac{1}{\gamma_1} + \frac{1}{\gamma_2}\right) \end{bmatrix}$$

$i,j = 1, 2,\ j \neq i$. Since the amount of fish caught by this fisherman now depends on the fish stock, and thus in particular on the realization of the disturbance factor w, in general these worst-case costs will differ from the actual revenues for him.

Notice that both $H_{1,j}$ and $H_{2,j}$ do not depend on the noise parameters and that $H_{1,j}$ is positive semi-definite. Consequently, $\frac{\gamma_j}{v_j - \gamma_j} H_{1,j}$ in equation (9.6.15) reflects the cost involved for the fisherman due to his worst-case expectations concerning the external factor w. This effect is almost negligible if the fisherman expects a modest influence of w on the fish growth (i.e. if v_i is at least several times larger than γ_i). In that case the costs J_i^* are close to the costs the fisherman has in the undisturbed case ($v_i = \infty$). In case the fisherman's worst-case expectations about w are large (i.e. v_i close to γ_i) these worst-case expected costs are completely dominated by $\frac{\gamma_j}{v_j - \gamma_j} H_{1,j}$. It seems reasonable to take the view that a fisherman will only go fishing (assuming that he wants to have a profit even under his worst-case expectations about w) if $-J_i^*$ is positive. This gives additional conditions on the parameters that have to be satisfied to consider this equilibrium outcome as a realistic one. Particularly when there is a very large initial fish stock, these conditions will usually not be satisfied. However, given our model assumptions this is a situation which we can rule out. Also, if the expected external factors w become generically dominating the revenues $-J_i^*$ become negative. This is because $H_{1,j}$ in equation (9.6.15) is positive semi-definite. So, in that case as well the equilibrium ceases to exist. Again this case is intuitively clear. If there is a large amount of 'external fishing', almost no fish will be left in the lake. So the fisherman is confronted with exceptional costs to catch the remainder of the fish. Since he gets a fixed price on the market for his fish he will quit fishing.

Finally, notice that both $\frac{\partial u_j^*}{\partial v_j} < 0$ and $\frac{\partial J_j^*}{\partial v_j} < 0$. From this, one conclusion is that the more fisherman j expects that the fish stock will be disturbed by external factors, the more fish he will catch himself. Another conclusion is that the expected returns under the worst-case scenario decrease for a fisherman if he expects more negative external impacts.

In the third equilibrium (9.6.14) the costs for fisherman j are:

$$
J_j^* = \hat{x}^T(0) \left(h_j H_{1,j} + \begin{bmatrix} \frac{\gamma_j(2\beta - r)}{3} & \frac{c\gamma_j}{3} \\ \frac{c\gamma_j}{3} & -\frac{p^2}{4r\gamma_j} - \frac{pc\gamma_j}{6\beta}\left(\frac{1}{\gamma_1} + \frac{1}{\gamma_2}\right) \end{bmatrix} \right) \hat{x}(0), \tag{9.6.16}
$$

$i,j = 1, 2,\ j \neq i$, where $h_j = \frac{2\gamma_i v_j - \gamma_j v_i + \gamma_1 \gamma_2}{3d}$. For this equilibrium a similar analysis can be performed as above, leading to similar conclusions. We will not elaborate those points here.

One way in which this equilibrium differs from the previous one is that the equilibrium action and expected worst-case revenues depend on the opposing fisherman's noise expectations. From equations (9.6.14) and (9.6.16) it follows that both $\frac{\partial u_j^*}{\partial v_i}$ and $\frac{\partial J_j^*}{\partial v_i}$ are negative, whereas both $\frac{\partial u_j^*}{\partial v_j}$ and $\frac{\partial J_j^*}{\partial v_j}$ are positive. This implies that each fisherman responds to an increase in worst-case expectations about the external factors of the other fisherman by catching more fish. Furthermore, an increase of the worst-case expectations of his opponent has a negative impact on his worst-case expected revenues. This, contrary to his own reaction to an increase in worst-case expectations with respect to external factors. If the fisherman himself expects more 'disturbances' he will react by catching less fish which has a positive impact on his worst-case expected revenues.

Finally, straightforward calculations show that not one of the three equilibria Pareto dominates another equilibrium. That is, comparing the worst-case revenues of any two of the above three equilibria one fisherman is always better off in one equilibrium whereas the other is better off in the other equilibrium.

In conclusion we observe that taking noise into account has a number of consequences. First, if the $r > 2\beta$, noise does not affect the outcome of the game. The fishermen keep fishing a fixed amount over time. If $r < 2\beta$ the noise expectations do play a role. Three different equilibria may occur. Two equilibria in which either one of the fishermen sticks to the noise-free optimal action and the other takes into account the current fish stock and his worst-case expectations about the external factors in the amount of fish he catches. At the other equilibrium both fishermen simultaneously take each other's noise expectations into account.

In all these equilibria we frequently observe a 'tragedy of common's' effect. That is, the fishermen react to an expected more disturbed fish stock growth by either himself or his opponent with increasing the number of fish they catch. Only in the last-mentioned equilibrium does a reverse reaction occurs – if a fisherman anticipates a more disturbed environment he will catch less himself. This effect is, however, counteracted if he also observes an increased sensitivity in his opponent w.r.t. the external factors.

9.7   A scalar numerical algorithm

Theorem 9.18 shows that the equations (9.5.5)–(9.5.8) play a crucial role in the question as to whether the game (9.5.1) and (9.5.2) will have a soft-constrained Nash equilibrium. Every soft-constrained Nash equilibrium has to satisfy these equations. So, the question arises under which conditions (9.5.5)–(9.5.8) will have one or more solutions and, if possible, to calculate these solutions. This is a difficult open question. Similar remarks apply here as were made in Section 8.5 for solving the corresponding set of algebraic Riccati equations to determine the feedback Nash equilibria. However, again for the scalar case, one can devise an algorithm to calculate all soft-constrained Nash equilibria. This algorithm is in the spirit of Algorithm 8.1 and will be discussed in this Section.

As in Section 8.5 we will consider the general N-player case under the simplifying assumptions that $b_i \neq 0$ and players have no direct interest in each others' control actions (i.e. $S_{ij} = 0$, $i \neq j$). Using again notation lower case, to stress the fact that we are dealing with the scalar case, the set of equations (9.5.5)–(9.5.8) become (see also the concluding remarks at the end of this chapter)

$$-2\left(a - \sum_{j \neq i}^{N} s_j x_j\right) x_i + (s_i - m_i)x_i^2 - q_i = 0, \quad i = 1, \ldots, N, \tag{9.7.1}$$

$$a - \sum_{j=1}^{N} s_j x_j < 0, \quad \text{and} \tag{9.7.2}$$

$$a - \sum_{j=1}^{N} s_j x_j + m_i x_i < 0, \quad i = 1, \ldots, N. \tag{9.7.3}$$

Some elementary calculation shows that equation (9.7.1) can be rewritten as

$$2\left(a - \sum_{j=1}^{N} s_j x_j\right)x_i + (s_i + m_i)x_i^2 + q_i = 0, \quad i = 1, \ldots, N. \tag{9.7.4}$$

For notational convenience next introduce, for $i = 1, \ldots, N$ and Ω some index set of the numbers $\{1, \ldots, N\}$[1], the variables

$$\tau_i := (s_i + m_i)q_i, \quad \tau_{max} := \max_i \tau_i, \quad \rho_i := \frac{s_i}{s_i + m_i}, \quad \gamma_i := -1 + 2\rho_i = \frac{s_i - m_i}{s_i + m_i}, \tag{9.7.5}$$

$$\gamma_\Omega := -1 + 2\sum_{i \in \Omega} \rho_i, \quad y_i := (s_i + m_i)x_i, \quad \text{and} \quad y_{N+1} := -a_{cl} := -\left(a - \sum_{i=1}^{N} s_i x_i\right).$$

With some small change of notation for a fixed index set Ω, γ_Ω will also be denoted without brackets and commas. That is, if e.g. $\Omega = \{1,2\}$, γ_Ω is also written as γ_{12}. Moreover, assume (without loss of generality) that $\tau_1 \geq \cdots \geq \tau_N$. Multiplication of equation (9.7.4) by $s_i + m_i$ shows that (9.7.1)–(9.7.3) has a solution if and only if

$$y_i^2 - 2y_{N+1}y_i + \tau_i = 0, \quad i = 1, \ldots, N, \tag{9.7.6}$$

$$y_{N+1} = -a + \sum_{i=1}^{N} \rho_i y_i, \tag{9.7.7}$$

$$-y_{N+1} + \frac{m_i}{m_i + s_i}y_i < 0, \quad i = 1, \ldots, N, \quad \text{and} \tag{9.7.8}$$

$$y_{N+1} > 0 \tag{9.7.9}$$

has a set of real solutions y_i, $i = 1, \ldots, N + 1$. Following the analysis of Section 8.5.1 one obtains the following lemma.

Lemma 9.20

The set of equations (9.7.6) and (9.7.7) has a solution such that $y_{N+1} > 0$ if and only if there exist $t_i \in \{-1, 1\}$, $i = 1, \ldots, N$, such that the equation

$$\left(-1 + \sum_{i=1}^{N} \rho_i\right)y_{N+1} + t_1\rho_1\sqrt{y_{N+1}^2 - \tau_1} + \cdots + t_N\rho_N\sqrt{y_{N+1}^2 - \tau_N} = a \tag{9.7.10}$$

has a solution $y_{N+1} > 0$. In fact all solutions of equations (9.7.6) and (9.7.7) are obtained by considering all possible sequences $(t_1, \ldots, t_N)$ in equation (9.7.10).

Obviously, a necessary condition for equation (9.7.10) to have a solution is that $y_{N+1}^2 \geq \tau_1$.   □

---

[1] So, for $N = 2$, Ω is either $\{1\}$, $\{2\}$ or $\{1,2\}$.

In the same way as in Section 8.5 we next define recursively the following functions for $n = 1, \ldots, N - 1$:

$$f_i^{n+1}(x) := f_i^n(x) + \rho_{n+1}x - \rho_{n+1}\sqrt{x^2 - \tau_{n+1}}, \quad i = 1, \ldots, 2^n,$$

$$f_{i+2^n}^{n+1}(x) := f_i^n(x) + \rho_{n+1}x + \rho_{n+1}\sqrt{x^2 - \sigma_{n+1}}, \quad i = 1, \ldots, 2^n,$$

with

$$f_1^1(x) := (-1 + \rho_1)x - \rho_1\sqrt{x^2 - \tau_1} \quad \text{and} \quad f_2^1(x) := (-1 + \rho_1)x + \rho_1\sqrt{x^2 - \tau_1}.$$

Each function f_i^N, $i = 1, \ldots, 2^N - 1$, corresponds to a function obtained from the left-hand side of equation (9.7.10) by making a specific choice of t_j, $j = 1, \ldots, N$, and substituting x for y_{N+1}. From Lemma 9.20 it is then obvious that equations (9.7.6) and (9.7.7) has a solution if and only if $f_i^N(x) = a$ has a solution for some $i \in \{1, \ldots, 2^N\}$. Or, stated differently, equations (9.7.6) and (9.7.7) have a solution if and only if the following function has a root

$$\prod_{i=1}^{2^N}(f_i^N(x) - a) = 0.$$

Denoting the function on the left-hand side of this equation, $\prod_{i=1}^{2^N}(f_i^N(x) - a)$, by $f(x)$ one obtains, using the same analysis as in Section 8.5, the following theorem.

Theorem 9.21

y_i is a solution of equations (9.7.6) and (9.7.7) if and only if y_{N+1} is a zero of $f(x)$ and there exist $t_i \in \{-1, 1\}$, such that $y_i = y_{N+1} + t_i\rho_i\sqrt{y_{N+1}^2 - \tau_i}$. Moreover, $f(x)$ is a polynomial of degree 2^N. $\square$

An immediate consequence of this theorem is the following corollary.

Corollary 9.22

The N-player scalar game has at most 2^N soft-constrained Nash equilibria. $\square$

In Corollary 8.15 it was shown that the N-player scalar undisturbed linear quadratic differential game always has at most $2^N - 1$ feedback Nash equilibria. The following example shows that we cannot draw a similar conclusion here, solely based on equations (9.7.6) and (9.7.7).

Example 9.6

Consider the two-player scalar game with $a = -2$, $s_i = 1$, $m_i = 9$, $i = 1, 2$, $q_1 = 0.1$ and $q_2 = 0.05$. For this case we plotted the four curves f_i^2 in Figure 9.1. From this graph we see that all curves are monotonically decreasing and they all have an inter Section point with -2 for a value $y_3 > 1$. Consequently, the set of equations (9.7.6) and (9.7.7) has four solutions. $\square$

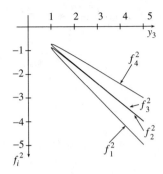

Figure 9.1   Curves f_i^2 in Example 9.6

Next, we develop a numerical algorithm similar to Algorithm 8.1 to find all solutions of equations (9.7.6) and (9.7.7). Again, for didactical reasons, the two-player case is considered first. Let p_1, p_2 be a possibly complex solution of equations (9.7.6) and (9.7.7). Denote the negative of the resulting closed-loop system parameter by

$$\lambda := -a + \rho_1 p_1 + \rho_2 p_2. \tag{9.7.11}$$

Then,

$$p_1^2 - 2\lambda p_1 + \tau_1 = 0, \tag{9.7.12}$$

and

$$p_2^2 - 2\lambda p_2 + \tau_2 = 0. \tag{9.7.13}$$

Consequently, using the definition of λ and equation (9.7.12), respectively,

$$\begin{aligned}
p_1\lambda &= -p_1 a + \rho_1 p_1^2 + \rho_2 p_1 p_2 \\
&= -p_1 a + \rho_1(2\lambda p_1 - \tau_1) + \rho_2 p_1 p_2.
\end{aligned}$$

From this it follows that

$$\gamma_1 p_1 \lambda = \rho_1 \tau_1 + a p_1 - \rho_2 p_1 p_2. \tag{9.7.14}$$

In a similar way using the definition of λ and (9.7.13), respectively, one obtains

$$\begin{aligned}
p_2\lambda &= -p_2 a + \rho_1 p_1 p_2 + \rho_2 p_2^2 \\
&= -p_2 a + \rho_1 p_1 p_2 + \rho_2(2\lambda p_2 - \tau_2).
\end{aligned}$$

Which gives rise to

$$\gamma_2 p_2 \lambda = \rho_2 \tau_2 + a p_2 - \rho_1 p_1 p_2. \tag{9.7.15}$$

Finally, using the definition of λ and both equations (9.7.12) and (9.7.13), respectively, we get

$$
\begin{aligned}
p_1 p_2 \lambda &= -p_1 p_2 a + \rho_1 p_1^2 p_2 + \rho_2 p_1 p_2^2 \\
&= -p_1 p_2 a + 2(\rho_1 + \rho_2)\lambda p_1 p_2 - \rho_1 \tau_1 p_2 - \rho_2 \tau_2 p_1,
\end{aligned}
$$

which gives

$$
(2(\rho_1 + \rho_2) - 1)p_1 p_2 \lambda = \rho_2 \tau_2 p_1 + \rho_1 \tau_1 p_2 + a p_1 p_2. \tag{9.7.16}
$$

So, using the notation (9.7.5), with

$$
\tilde{M} := \begin{bmatrix}
-a & \rho_1 & \rho_2 & 0 \\
\dfrac{\rho_1 \tau_1}{\gamma_1} & \dfrac{a}{\gamma_1} & 0 & -\dfrac{\rho_2}{\gamma_1} \\
\dfrac{\rho_2 \tau_2}{\gamma_2} & 0 & \dfrac{a}{\gamma_2} & -\dfrac{\rho_1}{\gamma_2} \\
0 & \dfrac{\rho_2 \tau_2}{\gamma_{12}} & \dfrac{\rho_1 \tau_1}{\gamma_{12}} & \dfrac{a}{\gamma_{12}}
\end{bmatrix}
$$

we conclude from equations (9.7.11) and (9.7.14)–(9.7.16) that, provided $\gamma_i \neq 0$ $i = 1, 2, 12$, every solution p_1, p_2 of (9.7.6) and (9.7.7) satisfies the equation

$$
\tilde{M} \begin{bmatrix} 1 \\ p_1 \\ p_2 \\ p_1 p_2 \end{bmatrix} = \lambda \begin{bmatrix} 1 \\ p_1 \\ p_2 \\ p_1 p_2 \end{bmatrix}.
$$

Using the fact that $p_i = (s_i + m_i)x_i$ an analogous reasoning to that used in Theorem 8.16 gives the next lemma.

Lemma 9.23

1. Assume that (x_1, x_2) solves expressions (9.7.1) and (9.7.2) and $\gamma_i \neq 0$, $i = 1, 2, 12$. Then $\lambda := -a + \sum_{i=1}^{2} s_i x_i > 0$ is an eigenvalue of the matrix

$$
M := \begin{bmatrix}
-a & s_1 & s_2 & 0 \\
\dfrac{\rho_1 q_1}{\gamma_1} & \dfrac{a}{\gamma_1} & 0 & -\dfrac{s_2}{\gamma_1} \\
\dfrac{\rho_2 q_2}{\gamma_2} & 0 & \dfrac{a}{\gamma_2} & -\dfrac{s_1}{\gamma_2} \\
0 & \dfrac{\rho_2 q_2}{\gamma_{12}} & \dfrac{\rho_1 q_1}{\gamma_{12}} & \dfrac{a}{\gamma_{12}}
\end{bmatrix}. \tag{9.7.17}
$$

Furthermore, $[1, x_1, x_2, x_1 x_2]^T$ is a corresponding eigenvector and $\lambda^2 \geq \tau_{max}$.

2. Assume that $[1, x_1, x_2, x_3]^T$ is an eigenvector corresponding to a positive eigenvalue λ of M, satisfying $\lambda^2 \geq \tau_{max}$, and that the eigenspace corresponding to τ has dimension one. Then, (x_1, x_2) solves expressions (9.7.1) and (9.7.2). $\qquad \square$

Lemma 9.20 and Lemma 9.23 then give rise to the following numerical algorithm.

Algorithm 9.1

Let $s_i := \frac{b_i^2}{r_i}$ and $m_i := \frac{e^2}{v_i}$. Assume that for every index set $\Omega \subset \{1, \ldots, N\}$, $\gamma_\Omega \neq 0$. Then, the following algorithm calculates all solutions of (9.7.1) and (9.7.2).

Step 1 Calculate matrix M in expression (9.7.17) and $\tau := \max_i(s_i + m_i)q_i$.

Step 2 Calculate the eigenstructure (λ_i, n_i), $i = 1, \ldots, k$, of M, where λ_i are the eigenvalues and n_i the corresponding algebraic multiplicities.

Step 3 For $i = 1, \ldots, k$ repeat the following steps:

    3.1. If (i) $\lambda_i \in \mathbb{R}$; (ii) $\lambda_i > 0$ and (iii) $\lambda_i^2 \geq \tau$ then proceed with Step 3.2 of the algorithm. Otherwise, return to Step 3.

    3.2. If $n_i = 1$ then

        (a) calculate an eigenvector z corresponding to λ_i of M. Denote the entries of z by $[z_0, z_1, z_2, \ldots]^T$. Calculate $x_j := \frac{z_j}{z_0}$. Then, $(x_1, \ldots, x_N)$ solve (9.7.1,9.7.2). Return to Step 3.

        If $n_i > 1$ then

        (b) Calculate $\tau_i := s_i q_i$.

        (c) For all 2^N sequences $(t_1, \ldots, t_N)$, $t_k \in \{-1, 1\}$,

        (i) calculate

$$y_j := \lambda_i + t_j \frac{s_i}{s_i + m_i}\sqrt{\lambda_i^2 - \sigma_j}, \quad j = 1, \ldots, N;$$

        (ii) if $\lambda_i = -a + \sum\limits_{j=1,\ldots,N} y_j$ then calculate $x_j := \frac{y_j}{s_j+m_j}$. Then, $(x_1, \ldots, x_N)$ solves (9.7.1) and (9.7.2).

Step 4 End of the algorithm. ☐

Example 9.7

Reconsider Example 9.6. That is, consider the two-player scalar game with $a = -2$, $b_i = e = 1$, $r_i = 1$, $v_i = \frac{1}{9}$, $i = 1, 2$, $q_1 = 0.1$ and $q_2 = 0.05$. To calculate the soft-constrained Nash equilibria of this game, we first determine all solutions of expressions (9.7.1) and (9.7.2). According to Algorithm 9.1, we first have to determine the eigenstructure of the following matrix

$$M := \begin{bmatrix} 2 & 1 & 1 & 0 \\ -1/80 & 5/2 & 0 & 5/4 \\ -1/160 & 0 & 5/2 & 5/4 \\ 0 & -1/120 & -1/60 & 10/3 \end{bmatrix}.$$

Using MATLAB, we find that M has the eigenvalues $\{2.0389, 2.4866, 2.5132, 3.2946\}$. Since all eigenvalues are larger than $\tau = 1$, we have to carryout Step 3 of the algorithm for all these eigenvalues. Since all the eigenvalues have a geometric multiplicity of one, there are four solutions satisfying expressions (9.7.1) and (9.7.2). From the corresponding eigenspaces one obtains the solutions show in Table 9.1 (with $a_{cl} = a - s_1 x_1 - s_2 x_2 = -$eigenvalue).

Table 9.1

Eigenvalue	(x_1, x_2)	$a_{cl} + m_1 x_1$	$a_{cl} + m_2 x_2$
2.0389	(0.0262,0.0127)	−1.8030	−1.9250
2.4866	(0.4763,0.0103)	1.8003	−2.3942
2.5132	(0.0208,0.4925)	−2.3265	1.9192
3.2946	(0.6434,0.6512)	2.4958	2.5666

From the last two columns of Table 9.1 we see that only the first solution satisfies the additional conditions (9.7.3). Since $q_i > 0$, and thus (9.5.9) and (9.5.10) are satisfied with $y_i = 0$, it follows that this game has one soft-constrained Nash equilibrium. The equilibrium actions corresponding to this equilibrium $(0.0262, 0.0127)$ are

$$u_1^*(t) = -0.0262x(t) \quad \text{and} \quad u_2^*(t) = -0.0127x(t).$$

Assuming that the initial state of the system is x_0, the worst-case costs expected by the players are

$$J_1^* = 0.0262x_0^2 \quad \text{and} \quad J_2^* = 0.0127x_0^2,$$

respectively. □

Note

If k of the γ_i parameters are zero, we obtain k linear equations in the variables $(1, p_1, p_2, p_1 p_2)$. Under some regularity conditions, k of these variables can then be explicitly solved as a function of the remaining $2^N - k$ variables. The solutions of the remaining $2^N - k$ equations can then be obtained using a similar eigenstructure algorithm.

As an example consider the case that in the two-player case described above $\gamma_1 = 0$ (and $\gamma_j \neq 0$, $j = 2, 12$). So, equations (9.7.11) and (9.7.14)–(9.7.16) reduce to

$$\lambda = -a + p_1 p_1 + p_2 p_2, \tag{9.7.18}$$
$$0 = p_1 \tau_1 + a p_1 - p_2 p_1 p_2 \tag{9.7.19}$$
$$\gamma_2 p_2 \lambda = p_2 \tau_2 + a p_2 - p_1 p_1 p_2 \tag{9.7.20}$$
$$\gamma_{12} p_1 p_2 \lambda = p_2 \tau_2 p_1 + p_1 \tau_1 p_2 + a p_1 p_2. \tag{9.7.21}$$

From equation (9.7.19) we can then solve, e.g., if $a \neq 0$,

$$p_1 = \frac{-\rho_1 \tau_1}{a} + \frac{\rho_2}{a} p_1 p_2. \tag{9.7.22}$$

Substitution of this into the remaining three equations (9.7.18), (9.7.20) and (9.7.21) gives

$$\lambda = -a - \frac{\rho_1^2 \tau_1}{a} + \rho_2 p_2 + \frac{\rho_1 \rho_2}{a} p_1 p_2, \tag{9.7.23}$$

$$\gamma_2 p_2 \lambda = \rho_2 \tau_2 + a p_2 - \rho_1 p_1 p_2 \tag{9.7.24}$$

$$\gamma_{12} p_1 p_2 \lambda = \frac{-\rho_1 \rho_2 \tau_1 \tau_2}{a} + \rho_1 \tau_1 p_2 + \left(\frac{\rho_2^2 \tau_2}{a} + a \right) p_1 p_2. \tag{9.7.25}$$

Or, stated differently,

$$\begin{bmatrix} -a - \frac{\rho_1^2 \tau_1}{a} & \rho_2 & \frac{\rho_1 \rho_2}{a} \\ \frac{\rho_2 \tau_2}{\gamma_2} & \frac{a}{\gamma_2} & -\frac{\rho_1}{\gamma_2} \\ \frac{-\rho_1 \rho_2 \tau_1 \tau_2}{\gamma_{12} a} & \frac{\rho_1 \tau_1}{\gamma_{12}} & \frac{\left(\frac{\rho_2^2 \tau_2}{a} + a \right)}{\gamma_{12}} \end{bmatrix} \begin{bmatrix} 1 \\ p_2 \\ p_1 p_2 \end{bmatrix} = \lambda \begin{bmatrix} 1 \\ p_2 \\ p_1 p_2 \end{bmatrix}.$$

By solving this eigenvalue problem, in a similar way to that described in Algorithm 9.1, one can determine the solutions p_2 and $p_1 p_2$ of the set of equations (9.7.23)–(9.7.25). Substitution of this result into equation (9.7.22) then gives p_1.   □

For the general N-player case we proceed as in Section 8.5.3. Let p_i, $i = 1, \ldots, N$, be a solution of (9.7.6) and (9.7.7). Denote the negative of the resulting closed-loop system parameter by

$$\lambda := -a + \sum_i \rho_i p_i. \tag{9.7.26}$$

Then,

$$p_i^2 - 2\lambda p_i + \tau_i = 0, \quad i = 1, \ldots, N. \tag{9.7.27}$$

Next we derive, again, for each index set $\Omega \subset \{1, \ldots, N\}$ a linear equation (linear in terms of products of p_i variables (Πp_i)). This gives us in addition to (9.7.26) another $2^N - 1$ linear equations. These equations, together with equation (9.7.26), determine our matrix M. In case Ω contains only one number we have, using the definition of λ and equation (9.7.27), respectively,

$$p_j \lambda = p_j \left(-a + \sum_{i=1}^{N} \rho_i p_i \right) = -a p_j + \rho_j p_j^2 + p_j \sum_{i \neq j} \rho_i p_i$$

$$= -a p_j + 2\lambda \rho_j p_j - \rho_j \tau_j + p_j \sum_{i \neq j} \rho_i p_i.$$

From which it follows that

$$p_j\lambda = \frac{\rho_j\tau_j}{\gamma_j} + \frac{a}{\gamma_j}p_j - p_j\sum_{i\neq j}\frac{\rho_i}{\gamma_j}p_i, \quad j = 1,\dots,N.$$

Next consider the general case $\prod_{j\in\Omega}p_j\lambda$. For notational convenience let Ω_{-i} denote the set of all numbers that are in Ω except number i. Then, following the lines of Section 8.5.3,

$$\prod_{j\in\Omega}p_j\lambda = -a\prod_{j\in\Omega}p_j + 2\lambda\sum_{i\in\Omega}\rho_i\prod_{j\in\Omega}p_j - \sum_{i\in\Omega}\rho_i\tau_i\prod_{j\in\Omega_{-i}}p_j + \sum_{i\notin\Omega}\prod_{j\in\Omega}p_j\rho_ip_i.$$

Therefore, with $\gamma_\Omega = -1 + 2\sum_{i\in\Omega}\rho_i$, it follows that

$$\prod_{j\in\Omega}p_j\lambda = \frac{1}{\gamma_\Omega}\left\{a\prod_{j\in\Omega}p_j + \sum_{i\in\Omega}\rho_i\tau_i\prod_{j\in\Omega_{-i}}p_j + \sum_{i\notin\Omega}\prod_{j\in\Omega}p_j\rho_ip_i\right\}. \tag{9.7.28}$$

Equations (9.7.26) and (9.7.28) determine the matrix $\tilde{M}$. That is, introducing

$$p := \left[1, p_1, \dots, p_N, p_1p_2, \dots, p_{N-1}p_N, \dots, \prod_{i=1}^{N}p_i\right]^T$$

$\tilde{M}p = \lambda p$. Since $p_i = (s_i + m_i)x_i$ and $\tau_i = (s_i + m_i)q_i$, matrix M is then easily obtained from $\tilde{M}$ by rewriting p as $p = Dx$, where $x := [1, x_1, \dots, x_N, x_1x_2, \dots, x_{N-1}x_N, \dots,$ $\prod_{i=1}^{N}x_i]^T$ and D is a diagonal matrix defined by $D := \mathrm{diag}\{1, s_1 + m_1, \dots, s_N + m_N,$ $(s_1 + m_1)(s_2 + m_2), \dots, (s_{N-1} + m_{N-1})(s_N + m_N), \cdots, \prod_{i=1}^{N}(s_i + m_i)\}$. Obviously, $M = D^{-1}\tilde{M}D$. Below we consider the case for $N = 3$.

Example 9.8

Consider the three-player case. With $p := [1, p_1, p_2, p_3, p_1p_2, p_1p_3, p_2p_3, p_1p_2p_3]^T$,

$$D = \mathrm{diag}\{1, s_1 + m_1, s_2 + m_2, s_3 + m_3, (s_1 + m_1)(s_2 + m_2), (s_1 + m_1)(s_3 + m_3),$$
$$(s_2 + m_2)(s_3 + m_3), (s_1 + m_1)(s_2 + m_2)(s_3 + m_3)\},$$

and

$$\tilde{M} = \begin{bmatrix} -a & \rho_1 & \rho_2 & \rho_3 & 0 & 0 & 0 & 0 \\ \frac{\rho_1\tau_1}{\gamma_1} & \frac{a}{\gamma_1} & 0 & 0 & -\frac{\rho_2}{\gamma_1} & -\frac{\rho_3}{\gamma_1} & 0 & 0 \\ \frac{\rho_2\tau_2}{\gamma_2} & 0 & \frac{a}{\gamma_2} & 0 & -\frac{\rho_1}{\gamma_2} & 0 & -\frac{\rho_3}{\gamma_2} & 0 \\ \frac{\rho_3\tau_3}{\gamma_3} & 0 & 0 & \frac{a}{\gamma_3} & 0 & -\frac{\rho_1}{\gamma_3} & -\frac{\rho_2}{\gamma_3} & 0 \\ 0 & \frac{\rho_2\tau_2}{\gamma_{12}} & \frac{\rho_1\tau_1}{\gamma_{12}} & 0 & \frac{a}{\gamma_{12}} & 0 & 0 & -\frac{\rho_3}{\gamma_{12}} \\ 0 & \frac{\rho_3\tau_3}{\gamma_{13}} & 0 & \frac{\rho_1\tau_1}{\gamma_{13}} & 0 & \frac{a}{\gamma_{13}} & 0 & -\frac{\rho_2}{\gamma_{13}} \\ 0 & 0 & \frac{\rho_3\tau_3}{\gamma_{23}} & \frac{\rho_2\sigma_2}{\gamma_{23}} & 0 & 0 & \frac{a}{\gamma_{23}} & -\frac{\rho_1}{\gamma_{23}} \\ 0 & 0 & 0 & 0 & \frac{\rho_3\tau_3}{\gamma_{123}} & \frac{\rho_2\tau_2}{\gamma_{123}} & \frac{\rho_1\tau_1}{\gamma_{123}} & \frac{a}{\gamma_{123}} \end{bmatrix}.$$

Which yields,

$$
M = \begin{bmatrix}
-a & s_1 & s_2 & s_3 & 0 & 0 & 0 & 0 \\
\frac{\rho_1 q_1}{\gamma_1} & \frac{a}{\gamma_1} & 0 & 0 & -\frac{s_2}{\gamma_1} & -\frac{s_3}{\gamma_1} & 0 & 0 \\
\frac{\rho_2 q_2}{\gamma_2} & 0 & \frac{a}{\gamma_2} & 0 & -\frac{s_1}{\gamma_2} & 0 & -\frac{s_3}{\gamma_2} & 0 \\
\frac{\rho_3 q_3}{\gamma_3} & 0 & 0 & \frac{a}{\gamma_3} & 0 & -\frac{s_1}{\gamma_3} & -\frac{s_2}{\gamma_3} & 0 \\
0 & \frac{\rho_2 q_2}{\gamma_{12}} & \frac{\rho_1 q_1}{\gamma_{12}} & 0 & \frac{a}{\gamma_{12}} & 0 & 0 & -\frac{s_3}{\gamma_{12}} \\
0 & \frac{\rho_3 q_3}{\gamma_{13}} & 0 & \frac{\rho_1 q_1}{\gamma_{13}} & 0 & \frac{a}{\gamma_{13}} & 0 & -\frac{s_2}{\gamma_{13}} \\
0 & 0 & \frac{\rho_3 q_3}{\gamma_{23}} & \frac{\rho_2 q_2}{\gamma_{23}} & 0 & 0 & \frac{a}{\gamma_{23}} & -\frac{s_1}{\gamma_{23}} \\
0 & 0 & 0 & 0 & \frac{\rho_3 q_3}{\gamma_{123}} & \frac{\rho_2 q_2}{\gamma_{123}} & \frac{\rho_1 q_1}{\gamma_{123}} & \frac{a}{\gamma_{123}}
\end{bmatrix}.
$$

Using this matrix M in Algorithm 9.1 one can determine all solutions of the three-player scalar equations (9.1.1) and (9.7.2). □

Example 9.9

Reconsider the EMU game from Example 8.15. Assume that in this game neither country expects severe external shocks to the economy, whereas the Central Bank is somewhat less optimistic in this respect. We model this by considering the game:

$$
\dot{s}(t) = -s(t) - f_1(t) + f_2(t) + \frac{1}{2}i_E(t) + w(t), \quad s(0) = s_0,
$$

with

$$
J_1 := \int_0^\infty \{2s^2(t) + f_1^2(t) - 4w^2(t)\}dt,
$$

$$
J_2 := \int_0^\infty \{2s^2(t) + 2f_2^2(t) - 4w^2(t)\}dt,
$$

and

$$
J_E := \int_0^\infty \{s^2(t) + 3i_E^2(t) - 2w^2(t)\}dt.
$$

With these parameters, $\rho_1 = 4/5$, $\rho_2 = 2/3$, $\rho_3 = 1/7$. Consequently, $\gamma_1 = 3/5$, $\gamma_2 = 1/3$, $\gamma_3 = -5/7$, $\gamma_{12} = -1 + 2\rho_1 + 2\rho_2 = 29/15$, $\gamma_{13} = -1 + 2\rho_1 + 2\rho_3 = 31/35$, $\gamma_{23} = -1 + 2\rho_2 + 2\rho_3 = 13/21$, and $\gamma_{123} = -1 + 2\rho_1 + 2\rho_2 + 3\rho_3 = 233/105$. To calculate the soft-constrained Nash equilibria of this game, we first determine all solutions of equations (9.7.1) and (9.7.2) using Algorithm 9.1. In this case matrix

$$
M := \begin{bmatrix}
1 & 1 & 1/2 & 1/12 & 0 & 0 & 0 & 0 \\
8/3 & -5/3 & 0 & 0 & -5/6 & -5/36 & 0 & 0 \\
4 & 0 & -3 & 0 & -3 & 0 & -1/4 & 0 \\
-1/5 & 0 & 0 & 7/5 & 0 & 7/5 & 7/10 & 0 \\
0 & 20/29 & 24/29 & 0 & -15/29 & 0 & 0 & -5/116 \\
0 & 5/31 & 0 & 56/31 & 0 & -35/31 & 0 & -35/62 \\
0 & 0 & 3/13 & 28/13 & 0 & 0 & -21/13 & -21/13 \\
0 & 0 & 0 & 0 & 15/233 & 140/233 & 168/233 & -105/233
\end{bmatrix}.
$$

Using MATLAB, we find the following eigenvalues for M

$$\{-2.1369, -1.7173, -1.9576 \pm 0.2654i, -1.267 \pm 0.5041i, 1.9543, 2.37\}.$$

Since the square of every positive eigenvalue is larger than $\tau = 5/2$ ($= \max\{5/2, 3/2, 7/12\}$), according to Algorithm 9.1, the equations (9.7.1) and (9.7.2) have two solutions.

From the corresponding eigenspaces we obtain the solutions given in Table 9.2 (with $a_{cl} = a - s_1 x_1 - s_2 x_2 - s_3 x_3$).

Table 9.2

Eigenvalue	(x_1, x_2, x_3)	$a_{cl} + m_1 x_1$	$a_{cl} + m_2 x_2$	$a_{cl} + m_3 x_3$
2.37	(0.4836,0.4546,7.909)	−2.2491	−2.2564	1.5845
1.9543	(0.6445,0.5752,0.2664)	−1.7932	−1.8105	−1.8211

From the last three columns of this table we see that only the second solution satisfies the additional conditions (9.7.3). Since $q_i > 0$, and thus (9.5.9) and (9.5.10) are satisfied with $y_i = 0$, this game therefore has one soft-constrained Nash equilibrium. The equilibrium actions corresponding to this equilibrium $(x_1^*, x_2^*, x_3^*) := (0.6445, 0.5752, 0.2664)$ are

$$f_1^*(t) = x_1^* s(t), \quad f_2^*(t) = -\frac{1}{2} x_2^* s(t) \quad \text{and} \quad i_E^*(t) = -\frac{1}{6} x_3^* s(t).$$

Assuming that the initial state of the system is x_0, the worst-case costs expected by the players are

$$J_1^* = 0.6445 s_0^2, \quad J_2^* = 0.5752 s_0^2 \quad \text{and} \quad J_E^* = 0.2664 s_0^2,$$

respectively. Taking a closer look at the equilibrium actions shows that all players use more control efforts than in the undisturbed case. The ratio of the increase in control efforts used by the fiscal player 1, fiscal player 2 and the Central Bank is approximately $6 : 4 : 3$. The expected increase in worst-case costs by these three players is approximately 3.7%, 2.5% and 1.8%, respectively. So we see that though, at first sight, it seems that the Central Bank is the most risk-averse player in this game, due to the model structure the Bank will suffer least from an actual realization of a worst-case scenario. Also, in coping with this uncertainty, the Bank deviates least from its original equilibrium action. Finally, we observe that if the players take uncertainty into account, the implemented equilibrium policies lead to a closed-loop system which adjusts faster towards its equilibrium value $s = 0$. That is, with $w = 0$, $a_{cl} = -1.9543$, whereas if the players do not take into account model uncertainty $a_{cl} = -1.9225$. □

We conclude this Section with two additional observations. The first point we should make is that the incorporation by players of noise into their decision making may result in the fact that a situation of no equilibrium changes into a situation in which an equilibrium does exist. Take, for example, $q_i = -1$, $b_i = r_i = v_i = e = 1$ and $a = -\frac{3}{2}$. For these parameters the undisturbed game has no equilibrium (see Theorem 8.10 part 2(b)) whereas the disturbed game has the equilibrium $x_i = -\frac{1}{2}$, $i = 1, 2$ (which can, for example, be verified by a direct substitution of x_i into (9.7.1)–(9.7.3) and taking $y_i = -1$ in (9.5.9) and (9.5.10)).

The second point we should make is that, using the implicit function theorem, one can analyse the consequences of a change in the m_i parameters on the solution set (x_1, x_2) of the algebraic Riccati equations (9.7.4) in the two-player case. To that end, rewrite $m_i =: \alpha_i s_i$, $i = 1, 2$. Then,

$$2(a - s_1 x_1 - s_2 x_2)x_1 + s_1(1 + \alpha_1)x_1^2 + q_1 = 0, \quad \text{and} \tag{9.7.29}$$

$$2(a - s_1 x_1 - s_2 x_2)x_2 + s_2(1 + \alpha_2)x_2^2 + q_2 = 0. \tag{9.7.30}$$

Assume that the solution (x_1^*, x_2^*) of equations (9.7.29) and (9.7.30) can be described locally as a function $h(\alpha_1, \alpha_2)$. Then, using the implicit function theorem, we obtain from (9.7.29) and (9.7.30) that

$$h' = \frac{-1}{2(p_1 p_2 - s_1 s_2 x_1^* x_2^*)} \begin{bmatrix} -p_2 & s_2 x_1^* \\ s_1 x_2^* & -p_1 \end{bmatrix} \begin{bmatrix} s_1 x_1^{*2} & 0 \\ 0 & s_2 x_2^{*2} \end{bmatrix}$$

where $p_i := -(a - s_1 x_1^* - s_2 x_2^* + m_i x_i^*) > 0$ (see equation (9.7.3)). From this it is immediately clear, for example, that at a positive equilibrium an increase in α_1 will have an opposite effect on the entries of x_i^*, $i = 1, 2$. One entry will increase, the other will decrease. Similarly, we see that the effect of such an increase on x_1^* and x_2^* are also opposite. If $r_i = b_i = 1$, $i = 1, 2$, and consequently the equilibrium strategies are given by $u_i^*(t) = -x_i^* x(t)$, this implies that the response to a more risk-averse behavior by one player is a more risk-seeking behavior by the other player. We do not undertake a more detailed analysis here since such an analysis can be carried out best in the context of a specific application, as we did in Section 9.6.

9.8 Stochastic interpretation

As is well known (Runolfsson (1994) and Başar (2003)), the deterministic formulation of the one-player optimization problem has, if $Q \geq 0$, an equivalent stochastic formulation. That is, consider the one-player linear noisy system (9.2.1) with w replaced by a stationary white gaussian noise with zero mean and covariance $E(w(t)w^T(\tau)) = \delta(t - \tau)$, and the cost function $L : \mathcal{F} \to \mathbb{R}$, defined by

$$\lim_{T \to \infty} \frac{1}{T} \log E\left\{ \exp\left(\frac{1}{2\gamma^2} \int_0^\infty \{x^T Q x(t) + u^T R u\} dt \right) \right\}.$$

Here, the number γ is positive, $Q \geq 0$ and $R > 0$. In Başar (2003) it has been shown that the question whether

$$\exists \bar{F} \in \mathcal{F} \text{ such that } L(\bar{F}) \leq L(F) \text{ for all } F \in \mathcal{F} \tag{9.8.1}$$

can be answered in terms of the stabilizing solution of the algebraic Riccati equation

$$Q + A^T X + XA - XSX + \gamma^{-2} XEE^T X = 0,$$

which also generates the saddle-point solution for the soft-constrained differential game related to the H_∞ control problem for γ sufficiently large (see Sections 9.3 and 9.4 or Başar and Bernhard (1995)). More precisely, if (Q, A) is observable, a positive number γ^* exists such that the above Riccati equation has a stabilizing solution $X \geq 0$ if and only if $\gamma > \gamma^*$. In that case matrix $A - SX$ is also stable. Runolfsson (1994) proved, under an additional controllability assumption, that this stabilizing solution also generates the solution for (9.8.1). We recall this result in the following theorem. The link between this theorem and Theorem 9.8 is obvious.

Theorem 9.24

Assume that (A, B) and (A, E) are controllable and that (Q, A) is observable. Let γ^* be as defined above. If $(A - SX, E)$ is controllable, then $\bar{F} := -R^{-1} B^T X$ is a solution of (9.8.1) and $L(\bar{F}) = \frac{1}{2} \gamma^{-2} \text{tr}(E^T XE)$. $\qquad\square$

The extension to the N-player case is straightforward. Consider system (9.1.1) and a cost function for player i

$$L_i(F_1, \ldots, F_N) := \lim_{T_1 \to \infty} \frac{1}{T_1} \log \mathrm{E} \left\{ \exp \frac{1}{2\gamma_i^2} \left(\int_0^{T_1} \left\{ x^T \left(Q_i + \sum_{j=1}^N F_j^T R_{ij} F_j \right) x(t) \right\} dt \right) \right\},$$

where the numbers γ_i are positive, the matrices R_{ij} are symmetric, $Q_i \geq 0$ and $R_{ii} > 0$. Consider the following analogue of Definition 9.2.

Definition 9.3

An N-tuple $\bar{F} = (\bar{F}_1, \ldots, \bar{F}_N) \in \mathcal{F}$ is called an **LEQG feedback Nash equilibrium** if for all i the following inequality holds: $\bar{L}_i(\bar{F}, x_0) \leq \bar{L}_i(\bar{F}_{-i}(F), x_0)$ for all initial states x_0 and for all $F \in \mathbb{R}^{m_i \times n}$ such that $\bar{F}_{-i}(F) \in \mathcal{F}$. $\qquad\square$

The generalization of Theorem 9.24 then reads as follows.

Theorem 9.25

Assume there exist N real symmetric matrices $X_i \geq 0$ satisfying

$$Q_i + A^T X_i + X_i A - \sum_{j \neq i}^N (X_i S_j X_j + X_j S_j X_i) - X_i S_i X_i + \sum_{j \neq i}^N X_j S_{ij} X_j + \gamma_i^{-2} X_i EE^T X_i = 0,$$

$$A_i := A - \sum_{j=1}^N S_j X_j + \gamma_i^{-2} EE^T X_i \text{ is stable } i = 1, \cdots, N,$$

$$\bar{A} := A - \sum_{j=1}^N S_j X_j \text{ is stable.}$$

Furthermore assume that the matrix pairs (A_i, B_i), $(\bar{A}, E)$ and (A_i, E) are controllable for all $i = 1, \ldots, N$, and the pair (Q_i, A_i) is observable. Under these conditions, the N-tuple of feedback matrices $(\bar{F}_1, \ldots, \bar{F}_N)$ with $\bar{F}_i := -R_{ii}^{-1}B_i^T X_i$ is an LEQG feedback Nash equilibrium, and $L_i(\bar{F}) = \frac{1}{2}\gamma_i^{-2}\mathrm{tr}(E^T X_i E)$. $\qquad\square$

9.9 Notes and references

For this chapter the papers by van den Broek, Engwerda and Schumacher (2003b,c) have been consulted. The latter reference also contains a third approach to deal with uncertainty, the so-called hard-bounded uncertainty approach. In that approach, again, a minmax problem is solved but the disturbance is not restrained by a cost term but simply by a direct norm bound (see also Bernhard and Bellec (1973) and Başar (2003)). An algorithm is indicated to calculate equilibria in such a game.

In this chapter we tried as much as possible to avoid the technicalities involved in setting up a rigorous stochastic framework. As a drawback, the stated results seem to be less general than they are. A more general framework and references under which the above presented statements concerning the stochastic finite-planning horizon equilibrium actions continue to hold can be found, for example, in Başar and Olsder (1999). For a recent general treatment of stochastic differential games refer to Buckdahn, Cardaliaguet and Rainer (2004).

If one does not have access to the full state at each point in time, the analysis becomes much more complicated. Some references dealing with various issues in this direction can be found in Başar and Olsder (1999).

Finally, we summarize the general N-player result of Theorem 9.18.

Theorem 9.26

(N-player analogue of Theorem 9.18)

Consider the differential game defined by (9.2.1) and (9.2.4)–(9.2.5). Assume there exist N real symmetric $n \times n$ matrices X_i and N real symmetric $n \times n$ matrices Y_i such that

$$-\left(A - \sum_{j\neq i}^N S_jX_j\right)^T X_i - X_i\left(A - \sum_{j\neq i}^N S_jX_j\right) + X_iS_iX_i - Q_i - \sum_{j\neq i}^N X_jS_{ij}X_j - X_iM_iX_i = 0,$$

$$A - \sum_{j=1}^N S_jX_j + M_iX_i \text{ is stable for } i = 1, \ldots, N,$$

$$A - \sum_{j=1}^N S_jX_j \text{ is stable,}$$

$$-\left(A - \sum_{j\neq i}^N S_jX_j\right)^T Y_i - Y_i\left(A - \sum_{j\neq i}^N S_jX_j\right) + Y_iS_iY_i - Q_i - \sum_{j\neq i}^N X_jS_{ij}X_j \leq 0.$$

Define the N-tuple $\bar{F} = (\bar{F}_1, \ldots, \bar{F}_N)$ by

$$\bar{F}_i := -R_{ii}^{-1}B_i^T X_i.$$

Then $\overline{F} \in \mathcal{F}$, and this N-tuple is a soft-constrained Nash equilibrium. Furthermore

$$\overline{J}_i^{SC}(\overline{F}_1, \ldots, \overline{F}_N, x_0) = x_0^T X_i x_0.$$

9.10 Exercises

1. Consider the scalar differential game (see Exercises 7.1 and 8.1)

$$\min_{u_i} J_i := E\left\{ \int_0^T \{q_i x^2(t) + r_i u_i^2(t)\} dt + q_{iT} x^2(T) \right\}, \quad i = 1, 2, \text{ subject to}$$
$$\dot{x}(t) = ax(t) + b_1 u_1(t) + b_2 u_2(t) + w(t),$$

where w is a white noise zero mean random variable with covariance $3\delta(t - \tau)$ and $x(0)$ is a Gaussian random variable (independent of w) with mean 2 and covariance 4. Determine which of the following differential games has a linear feedback Nash equilibrium. If an equilibrium exists, compute the equilibrium actions and involved cost.

(a) $T = 2$, $a = 0$, $b_1 = 1$, $b_2 = 1$, $q_1 = 3$, $q_2 = 1$, $r_1 = 4$, $r_2 = 4$, $q_{1T} = 0$ and $q_{2T} = 0$.

(b) $T = 1$, $a = 1$, $b_1 = 2$, $b_2 = 1$, $q_1 = 1$, $q_2 = 4$, $r_1 = 1$, $r_2 = 1$, $q_{1T} = 0$ and $q_{2T} = 0$.

(c) $T = 1$, $a = 0$, $b_1 = 1$, $b_2 = 1$, $q_1 = -1$, $q_2 = 2$, $r_1 = 1$, $r_2 = 1$, $q_{1T} = 0$ and $q_{2T} = 1$.

2. Consider the zero-sum differential game

$$\min_{u_1} J := E\left\{ \int_0^T \{x^T(t)Q(t) + u_1^T(t)R_1 u_1(t) - u_2^T R_2 u_2(t)\} dt + x^T(T)Q_T x(T) \right\}, \text{ subject to}$$
$$\dot{x}(t) = Ax(t) + B_1 u_1(t) + B_2 u_2(t) + w(t),$$

where w is a white noise zero mean random variable with covariance $V\delta(t - \tau)$ and $x(0)$ is a Gaussian random variable (independent of w) with mean m_0 and covariance P_0, and player 2 likes to maximize J w.r.t. u_2. Here $R_i > 0$, $i = 1, 2, V > 0$ and $P_0 > 0$.

(a) Provide sufficient conditions under which this game has a linear feedback Nash equilibrium.

(b) Consider the scalar case (see also Exercises 7.3 and 8.2) $T = A = B_1 = B_2 = Q = R_1 = 1$, $R_2 = \frac{1}{2}$, $Q_T = -\frac{1}{4}$, $m_0 = -4$, $V = 1$ and $P_0 = 2$. Find the corresponding equilibrium strategies and cost.

3. Consider the scalar differential game

$$\min_{u_i} J_i := \lim_{T \to \infty} \frac{1}{T} \mathrm{E} \left\{ \int_0^T \{q_i x^2(t) + r_i u_i^2(t)\} dt \right\}, \quad i = 1, 2, \text{ subject to}$$

$$\dot{x}(t) = ax(t) + b_1 u_1(t) + b_2 u_2(t) + w(t),$$

where w is a white noise zero mean random variable with covariance $3\delta(t - \tau)$ and $x(0)$ is a Gaussian random variable (independent of w) with mean 1 and covariance 2. Determine which of the following differential games has a stochastic variance-independent feedback Nash equilibrium. If an equilibrium exists, compute the equilibrium actions and involved cost.

(a) $a = 0$, $b_1 = 1$, $b_2 = 1$, $q_1 = 3$, $q_2 = 1$, $r_1 = 4$ and $r_2 = 4$.
(b) $a = 1$, $b_1 = 2$, $b_2 = 1$, $q_1 = 1$, $q_2 = 4$, $r_1 = 1$ and $r_2 = 2$.
(c) $a = -1$, $b_1 = 1$, $b_2 = -1$, $q_1 = -1$, $q_2 = 2$, $r_1 = 2$ and $r_2 = 1$.

4. Consider the scalar differential game

$$\min_{u_i} J_i := \min_{u_i} \sup_{w \in L_2} \left\{ \int_0^\infty \{q_i x^2(t) + u_i^2(t) - w^2(t)\} dt \right\}, \quad i = 1, 2, \text{ subject to}$$

$$\dot{x}(t) = ax(t) + u_1(t) + u_2(t) + w(t),$$

where w is some deterministic noise. Assume that $q_1 = 8$ and $q_2 = 6\frac{3}{4}$ (see also Exercise 8.7).

(a) Use Algorithm 9.1 to calculate all soft-constrained feedback Nash equilibria of this game if
    (i) $a = -1$, (ii) $a = 3\frac{1}{2}$ and (iii) $a = 4$, respectively.

(b) Compare your answers with those of Exercise 8.7 and conclude.

5. Consider the scalar differential game

$$\min_{u_i} J_i := \min_{u_i} \sup_{w \in L_2} \left\{ \int_0^\infty \{q_i x^2(t) + u_i^2(t) - v_i w^2(t)\} dt \right\}, \quad i = 1, 2, \text{ subject to}$$

$$\dot{x}(t) = ax(t) + u_1(t) + u_2(t) + w(t),$$

where w is some deterministic noise. Assume that $q_1 = -5$, $q_2 = -32$, $v_1 = 4$ and $v_2 = 2$ (see also Exercise 8.8).

(a) Let $a = -10$. Use Algorithm 9.1 to calculate all solutions of (9.7.1) and (9.7.2).
(b) Use the results of part (a) to verify whether condition (9.5.7) is satisfied.
(c) Use the results of part (a) to verify whether there exist y_i satisfying (9.5.9) and (9.5.10).

(d) Determine the soft-constrained feedback Nash equilibria of this game.

(e) Answer the same questions in case (ii) $a = -5$, (iii) $a = 0$, (iv) $a = 5$ and (v) $a = 10$, respectively.

(f) Compare your answers with those of Exercise 8.8 and conclude.

6. In this exercise you are asked to formulate some numerical algorithms to calculate a solution of the set of equations (9.5.5)–(9.5.8).

   (a) Reconsider the numerical algorithms we formulated in Exercises 8.10 and 8.11. Formulate the analogues of both these algorithms to determine a solution of equations (9.5.5), (9.5.6) and (9.5.8).

   (b) Formulate the analogue of the algorithm considered in Exercise 8.11 to determine a solution of equations (9.5.5)–(9.5.7).

   (c) Calculate the solutions using your algorithms for the games considered in Exercise 4.

   (d) Calculate the solutions also using your algorithms for a number of games with more than one state variable. Discuss some (dis)advantages of the various algorithms.

7. Reconsider the EMU differential game of Example 9.9. Assume now that in this game both countries are less optimistic than the Central Bank concerning the occurrence of external shocks. We model this by considering the game:

$$\dot{s}(t) = -s(t) - f_1(t) + f_2(t) + \frac{1}{2}i_E(t) + w(t), \quad s(0) = s_0,$$

with

$$J_1 := \int_0^\infty \{2s^2(t) + f_1^2(t) - 2w^2(t)\}dt,$$

$$J_2 := \int_0^\infty \{2s^2(t) + 2f_2^2(t) - 2w^2(t)\}dt,$$

and

$$J_E := \int_0^\infty \{s^2(t) + 3i_E^2(t) - 4w^2(t)\}dt.$$

(a) Determine all soft-constrained feedback Nash equilibria of this game.

(b) Compare your answer with the results of Example 9.9 and conclude.

8. Reconsider the lobby-game from Exercise 8.6. However, now assume that the future dynamics are not completely known to the players and that the system dynamics are corrupted by some deterministic noise. Both the industry and environmental lobbyist

take this uncertainty into account by considering the following optimization problems, respectively.

$$\min_{u_i} \int_0^\infty e^{-rt}\{-c(t) + \rho u_i^2(t) - u_e^2(t) + v_1 w^2(t)\}dt \quad \text{and}$$

$$\min_{u_e} \int_0^\infty e^{-rt}\{c(t) - \rho u_i^2(t) + u_e^2(t) + v_2 w^2(t)\}dt$$

subject to $\dot{c}(t) = -\delta c(t) + u_i(t) - u_e(t) + w(t), \ c(0) = c_0$.

(a) Interpret this problem setting.

(b) Reformulate the problem into the standard framework.

(c) Consider the case $r = 0.1$, $\delta = 0.95$, $\rho = 7/36$, $v_1 = 7/15$ and $v_2 = 7$. Show that

with $X_1 := \begin{bmatrix} -1/3 & 0.2063 \\ 0.2063 & -8.3962 \end{bmatrix}$ and $X_2 := \begin{bmatrix} 1 & -1.0749 \\ -1.0749 & 10.7163 \end{bmatrix}$ the equa-

tions (9.5.5)–(9.5.8) are satisfied.

9. Assume that the state of the system is generated by the linear noisy system,

$$\dot{x}(t) = Ax(t) + B_1 u_1(t) + \cdots + B_N u_N(t) + Ew(t).$$

The noise w is white, Gaussian, of zero mean and has covariance $\delta(t - \tau)$. The initial state at time $t = 0$, x_0, is a Gaussian random variable of mean m_0 and covariance P_0. This random variable is independent of w. The strategy spaces considered by the players are assumed to be the set of linear feedback actions Γ_i^{lfb}, as defined in Section 8.2. Consider the cost function for player i

$$L_i(F_1, \ldots, F_N) := \lim_{T_1 \to \infty} \frac{1}{T_1} \log \text{E} \left\{ \exp \frac{1}{2\gamma_i^2} \left(\int_0^{T_1} \left\{ x^T \left(Q_i + \sum_{j=1}^N F_j^T R_{ij} F_j \right) x(t) \right\} dt \right) \right\}.$$

The numbers γ_i are positive, the matrices R_{ij} are symmetric, $Q_i \geq 0$ and $R_{ii} > 0$.

(a) Show that the conditions from Theorem 9.25 under which this game has an LEQG feedback Nash equilibrium coincide with the conditions under which the game considered in Theorem 9.18 has a solution.

(b) Consider the two-player scalar game $A = -1$, $B_1 = B_2 = E = 1$, $R_{11} = 2$, $R_{22} = \frac{1}{2}$, $R_{12} = R_{21} = 0$, $Q_1 = Q_2 = 1$, $\gamma_1 = 1$ and $\gamma_2 = 2$. Show that all conditions mentioned in Theorem 9.25 are satisfied. Determine the LEQG feedback Nash equilibrium of this game and the associated cost for both players.

(c) Answer the same question as in part (b) but now with $\gamma_1 = 3$ and $\gamma_2 = 6$.

(d) Answer the same question as in part (b) but now with $\gamma_1 = 1/3$ and $\gamma_2 = 2/3$.

(e) Compare your answers in parts (b)–(d) and conclude. Interpret your answer also using a deterministic framework.

10. Consider two countries. Country 2 is a developing country, whereas country 1 contributes a fixed amount (1%) of its national product to the development of

country 2. Utility in both countries is assumed to be a quadratic function of its national product. The government's budget in both countries is ideally assumed to be a fixed fraction (γ_i) of its national product. The goal of both governments is to maximize the total discounted utility in their countries. The problem is formalized as follows

$$\min_{f_i} \int_0^\infty e^{-2rt}\{-y_i^2(s) + \theta_i(f_i(s) - \gamma_i y_i(s))^2\}ds$$

subject to

$$\dot{y}_1(t) = \alpha_1 y_1(t) + b_1 f_1(t), \quad y_1(0) = \bar{y}_1$$
$$\dot{y}_2(t) = \beta_1 y_2(t) + 0.01\beta_2 y_1(t) + b_2 f_2(t), \quad y_2(0) = \bar{y}_2.$$

(a) Interpret the above model and, in particular, the variables and parameters involved.

(b) Rewrite the model into the standard framework. For that purpose introduce the new control variables $u_i(t) := f_i(t) - \gamma_i y_i(t)$.

Introduce

$$s_i = \frac{b_i^2}{\theta_i}, \ i = 1,2, \ A := \begin{bmatrix} a_1 & 0 \\ a_2 & a_3 \end{bmatrix} := \begin{bmatrix} \alpha_1 - r + b_1\gamma_1 & 0 \\ 0.01\beta_2 & \beta_1 - r + b_2\gamma_2 \end{bmatrix},$$

$$S_1 := \begin{bmatrix} s_1 & 0 \\ 0 & 0 \end{bmatrix}, S_2 := \begin{bmatrix} 0 & 0 \\ 0 & s_2 \end{bmatrix}, Q_1 = \begin{bmatrix} -1 & 0 \\ 0 & 0 \end{bmatrix} \text{ and } Q_2 = \begin{bmatrix} 0 & 0 \\ 0 & -1 \end{bmatrix}.$$

(c) Consider the matrix M associated with the open-loop Nash equilibria (see 7.4.3). Show that the characteristic polynomial of M is $(-a_3 - \lambda)(-a_1 - \lambda)$ $(\lambda^2 - (a_1^2 - s_1))(\lambda^2 - (a_3^2 - s_2))$. Prove that a basis for the eigenspace of M corresponding to the eigenvalue $\lambda = -a_1$ is $[0\ 0\ 0\ 0\ 1\ 0]^T$.

(d) Show that the game has

    (i) no open-loop equilibrium if both $a_1^2 < s_1$ and $a_3^2 < s_2$,

    (ii) one open-loop equilibrium that permits a feedback synthesis if $a_1^2 > s_1$ and $a_3^2 < s_2$.

    (iii) more than one open-loop equilibrium that permits a feedback synthesis if both $a_1^2 > s_1$ and $a_3^2 > s_2$.

Can you find an intuitive explanation for these results?

(e) Consider the case that both $a_1^2 > s_1$ and $a_3^2 > s_2$. Show that the game has a feedback Nash equilibrium with $K_1 = \begin{bmatrix} k_1 & 0 \\ 0 & 0 \end{bmatrix}$ and $K_2 = \begin{bmatrix} x_1 & x_2 \\ x_2 & x_3 \end{bmatrix}$ for appropriately chosen numbers k_1 and x_i, $i = 1,2,3$.

(f) Choose $a_1 = 0.25$, $a_2 = 0.002$, $a_3 = 0.02$, $s_1 = 0.016$ and $s_2 = 0.0002$. Calculate the feedback Nash equilibrium in part (e).

(g) Consider the above model where both differential equations are corrupted by some deterministic input. Formulate the corresponding soft-constrained optimization problem. Next take as a starting point the feedback Nash equilibrium you calculated in part (e) to assess the influence of different risk attitudes by both governments on the feedback Nash equilibrium outcome. To calculate the various soft-constrained equilibria use your algorithm from Exercise 6. Interpret your results.

11. Consider the following game on government debt stabilization (for notation, see Section 7.8.1):

$$\min_{f(.)} L^F = \int_0^\infty e^{-\delta t}\{f^2(t) + \eta m^2(t) + \lambda d^2(t)\}dt \quad \text{and} \quad \min_{m(.)} L^M = \int_0^\infty e^{-\delta t}\{m^2(t)$$
$$+ \kappa d^2(t)\}dt,$$

subject to the dynamic constraint

$$\dot{d}(t) = r(1 + \sin te^{-t})d(t) + f(t) - m(t), \quad d(0) = d_0. \tag{9.10.1}$$

(a) Assume the monetary and fiscal authorities do not know the exact dynamic evolution (9.10.1) of the debt, but are aware of this fact, and consider instead the dynamic game

$$\min_{f(.)} \sup_{w(.)\in L_2} L^F = \int_0^\infty e^{-\delta t}\{f^2(t) + \eta m^2(t) + \lambda d^2(t) - v_F w^2(t)\}dt \quad \text{and}$$

$$\min_{m(.)} \sup_{w(.)\in L_2} L^M = \int_0^\infty e^{-\delta}\{m^2(t) + \kappa d^2(t) - v_M w^2(t)\}dt,$$

subject to the dynamic constraint

$$\dot{d}(t) = rd(t) + f(t) - m(t) + w(t), \quad d(0) = d_0.$$

Determine the equilibrium actions for this game. Do not solve the involved equations explicitly.

(b) Let $d_0 = 1$, $r = \delta = 0.02$, $\eta = 0.25$, $\lambda = 0.6525$, $\kappa = 0.7150$ and $v_F = v_M = 10$. Calculate numerically the equilibrium actions. Calculate numerically the real evolution of the debt using these equilibrium actions.

(c) Answer the same questions as in part (b), but now with $v_F = 5$. Compare the results with those obtained in part (b) and conclude.

(d) Answer the same questions if $v_F = 2$.

9.11 Appendix

Proof of Theorem 9.8

The matrices $A - SX$ and $A - SX + MX$ are stable by assumption. This implies that $\bar{F} \in \mathcal{F}$ and $\bar{w} \in L_2^q(0, \infty)$, respectively. By Lemma 9.7

$$J(F, w, x_0) = x_0^T X x_0 + \int_0^\infty \{\|(F - \bar{F})x\|_R^2 - \|w - V^{-1}E^T Xx\|_V^2\}dt.$$

From this it follows that

$$J(\bar{F}, w, x_0) = x_0^T X x_0 - \int_0^\infty \{\|w - V^{-1}E^T X\tilde{x}\|_V^2\}dt \le x_0^T X x_0,$$

where $\tilde{x}$ is generated by $\dot{\tilde{x}} = (A + B\bar{F})\tilde{x} + Ew$, $\tilde{x}(0) = x_0$. Furthermore, if $J(\bar{F}, w, x_0) = x_0^T X x_0$ then $w = \bar{w}$. Hence $J(\bar{F}, w, x_0) < x_0^T X x_0$ for all $w \ne \bar{w}$, and $J(\bar{F}, \bar{w}, x_0) = x_0^T X x_0$. This, obviously, also implies that $\bar{F} \in \bar{\mathcal{F}}$.

Next, we show that $J(F, \bar{w}, x_0) \ge J(\bar{F}, \bar{w}, x_0)$ for all $F \in \mathcal{F}$. Let $\hat{x}$ and $\bar{x}$ be generated by

$$\dot{\hat{x}} = (A + BF)\hat{x} + E\bar{w}, \quad \hat{x}(0) = x_0,$$

and

$$\dot{\bar{x}} = (A + B\bar{F})\bar{x} + E\bar{w}, \quad \bar{x}(0) = x_0,$$

respectively. Define furthermore

$$\nu := (\bar{F} - F)\hat{x}, \quad \zeta := \bar{w} - V^{-1}E^T X\hat{x}.$$

Then $J(F, \bar{w}, x_0) - J(\bar{F}, \bar{w}, x_0) = \int_0^\infty \{\|\nu\|_R^2 - \|\zeta\|_V^2\}dt$. Introducing $\xi := \bar{x} - \hat{x}$ we have that

$$\dot{\xi} = (A + B\bar{F})\xi + B\nu \tag{9.11.1}$$

with $\xi(0) = 0$, and $\zeta = V^{-1}E^T X\xi$. Since both $\hat{x}$ and $\bar{x}$ belong to $L_2^n(0, \infty)$ it follows that ξ and ν are quadratically integrable as well, which implies that $\xi(t) \to 0$ for $t \to \infty$. So, $\int_0^\infty \frac{d}{dt}\xi^T X\xi dt = 0$. Hence

$$\begin{aligned}
J(F, \bar{w}, x_0) - J(\bar{F}, \bar{w}, x_0) &= \int_0^\infty \left\{(\|\nu\|_R^2 - \|\zeta\|_V^2) - \frac{d}{dt}\xi^T X\xi\right\}dt \\
&= \int_0^\infty \{\|\nu\|_R^2 - 2\nu^T B^T X\xi - \xi^T(A^T X + XA - 2XSX + XMX)\xi\}dt \\
&= \int_0^\infty \{\|\nu - R^{-1}B^T X\xi\|_R^2 - \xi^T(A^T X + XA - XSX + XMX)\xi\}dt \\
&= \int_0^\infty \{\|\nu + \bar{F}\xi\|_R^2 + \xi^T Q\xi\}dt.
\end{aligned}$$

Next, define $w := v + \bar{F}\xi = \bar{F}x - F\hat{x}$. Then, equation (9.11.1) shows that $\dot{\xi} = A\xi + Bw$. Since $\xi(0) = 0$ and $\xi(t) \to 0$ for $t \to \infty$ also $\int_0^\infty \left(\frac{d}{dt}\xi^T Y\xi\right) dt = 0$. Hence

$$J(F, \bar{w}, x_0) - J(\bar{F}, \bar{w}, x_0) = \int_0^\infty \left\{ (\|w\|_R^2 + \|\xi\|_Q^2) + \frac{d}{dt}\xi^T Y\xi \right\} dt$$

$$= \int_0^\infty \{w^T Rw + 2w^T B^T Y\xi + \xi^T (Q + A^T Y + YA)\xi\} dt$$

$$= \int_0^\infty \{\|w + R^{-1}B^T Y\xi\|_R^2 + \xi^T (Q + A^T Y + YA - YSY)\xi\} dt \geq 0,$$

where the last inequality follows by assumption. $\qquad\square$

Proof of Theorem 9.10

By definition $\bar{F} \in \mathcal{F}$ is such that the Hamiltonian matrix $H_{\bar{F}}$ defined in equation (9.3.6) has no eigenvalues on the imaginary axis. This implies that there is an open neighborhood $O_{\bar{F}} \subset \mathcal{F}$ of $\bar{F}$ such that for all $F \in O_{\bar{F}}$, H_F has no eigenvalues on the imaginary axis. Let $F \in O_{\bar{F}}$ be an arbitrary element. Then Lemma 9.6 part (ii) applies with A, Q and $\phi(w, x_0)$ replaced by $A + BF$, $Q + F^T RF$ and $J(F, w, x_0)$, respectively. By this lemma $\bar{J}(F, x_0) := \max_{w \in L_2^q(0,\infty)} J(F, w, x_0) = x_0^T \psi(F) x_0$ where $\psi : O_{\bar{F}} \to \mathbb{R}^{n \times n}$ is defined by $\psi(F) := X$, and X is the stabilizing solution of

$$Q + F^T RF + (A + BF)^T X + X(A + BF) + XMX = 0.$$

In Lancaster and Rodman (1995) it is shown that the maximal solution of

$$\tilde{X}(\mu)\tilde{D}(\mu)\tilde{X}(\mu) - \tilde{X}(\mu)\tilde{A}(\mu) - \tilde{A}^T(\mu)\tilde{X}(\mu) - \tilde{C}(\mu) = 0 \qquad (9.11.2)$$

is a real-analytic function of k real variables $\mu \in \Omega$, where Ω is an open connected set in $\mathbb{R}^k$ if (i) $\tilde{A}(\mu)$, $\tilde{C}(\mu)$ and $\tilde{D}(\mu)$ are real-analytic functions of μ, (ii) $\tilde{D}(\mu) \geq 0$, (iii) $(\tilde{A}(\mu), \tilde{D}(\mu))$ is stabilizable, and (iv) the matrix

$$\begin{bmatrix} -\tilde{A}(\mu) & \tilde{D}(\mu) \\ \tilde{C}(\mu) & \tilde{A}^T(\mu) \end{bmatrix}$$

has no eigenvalues on the imaginary axis for all $\mu \in \Omega$. Under the conditions (ii) and (iii), the maximal solution of (9.11.2) coincides with the unique solution of (9.11.2) for which the eigenvalues of $\tilde{A}(\mu) - \tilde{D}(\mu)\tilde{X}(\mu)$ lie in the closed left-half plane (for example, Lancaster and Rodman (1995)). Note that $-X$ is the maximal solution of (9.11.2) with $\tilde{A}(\mu) = A + BF$, $\tilde{C}(\mu) = -Q - F^T RF$, $\tilde{D}(\mu) = M$ and $\mu = \text{vec}\, F$ (vec F denotes the vector obtained from F by stacking the columns of F). Clearly, condition (i) and (ii) hold; condition (iii) follows from the stability of $A + BF$ and condition (iv) follows from the easily verifiable fact that the matrices H_F and

$$\begin{bmatrix} -A - BF & M \\ -Q - F^T RF & (A + BF)^T \end{bmatrix}$$

have the same eigenvalues. Hence, ψ is an analytic function of F in any open connected subset of $\bar{\mathcal{F}}$. In particular $\bar{J}$ is differentiable with respect to F in such a set. Since $\bar{J}$ attains its minimum at $\bar{F} \in \mathcal{F}$, for each $x_0 \in \mathbb{R}^n$, a differentiation argument shows that the Fréchet derivative $\partial \psi(\bar{F}) = 0$. Next, define the transformation $\Psi : \bar{\mathcal{F}} \times \mathbb{R}^{n \times n} \to \mathbb{R}^{n \times n}$ by

$$\Psi(F, X) := Q + F^T R F + (A + BF)^T X + X(A + BF) + XMX.$$

By definition, $\Psi(F, \psi(F)) = 0$ for all $F \in O_{\bar{F}}$. Taking the derivative of this equality at $F = \bar{F}$ shows that $\bar{F} = -R^{-1} B^T \psi(\bar{F})$. Substituting this in $\Psi(\bar{F}, \psi(\bar{F})) = 0$ gives

$$Q + A^T \psi(\bar{F}) + \psi(\bar{F})A - \psi(\bar{F})S\psi(\bar{F}) + \psi(\bar{F})M\psi(\bar{F}) = 0.$$

This shows that $\psi(\bar{F})$ satisfies equation (9.3.9) and furthermore, since it is the stabilizing solution of the equation $\Psi(\bar{F}, X) = 0$ it follows that $A + B\bar{F} + M\psi(\bar{F}) = A - S\psi(\bar{F}) + M\psi(\bar{F})$ is stable. Finally, since $\bar{F} \in \bar{\mathcal{F}}$, the matrix $A - S\psi(\bar{F})$ is stable. $\square$

References

Aarle, B. van, Bovenberg, L. and Raith, M. (1995) Monetary and fiscal policy interaction and government debt stabilization. *Journal of Economics*, **62**, 111–140.

Aarle, B. van, Engwerda, J.C. and Plasmans, J. (2002) Monetary and fiscal policy interaction in the EMU: a dynamic game approach. *Annals of Operations Research*, **109**, 229–264.

Aarle, B. van, *et al.* (2001) Macroeconomic policy interaction under EMU: a dynamic game approach. *Open Economies Review*, **12**, 29–60.

Abou-Kandil, H. and Bertrand, P. (1986) Analytic solution for a class of linear quadratic open-loop Nash games. *International Journal of Control*, **43**, 997–1002.

Abou-Kandil, H., Freiling, G. and Jank, G. (1993) Necessary and sufficient conditions for constant solutions of coupled Riccati equations in Nash games. *Systems and Control Letters*, **21**, 295–306.

Abou-Kandil, H. *et al.* (2003) *Matrix Riccati Equations in Control and Systems Theory*, Birkhäuser, Berlin.

Ambartsumian, V.A. (1943) The problem of diffuse reflection of light by a turbid medium, reprinted, in *A Life in Astrophysics: Selected Papers of Viktor A. Ambartsumian*, Allerton Press, New York (1998). pp. 83–87.

Anderson, B.D.O. and Moore, J.B. (1989) *Optimal Control: Linear Quadratic Methods*, Prentice-Hall, Englewood Cliffs, New Jersey.

Arrow, K.J. (1968) Applications of control theory to economic growth, in *Mathematics of the Decision Sciences 2*, (eds G.B. Dantzig and A.F. Veinott (Jr.)) American Mathematical Society, Providence, RI, p. 92.

Bagchi, A. and Olsder, G.J. (1981) Linear pursuit evasion games. *Journal of Applied Mathematics and Optimization*, **7**, 95–123.

Başar, T. (1974) A counterexample in linear-quadratic games: existence of nonlinear Nash strategies. *Journal of Optimization Theory and Applications*, **14**, 425–430.

Başar, T. (1975) Nash strategies for *M*-person differential games with mixed information structures. *Automatica*, **11**, 547–551.

Başar, T. (1977) Informationally nonunique equilibrium solutions in differential games. *SIAM, J. Control and Optimization*, **15**, 636–660.

Başar, T. (1981) On the saddle-point solution of a class of stochastic differential games. *Journal of Optimization Theory and Applications*, **33**, 539–556.

Başar, T. (2003) Paradigms for robustness in controller and filter design, in *Modeling and Control of Economic Systems 2002*, (ed R. Neck), Elsevier, pp. 1–13.

Başar, T. and Bernhard, P. (1995) H_∞-*Optimal Control and Related Minimax Design Problems*, Birkhäuser, Boston.

Başar, T. and Li, S. (1989) Distributed computation of Nash equilibria in linear-quadratic differential games. *SIAM J. Control and Optimization*, **27**, 563–578.

Başar, T. and Olsder, G.J. (1999) *Dynamic Noncooperative Game Theory*, SIAM, Philadelphia.

Bellman, R.E. (1956) *Dynamic Programming*, Princeton University Press, Princeton, New Jersey.

Bellman, R.E. and Dreyfus, S.E. (1962) *Applied Dynamic Programming*, Princeton University Press, Princeton, New Jersey.

Berkovitz, L.D. (1961) A variational approach to differential games, RM-2772, RAND Corporation, Santa Monica, USA. Also (1964) *Annals of Mathematics Studies*, **52**, 127–174.

Berkovitz, L.D. and Fleming, W.H. (1957) On differential games with integral payoff, in *Contributions to the Theory of Games, Vol.3* (eds M. Dresher, A.W. Tucker, and P. Wolfe), Princeton University Press, Princeton, pp. 413–435.

Bernhard, P. (1979) Linear-quadratic, two-person, zero-sum differential games: necessary and sufficient conditions. *Journal of Optimization Theory and Applications*, **27**, 51–69.

Bernhard, P. and Bellec, G. (1973) On the evaluation of worst-case design with an application to the quadratic synthesis technique, in *Proceedings of the 3rd IFAC Symposium on Sensitivity, Adaptivity, and Optimality*, Ischia, Italy, pp. 349–352.

Bittanti, S. (1991) Count Riccati and the early days of the Riccati equation, in *The Riccati Equation*, (eds S. Bittanti, A.J. Laub and J.C. Willems), Springer-Verlag, Berlin, pp. 1–10.

Bittanti, S., Laub, A.J. and Willems, J.C. (eds) (1991) *The Riccati Equation*, Springer-Verlag, Berlin.

Blaquiere, A. Gerard, F. and Leitmann, G. (1969) *Quantitative and Qualitative Games*, Academic Press, New York.

Breitner, M.H. (2002) Rufus P. Isaacs and the early years of differential games: a survey and discussion paper, in *Proceedings of the 10th International Symposium on Dynamic Games and Applications* (eds Petrosjan and Zenkevich), St. Petersburg, Russia, pp. 113–129.

Brock, W.A. and Haurie, A. (1976) On existence of overtaking optimal trajectories over an infinite time horizon. *Mathematics of Operations Research*, **1**, 337–346.

Broek, W.A. van den (2001) *Uncertainty in Differential Games*, PhD. Thesis, CentER Dissertation Series, Tilburg University, The Netherlands.

Broek, W.A. van den, Engwerda, J.C. and Schumacher, J.M. (2003a) Robust equilibria in indefinite linear-quadratic differential games, in *Proceedings of the 10th International Symposium on Dynamic Games and Applications* (eds Petrosjan and Zenkevich), St. Petersburg, Russia, pp. 139–158.

Broek, W.A. van den, Engwerda, J.C. and Schumacher, J.M. (2003b) An equivalence result in linear-quadratic theory. *Automatica*, **39**, 355–359.

Broek, W.A. van den, Engwerda, J.C. and Schumacher, J.M. (2003c) Robust equilibria in indefinite linear-quadratic differential games. *Journal of Optimization Theory and Applications*, **119**, 565–595.

Buckdahn, R. Cardaliaguet, P. and Rainer, C. (2004) Nash equilibrium payoffs for nonzero-sum stochastic differential games, Internal Report Université de Bretagne Occidentale, Brest cedex, France.

Bushnell, L (ed) (1996) *Special Issue on the History of Control*, IEEE Control Systems Magazine, **16**(3).

Callier, F. and Willems, J.L. (1981) Criterion for the convergence of the solution of the Riccati differential equation. *IEEE Trans. Automat. Contr.*, **26**, 1232–1242.

Carlson, D.A. and Haurie, A. (1991) *Infinite Horizon Optimal Control*, Springer-Verlag, Berlin.

Case, J.H. (1967). *Equilibrium points of N-person differential games*, PhD thesis, University of Michigan, Ann Arbor, MI, Dept. of Industrial Engineering, Tech. report no. 1967–1.

Chiang, A. (1992) *Elements of Dynamic Optimization*, McGraw-Hill, New York, Section 2.4.

Chintagunta, P.K. (1993) Investigating the sensitivity of equilibrium profits to advertising dynamics and competitive effects. *Management Science*, **39**, 1146–1162.

Clements, D.J. and Anderson, B.D.O. (1978) *Singular Optimal Control: The Linear-Quadratic Problem*, Springer-Verlag, Berlin.

Coddington, E.A. and Levinson, N. (1955) *Theory of Ordinary Differential Equations*, McGraw Hill, New York.

Cohen, D. and Michel, P. (1988) How should control theory be used to calculate a time-consistent government policy? *Review of Economic Studies*, **55**, 263–274.

Coppel, W.A. (1966) A survey of quadratic systems. *Journal of Differential Equations*, 2,293–304.

Cournot, A.A. (1838) *Recherches sur les Principes Mathematiques de la Theorie des Richesses*, Hachette, Paris.

Currie, D. Holtham, G. and Hughes Hallett, A. (1989) The theory and practice of international policy coordination: does coordination pay? in *Macroeconomic Policies in an Interdependent World.* (ed Byrant), IMF, Washington.

Cramer, G. (1750) Introduction à l'Analyse des Lignes Courbes Algébraique.

Cruz, J.B. and Chen, C.I. (1971) Series Nash solution of two person nonzero-sum linear-quadratic games. *Journal of Optimization Theory and Applications*, **7**, 240–257.

Damme, E. van (1991) *Stability and Perfection of Nash Equilibria*, Springer-Verlag, Berlin.

Davis, M.H.A. (1977) *Linear Estimation and Stochastic Control*, Chapman and Hall, London.

Dockner, E., Feichtinger, G. and Jørgensen, S. (1985) Tracktable classes of nonzero-sum open-loop Nash differential games. *Journal of Optimization Theory and Applications*, **45**, 179–197.

Dockner, E. *et al.* (2000) *Differential Games in Economics and Management Science*, Cambridge University Press, Cambridge.

Dooren, P. van (1981). A generalized eigenvalue approach for solving Riccati equations. *SIAM J. Scientific Statistical Computation*, **2**, 121–135.

Douven, R.C. (1995) *Policy Coordination and Convergence in the EU*, Ph.D. Thesis, Tilburg University, The Netherlands.

Douven, R.C. and Engwerda, J.C. (1995) Is there room for convergence in the E.C.? *European Journal of Political Economy*, **11**, 113–130.

Douven, R.C. and Plasmans, J.E.J. (1996) SLIM, a small linear interdependent model of eight EU-member states, the USA and Japan. *Economic Modelling*, **13**, 185–233.

Dresher, M. Tucker, A.W. and Wolfe, P. (eds) (1957). Special Issue: *Contributions to the Theory of Games, Vol.3*, Annals of Mathematics Studies, **39**.

Edgeworth, F.Y. (1881) *Mathematical Psychics: An Essay on the Application of Mathematics to the Moral Sciences*, C. Kegan Paul and Co., London.

Eisele, T. (1982) Nonexistence and nonuniqueness of open-loop equilibria in linear-quadratic differential games. *Journal of Optimization Theory and Applications*, **37**(4), 443–468.

Engwerda, J.C. (1998a) On the open-loop Nash equilibrium in LQ-games. *Journal of Economic Dynamics and Control*, **22**, 729–762.

Engwerda, J.C. (1998b) Computational aspects of the open-loop Nash equilibrium in LQ-games. *Journal of Economic Dynamics and Control*, **22**, 1487–1506.

Engwerda, J.C. (2000a) Feedback Nash equilibria in the scalar infinite horizon LQ-game. *Automatica*, **36**, 135–139.

Engwerda, J.C. (2000b) The solution set of the *N*-player scalar feedback Nash algebraic Riccati equations. *IEEE Trans. Automat. Contr.*, **45**, 2363–2368.

Engwerda, J.C. (2003) Solving the scalar feedback Nash algebraic Riccati equations: an eigenvector approach. *IEEE Trans. Automat. Contr.*, **48**, 847–853.

Engwerda, J.C. and Weeren, A.J.T.M. (1994) On the inertia of the matrix $M = \begin{bmatrix} -A & S_1 & S_2 \\ Q_1 & A^T & 0 \\ Q_2 & 0 & A^T \end{bmatrix}$, in

Proceedings of the Sixth International Symposium on Dynamic Games and Applications (eds Breton and Zaccour), GERAD Ecole des H.E.C. Montréal, Canada, pp. 565–571.

Engwerda, J.C. Aarle, B. van and Plasmans, J. (1999) The (in)finite horizon open-loop Nash LQ game: An application to the EMU. *Annals of Operations Research*, **88**, 251–273.

Engwerda, J.C. Aarle, B. van and Plasmans, J. (2002) Cooperative and non-cooperative fiscal stabilisation policies in the EMU. *Journal of Economic Dynamics and Control*, **26**, 451–481.

Euler, L. (1744) *Methodus inveniendi lineas curvas maximi minimi proprietate gaudentes sive solutio problematis isoperimetrici latissimo sensu accepti (Method for finding plane curves that show some property of maxima and minima)*, Lausanne.

Fan, K. Glicksberg, I. and Hoffman, A.J. (1957) Systems of inequalities involving convex functions. *American Mathematical Society Proceedings*, **8**, 617–622.

Feichtinger, G. and Hartl, R.F. (1986) *Optimale Kontrolle Ökonomischer Prozesse. Anwendungen des Maximumprinzips in den Wirtschaftswissenschaften*, De Gruyter, Berlin.

Fershtman, C. and Kamien, I. (1987) Dynamic duopolistic competition with sticky prices. *Econometrica*, **55**, 1151–1164.

Feucht, M. (1994) *Linear-quadratische Differentialspiele und gekoppelte Riccatische Matrixdifferentialgleichungen*, Ph.D. Thesis, Universität Ulm, Germany.

Fleming, W.H. (1957) A note on differential games of prescribed duration, in *Contributions to the Theory of Games, Vol.3*, (eds Dresher, Tucker and Wolfe), Princeton University Press, Princeton, pp. 407–412.

Fleming, W.H. and Rishel, R.W. (1975) *Deterministic and Stochastic Optimal Control*, Springer-Verlag, Berlin.

Foley, M.H. and Schmitendorf, W.E. (1971) On a class of non-zero-sum linear-quadratic differential games. *Journal of Optimization Theory and Applications*, **5**, 357–377.

Foreman, J.G. (1977) The princess and the monster on the circle. *SIAM J. Control and Optimization*, **15**, 841–856.

Francis, B.A. (1987) *A course in H_∞ Control Theory*, Lecture Notes in Control and Information Sciences Vol. 88, Springer-Verlag, Berlin.

Freiling, G. Jank, G. and Abou-Kandil, H. (1996) On global existence of solutions to coupled matrix Riccati equations in closed-loop Nash games. *IEEE Trans. Automat. Contr.*, **41**, 264–269.

Friedman, A. (1971) *Differential Games*, John Wiley and Sons, New York.

Fudenberg, D. and Tirole, J. (1991) *Game Theory*, MIT Press, Cambridge, Massachusetts.

Geerts, T. (1989) *Structure of Linear-Quadratic Control*, PhD. Thesis, Eindhoven Technical University, The Netherlands.

Glover, K. and Doyle, J.C. (1988) State-space formulae for all stabilizing controllers that satisfy an H_∞ norm bound and relations to risk sensitivity. *Systems and Control Letters*, **11**, 167–172.

Grabowski, P. (1993) The LQ controller synthesis problem. *IMA Journal of Mathematical Control and Information*, **10**, 131–148.

Greub, W.H. (1967) *Linear Algebra*, Springer-Verlag, Berlin.

Halkin, H. (1974) Necessary conditions for optimal control problems with infinite horizons. *Econometrica*, **42**, 267–272.

Halmos, P.R. (1966) *Measure Theory*, D. van Nostrand Company, Inc., Princeton, New Jersey.

Hansen, L. Epple, D. and Roberts, W. (1985) Linear-quadratic models of resource depletion, in *Energy, Foresight and Strategy* (ed. Sargent), RFF Press, Baltimore, pp. 102–142.

Hansen, L.P. Sargent, T.J. and Tallarini, Jr. T.D. (1999) Robust permanent income and pricing. *Review of Economic Studies*, **66**, 873–907.

Harsanyi, J.C. (1956) Approaches to the bargaining problem before and after the theory of games: a critical discussion of Zeuthen's, Hicks' and Nash's theories. *Econometrica*, **24**, 144–157.

Haurie, A. (2001) A historical perspective on cooperative differential games, in *Annals of the International Society of Dynamic Games: Advances in Dynamic Games and Applications*, (eds Altman and Pourtallier), Birkhäuser, Berlin, pp. 19–31.

Haurie, A. and Tolwinski, B. (1984) Acceptable equilibria in dynamic bargaining games. *Large Scale Systems*, **6**, 73–89.

Haurie, A. and Tolwinski, B. (1985) Definition and properties of cooperative equilibria in a two-player game of infinite duration. *Journal of Optimization Theory and Applications*, **46**, 525–534.

Hautus, M.L.J. (1980) *Optimalisering van Regelsystemen*, Onderafdeling der Wiskunde, Technical University Eindhoven, The Netherlands.

Hestenes, M.R. (1949) A general problem in the calculus of variations with applications to paths of least time. *RAND Corporation Report*.

Hestenes, M.R. (1966) *Calculus of Variations and Optimal Control Theory*, John Wiley and Sons, New York.

Ho, Y.C. Bryson, E. and Baron, S. (1965) Differential games and optimal pursuit-evasion strategies. *IEEE Trans. Automat. Contr.*, **10**, 385–389.

Horn, R.A. and Johnson, C.A. (1985) *Matrix Analysis*, Cambridge University Press, Cambridge.

Howard, J.V. (1992) A social choice rule and its implementation in perfect equilibrium. *Journal of Economic Theory*, **56**, 142–159.

Hughes Hallett, A. (1984) Non-cooperative strategies for dynamic policy games and the problem of time inconsistency. *Oxford Economic Papers*, **36**, 381–399.

Hughes Hallett, A. and Petit, M.L. (1990) Cohabitation or forced marriage? A study of the costs of failing to coordinate fiscal and monetary policies. *Weltwirtschaftliches Archiv*, **126**, 662–689.

Isaacs, R. (1951) Games of pursuit, Paper P-257, RAND Corporation, Santa Monica, USA.

Isaacs, R., (1954–1955), Differential games I,II,III,IV. Research Memoranda RM-1391, RM-1399, RM-1411, RM-1486, RAND Corporation, Santa Monica, USA.

Isaacs, R. (1965) *Differential Games - A Mathematical Theory with Applications to Warfare and Pursuit, Control and Optimization*, John Wiley and Sons, New York.

Jacobson, D.H. (1971) Totally singular quadratic minimization problems. *IEEE Trans. Autom. Contr.*, **16**, 578–589.

Jacobson, D.H. (1973) Optimal stochastic linear systems with exponential performance criteria and their relation to deterministic differential games. *IEEE Trans. Automat. Contr.*, **18**, 124–131.

Jacobson, D.H. (1977a) On values and strategies for infinite-time linear quadratic games. *IEEE Trans. Autom. Contr.*, **22**, 490–491.

Jacobson, D.H. (1977b) *Extensions of Linear-Quadratic Control, Optimization and Matrix Theory*, Acadamic Press, London.

Jank, G. and Kun, G. (2002) Optimal control of disturbed linear-quadratic differential games. *European Journal of Control*, **8**, 152–162.

Jódar, L. (1990) Explicit solutions of Riccati equations appearing in differential games. *Applied Mathematics Letters*, **3**, 9–12.

Jódar, L. and Abou-Kandil, H. (1988) A resolution method for Riccati differential systems coupled in their quadratic terms. *SIAM Journal Mathematical Analysis*, **19**, 1425–1430.

Jódar, L. and Abou-Kandil, H. (1989) Kronecker products and coupled matrix Riccati differential equations. *Linear Algebra and its Applications*, **121**, 39–51.

Jódar, L. and Navarro, E. (1991a) Exact computable solution of a class of strongly coupled Riccati equations. *Journal of Computational and Applied Mathematics*, **36**, 265–271.

Jódar, L. and Navarro, E. (1991b) Exact solution of a class of strongly coupled Riccati matrix differential equations. *Applied Mathematics and Computation*, **43**, 165–173.

Jódar, L., Navarro, E. and Abou-Kandil, H. (1991) Explicit solutions of coupled Riccati equations occuring in Nash games – the open loop case. *Control and Cybernetics*, **20**, 59–66.

Jørgensen, S. and Zaccour, G. (1999) Equilibrium pricing and advertising strategies in a marketing channel. *Journal of Optimization Theory and Applications*, **64**, 293–310.

Jørgensen, S. and Zaccour, G. (2003) *Differential Games in Marketing*, Kluwer, Deventer.

Kailath, T. (1980) *Linear Systems*, Prentice-Hall, Englewood Cliffs, New Jersey.

Kaitala, V., Pohjola, M. and Tahvonen, O. (1992) Transboundary air polution and soil acidification: a dynamic analysis of an acid rain game between Finland and the USSR. *Environmental and Resource Economics*, **2**, 161–181.

Kalai, E. (1977) Proportional solutions to bargaining problems: interpersonal utility comparisons. *Econometrica*, **45**, 1623–1630.

Kalai, E. and Smorodinsky, M. (1975) Other solutions to Nash's bargaining problem. *Econometrica*, **43**, 513–518.

Kelendzeridze, D.L. (1961) Theory of an optimal pursuit strategy. *Doklady Akademii Nauk SSRS*, **138**, 529–532. English translation: *Sov. Math. Dokl.*, **2**, 654–656.

Kleinman, D. (1968) On an iterative technique for Riccati equation computations. *IEEE Trans. Automat. Contr.*, **13**, 114–115.

Klompstra, M.B. (2000) Nash equilibria in risk-sensitive dynamic games. *IEEE Trans. Automat. Contr.*, **45**, 1397–1401.

Knobloch, H.W. and Kwakernaak, H. (1985) *Lineare Kontrolltheorie*, Springer-Verlag, Berlin.

Krasovskii, N.N. and Subbotin, A.I. (1988) *Game-Theoretical Control Problems*, Springer-Verlag, Berlin.

Kremer, D. (2002) *Non-Symmetric Riccati Theory and Noncooperative Games*, Ph.D. Thesis, RWTH-Aachen, Germany.

Krikelis, N. and Rekasius, Z. (1971) On the solution of the optimal linear control problems under conflict of interest. *IEEE Trans. Automat. Contr.*, **16**, 140–147.

Kucera, V. (1991) Algebraic Riccati equation: Hermitian and definite solutions, in *The Riccati Equation*, (eds Bittanti, Laub and Willems), Springer-Verlag, Berlin, pp. 53–88.

Kumar, P.R. and Schuppen, J.H. (1980) On Nash equilibrium solutions in stochastic dynamic games. *IEEE Trans. Automat. Contr.*, **25**, 1146–1149.

Kun, G. (2001) *Stabilizability, Controllability, and Optimal Strategies of Linear and Nonlinear Dynamical Games*, Ph.D. Thesis, RWTH-Aachen, Germany.

Kwakernaak, H. and Sivan, R. (1972) *Linear Optimal Control Systems*, John Wiley and Sons, New York.

Kydland, F. (1976) Decentralized stabilization policies: optimization and the assignment problem. *Annals of Economic and Social Measurement*, **5**, 249–261.

Lay, D.C. (2003) *Linear Algebra and its Applications*, Pearson Education, New York.

Lancaster, P. and Rodman, L. (1995) *Algebraic Riccati Equations*, Clarendon Press, Oxford.

Lancaster, P. and Tismenetsky, M. (1985) *The Theory of Matrices*, Academic Press, London.

Laub, A.J. (1979) A Schur method for solving algebraic Riccati equations. *IEEE Trans. Automat. Contr.*, **24**, 913–921.

Laub, A.J. (1991) Invariant subspace methods for the numerical solution of Riccati equations, in *The Riccati Equation*, (eds Bittanti, Laub and Willems). Springer-Verlag, Berlin, pp. 163–199.

Lee, E.B. and Markus, L. (1967) *Foundations of Optimal Control Theory*, John Wiley and Sons, Inc., New York.

Leitmann, G. (1974) *Cooperative and Non-cooperative Many Players Differential Games*, Springer-Verlag, Berlin.

Levine, P. and Brociner, A. (1994) Fiscal policy coordination and EMU: a dynamic game approach. *Journal of Economic Dynamics and Control*, **18**, 699–729.

Li, T-Y and Gajic, Z. (1994) Lyapunov iterations for solving couple algebraic Riccati equations of Nash differential games and algebraic Riccati equations of zero-sum games. *Annals of Dynamic Games*, **3**, 333–351.

Lockwood, B. (1996) Uniqueness of Markov-perfect equilibrium in infinite-time affine-quadratic differential games. *Journal of Economic Dynamics and Control*, **20**, 751–765.

Lockwood, B. and Philippopoulos, A. (1994) Inside power, unemployment dynamics, and multiple inflation equilibria. *Economica*, **61**, 59–77.

Luenberger, D.G. (1969) *Optimization by Vector Space Methods*, John Wiley and Sons, New York, Chapter 7.

Lukes, D.L. (1971) Equilibrium feedback control in linear games with quadratic costs. *SIAM J. Control and Optimization*, **9**, pp. 234–252.

Lukes, D.L. and Russell, D.L. (1971) Linear-quadratic games. *Journal of Mathematical Analysis and Applications*, **33**, 96–123.

MacFarlane, A.G.J. (1963) An eigenvector solution of the optimal linear regulator problem. *Journal of Electronical Control*, **14**, 496–501.

MacTutor History of Mathematics Archive (2003). University of St. Andrews, St. Andrews, Fife, Scotland, http://www-groups.dcs.st-andrews.ac.uk/ history/index.html.

McShane, E.J. (1939) On multipliers for Lagrange problems. *American Journal Mathematics*, **61**, 809–819.

Mageirou, E.F. (1976) Values and strategies for infinite time linear quadratic games. *IEEE Trans. Aut. Cont.*, **21**, 547–550.

Mageirou, E.F. (1977) Iterative techniques for Riccati game equations. *Journal of Optimization Theory and Applications*, **22**, 51–61.

Mäler, K.-G. (1992) Critical loads and international environmental cooperation, in *Conflicts and Cooperation in Managing Environmental Resources*, (ed R. Pethig), Springer-Verlag, Berlin.

Mäler, K.-G. and Zeeuw, A.J. de (1998). The acid rain differential game. *Environmental and Resource Economics*, **12** 167–184.

Mangasarian, O.L. (1966) Sufficient conditions for the optimal control of nonlinear systems. *SIAM Journal on Control*, **4**, 139–152.

Mehlmann, A. (1988) *Applied Differential Games*, Plenum Press, New York.

Mehrmann, V.L. (1991) The Autonomous Linear Quadratic Control Problem: Theory and Numerical Solution, in *Lecture Notes in Control and Information Sciences 163*, (eds Thoma and Wyner), Springer-Verlag, Berlin.

Miller, M. and Salmon, M. (1985a) Policy coordination and dynamic games, in *International Economic Policy Coordination*, (eds Buiter and Marston), Cambridge University Press, Cambridge.

Miller, M. and Salmon, M. (1985b) Dynamic games and the time-inconsistency of optimal policy in open economies. *Economic Journal*, **95**, 124–213.

Miller, M. and Salmon, M. (1990) When does coordination pay? *Journal of Economic Dynamics and Control*, **14**, 553–569.

Molinari, B.P. (1975) Nonnegativity of a quadratic functional. *SIAM Journal on Control*, **13**, 792–806.

Molinari, B.P. (1977) The time-invariant linear-quadratic optimal control problem. *Automatica*, **13**, 347–357.

Morgan, A.P. (1987) *Solving Polynomial Systems Using Continuation for Scientific and Engineering Problems*, Prentice-Hall, Englewood Cliffs, New Jersey.

Moulin, H. (1984) Implementing the Kalai–Smorodinsky bargaining solution. *Journal of Economic Theory*, **33**, 32–45.

Nash, J.F. (1950a) Equilibrium points in N-person games, in *Proceedings of the National Academy of Sciences of the United States of America*, **36**, 48–49.

Nash, J. (1950b) The bargaining problem. *Econometrica*, **18**, 155–162.

Nash, J.F. (1951) Non-cooperative games. *Annals of Mathematics*, **54**, 286–295.

Nash, J. (1953) Two-person cooperative games. *Econometrica*, **21**, 128–140.

Neck, R. and Dockner, E.J. (1995) Commitment and coordination in a dynamic-game model of international economic policy-making. *Open Economies Review*, **6**, 5–28.

Neck, R. Behrens, D.A. and Hager, M. (2001) Solving dynamic macroeconomic policy games using the algorithm OPTGAME 1.0. *Optimal Control Applications and Methods*, **22**, 301–332.

Neese, J.W. and Pindyck, R. (1984) Behavioural assumptions in decentralized stabilization policies, in *Applied Decision Analysis and Economic Behaviour*, (ed Hughes Hallett), Martinus Nijhof, Dordrecht, The Netherlands, pp. 251–270.

Nerlove, M. and Arrow, K.J. (1962) Optimal advertising policy under dynamic conditions. *Economica*, **29**, 129–142.

Neustadt, L.W. (1976) *Optimization. A Theory of Necessary Conditions*, Princeton University Press, Princeton.

Obstfeld, M. (1991) *Dynamic Seignorage Theory: an Exploration*, Center for Economic Policy Research, London, Discussion Paper 519.

Osborne, M. and Rubinstein, A. (1991) *Bargaining and Markets*, Academic Press, Boston.

Özgüner, U. and Perkins, W.R. (1977) A series solution to the Nash strategy for large scale interconnected systems. *Automatica*, **13**, 313–315.

Paige, C. and Van Loan, C. (1981) A Schur decomposition for Hamiltonian matrices. *Linear Algebra and its Applications*, **41**, 11–32.

Papavassilopoulos, G.P. and Cruz, J. (1979) On the uniqueness of Nash strategies for a class of analytic differential games. *Journal of Optimization Theory and Applications*, **27**, 309–314.

Papavassilopoulos, G.P. and Olsder, G.J. (1984) On the linear-quadratic, closed-loop, no-memory Nash game. *Journal of Optimization Theory and Applications*, **42** 551–560.

Papavassilopoulos, G.P. Medanic, J. and Cruz, J. (1979) On the existence of Nash strategies and solutions to coupled Riccati equations in linear-quadratic games. *Journal of Optimization Theory and Applications*, **28**, pp. 49–75.

Pareto, V. (1896) *Cours d'economie politique*, F. Rouge, Lausanne.

Pars, L.A. (1962) *An Introduction to the Calculus of Variations*, Heinemann, London.

Peeters, R. (2002) *Computation and Selection of Equilibria in Noncooperative Games*, PhD. Thesis, Universitaire Pers Maastricht, University of Maastricht, The Netherlands.

Perko, L. (2001) *Differential Equations and Dynamical Systems*, Springer-Verlag, Berlin.

Petit, M.L. (1989) Fiscal and monetary policy co-ordination: a differential game approach. *Journal of Applied Econometrics*, **4**, 161–179.

Petrosjan, L.A. (1965) *On a Class of Pursuit-Evasian Games*, PhD. Thesis, St. Petersburg/Leningrad University, Russia.

Petrosjan, L.A. and Zenkevich, N.A. (eds) (2002) *Proceedings of the Tenth International Symposium on Dynamic Games and Applications*, St. Petersburg State University, Russia.

Pindyck, R. (1976) The cost of conflicting objectives in policy formulation. *Annals of Economic and Social Measurement*, **5**, 239–248.

Polderman, J.W. and Willems, J.C. (1998) *Introduction to Mathematical Systems Theory: A Behavioral Approach*, Springer-Verlag, Berlin.

Pontrjagin, L.S. (1964) On some differential games. *Doklady Akademii Nauk SSRS*, **156**, 738–741. English translation: *Sov. Math. Dokl.*, **5**, 712–716.

Pontrjagin, L.S. *et al.* (1961) *The Mathematical Theory of Optimal Processes*, Fizmatgiz, Moscov. English translation: Interscience, New York, 1962.

Potter, J.E. (1966) Matrix quadratic solutions. *SIAM Journal of Applied Mathematics*, **14**, 496–501.

Pratt, J. (1964) Risk aversion in the small and in the large. *Econometrica*, **32**, 122–136.

Reid, W.T. (1972) *Riccati Differential Equations*, Academic Press, London.

Reinganum, J.F. and Stokey, N.L. (1985) Oligopoly extraction of a common property natural resource: the importance of the period of commitment in dynamic games. *International Economic Review*, **26**, 161–173.

Reyn, J.W. (1987) Phase portraits of a quadratic system of differential equations occurring frequently in applications. *Nieuw Archief voor Wiskunde*, **5**, 107–154.

Reynolds, S. (1987) Capacity investment, preemption and commitment in an infinite horizon model. *International Economic Review*, **28**, 69–88.

Riccati, J.F. (1724) Animadversationes in aequationes differentiales secundi gradus. *Actorum Eruditorum quae Lipsae publicantur. Supplementa 8*, 66–73.

Riesz, F. and Sz.-Nagy, B. (1972) *Functional Analysis*, Frederick Ungar Publishing Co., New York.

Roberts, A.W. and Varberg, D.E. (1973) *Convex Functions*, Academic Press, New York.

Rudin, W. (1964) *Principles of Mathematical Analysis*, McGraw-Hill, New York.

Runolfsson, T. (1994) The equivalence between infinite-horizon optimal control of stochastic systems with exponential-of-integral performance index and stochastic differential games. *IEEE Trans. Automat. Contr.*, **39**, 1551–1563.

Scalzo, R. (1974) N-person linear-quadratic differential games with constraints. *SIAM Journal of Control*, **12**, 419–425.

Scarf, H.E. (1957) On differential games with survival payoff, in *Contributions to the Theory of Games, Vol.3*, (eds Dresher, Tucker and Wolfe), Princeton University Press, Princeton, pp. 393–405.

Schmeidler, D. (1969) The nucleolus of a characteristic function game. *SIAM Journal of Applied Mathematics*, **17**, 1163–1170.

Schmitendorf, W.E. (1970) Existence of optimal open-loop strategies for a class of differential games. *Journal of Optimization Theory and its Applications*, **5**, 363–375.

Schmitendorf, W.E. (1988) Designing stabilizing controllers for uncertain systems using the Riccati equation approach. *IEEE Trans. Automat. Contr.*, **33**, 376–378.

Schumacher, J.M. (2002) *Course Notes on Dynamic Optimization*, Department of Econometrics, Tilburg University, The Netherlands.

Seierstad, A. and Sydsaeter, K. (1987) *Optimal Control Theory with Economic Applications*, North-Holland, Amsterdam.

Shapley, L. (1953) A value for *n*-person games, in *Contributions to the Theory of Games II*, (eds Tucker and Kuhn), Princeton University Press, Princeton, pp. 307–317.

Shayman, M.A. (1986) Phase portrait of the matrix Riccati equation. *SIAM J. Control and Optimization*, **24**, 1–64.

Simaan, M. and Cruz, J.B. Jr. (1973) On the solution of the open-loop Nash Riccati equations in linear quadratic differential games. *International Journal of Control*, **18**, 57–63.

Simon, C.P. and Blume, L. (1994) *Mathematics for Economists*, W.W. Norton and Company, New York.

Soethoudt, J.M. and Trentelman, H.L. (1989) The regular indefinite linear-quadratic problem with linear endpoint constraints. *Systems and Control Letters*, **12**, 23–31.

Stackelberg, H. von (1934) *Marktform und Gleichgewicht*, Springer-Verlag, Berlin.

Starr, A.W. (1969) Nonzero-sum differential games: concepts and models, *Technical Report No. 590*, Division of Engineering and Applied Physics, Harvard University.

Starr, A.W. and Ho, Y.C. (1969a) Nonzero-sum differential games. *Journal of Optimization Theory and Applications*, **3**, 184–206.

Starr, A.W. and Ho, Y.C. (1969b) Further properties of nonzero-sum differential games. *Journal of Optimization Theory and Applications*, **3**, 207–219.

Stoer, J. and Bulirsch, R. (1980) *Introduction to Numerical Analysis*, Springer-Verlag, New York.

Tabak, D. (1975) Numerical solution of differential game problems. *Int. J. Syst. Science*, **6**, 591–599.

Tabellini, G. (1986) Money, debt and deficits in a dynamic game. *Journal of Economic Dynamics and Control*, **10**, 427–442.

Takayama, A. (1985) *Mathematical Economics*, Cambridge University Press, Cambridge.

Teo, K.L., Goh, C.J. and Wong, K.H. (1991) *A Unified Computational Approach to Optimal Control Problems*, Longman Scientific and Technical, New York.

Thomson, W. (1994) Cooperative models of bargaining, in *Handbook of Game Theory, Vol.2*, (eds Aumann and Hart), Elsevier Science, pp. 1238–1277.

Tijs, S. (1981) Bounds for the core and the τ-value, in *Game Theory and Mathematical Economics*, (eds Moeschlin and Pallaschke), North-Holland, Amsterdam, pp. 123–132.

Tijs, S. (2004) *Introduction to Game Theory*, Hindustan Book Agency, New Delhi.

Tolwinski, B., Haurie, A. and Leitmann, G. (1986) Cooperative equilibria in differential games. *Journal of Mathematical Analysis and Applications*, **119**, 182–202

Trentelman, H.L. (1989) The regular free-endpoint linear quadratic problem with indefinite cost. *SIAM J. Control and Optimization*, **27**, 27–42.

Trentelman, H.L. and Willems, J.C. (1991) The dissipation inequality and the algebraic Riccati equation, in *The Riccati Equation*, (eds Bittanti, Laub and Willems), Springer-Verlag, Berlin, pp. 197–242.

Trentelman, H.L., Stoorrogel, A.A. and Hautus, M.L.J. (2001) *Control Theory for Linear Systems*, Springer-Verlag, Berlin.

Tsutsui, S. and Mino, K. (1990) Non-linear strategies in dynamic duopolistic competition with sticky prices. *Journal of Economic Theory*, **52**, 136–161.

Turnovsky, S.J. (1977) *Macroeconomic Analysis and Stabilization Policies*, Cambridge University Press, Cambridge.

Valentine, F.A. (1937) The problem of Lagrange with differential inequalities as added side conditions, in *Contributions to the Theory of the Calculus of Variations*, University of Chicago Press, Chicago, Illinois, 1933–1937.

Van Hentenryck, P., McAllester, D. and Kapur, D. (1997) Solving polynomial systems using a branch and prune approach. *SIAM J. Num. Anal.*, **34**, 797–827.

Verkama, M. (1994) *Distributed Methods and Processes in Games of Incomplete Information*, Ph.D. Thesis, Helsinki University of Technology, Finland.

Verschelde, J., Verlinden, P. and Cools, R. (1994) Homotopies exploiting Newton polytopes for solving sparse polynomial systems. *SIAM J. Num. Anal.*, **31**, 915–930.

Von Neumann, J. (1928) Zur Theorie der Gesellschaftsspiele. *Mathematische Annalen*, **100**, 295–320.

Von Neumann, J. and Morgenstern, O. (1944) *Theory of Games and Economic Behavior*, Princeton University Press, Princeton.

Walker, P (2001) A chronology of game theory, University of Canterbury, Christchurch, New Zealand, http://www.econ.canterbury.ac.nz/hist.htm.

Weeren, A.J.T.M. (1995) *Coordination in Hierarchical Control*, Ph.D. Thesis, Tilburg University, The Netherlands.

Weeren, A.J.T.M. Schumacher, J.M. and Engwerda, J.C. (1999) Asymptotic analysis of linear feedback Nash equilibria in nonzero-sum linear-quadratic differential games. *Journal of Optimization Theory and Applications*, **101**, 693–723.

Willems, J.L. (1970) *Stability Theory of Dynamical Systems*, Nelson and Sons, London.

Willems, J.C. (1971) Least squares stationary optimal control and the algebraic Riccati equation. *IEEE Trans. Automat. Contr.*, **16**, 621–634.

Zeeuw, A.J. de (1984). *Difference Games and Linked Econometric Policy Models*, Ph.D. Thesis, Tilburg University, The Netherlands.

Zeeuw, A.J. de and Ploeg, F. van der (1991). A differential game of international pollution control. *Systems and Control Letters*, **17**, 409–414.

Zelikin, M.I. and Tynyanskij, N.T. (1965) Deterministic differential games. *Uspeki Mat. Nauk*, **20**, 151–157 (in Russian).

Zermelo, E. (1913) Über die Anwendung der Mengenlehre auf die Theorie des Schachspiels, in *Proceedings of the Fifth International Congress of Mathematicians, Vol.II*, (eds Hobson and Love), Cambridge University Press, Cambridge, pp. 501–504.

Zeuthen, F. (1930) *Problems of Monopoly and Economic Warfare*, George Routledge and Sons, London.

Zhou, K., Doyle, J.C. and, Glover, K. (1996). *Robust and Optimal Control*, Prentice Hall, New Jersey.

Index